Vishvanath Nene, Chittaranjan Kole
(Editors)

Genome Mapping and Genomics in Animal-Associated Microbes

Springer

VISHVANATH NENE
Institute for Genome Sciences and
Department of Microbiology &
Immunology
University of Maryland School of Medicine
Baltimore, MD 21021
USA

e-mail: vnene@som.umaryland.edu

CHITTARANJAN KOLE
Department of Genetics & Biochemistry
Clemson University, 111 Jordan Hall
Clemson, SC 29634
USA

e-mail: ckole@clemson.edu

ISBN 978-3-540-74040-7 ISBN 978-3-540-74042-1 (eBook)
DOI 10.1007/978-3-540-74042-1

Library of Congress Control Number: 2008932285

Cover design: WMXDesign GmbH, Heidelberg, Germany

Printed on acid-free paper

9 8 7 6 5 4 3 2 1

springer.com

Preface

The ability to cost-effectively and rapidly sequence genomes, advances in computational biology and development of high-throughput technologies that facilitate dissection of the ever-expanding list of "omics" has revolutionized research approaches within the biological sciences. For example, the concept of reverse vaccinology has brought antigen identification into the genomics era, and bioinformatics tools can now be used to select and prioritize a list of candidate vaccine antigens from predicted pathogen proteomes for further testing. In addition, the ability to infer metabolic capacity based on pathogen genome sequences has resulted in the identification of several targets for chemogenomics, a discipline that may lead to the generation of novel chemotherapeutics. These two tangible outputs, candidate vaccine antigens and drug targets, have energized efforts in developing improved methods of pathogen control. In addition, development of genome-wide molecular diagnostic tools provides an opportunity to study pathogen genotypes and population dynamics, and should allow those all-important correlates with phenotype to be made. Thus, genomics technologies are now being used to develop a holistic approach to study the genetics of host and pathogen populations and their molecular responses to each other at a level of detail not previously possible. Application of genomics technologies to study the process of infection, disease, vaccination, and interventions leading to immunity is likely to result in building databases, which will help realize the holy grail of developing rational approaches for pathogen and disease control.

The scope for a book on genome mapping and genomics of animal-associated microbes is too huge. We decided to exclude viruses and nonpathogenic microbial associations and chose livestock as the target animals. We concentrated on selecting bacterial or protozoan pathogens of global and developing country significance. We also picked pathogens whose genome sequence has been available for a few years so that the impact of genomics in driving research would be more apparent to the reader. Of the six pathogens chosen, most affect ruminants, and three are vector-transmitted, increasing the complexity of their "animal" associations but providing another target for potential intervention.

We are grateful to the senior authors and co-authors of the chapters presented in this volume, as they have done a marvelous job in summarizing complex topics. They have provided excellent overviews of their pet pathogens as well as comparative aspects with related pathogens. Remarkable progress has been made in studying the organisms reported on here and the next few years promise exciting scientific breakthroughs in both basic and applied pathogen biology.

Baltimore, MD
Clemson, SC
June 2008

Vishvanath Nene
Chittaranjan Kole

Contents

Index ... 2[illegible]

Contributors

Mitchell S. Abrahamsen
Department of Veterinary
and Biomedical Science
College of Veterinary Medicine
University of Minnesota
St. Paul, MN 55108, USA
abe@umn.edu

Jabbar S. Ahmed
Division of Veterinary Infection
Biology and Immunology
Parkallee 22, Research Center Borstel
23845 Borstel, Germany
jahmed@fz-borstel.de

Basil A. Allsopp
Department of Veterinary
Tropical Diseases
Faculty of Veterinary Science
University of Pretoria
Private Bag X04
Onderstepoort 0110, South Africa
Basil.Allsopp@up.ac.za

John Bannantine
National Animal Disease Center
USDA-ARS, Ames
IA 50010-0000, USA
john.bannantine@ars.usda.gov

Richard Bishop
The International Livestock
Research Institute (ILRI)
P.O. Box 30709
Nairobi 00100, Kenya
r.bishop@cgiar.org

Stephen M. Boyle
Center for Molecular Medicine
and Infectious Diseases
Virginia-Maryland Regional College
of Veterinary Medicine
Virginia Tech, Blacksburg
VA 24061-0342, USA
smboyle@vt.edu

Kelly A. Brayton
Programs in Genomics
and Vector-borne Diseases
Department of Veterinary
Microbiology and Pathology
Washington State University
Pullman
WA 99164-7040, USA
kbrayton@vetmed.wsu.edu

Betsy Bricker
National Animal Disease Center
ARS, USDA, 2300 Dayton Road
Ames, IA 50010, USA
bbricker@nadc.ars.usda.gov

Oswald R. Crasta
Cyberinfrastructure Group
Virginia Bioinformatics Institute
Virginia Polytechnic
and State University
Blacksburg VA 24061, USA
ocrasta@vbi.vt.edu

Michael J. Dark
Programs in Genomics
and Vector-borne Diseases
Department of Veterinary
Microbiology and Pathology
Washington State University
Pullman
WA 99164-7040, USA
darkmich@vetmedu.wsu.edu

Paul deFigueiredo
Plant Pathology
Texas A&M University
College Station
TX 77843, USA
pjdefigueiredo@tamu.edu

Etienne P. de Villiers
The International Livestock
Research Institute (ILRI)
P.O. Box 30709
Nairobi 00100, Kenya
e.villiers@cgiar.org

Shinichiro Enomoto
Department of Veterinary
and Biomedical Science
College of Veterinary Medicine
University of Minnesota
St. Paul, MN 55108, USA
enomo001@tc.umn.edu

Thomas A. Ficht
Veterinary Pathobiology
Texas A&M University
4467 TAMU, College Station
TX 77843-4467, USA
tficht@cvm.tamu.edu

Jason M. Fritzler
Department of Veterinary
Pathobiology
College of Veterinary Medicine
and Biomedical Sciences
Texas A&M University
College Station, TX 77843-4467, USA
jfritzler@cvm.tamu.edu

Elizabeth Glass
Division of Genetics and Genomics
Roslin Institute, Roslin
Midlothian EH25 9PS, UK
liz.glass@bbsrc.ac.uk

Simon P. Graham
Virology Department
Veterinary Laboratories Agency
Woodham Lane, New Haw, Addlestone
Surrey KT15 3NB, UK
s.graham@cgiar.org

Lee R. Haines
Liverpool School of
Tropical Medicine
Pembroke Place
Liverpool L3 5QA UK
l.haines@liv.ac.uk

Shirley M. Halling
National Animal Disease Center
ARS, USDA, 2300 Dayton Road
Ames, IA 50010, USA
shalling@nadc.ars.usda.gov

Oliver He
Unit for Laboratory
Animal Medicine
School of Medicine
University of Michigan
Ann Arbor, MI 48105, USA
yongqunh@umich.edu

Kirsty Jensen
Division of Genetics and Genomics
Roslin Institute, Roslin
Midlothian EH25 9PS, UK
kirsty.jensen@bbsrc.ac.uk

Sagarika Kanjilal
Department of Veterinary
and Biomedical Sciences
Pennsylvania State University
University Park, PA, USA

and

Milton S. Hershey Medical Center
Pennsylvania State University
Hershey, PA 16802, USA
kanjilal@psu.edu

Vivek Kapur
Department of Veterinary
and Biomedical Sciences
Center for Infectious
Disease Dynamics
Pennsylvania State University
University Park
PA 16802, USA
vkapur@psu.edu

Ling-Ling Li
Department of Veterinary
and Biomedical Sciences
Pennsylvania State University
University Park
PA 16802, USA
lul17@psu.edu

David J Mann
Department of Biochemistry
Imperial College
South Kensington
London SW7 2AZ, UK
d.mann@imperial.ac.uk

Jere W. McBride
Departments of Pathology and
Microbiology and Immunology
Center for Biodefense and Emerging
Infectious Diseases
Sealy Center for Vaccine Development
Institute for Human Infections
and Immunity
University of Texas Medical Branch
Galveston
TX 77555-0609, USA
jemcbrid@utmb.edu

David O'Callahan
INSERM U431, Faculté de Médecine
Avenue Kennedy
F-30900 Nîmes France
david.ocallaghan@univ-montp1.fr

David O. Odongo
The International Livestock
Research Institute (ILRI)
P.O. Box 30709
Nairobi 00100, Kenya
d.odongo@cgiar.org

Steven C. Olsen
National Animal Disease Center
USDA/ARS, 2300 Dayton Avenue
Ames, IA 50010, USA
solsen@nadc.ars.usda.gov

Guy H. Palmer
Programs in Genomics
and Vector-borne Diseases
Department of Veterinary
Microbiology and Pathology
Washington State University
Pullman
Washington, 99164-7040, USA
gpalmer@vetmed.wsu.edu

Terry W. Pearson
Department of Biochemistry
and Microbiology Petch Building
University of Victoria, Victoria
British Columbia, Canada V8W 3P6
parasite@uvic.ca

Anjan Purkayastha
Cyberinfrastructure Group
Virginia Bioinformatics Institute
Virginia Polytechnic
and State University
Blacksburg VA 24061, USA
anjan@vbi.vt.edu

C. Marty Roop II
Department of Microbiology
and Immunology
East Carolina University School
of Medicine Greenville
NC 27858-4354, USA
ROOPR@mail.ecu.edu

Luis E. Samartino
INTA, CICVyA
Instituto Patobiología
Buenos Aires, Argentina
lsanma@correo.inta.gov.ar

Ulrike Seitzer
Division of Veterinary Infection
Biology and Immunology
Parkallee 22, Research Center Borstel
23845 Borstel, Germany
useitzer@fz-borstel.de

Mohamed N. Seleem
Center for Molecular Medicine
and Infectious Diseases
Virginia-Maryland Regional
College of Veterinary Medicine
Virginia Tech, Blacksburg
VA 24061-0342, USA
naguieb@vt.edu

Sushmita Singh
Biomedical Genomics Center
University of Minnesota
St. Paul, MN 55108, USA
singh056@umn.edu

Bruno W. Sobral
Cyberinfrastructure Group
Virginia Bioinformatics Institute
Virginia Polytechnic
and State University
Blacksburg
VA 24061, USA
sobral@vt.edu

Eric E. Snyder
Cyberinfrastructure Group
Virginia Bioinformatics Institute
Virginia Polytechnic
and State University
Blacksburg
VA 24061, USA
snyderee@vt.edu

Nammalwar Sriranganathan
Center for Molecular Medicine
and Infectious Diseases
Virginia-Maryland Regional
College of Veterinary Medicine
Virginia Tech, Blacksburg
VA 24061-0342, USA
nathans@vt.edu

Chihiro Sugimoto
Department of Veterinary Science
University of Hokkaido
Sapporo, Japan
sugimoto@czc.hokudai.ac.jp

Thomas J. Templeton
Department of Microbiology
and Immunology
Weill Cornell Medical College
Weill Graduate School
of Medical
Sciences of Cornell University
New York
NY 10021, USA
tjt2001@med.cornell.edu

Rene M. Tsolis
Medical Microbiology
and Immunology
GBSF, Room 5312
University of California
Davis
CA 95616-8732, USA
rmtsolis@ucdavis.edu

Rebecca A. Wattam
Cyberinfrastructure Group
Virginia Bioinformatics Institute
Virginia Polytechnic and
State University
Blacksburg
VA 24061, USA
wattam@vbi.vt.edu

Kelley Williams
Cyberinfrastructure Group
Virginia Bioinformatics Institute
Virginia Polytechnic and
State University
Blacksburg
VA 24061, USA
kellwill@vt.edu

Adrian M. Whatmore
Department of Statutory
and Exotic Bacterial Diseases
Veterinary Laboratories Agency
Addlestone
Surrey, KT15 3NB, UK
a.whatmore@vla.defra.gsi.gov.uk

Gong-Xin Yu
Boise State University
1910 University Drive
Boise
ID 83725-1555, USA
gongxinyu@boisestate.edu

Guan Zhu
Department of Veterinary
Pathobiology
College of Veterinary Medicine
and Biomedical Sciences,
Faculty of Genetics Program
Texas A&M University
College Station
TX 77843-4467, USA
gzhu@cvm.tamu.edu

Abbreviations

2D	Two-dimensional
ABC	Adenosine 5-triphosphate-binding cassette
ADA	Adenylate deaminase
ADH	Alcohol dehydrogenase
ADP	Adenosine diphosphate
AFLP	Amplified fragment length polymorphism
AK	Adenosine kinase
ANI	Average nucleotide identity
AOX	Alternative oxidase
APHIS	Animal and Plant Health Inspection Service
ATP	Adenosine tri-phosphate
ATPase	AAA-adenosine triphosphatase
BAC	Bacterial artificial chromosome
BBP	*Brucella* Bioinformatics Portal
BCG	Bacillus Calmette Guérin (attenuated *Mycobacterium bovis*)
BLAST	Basic local alignment search tool
BLASTN	BLAST for nucleotide (database)
BLASTP	BLAST for protein (database)
BoLA	Bovine major histocompatibility complex
bp	base pair
BRC	Bioinformatics Resource Center
Bv	Biovariety
cdks	Cyclin-dependent kinases
CDS	Coding sequence
CFU	Colony forming unit
CHEF	Contour-clamped homogeneous electric field
Chr I	Chromosome 1
Chr II	Chromosome 2
CME	Canine monocytic ehrlichosis
CML	Chronic myelogenous leukemia
CMR	Comprehensive Microbial Resource
COG	Clusters of orthologous groups of proteins
COWP	*Cryptosporidium* oocyst wall protein
CTL	Cytotoxic T lymphocytes
CtrA	Conserved global response regulator A
CXCR 4	Chemokine receptor 4
DHF	Di-hydrofolate
dN	Number of nonsynonymous (causing amino acid change) nucleotide substitutions in a gene
DNA	Deoxyribonucleic acid
dS	Number of synonymous (causing no amino acid change) nucleotide substitutions in a gene
EC	Enzyme commission

ECF	East Coast fever
ERIC-PCR	Enterobacterial repetitive intergenic consensus-PCR
ES	Expression site
EST	Expressed sequence tag
e-value	Expect value
FAINT	Frequently associated in *Theileria*
FALP	Fluorescent amplified fragment length polymorphism
FAO	Food and Agriculture Organization of the United Nations
FAS	Fatty acid synthase
G+C	Guanine + cytosine
GenVar	Genome context based and sequence variant-oriented analysis pipeline
GFP	Green fluorescent protein
GLTS	Glutamate synthase
GMPS	GMP synthetase
GPI	Glycosylphosphatidylinositol
GS	Glutamine synthase
GSS	Genomic sequence survey
HEYM	Herrold's egg yolk medium
HME	Human monocytic ehrlichiosis
HOOF	Hypervariable octameric oligonucleotide fingerprints
HPLC	High-performance liquid chromatography
HVR	Hypervariable region
HXGPRT	Hypoxanthine xanthine guanine phosphoribosyl transferase
ICZN	International Commission on Zoological Nomenclature
IFA	Immunofluorescence assay
IHF	Integration host factor
IL	Interleukin
IMPDH	Inosine monophosphate dehydrogenase
INF-γ	Interferon gamma
IRF-1	Interferon regulatory factor 1
IRS	Interspersed repetitive sequence
IS	Insertion sequence
ITM	Infection and treatment method
ITS	Internal transcribed spacer (in rRNA transcripts)
IVET	In vitro expression technology
JD	Johne's disease
kb	Kilobase
KEGG	Kyoto Encyclopedia of Genes and Genomes
LC-MS/MS	Liquid chromatography – tandem mass spectrometry
LD50	Lethal dose of drug or pathogen for 50% of recipients
LDH	Lactate dehydrogenase
LPS	Lipopolysaccharide
LSP	Large secreted protein
LTR	Large tandem DNA repeat
MAC	*Mycobacterium avium* complex

MALDI-TOF	Matrix-assisted laser desorption ionization time of flight
Map	*Mycobacterium avium* subspecies *paratuberculosis*
map1	Major outer membrane protein 1 gene of *Ehrlichia ruminantium*
MAP1	Major outer membrane protein 1 of *Ehrlichia ruminantium*
MAUVE	Collaborative project whose goal is to create a free suite of functional, black box, tests for the core Java(tm) libraries
Mav	*Mycobacterium avium*
Mb	Megabase
MDH	Malate dehydrogenase
MESH	Medical Subjects Headings page
MHC	Major histocompatibility complex
MINet	Molecular interaction network
MIRU	Mycobacterial interspersed repetitive unit
MLEE	Multi-locus enzyme electrophoresis
MLGs	Multi-locus genotypes
MLSA	Multi-locus sequence analysis
MLSSR	Multi-locus short sequence repeats
MLVA	Multiple-locus variable number tandem repeat analysis
MPA	Mycophenlic acid
MPIL	Multiplex PCR specific for an IS900 integration loci
MPSS	Massively parallel signature sequencing
mRNA	Messenger RNA
msp	Major surface protein
Mtb	*Mycobacterium tuberculosis*
MTX	Metronidazole
NADH	Nicotinamide adenine dinucleotide, reduced form
NAHMS	National Animal Health Monitoring System
NCBI	National Center for Biotechnology Information
NMR	Nuclear magnetic resonance
N-terminus/N-terminal	Amino-terminus/terminal
NTZ	Nitazoxanide
OMP	Outer membrane protein
ORF	Open reading frame
OWP	Oocyst wall protein
PATRIC	Pathosystems Resource Integration Center
PBMC	Peripheral blood mononuclear cell
PCR	Polymerase chain reaction
pCS20	Genomic region of *E. ruminantium* containing overlapping genes *rnc* and *ctaG*
PE	Proline–glutamic acid
PEP	Phosphoenol pyruvate
PFAM	A database of protein families
PFGE	Pulse-field gel electrophoresis
PFO	Pyruvate ferredoxin oxidoreductase

PHYML	A maximum likelihood phylogenetic tree inferring program
PKB	Protein kinase B
PKS	Polyketide synthase
PMF	Peptide mass fingerprinting
PNO	Pyruvate:NADP + oxidoreductase
PPE	Proline–proline–glutamic acid
PPi-PFK	Pyrophosphate-dependent phosphofructokinase
pRb	Retinoblastoma protein
PRT	Phosphoribosyltransferase
PUBMED	On line search engine providing access to citations from biomedical literature
PV	Parasitophorous vacuole
PVM	Parasitophorous vacuolar membrane
RAPD	Random amplified polymorphic DNA
RAPD-PCR	Random amplified polymorphic DNA-PCR
RB51	*Brucella abortus* strain RB51
RBCs	Red blood cells
RDP	Ribosomal Database Project
REA	Restriction endonuclease analysis
REP-PCR	Repetitive extragenic palindromic-PCR
RFLP	Restriction fragment length polymorphism
RNA	Ribonucleic acid
rRNA	Ribosomal RNA
rrs	Small subunit ribosomal RNA gene
S19	*Brucella abortus* strain 19, cattle vaccine strain
SDS	Sodium dodecyl sulfate
Si	Plasmid in *Rhizobium* responsible for symbiosis
SIRPA	Signal inhibitory-regulatory protein alpha
SNP	Single nucleotide polymorphism
SOCS-1	Suppressor of cytokine signaling 1
SSR	Simple sequence repeat
SSU rRNA	Small subunit ribosomal RNA
T4SS	Type four secretion system
TAP	Transcription-associated proteins
Tar	*Theileria annulata* repeat
TaSR	*Theileria annulata* subtelomeric repeat
TBD	Tick-borne diseases
TBP	TATA binding protein
TCA	Tricarboxylic acid
thy1	Flavin-dependent thymidylate synthase
thyA	Thymidylate synthase
Ti	Plasmid in *Agrobacterium* responsible for tumorigenesis
TIGR	The Institute for Genomic Research
TIGRFAM	Collection of protein families featuring curated multiple sequence alignments
TK	Thymidine kinase
TLR	Toll-like receptor
TNF	Tissue necrotizing factor

Tpr	*Theileria parva* repeat
TpSR	*Theileria parva* subtelomeric repeat
TRAP	Thrombospondin-related adhesive protein
tRNA	Transfer RNA
VLPT	Variable length PCR target
VNTR	Variable number of tandem repeats
XML	Extensible Markup Language

1 *Brucella*

Nammalwar Sriranganathan[1], Mohamed N. Seleem[1], Steven C. Olsen[2], Luis E. Samartino[3], Adrian M. Whatmore[4], Betsy Bricker[2], David O'Callaghan[5], Shirley M. Halling[6], Oswald R. Crasta[7], Alice R. Wattam[7], Anjan Purkayastha[7], Bruno W. Sobral[7], Eric E. Snyder[7], Kelley P. Williams[7], Gong-Xi Yu[8], Thomas A. Ficht[9], R. Martin Roop II[10], Paul deFigueiredo[11], Stephen M. Boyle[1] (✉), Yongqun He[12], and Renée M. Tsolis[13]

[1] Center for Molecular Medicine and Infectious Diseases, Virginia-Maryland Regional College of Veterinary Medicine, Virginia Polytechnic Institute and State University, Blacksburg, VA 24061-0342, USA, smboyle@vt.edu

[2] National Animal Disease Center, ARS, USDA, Ames, IA 50010, USA

[3] INTA, CICVyA, Instituto Patobiología, Buenos Aires, Argentina

[4] Department of Statutory and Exotic Bacterial Disease, Veterinary Laboratories Agency, Woodham Lane, Addlestone, Surrey KT15 3NB, UK

[5] INSERM ESPRI 26, UFR Medecine, Avenue Kennedy, 30908 Nimes, France

[6] Bacterial Diseases of Livestock Research Unit, Agricultural Research Service, United States Department of Agriculture, Ames, Iowa, USA

[7] Cyberinfrastructure Group, Virginia Bioinformatics Institute, Virginia Polytechnic Institute and State University, Blacksburg, VA 24061, USA

[8] Department of Biology and Computer Science, Boise State University, Boise, ID 83725, USA

[9] Veterinary Pathobiology, Texas A&M University, College Station, TX 77843, USA

[10] Department of Microbiology and Immunology, East Carolina University School of Medicine, Greenville, NC 27858, USA

[11] Plant Pathology, Texas A&M University, College Station, TX 77843, USA

[12] Unit for Laboratory Animal Medicine and Department of Microbiology and Immunology, School of Medicine, University of Michigan, Ann Arbor, MI 48109, USA

[13] Medical Microbiology and Immunology, University of California, Davis, CA 95616, USA

1.1.1 Discovery of the *Brucella*

Examination of the skeletal remains of the Roman residents of Herculaneum (Naples, Italy) killed by the catastrophic volcanic eruption of Mt. Vesuvius in the late August AD 79 revealed vertebral bone lesions typical of brucellosis in more than 17% of the residents. Scanning electron microscopy of recovered cheese provided a likely explanation for the high incidence of the disease. The buried carbonised cheese, made from sheep's milk and found with the bones, revealed the presence of cocco-bacillary forms that were morphologically similar to *Brucella spp.* (Capasso 2002). Eighteen centuries later, Sir David Bruce isolated *Micrococcus melitensis* (now *Brucella melitensis*) from the spleen of a British soldier who died from a febrile illness (Malta fever) common among military personnel stationed on Malta, an island not far away from Herculaneum (Godfroid et al. 2005). For almost 20 years, brucellosis was thought to be a vector-borne disease. The zoonotic nature of the brucellosis was accidentally demonstrated in 1905 by isolating *B. melitensis* from goat's milk used for the production of soft cheese in Malta (Nicoletti 2002; Godfroid et al. 2005). It was believed that goats were not the source of infection since they did not become ill when inoculated with *Brucella* cultures. Although raw goat's milk had been used as an essential nutritional meal for hospitalized patients suffering from Malta fever, it was decided to ban it from hospitals. The public did not follow the same recommendation and consumed infected dairy products and remained exposed to the disease (Nicoletti 2002).

1.1.2 Species Discovery

In 1897, a Danish veterinarian, L.F. Benhard Bang, discovered Bang's bacillus or bacillus of abortion

Genome Mapping and Genomics in Animal-Associated Microbes
V. Nene, C. Kole (Eds.)

(*B. abortus*) the causative agent of Bang's disease (brucellosis in cattle). Bang's bacillus was not recognised as being related to *Micrococcus melitensis* (isolated by Bruce) until 1918, when Alice Evans in the Hygiene Laboratory of the U.S. Public Health Service (now the National Institutes of Health) showed the close relationship between the two organisms and renamed the genus *Brucella* to honour Bruce (Meyer and Show 1920; Bang 1933; Nicoletti 2002).

In 1914, Traum isolated *B. suis* from an aborted pig foetus in U.S. (Traum 1914; Nicoletti 2002). The description of isolates from cattle and swine led to the recognition of widespread distribution of the disease.

In 1953, a different strain, thought to be a rough *Brucella* mutant, was described in sheep in New Zealand by Buddle and in Australia by Simmons (Simmons and Hall 1953; Buddle 1956; Diaz et al. 1967). Although the Subcommittee on the Taxonomy of *Brucella* of the International Committee on Bacteriological Nomenclature was not satisfied that the organism was a member of the genus *Brucella* and advised further study, the species was eventually recognized as *B. ovis* (Diaz et al. 1967).

In 1957, Stoenner and Lackman isolated *B. neotomae* from desert wood rat (*Neotoma lepida*) in Utah, U.S. (Stoenner and Lackman 1957). Carmichael isolated *B. canis* in 1966 from beagles in the U.S. (Carmichael and Bruner 1968).

Brucellosis in marine mammals was first described in 1994 in the U.S. when a bacterial isolate from the aborted foetus of a bottlenose dolphin (*Tursiops truncatus*) was characterized as a nontypical *Brucella spp.* (Ewalt et al. 1994). Since 1994, several new *Brucella* species have been isolated from marine mammals (Ross et al. 1994; Foster et al. 1996). The zoonotic nature of marine brucellae and its ability to cause abortion in cattle were documented (Brew et al. 1999; Rhyan et al. 2001). The discovery of the marine *Brucella* has changed the concept of a land-based distribution of brucellosis and associated control measures to that of a land- and ocean-based approach for control and eradication.

As of 2006, eight *Brucella* species are recognized. Six of them infect terrestrial animals: *B. abortus*, *B. melitensis*, *B. suis*, *B. ovis*, *B. canis* and *B. neotomae* (Verger et al. 1987) and two infect marine mammals: *B. cetaceae* and *B. pinnipediae* (Verger et al. 2000). Within these species, seven biovars are recognized for *B. abortus*, three for *B. melitensis* and five for *B. suis* (Verger et al. 2000); the remaining species have not been differentiated into biovars.

1.1.3 Zoonoses

Although *Brucella* was first isolated by Bruce in the nineteenth century, clinical conditions characteristic of brucellosis have been described by Hippocrates in 450 BC (Evans 1950). In 1751, Cleghorn, a British army surgeon stationed on the Mediterranean island of Minorca, described cases of chronic, relapsing febrile illness and cited Hippocrates's description of a similar disease (Hoover and Friedlander 1997). Marston, a British army surgeon working on the island of Malta, described the clinical characteristics (Malta fever) of his own infection in 1861 (Hoover and Friedlander 1997).

Brucella was discovered and isolated for the first time from humans in 1887 before it was recognised as an animal pathogen in 1905. The first recognised human case of brucellosis in the USA was in an army officer based in Puerto Rico in 1898 (Brown 1977; Nicoletti 2002). The zoonotic nature of *B. canis* was reported in 1975 in US (Blankenship and Sanford 1975; Munford et al. 1975). The zoonotic nature of marine brucellae was documented in 1999 in a case of a laboratory-acquired human infection (Brew et al. 1999).

B. suis was the first biological agent to be weaponised by the US in 1942 during its offensive biological warfare program. The agent was formulated to maintain long-term viability, placed into bombs and tested in field trials during 1944–1945 using animal targets (Hoover and Friedlander 1997). By 1967, the USA terminated its offensive program for the development and deployment of *Brucella* and other pathogens as biological weapons (Hoover and Friedlander 1997).

B. melitensis, *B. suis* and *B. abortus* are listed as potential bioweapons by the Centers for Disease Control and Prevention (Kaufmann et al. 1997; Kortepeter and Parker 1999), because of their virulence in humans. This is due to the highly infectious nature of all three species as they can be readily aerosolized. Moreover, an outbreak of brucellosis would be difficult to detect because the initial symptoms are easily confused with those of influenza (Chain et al. 2005).

In comparison to abortions, orchitis, followed by persistent infections of supra-mammary lymph nodes and reticuloendothelial system in animals (Adams 2002), humans develop symptoms that start out as flu-like symptoms followed by undulant fever with severe cold sweats in between (Pappas et al. 2006a, b). In some affected individuals the disease could be fatal if untreated, while others can become permanently infected and suffer from fever and cold sweats, particularly when they are stressed. Brucellosis has also been associated with mild to severe cases of arthritis in adults and childen (Pourbagher et al. 2006).

1.1.4 Eradication Program

By the year 1922, several states in the USA had passed laws and regulations in an attempt to prevent introduction of the disease by cattle purchased from other states (Nicoletti 2002). The Cooperative State-Federal Brucellosis Eradication Program began in 1934 and cost about $3.5 billion by 1997. The program's Uniform Methods and Rules set forth the minimum standards for states to achieve eradication. A state is designated as brucellosis-free when none of the cattle in that state are found infected for 12 consecutive months under an active surveillance program. In 1956, there were 124,000 affected herds in the U.S.A., which corresponds to one in every eight cattle herds. By 1992, this number had dropped to 700 herds, and by 2000 there were only six known infected herds remaining in the entire U.S.A. Consequently, the number of human brucellosis cases in the USA has dropped from 6,321 in 1947 to about 100 per year by 1998, mostly acquired overseas or due to consumption of infected milk products from Mexico (Cook et al. 2002).

Infected wild life (bison, elk and feral swine) still remains a source of infection to domestic livestock three of the five brucellosis affected cattle herds disclosed in FY 2005 are due to wild life (Olsen and Stoffregen 2005).

In spite of the availability of very effective vaccines like *B. abortus* strains 19 and RB51, eradication of cattle brucellosis has not been accomplished in all the countries of the world. Most of North America has essentially eradicated the disease from their cattle. In Mexico, South America, Asia, Africa, Middle East and Caucuses States, the disease is highly prevalent, even though in many of these countries, there are 'test and slaughter' programs in place. Because of lack of indemnification, the programs have not been very effective. In countries like India, penal codes that prohibit slaughter of cows complicate the issue even further. From the long and successful efforts in the USA, one can conclude that eradication of brucellosis in cattle can be accomplished only when all the concerned parties get involved in finding a solution. It is basically a political disease; unless there is a strong political support for the indemnification of the farmer for the loss due to removal of infected animals, it is almost impossible to control this important zoonosis. The farmers, milk and milk products industry, breeding companies, consumers and the politicians must work together and find a practical eradication effort that is suitable for each country.

In this regard, the recent brucellosis control effort in Iran is worth careful consideration by other countries. The Iranian investigators effectively eradicated the disease from a set of large commercial dairy farms by monthly serological testing and slaughtering of all positive cows for a period of a year (personal communication, Kamran Afshar Pad, Veterinary Organization of Iran, 2005). This was followed by mass vaccination of the entire herd with regular doses of strain RB51. The investigators educated the commercial farmers to expect a certain level of abortion in the vaccinated animals. They established standard operating procedures for the safe removal of all aborted materials and the cows. They were able to attain what could be considered as disease-free status in these large commercial farms within 10–18 months; this process took 20 years for USA to accomplish. The commercial farms were willing to withstand the losses by having increased productivity and the lack of abortions. This approach would certainly have to be modified to address the very different economic conditions of a small-scale farmer, who would also be the primary target of such an eradication effort in many of the above countries having endemic brucellosis.

1.1.5 Vaccination

The first attempt at using a *Brucella* vaccine was performed in 1906 by Bang (Bang 1906). He demonstrated that the injection of live *B. abortus* protected cattle against brucellosis, whereas killed organisms were ineffective. It has since become evident that a live vaccine is superior to a killed vaccine because of the former's ability to induce a strong cellular immunity (Nicoletti 2002).

Strain 19, a laboratory-derived strain of *B. abortus* attenuated by an unknown process during subculture, was first described in 1930 and was originally isolated from the milk of a Jersey cow as a virulent strain in 1923. After being kept in the laboratory at room temperature for over a year, it was found to have become attenuated (Buck 1930; Schurig et al. 1991). In 1941, strain 19 was introduced and used in the US (Nicoletti 2002).

B. abortus smooth strain 45/20 was isolated from a cow in 1922 and a rough derivative was obtained after 20 passages in guinea pigs. The strain was able to protect guinea pigs and cattle from *Brucella* infection (McEwen 1940). When used as a live vaccine, strain 45/20 was not stable and tended to revert to the smooth virulent form (Schurig et al. 1991). The instability of the strain prevented its extensive use as live vaccines (Schurig et al. 1991).

In the mid-1950s, Elberg and Herzberg created a streptomycin-dependent strain of *B. melitensis*, assuming that controlling its growth in the host might be achieved by co-injection of the antibiotic. The vaccine protected mice and guinea pigs against brucellosis. However, in monkeys and goats, the results were unsatisfactory as attributed to the fact that the strain survived in the animals for a shorter period than was necessary to stimulate a strong immune response (Elberg and Faunce 1957; Banai 2002). This led to reselection of a non-streptomycin requiring strain that was used to successfully immunise mice and guinea pigs (Herzberg and Elberg 1955; Banai 2002). The new strain, designated Rev.1, was found sufficiently attenuated in mice and guinea pigs, while maintaining the capacity to persist in the spleens of different mice strains for 6–12 weeks (Banai 2002) and induce protection.

In 1991, Schurig et al. (1991) developed *B. abortus* strain RB51, a stable rough mutant derived from the strain 2308. Being a rough strain, vaccination with strain RB51 does not result in O antigen-specific antibodies, thereby greatly facilitating the serological differentiation of infected from vaccinated animals (Schurig et al. 1991). Strain RB51 has replaced strain 19 as the official vaccine for cattle in the US (Stevens et al. 1997).

1.1.6 Molecular Study and Diagnoses

In 1952, the milk ring test was introduced to the Cooperative State-Federal Brucellosis Eradication Pro-gram in the US. It remains the primary method of surveillance for brucellosis among dairy cattle herds (Nicoletti 2002). The study of the *Brucella* genome began as early as the 1960s, when DNA homology was used to show that the different strains of *Brucella* are related (Hoyer and McCullough 1968a, b). In the early 1970s, Altenberg used chemical mutagenesis to create auxotrophic mutants in *B. abortus* S19 and then used marker frequency analysis to map certain genes (Altenbern 1973). In the 1980s, the analysis of the genetics of *Brucella* and the structure and organisation of its genome began (Michaux-Charachon et al. 2002). The sequence of the *B. melitensis* 16M and *B. suis* genome was published in 2002 (DelVecchio et al. 2002c; Paulsen et al. 2002) and that of *B. abortus* in 2005 (Halling et al. 2005).

In 1990, Fekete reported the first PCR-based assays developed for *Brucella* diagnoses (Fekete et al. 1990). In 1994, Bricker and Halling (1994) introduced AMOS-PCR assay to differentiate *B. abortus* biovars 1, 2 and 4; *B. melitensis*, *B. ovis* and *B. suis* biovar 1.

1.2 Economic and Zoonotic Implications of *Brucella*

1.2.1 Distribution of *Brucella spp.* Worldwide

The genus *Brucella* is composed of facultative bacteria that infect numerous mammalian hosts with localization predominantly in intracellular environments, for example macrophages. The genus is generally considered to be composed of four zoonotic species, which in order of pathogenicity to humans are *B. melitensis*, *B. suis*, *B. abortus* and *B. canis*. The genus also includes *B. ovis*, *B. neotomae* and recently discovered marine *Brucella* isolates. The zoonotic potential of marine *Brucella* strains remains uncharacterized, although they have been identified in two human patients with neurobrucellosis, and a laboratory worker was infected and developed clinical brucellosis after working with a seal isolate (Brew et al. 1999; Sohn et al. 2003). Although species of *Brucella* tend to have preferred hosts, most pathogenic species are capable of infecting numerous mammalian hosts. Infections of domestic livestock with *Brucella* are generally associated with economic losses due to infertility, fetal loss and reduced milk production.

Brucella melitensis, *B. suis* and *B. abortus* infections occur in humans and domestic livestock worldwide, but prevalence tends to be higher in parts of the Mediterranean basin, Central and South America, the Middle East, sub-Saharan Africa and Asia (Pappas et al. 2005b, 2006a, b). As a general rule, prevention of human brucellosis depends predominantly on the control of the disease in animals (Ragan 2002; Godfroid et al. 2005). Although brucellosis has been eradicated or nearly eradicated from domestic cattle, swine or small ruminants in Europe, Canada and the United States, reservoir hosts in other species, such as wildlife, remain of concern for re-emergence in domestic livestock. Other countries, particularly in the developing world, still live with a huge disease burden (Cutler et al. 2005). Over the last 10–15 years, economic instability, socioeconomic factors and insufficient regulatory programs in domestic livestock have led to re-emergence of brucellosis in both livestock and humans in a number of countries (Pappas et al. 2006a, b).

1.2.2 Zoonotic Characteristics of *Brucella spp.*

Human brucellosis is a worldwide zoonosis and is estimated to infect over 500,000 people yearly in non-industrialized countries of the world (Alp et al. 2006). Estimates of the prevalence of brucellosis range from <1 per 100,000 people in the United Kingdom, United States and Australia, to >70 per 100,000 people in some countries in the Middle East (Cutler et al. 2005). Infections in humans primarily result from direct contact with infected livestock or consumption of non-pasturized dairy products (Roth et al. 2003). Human brucellosis can also occur through handling or consumption of raw meat from infected animals, inadvertant exposure in laboratories or accidental infection with live vaccines intended for domestic livestock. Although reports are rare, human-to-human transmission of brucellosis through veneral transmission (Ruben et al. 1991) or through breast milk (Palanduz et al. 2000) has been reported.

Cultural habits are a very important factor for influencing human infections with *Brucella spp.* As an example, Mexico, Peru, Argentina and Paraguay have a large number of people infected with *B. melitensis* due to the consumption of non-pasteurized cheese produced from milk of infected goats. In a similar manner, in the United States, which is considered to be free of *B. melitensis*, human infection continues to occur in certain ethnic groups due to consumption of non-pasturized cheese products transported into the country from areas endemic with caprine brucellosis (Fosgate et al. 2002). Close association with livestock also contributes to infection in countries where caprine brucellosis is endemic, as all family members participate in the work of raising goats.

In humans, the incubation period is variable and can range from less than a week to several months (Yagupsky and Baron 2005). Frequent bacteremia in the early stages of human infection leads to widespread in vivo distribution primarily to sites rich in reticuloendothelial tissue. Clinical symptoms in humans are not pathognomonic and can include recurrent pyrexia (undulant fever), cephalagia, malaise, joint and muscle pain, night sweats and even neurologic manifestations. *Brucella* can be distributed to almost any tissue or site in vivo, leading to related clinical symptoms caused by inflammatory lesions associated with bacterial localization. Osteoarticular disease is the most common complication and can include peripheral arthritis, sacroilitis and spondylitis (Pappas et al. 2005b; Alp et al. 2006). Like domestic livestock, human brucellosis is generally associated with low mortality. Endocarditis remains the principal cause of human mortality in the course of brucellosis (Pappas et al. 2005b).

1.2.3 Ecomonic Losses to Producers and Costs for Human Clinical Care

Economic losses vary from country to country, depending upon numbers and types of livestock, prevalence of disease and differences in livestock production methods. Economic losses to producers are primarily associated with loss of offspring and reduction in milk production, whereas economic impact on individual countries may relate to the cost of control or surveillance programs, loss of international markets, increased public health costs and lost productivity. Although worldwide estimates are generally unavailable, in 2002 it was estimated that $25 million in economic losses occurred in Central America per year due to brucellosis (Moreno 2002). Although compensation for elimination of *Brucella*-infected livestock is provided in some countries, many countries do not compensate producers for removal of infected livestock. Producers with infected herds may also suffer economic losses from reduced prices for milk, inability to sell livestock due to quarantine procedures and loss of available markets.

The poor in every society, and particularly in developing countries, bear a disproportionately high share of the burden of disease (World Health Organization 2005). Not only are they at greater risk for contracting zoonotic diseases due to close contact with animal reservoirs of disease, but secondly, once infected, are least likely to get proper treatment (World Health Organization 2005). Furthermore, the impact of disease is worse in poor households where a dual burden is borne, because it affects both people and animals (World Health Organization 2005). Although often overlooked, the impact of brucellosis in pastoral production systems may have greater relative impact than the economic losses associated with livestock production in developed countries. Associated fetal losses and reduced milk consumption may threaten the welfare of subsistence producers, in addition to the potential for greater transmission of brucellosis from livestock to humans due to the close association in pastoral production systems.

Clinical care of human cases is a significant cost to countries where brucellosis is endemic. Economic costs not only include hospital care and long-term medications, but also reductions in productivity during clinical disease. The costs of surgical intervention, 25% in one study of spinal brucellosis (Alp et al. 2006), and frequency of relapse (10–15%) in human patients (Solera et al. 1998; Pappas et al. 2005a; Alp et al. 2006) also contribute to the cost of human medical care.

Brucellosis continues to have greater impact on rural populations in developing countries. In the 1980s, in Argentina, approximately 1,800 cases per year were reported, with 60% occurring in rural areas (Garcia 1990). Between 1993 and 1995, the number of cases in Argentina had been reduced to approximately 100 per year (212 cases in total). Although primarily associated with rural areas (85%), infections were primarily caused by *B. abortus* (42%), followed by *B. melitensis* (27%) and *B. suis*, (15%) (Samartino 2002). In the late 1990s, Argentina estimated its losses due to brucellosis to be $66 million per year in economic losses in the livestock industry and 24 million dollars per year due to human brucellosis (Garcia et al. 1990). In a similar manner, Mexico reported 37,807 cases of human brucellosis, primarily due to *B. melitensis*, between 1990 and 2000 (6.4–13.8 per 100,000 population) with an estimated cost of $150,000 dollars per year for treatment (Gil 2000; Luna-Martinez and Mejia-Teran 2002). In Peru, where human brucellosis cases ranged from 1,800 cases per year in 1998 to 2,560 cases in 2002 (Gil 2000), based on a cost of treatment of $255, the Nacional Program of Zoonosis estimated human brucellosis treatment costs exceeded $650,000 dollars in 2002. In many developing countries, a lack of surveillance activities in both public health and animal health sector combined with poor coordination and limited medical care may underestimate brucellosis prevalence and economic costs (Pan American Health Organization 1998).

In particular, children appear to represent a high proportion of human brucellosis cases. In northern Argentina, 57% of *B. melitensis* infections were in children (Samartino 2002). In Kyrgyzstan in Central Asia, while prevalence of brucellosis in humans is increasing (>150 per 100,000), 40% of infections are in children. Although reports suggest that clinical appearance and response to treatment are similar to those in adults, *Brucella* infections in children may cause losses from reduced parental productivity and may also be associated with other unidentified economic costs.

1.2.4 Control Programs and Associated Economic Costs

The most effective method to resolve human brucellosis is to address the disease in domestic livestock. Control or eradication programs generally utilize combinations of sanitation, isolation procedures, vaccination and test and removal programs dependent upon economic resources. Control of brucellosis can be expensive as the United States spends approximately $40 million annually for their surveillance and eradication programs in cattle and swine. Brucellosis programs generally require long-term commitment of resources and personnel. In general, vaccination with *B. abortus* or *B. melitensis* vaccines is the most economical approach for preventing brucellosis in domestic livestock caused by these two species. There are several well-documented instances in which reductions in vaccination programs were associated with resurgence of brucellosis in domestic livestock and/or human brucellosis (Taleski et al. 2002; Minas et al. 2004).

Although expensive, money spent on livestock programs can give favorable economic returns by preventing human brucellosis. For example, in Mongolia, where prevalence of brucellosis is high, it was estimated that the cost benefit ratio for mass vaccination against brucellosis would be 3.2 and, with an overall benefit of $26.6 million dollars, would benefit both public health and livestock sectors (Roth et al. 2003). Ten years after eradication of brucellosis in the Czech Republic, it was estimated that the cumulative benefit/eradication cost ratio was 7:1 and eradication had averted losses of approximately $700 million and prevented infection in 2,000 people (Kouba 2003).

1.2.5 Conclusions

Brucellosis continues to remain a significant pathogen in both humans and domestic livestock in many parts of the world. Although estimates of the costs associated with brucellosis infections remain limited to specific countries or areas, all data suggest that worldwide economic losses due to brucellosis are extensive not only in animal production, but also in the area of public health. Success or failure to control brucellosis in domestic livestock strongly correlates to the incidence of human infection with brucellosis. The continued high prevalence of brucellosis in some countries–and resurgence in others–emphasises the numerous factors that contribute to failure in elimination of this pathogen from domestic livestock reservoirs.

1.3 *Brucella* Taxonomy

1.3.1 Morphology/Life Cycle

Brucellae are gram-negative, aerobic, non-motile, non-spore-forming coccobacilli that infect many animal species as well as cause human brucellosis, still one of the major zoonotic diseases with a worldwide distribution (Pappas et al. 2005b). Although classically considered facultatively intracellular since the organism can survive on bacteriological media and, to some extent, in the open environment, their preferred niche is intracellular. Thus, in practice, they behave as strict parasites relying for their survival, reproduction and persistence on transmission between animal hosts. The usual route of infection is through the digestive tract, although conjunctival infection may also occur. Tissue and fluids aborted as a result of *Brucella* infection are heavily contaminated with *Brucella* and are thus considered the major route of transmission. However, vaginal secretions and milk are also likely sources of infection following close contact, and venereal transmission is also relevant particularly in swine, ovine and canine brucellosis. Six species are classically recognized within the genus *Brucella* based largely on differences in pathogenicity and primary host preference. The main pathogenic species are *B. melitensis*, responsible for ovine and caprine brucellosis and considered the most pathogenic species for man, *B. abortus* and *B. suis* (bovine and swine brucellosis, respectively). The three remaining *Brucella* species, *B. ovis* (ram epididymitis), *B. canis* (canine brucellosis) and *B. neotomae* (only ever found in the desert wood rat), have rarely or never been associated with human infections.

1.3.2 Phylogenetic Position of the Genus

The true phylogenetic position of the genus *Brucella* was uncertain for many years, but it is now clear that the genus belongs to the order *Rhizobiales* within the class alpha-proteobacteria as shown in Fig. 1 (Gupta 2005). The unexpected phylogenetic

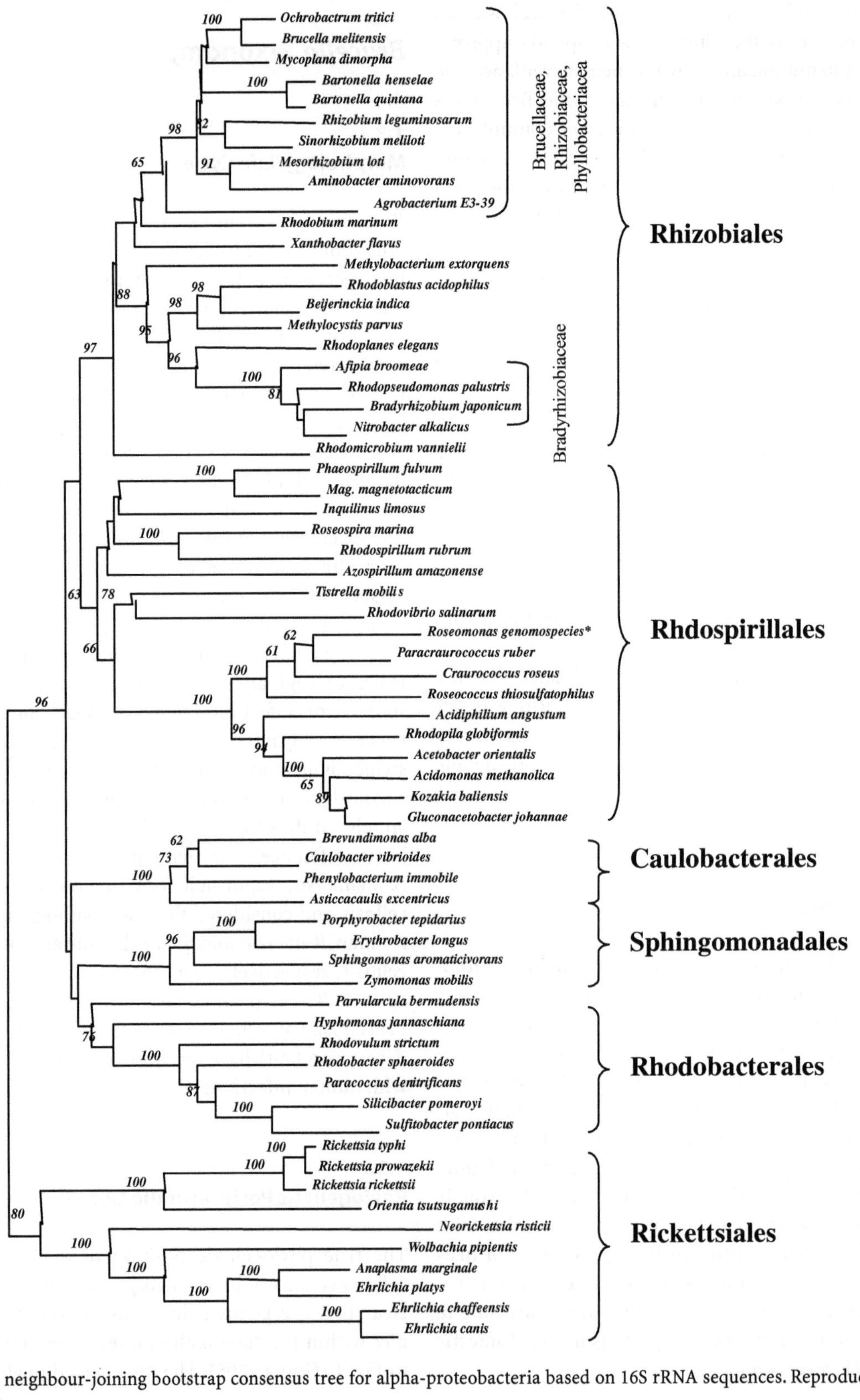

Fig. 1 A neighbour-joining bootstrap consensus tree for alpha-proteobacteria based on 16S rRNA sequences. Reproduced with permission

position of *Brucella* within the alpha-proteobacteria was confirmed on the basis of ribosomal cistron similarities and 16S rRNA sequence comparisons (De Ley et al. 1987; Moreno et al. 1990; Yanagi and Yamasato 1993). Thus, although *Brucella spp.* are animal pathogens, they share close relationships with organisms that inhabit soil (e.g., *Ochrobactrum spp.*), that establish symbiotic relationships with plants (e.g., *Rhizobium spp.*) or that are phytopathogens (e.g., *Agrobacterium spp.*). The family *Brucellaceae* consists of the genera *Brucella*, *Mycoplana* and *Ochrobactrum* with *Ochrobactrum spp.* being phenotypically and genetically most closely related to *Brucella* (De Ley et al 1987; Holmes et al. 1988; Velasco et al. 1998; Lebuhn et al. 2000). The *Ochrobactrum* constitute, in contrast to *Brucella*, a rather diverse genus (Whatmore et al. 2005; Lebuhn et al. 2006) that currently consists of six species. Among these *O. intermedium* has been described as most phylogenetically and taxonomically related to *Brucella* (Velasco et al. 1998; LealKlevezas et al. 2005; Lebuhn et al. 2006). Indeed, based on internal spacer region 1 (ITS1) sequence, it has recently been suggested that the genus *Brucella* actually falls within the *Ochrobactrum* clade. This led to the suggestion that *Brucella* should be regarded as *Ochrobactrum* species from a scientific point of view, although the need to retain the name *Brucella* to distinguish a group of clinically important organisms was acknowledged (Lebuhn et al. 2006)

1.3.3 *Brucella* Taxonomy

The internal taxonomy of the genus has been an area of some controversy in recent years (Moreno et al. 2002), with the status of the six classical species described above being questioned. The species have classically been identified on the basis of distinct host specificity and a biotyping approach based on serotyping, phage typing, dye sensitivity, CO_2 requirement, H_2S production and metabolic properties (Alton et al. 1988). This scheme identifies both the six species and further subdivides *B. abortus*, *B. melitensis* and *B. suis* into biovars (seven, three and five, respectively). However, DNA–DNA hybridization studies carried out within the genus revealed a high degree of homology (>90%) between the six species (Hoyer and McCullough 1968a, b; Verger et al. 1985) and, on this basis, it was proposed that *Brucella* should constitute a monospecific genus (Verger et al. 1985; 1987). This proposal was supported by the Subcommittee on the Taxonomy of *Brucella*, with *B. melitensis* becoming the sole representative species and the other species being considered biovars of *B. melitensis* (Anonymous 1988).

However, the move to a monospecific genus failed to find widespread support among the scientific community largely for practical reasons, with most preferring to use the nomenspecies designation. Furthermore, in the years following this decision, the development of molecular approaches allowed the identification of many species-specific markers that largely supported the classical species divisions (Moreno et al. 2002). While low resolution approaches such as 16S rRNA gene sequencing (Gee et al. 2004) confirmed the high degree of genetic conservation within the genus, approaches such as pulsed-field gel electrophoresis (PFGE) (Michaux-Charachon et al. 1997), examination of IS*711* insertion sites (Bricker and Halling 1994, Ouahrani et al. 1996), restriction fragment length polymorphism (RFLP) analysis of various *omp* gene fragments (summarized in Vizcaíno et al. 2000), amplified fragment length polymorphism (AFLP) (Whatmore et al. 2005), infrequent restriction site (IRS) PCR (Cloeckaert et al. 2003), variable number of tandem repeat (VNTR) analysis (Le Fleche et al. 2006; Whatmore et al. 2006) and multi-locus sequence analysis (MLSA) (Whatmore et al. 2007) all provide broad support for the classical species groups. However, there are some exceptions that should be noted particularly with regard to the *B. suis* group in which the failure to identify species-specific markers has been highlighted previously (Moreno et al. 2002). This reflects the very close genetic relationship between *B. canis* and *B. suis* and the inclusion of *B. suis* biovar 5, that by a variety of methods appears not to cluster with other members of the species, within the group. This overall picture of molecular differences consistent with classical species, in conjunction with the clear host preference relating to each of the classical species, and the practical value of designations that relate to differential pathogenicity for man led to recent moves to reverse the monospecies designation. These culminated with a decision to return to the pre-1986 taxonomic opinion on the genus *Brucella* (Osterman 2006).

The taxonomic situation has been further complicated by the identification of novel *Brucella* isolates from marine mammals in recent years. These isolates clearly belong to the genus *Brucella*, again showing high DNA:DNA hybridization values with other members of the genus (Verger et al. 2000) but are phenotypically and molecularly distinct from all the existing terrestrial *Brucella* species. A number of controversial names have been suggested for these new isolates though none has yet been formally accepted. Initially the name *B. maris* was suggested (Jahans et al. 1997) but as it became clear that there was considerable internal diversity within the group, subdivision into two species, *B. pinnipediae* (seals) and *B. cetaceae* (dolphins and porpoises), named according to classical criteria of host preferentialism was suggested (Cloeckaert et al. 2001). However, it is becoming clear that there is further subdiversity within these groups that likely corresponds to host specificity (Boschiroli et al. 2001; Groussaud et al. 2007). This is consistent with the division into at least three species congruent with *B. phocae* (seals), *B. phoecoenae* (porpoises) and *B. delphini* (dolphins) as described by Corbel and Banai (2005).

Future work will be required to clarify the status of anomalous isolates, such as *B. suis* biovar 5 described earlier, to confirm the placement of a number of newly described *Brucella*-like isolates (both the marine mammal isolates described earlier and other isolates identified in terrestrial hosts), to clarify the phylogenetic relationship between *Ochrobactrum* and *Brucella* and to update the current minimal standards for species designation within the group (Corbel 1975).

1.4 Molecular Epidemiology and Population Dynamics of *Brucella*

The remarkable genetic homogeneity among *Brucella* species has hindered the discovery of unique DNA sequence targets exploitable for the identification and differentiation of *Brucella* species and strains. Such targets have been sought for a variety of purposes. Stable genetic markers are needed for unequivocal identification of *Brucella* pathogens, at the genus or preferably at the species level, for diagnostic application. Other genetic markers are wanted to clarify the controversial and complex taxonomy of the *Brucella* family. Recently, there have been renewed efforts to find highly polymorphic targets for subtyping *Brucella* strains beyond the species level, to aid in epidemiological investigations of *Brucella* outbreaks.

1.4.1 Tools for Molecular Epidemiology

There are now many techniques and applications that use *Brucella* molecular markers for diagnostic and taxonomic purposes, a subject that has recently been reviewed elsewhere (Vizcaino et al. 2000; Moreno et al. 2002; Bricker et al. 2003; Cloeckaert 2004; Al Dahouk et al. 2005). However, until very recently, little had been published about molecular targets that could differentiate *Brucella* strains at a level suitable for epidemiological use. The highly conserved and clonal nature of the brucellae rendered useless many of the conventional methods and targets for strain discrimination.

Historically, *Brucella* typing has been based on a range of phenotypic traits resulting in the current biovar (biotype) system (Alton et al. 1975). However, this methodology is very time consuming, involves handling of the live pathogen and results in only a very small number of subtypes. With the advent of molecular based methods for strain typing, this technology was quickly applied towards typing *Brucella*. Early on, molecular strain typing was based on the identification of genetic polymorphisms in randomly or semi-randomly selected DNA fragments. This approach was attractive because little needed to be known about the specific DNA sequences of the fragments under study, since the specificity came from how the fragments were generated. Some of the methods that were successful in subtyping *Brucella* into groups at or below the species level include PFGE (Allardet-Servent 1988); RFLP (Ficht et al. 1990; Bricker et al. 2000); AP/RAPD-PCR (Fekete et al. 1992; Tcherneva et al. 2000); IS6501(IS711)-anchored PCR (Ouahrani-Bettache et al. 1996); ERIC-PCR and REP-PCR (Mercier et al. 1996; Tcherneva et al. 1996) and most recently IRS-PCR and AFLP (Cloeckaert et al. 2003; Whatmore et al. 2005). In general, the major disadvantages of these methods are that they are either time consuming (e.g., PFGE, RFLP and

AFLP) or are difficult to standardise from laboratory to laboratory (e.g., AP/RAPD-PCR and ERIC/REP-PCR). Another drawback of this approach is that it can be difficult to put the resulting data into proper genetic context, because much of the data is based on variations in unknown DNA sequence. Over time, as some *Brucella* DNA sequence data became available and PCR revolutionized DNA methodology, typing techniques became more specific and more reproducible but not highly discriminating (e.g., PCR-RFLP (reviewed by Al Dahouk et al. 2005).

Beginning in 2002, the complete genome sequences of the three main *Brucella* species: *B. melitensis*, *B. suis* and *B. abortus*, were published (Paulsen et al. 2002; DelVecchio et al. 2002a–c; Halling et al. 2005). The complete sequence of a second *B. abortus* strain, strain 2308, followed in 2005 (Chain et al. 2005). The opportunity to compare genome sequences has accelerated marker discovery and several new methods have been published.

One new approach exploits the accelerated mutation rates associated with reiterated sequences known as microsatellites or VNTRs. The mutations associated with VNTRs involve the loss or gain of complete repeat units. These incremental mutations principally arise from replication errors as the polymerase passes through a repeat region that has developed slipped-strand mispairing (Levinson and Gutman 1987). The mutation rates at tandem repeat loci are variable, influenced by a range of factors including the size of the repeat sequence, the number of repeat units per locus, the regional DNA topography and the flanking DNA sequence.

Because of the potential for high mutation rates, VNTR loci can be highly polymorphic, even in highly conserved genomes (Renders et al. 1999; Keim et al. 2000; Farlow et al. 2001). To maximise strain discrimination potential, the genetic fingerprints, or genotypes, are typically constructed from multiple VNTR loci. Following publication of the first fully sequenced *Brucella* genome (DelVecchio et al. 2002a–c), VNTR markers were used to genotype *Brucella* strains (Bricker et al. 2003). The method, called HOOF-Prints, was based on independent loci consisting of tandem repeat units of eight nucleotides. Initially, the HOOF-Print assay examined eight VNTR loci, though currently 12 loci are analyzed. The loci selected for this assay range from highly polymorphic, with many alleles identified among *Brucella* strains, to slightly polymorphic, with only a few alleles identified so far (Bricker et al. 2003; Bricker and Ewalt 2005). Some loci are non-polymorphic for a specific group or species. A second multi-locus VNTR analysis (MLVA) typing system was developed by Le Fleche and colleagues (Le Fleche et al. 2006). They examined 80 potential tandem repeat loci from which 15 loci were selected for routine use. A few months later, a third MLVA typing system was published (Whatmore et al. 2006). This assay evaluated 21 polymorphic loci, including the loci used in the HOOF-Print assay. All three assays have been very successful in discriminating among *Brucella* strains and will undoubtedly prove helpful for epidemiological investigations of brucellosis outbreaks.

1.4.2 Population Dynamics and How They Can Impact Molecular Epidemiology

To date, much of the molecular typing research has focused on achieving maximum discrimination among strains. However, the goal of molecular epidemiology is to identify genetic linkages. Therefore, to be useful, the molecular markers selected must be able to provide genetic linkage information as well as being able to discriminate strains. But, evolution is a continuous process, and genetic relationships may be obscured if only the terminal nodes of bacterial lineages are considered. Critical to untangling the genetic relationships is the understanding how population ecology and evolutionary processes shaped the specific *Brucella* population that was present in the host at the time when the isolation was made.

Very little research has been reported about the ecology and evolution of *Brucella* populations during infection (Moreno 1997, 1998; Moreno et al. 2002). Evidence indicates that *Brucella* mainly propagate clonally (Moreno 1997). Nevertheless, genetic changes caused by mutations, rearrangements, gene conversions and, to a lesser extent, horizontal gene transfer appear to be evident in the *Brucella* genomes, and each of these processes contributes to the genetic evolution of each organism. External forces such as immune responses, the intracellular environment, host behavior and transient exposure to the environment outside the host impact the fitness and survival of every organism within the population. Combined, these internal and external factors create the context in which each genotype develops.

The following factors are particularly important to consider from an epidemiology perspective and warrant further investigation:

Homoplasy is the convergent evolution of similar traits by unrelated individuals or populations. In the context of *Brucella* epidemiology, this can confound interpretation when unrelated populations evolve similar or identical VNTR fingerprints. Homoplasy can be minimized by considering a number of independent loci, as each additional locus decreases the probability of a random match. All three current VNTR assays incorporate multiple loci, diminishing the likelihood of homoplasy; however, decreasing the number of loci examined by any of these tests could compromise the results.

Population bottlenecks and Founder effects are well known evolutionary forces in nature. They describe how a major reduction in population size will result in the loss of genetic diversity within the surviving population. The genetic composition of this new population (the founder population) provides a fresh starting point for genetic evolution during subsequent population expansion. In the case of *Brucella*, these forces are relevant during infection of the host in a dose-dependent manner. If the host is exposed to a large number of bacteria that survive the initial transmission and invasion process, the loss of genetic variability in the founding population will be small. However, if the host is exposed to a small number of organisms or a small number survives the infection process, then the loss of genetic variability in the founding population will be substantial, thus limiting the genetic potential of the expanded population. No matter how large the initial dose may be, it will always result in a downsized population and some genetic variation will likely be lost. Thus, each transmission event to another host or environment creates a population bottleneck, with a new founding population that follows its own evolutionary path.

Multiple infections by two distinct strains either simultaneously (dual infections) or chronologically (superinfections) has rarely been observed or reported in *Brucella* outbreaks (Ewalt and Harrington 1979; Deqiu et al. 2002; Etter and Drew 2006). However, to be detected, strain differences would have to be obvious, such as an alternate species or biovar or a significant change in phenotype, but at the same time exhibit an equal fitness in the host. There are practical reasons why this would rarely be observed. Recently, the powerful discriminatory power of the MLVA typing has revealed that *Brucella* populations are more genetically diverse than previously thought. Of particular note, archived stocks of the *Brucella* type strains in culture collections have been found to be genetically variable. Not surprisingly, strain populations isolated from regions with a history of enzootic brucellosis appear to exhibit greater genetic diversity than *Brucella* strains isolated from spontaneous outbreaks (unpublished observation). High genetic diversity within a population can confound the results of MLVA typing as multiple alleles are simultaneously detected. Therefore, it is prudent to examine several bacterial colonies cultured from each outbreak to create a more complete picture of the microbial population.

Mode of transmission can impact the evolution of pathogenic bacterial populations. For example, *B. abortus* and *B. melitensis* are mainly transmitted vertically from mother to offspring or from exposure to aborted tissues. In both cases, the opportunity for transmission is limited and sporadic. The chance of population bottlenecks and Founder effect is high. *B. canis* and *B. suis*, on the contrary, are shed in urine, sometimes for years, as well as through the reproductive route, and so there can be an opportunity for the host to have repeated and prolonged exposure to the *Brucella* population. This may decrease the likelihood of the Founder effect in areas with high infection levels, and it may increase the likelihood of multiple strain infections. Consistent with this, we routinely detect greater genetic diversity among isolates of *B. suis*, *B. canis* and the brucellae from marine mammal hosts than is seen from *B. abortus* isolates (unpublished observation).

Host behaviour and ecology have the ability to affect the composition of *Brucella* populations. In general, non-migratory gregarious species are confined to a local region. The associated enzootic *Brucella* populations are likely to be genetically stable and spatially limited. Hosts with an expanded home range, such as bison, feral swine, canines and caribou, have the potential to disperse their *Brucella* populations (and consequently genotypes) over a larger area and are also more likely to encounter new *Brucella* populations. In these hosts, the *Brucella* populations are predicted to have greater genetic diversity. Many marine mammals are highly migratory and very social animals, favoring their exposure to many *Brucella* populations.

Brucellosis in American elk illustrates how a change in behaviour can affect the transmission of brucellosis and potentially the dynamics of the resident *Brucella* population. Elk are migratory but typically solitary animals. Under normal conditions, the incidence of brucellosis in elk is very low. However, in areas where artificial winter feeding grounds have been established, the normally solitary animals become highly concentrated, and are still found at high densities during calving season. This has resulted in an artificially high brucellosis infection rate (Etter and Drew 2006). Mixing of animals from different home ranges on the feeding grounds provides the opportunity for dispersal of many different *Brucella* populations. More research is needed to determine the full impact on *Brucella* population diversity.

1.4.3 Conclusions

The future of molecular typing of *Brucella* strains for epidemiological exploitation appears to be bright. The new methods are highly discriminating, simple to perform and highly reproducible. It is anticipated that they will soon be routinely incorporated into epidemiological investigation protocols. However, with the advantages these new technologies offer come the responsibility to correctly interpret the data that they produce. These data are a reflection of a single time point in population dynamics and evolutionary processes that are continuous. More research will be necessary to fully appreciate how the population at that time point came to be and how it relates to other *Brucella* populations.

1.5 Physical and Genetic Map of Genomes

1.5.1 Bacterial Genomes and Their Analyses

Bacteria were centre stage in the confirmation of DNA as genetic material. Avery et al. (1944) showed that the fraction transforming *Pneumococcus* contained deoxyribonucleic acid. Just over 50 years later, the genome of the first living organism, the bacterium *Haemophilus influenzae*, was sequenced (Fleischmann et al. 1995). And, a dozen years after that, over 300 bacterial genomes including four *Brucella* genomes have been sequenced. Sizes of bacterial genomes vary greatly. *Mycoplasma*, which are amongst the smallest free living bacteria, have a genome of 0.58 Mb (Hutchison et al. 1999), while the plant symbiont *Bradyrhizobium japonicum* has a genome of 10.2 Mbp. The *Brucella* genome is 3.3 Mbp.

The first investigations of the *Brucella* genome began in the late 1960s when DNA homology was used to show that the different classical species were related (Hoyer and McCullough 1968a, b). Later, Altenbern used chemical mutagenesis to induce auxotrophic mutants in *B. abortus* S19 followed by marker frequency analysis to map specific genes (Altenbern 1973). In the 1980s, the development of new molecular biology techniques allowed the investigation of the structure and organization of the *Brucella* genome to begin in earnest.

1.5.2 Global Organization–Physical Maps

A prototype physical map with a single circular chromosome was proposed by Allardet-Servent and colleagues (Allardet-Servent et al. 1991). After it was found that the enzyme *Pac*I, which recognizes an eight bp sequence containing only A's and T's, cut the *B. melitensis* 16M genome into eight fragments, a new map with two circular replicons was proposed (Michaux et al. 1993). When undigested genomic DNA is analysed by electrophoresis, two replicons, one of 1.1 Mbp and the other of 2.2 Mbp giving a genome of approximately 3.2 Mbp, are visualized by pulsed field gel electrophoresis (PFGE). Jumas-Bilak and colleagues (Jumas-Bilak et al. 1995) used a Tn5 derivative, Tn5-Map, to introduce a unique restriction site (I-*Sce*I, meganuclease) into the replicons of *B. melitensis*, leading to the demonstration of their circularity by linearization. The ability to linearize an erstwhile circular molecule allowed the development of a rapid physical mapping technique based on the use of partial digestions and the completion of the physical maps of the reference strains of all the recognized species and biovars of the genus (Michaux-Charachon et al. 1997). Comparison of these maps shows that the organization of the genomes amongst different strains is very similar, with the major differences being a 600 kbp inversion in the small chromosome of *B. abortus* biovar 1, 2 and 4, a variety of small insertions and deletions and a 21 kbp fragment unique to *B. ovis*. Similarly, large inversions and strain-specific restriction fragments were observed among physical maps of three marine mammal strains (Bourg et al. 2007).

1.5.3 The *Brucella* Genome is Highly Conserved

The observation that each strain had a unique restriction fragment length polymorphism profile (RFLP) when subjected to PFGE has been extended by more intensive analysis of strains. It is now clear that each biovar has a unique and conserved profile and that PFGE can be used to type clinical isolates. Because in nature *Brucella* strains multiply principally within their preferential hosts, the surprising degree of conservation seen among the isolates of a given biovar may have arisen from their being genetically isolated. A constant feature of the *Brucella* genome is the absence of plasmids and temperate phage. As broad host range plasmids can be stably maintained in *Brucella* (Rigby and Fraser 1989; Verger et al. 1993; Elzer et al. 1995), it suggests that *Brucella* does not encounter other bacteria during its natural life, and thus, has little chance of incorporating new genetic material through horizontal transfer. For *Brucella*, there is no evidence of transfer of genetic material in nature. This is consistent with different strains of *Brucella* being clones, evolving with their mammalian hosts (Michaux-Charachon et al. 1997; Boschiroli et al. 2002a, b).

1.5.4 The *Brucella* Genome and α-Proteobacteria Evolution

Brucella are members of the alpha-subdivision of the proteobacteria, a very large and diverse group of gram-negative microbes (Batut et al. 2004). The α-proteobacteria show great variability in metabolic capacities, morphologies and life cycles. They are found in a wide range of ecological niches, ranging from water to soil to both extra- and intracellular associations with eukaryotes, which spans unicellular organisms to multi-cellular plants and arthropods to mammals. Complete or draft genomic sequences for 95 α-proteobacteria are available, covering the full spectrum of lifestyles from pericellular parasites of plants (*Agrobacterium*) to facultative intracellar pathogens of mammals (*Brucella* and *Bartonella*) and symbionts of plants (*Sinorhizobium*) to obligate intracellular pathogens of mammals and arthropods (*Rickettsia*, *Erlichia*, and *Wolbachia*). *Caulobacter crescentus* stands apart from these in that it is free-living (Tsolis 2002).

Bioinformatical analysis of the frequencies of deletions, duplications and gene genesis events underlying the observed genome size variations in the α-proteobacteria subdivision suggest that the genome of the common ancestor of the α-proteobacteria contained between 3,000–5,000 genes (Boussau et al. 2004). Two major trends associated with the conversion of this ancestral gene pool into the genomes of the actual species are recognised: intracellular bacteria associated with invertebrates, animals and humans have evolved by gene loss, whereas the soil growing, plant-associated bacteria have evolved through genome size expansion. This results in a nine-fold difference in genome sizes (from 1 to 9 Mb). The dramatic variations in gene content over time have primarily affected the number of genes coding for proteins involved in regulation, transport and small molecule metabolism. Despite these genome size differences and the vast phenotypic diversity, the coherence of the α-proteobacteria subdivision is supported by homologies of rRNA and other genes. The obligate intracellular α-proteobacteria with very small genomes (*Rickettsia, Wolbachia*) cannot replicate outside their eukaryotic host, and *Bartonella* replicates within cells far faster than on synthetic media. *Brucella*, however, can grow axenically in defined medium and do survive (even if they do not significantly multiply) in this environment. The *Brucella* genome, with just over 3,000 genes encoded by 3.3 Mbp, is clearly in the early stages of genome reduction as it adapts to life in its intracellular niche.

1.5.5 Multiple Genomes and Genome Rearrangements

A common theme amongst the α-proteobacteria is the occurrence of complex genomes, with a large number of species having multiple circular and, sometimes, linear replicons (Jumas-Bilak et al. 1998a). Following the discovery of two replicons in *B. melitensis*, a fascinating observation was made when the intact genomes of the four *B. suis* biovars were separated by PFGE. The genome of *B. suis* 1330 (bv 1) had a similar structure to those of *B. melitensis* and *B. abortus*, while those of strains Thomsen (bv2) and 40 (bv4) differed. In these two strains, the small chromosome was larger (1.35 vs. 1.2 Mbp) and the large chromosome smaller (1.85 vs. 2.1 Mbp). However, in biovar 3, strain 686, a single circular replicon of 3.3 Mbp was observed. Analysis of the physical maps showed that these three possible genomic structures are most probably the products of recombination events between *rrn* loci. Two of the three *rrn* loci occur on Chr I. At the *rrn* loci on Chr II, there is a transposon flanked on each side by a t-RNA gene.

It was proposed that *Brucella* strains evolved from an ancestor with a single circular chromosome (Jumas-Bilak et al. 1998b). However, so far only one strain (*B. suis* 686, bv3) has been found to contain a single circular replicon. Considering the recent theory of the evolution of the genomes of intracellular α-proteobacteria by gene loss and the fact that the closely related *Ochrobactrum* also has two circular chromosomes of similar sizes to those found in *Brucella*, it is more likely that the ancestor of the genus had two chromosomes and that strain 686 is a unique occurrence.

1.5.6 Mobile Genetic Elements, Repeat Elements and Genomic Rearrangements

Mobile genetic elements such as transposons (Tn) and insertion sequences (IS) confer instability on genomic structures by generating chromosomal rearrangements such as insertions, deletions and inversions. However, they have not greatly affected the physical maps of *Brucella*. Only two genetic elements from *Brucella* are clearly recognizable as transposons, Tn*1953* and Tn*2020* (Halling and Zuerner 2002), which is bordered by IS*2020* and tIS*2020A*. Both of these transpsons have been associated with deletions. As discussed earlier, Tn*1953* is a large transposon common to all *Brucella* analysed except *B. ovis* for which most of the transposon is missing. The loci containing Tn*2020* is consistent, with a copy of Tn*2020* having transposed into the IS*2020* of a second Tn*2020* oriented in the same direction, deleting most of the other transpson copy and all except for a short portion of its IS*2020*. Of course if the two copies were distant from each other when this occurred, any sequences between them were deleted as well. A further deletion occurred in *B. melitensis* 16M, likely by recombination between IS*2020* copies thereby deleting sequences internal to the transposon, leaving a single IS*2020* copy in *B. melitensis* 16M.

The insertion sequence IS*711* (Halling et al. 1993), also designated IS*6501* (Ouahrani et al. 1993), appears to be ancestral as copies are found at identical loci among the *Brucella spp*. Further, the sequences of these IS*711* elements are polymorphic, supporting genetic drift in situ (Halling et al. 1993). IS*711* can transpose as sites of insertion and number of copies varies widely amongst the *Brucella* classical species. (Ouahrani et al. 1993; Bricker and Halling 1994). Though IS*711* and other insertion sequences appear to have had a minimal affect on the physical map with regards to deletions and rearrangements, a 2,774 bp fragment containing IS*711* and IS*BmI* related sequences found in *B. melitensis* and *B. suis* is missing from *B. abortus* (Halling et al. 2005). Both *B. ovis* and the *Brucella* marine mammal isolates have many copies of IS*711* (Ouahrani et al. 1993; Bricker and Halling 1994; Bricker et al. 2000) relative to *B. abortus*, *B. suis* or *B. melitensis*. IS*711* appears to be the only active transposable element in *Brucella* and, hence, has the potential to participate in evolution.

Repeated sequences are another common feature of bacterial chromosomes. Even though REPs or repetitive extragenic palindromic DNA sequences of 21–65 bp constituting more than 0.5% of the extragenic sequences have been identified in several α-proteobacteria, they were not observed in *B. melitensis* (Tobes and Ramos 2005). There are, however, a total of 54 copies of two longer (103–105 bp) small repetitive palindromic elements, Bru-RS1 and a related (65% similarity) element Bru-RS2 (Halling and Bricker 1994; Halling et al. 2005) in the *Brucella* genome. These elements are mostly extragenic. Only two of the Bru-RS1 copies are found within open reading frames. One is within one of two proline racemases, namely BruAb1_0363, and the other within the probable transcription regulator BruAb1_1398. Bru-RS1 and Bru-RS2 appear to have transposed throughout the genome to sites similar in sequence to those of IS*711*, and, indeed, these elements have been found side by side in several instances. As the marine isolates have many more copies of IS*711*, determination of the IS*711* insertion sites relative to Bru-RS1 and Bru-RS2 would be informative.

1.5.7 Genomic Polymorphisms: Transposons, Phage and Plasmid Associated Sequences

Other than the occasional rearrangement, the genomes appear to be relatively stable. There are a number of insertion sequences and remnants of phages and plasmids (Paulsen et al. 2002), suggesting that the genome has evolved in several steps. First, there was a period during which a great deal of horizontal transfer occurred. This period was followed by genome reduction as described for intracellular α-proteobacteria during which transposon, phage and plasmid sequences that conferred no selective advantage were lost. More recent evolution caused by adaptation to different mammalian hosts has led to *Brucella* speciation.

The two largest regions of DNA that have clearly been acquired by horizontal transfer are Tn*1953* described earlier and an 18 kbp region found in the large chromosome of *B. suis* 1330 that shows similarity to IncP plasmids and bacteriophage. The 18 kbp region is present in all biovars of *B. suis, B. canis, B. neotomae* and the marine isolates but absent from *B. melitensis, B. abortus* and *B. ovis*. The region is integrated into a hotspot at the *guaA* gene and appears to have been acquired by site specific recombination. Interestingly, the region was seen to actively excise in *B. suis* 1330 (Lavigne et al. 2005); however, clones having lost the region were not isolated, suggesting a selective advantage, possibly an anti-toxin. To date, none of the regions believed to have been acquired by horizontal transfer have been shown to play any role in virulence; however, several genes involved in virulence, such as MgtC (Lavigne et al. 2005), and genes encoding the twin arginine translocator system (Lavigne and O'Callaghan unpublished data) have low GC content and are next to tRNA loci.

1.5.8 Genetic Maps: Synteny

Conservation of gene order, or synteny, is a clear indication of a close evolutionary relationship. Bacteria within a taxonomic family, that is having a common ancestor, would be expected to have higher synteny amongst themselves than to bacteria of another family. During evolution, synteny would be disrupted by molecular rearrangements caused by recombination leading to inversions and deletions, acquisition of horizontally transferred DNA, illegitimate recombination via transposable elements, selection during niche adaptation and genetic drift. Analyses of the sequenced genomes of the *Rhizobiales*, including that of *Brucella*, show considerable regions of synteny. Paulsen and colleagues (Paulsen et al. 2002) reported that of the 3,388 orfs that they predicted in the *B. suis* 1330 genome, 1,902 had close homologs in the genomes of *A. tumefaciens*, *S. meliloti* and *M. loti*, while 2,408 had a close homolog in at least one of the three. Guerrero and colleagues (Guerrero et al. 2005) have performed a detailed bioinformatic analysis of syntenic genes in the *Rhizobiales*. Comparison of the *Brucella* genome with that of *S. meliloti* shows that 76% of genes are spread over 173 short regions of synteny. Further, syntenic gene products were found to have higher levels of identity than the non-syntenic genes. Interestingly, comparison with the *Enterobacteriales* showed similar levels of syntenic conservation but with different gene orders.

1.6 Genome Sequencing and Bioinformatics Resources

1.6.1 Introduction

Brucellae are gram-negative bacteria that are intracellular pathogens. Although their hosts are primarily animals, humans can also become infected (Godfroid et al. 2005). The bacteria invade their target host though the mucosa of nasopharyngeal cavities or the gastrointestinal tract and are then phagocytized by host macrophages (Cardoso et al. 2006). For most bacteria, being engulfed by a macrophage is a situation to avoid, as these eukaryotic cells use a variety of microbiocidal functions to destroy invaders that they capture (Celli 2006). *Brucella*, in contrast, actually require this process in order to replicate and survive. This makes them members of an exclusive bacterial class that have evolved unique methods to survive in extremely hostile environments. For *Brucella*, survival relies on its membrane-bound compartment (called the *Brucella*-containing vacuole or BCV) to avoid fusion with lysosomes (Celli 2006).

The availability of the genomic data has created abundant opportunities to understand pathogens and their virulence, pathogenicity and host-specificity. This chapter provides a detailed account of the high-throughput technologies currently available, their use in sequencing the genomes of *Brucella* species and the availability of and the need for bioinformatics resources to make use of the data. The generation of vast amount of genomic and post-genomic data also has created enormous challenges for curation, integration, interoperability and data-mining, which are necessary to draw meaningful inferences, to further the pathosystems biology field and to facilitate developing safer and more specific countermeasures in human and animals.

1.6.2 Genome Sequencing

High-throughput sequencing is now a powerful technique and has become part of biological research. Whole-genome sequencing enables the identification of novel genes and gene variants, which would help in understanding the mechanisms of host–pathogen interactions, host preference, pathogenicity, virulence; and also provides the basis for identification of targets for drugs, diagnostics and vaccines. The concept of personalized medicine is becoming a reality due to the possibility of sequencing individual genomes rapidly and at low cost. There is little doubt that genetic screening will become routine in future medical care.

1.6.3 High-Throughput Sequencing Technologies

The DNA sequencing process dates back to 1970s (Sanger et al. 1977). However, it is in the last decade that this field has gone through the most amazing transformation (Service 2006). At the time of writing this chapter, there are approximately 400 published annotated, whole-genome sequences with many more in the pipeline (Fig. 2).

The worldwide market of sequencing and analysis products in 2003 alone was US $2.9B. Currently, the cost of sequencing for an average user of a core facility varies from $6–23, with an average cost of $0.01 per base. Although this has come down exponentially over the last decade, some of the

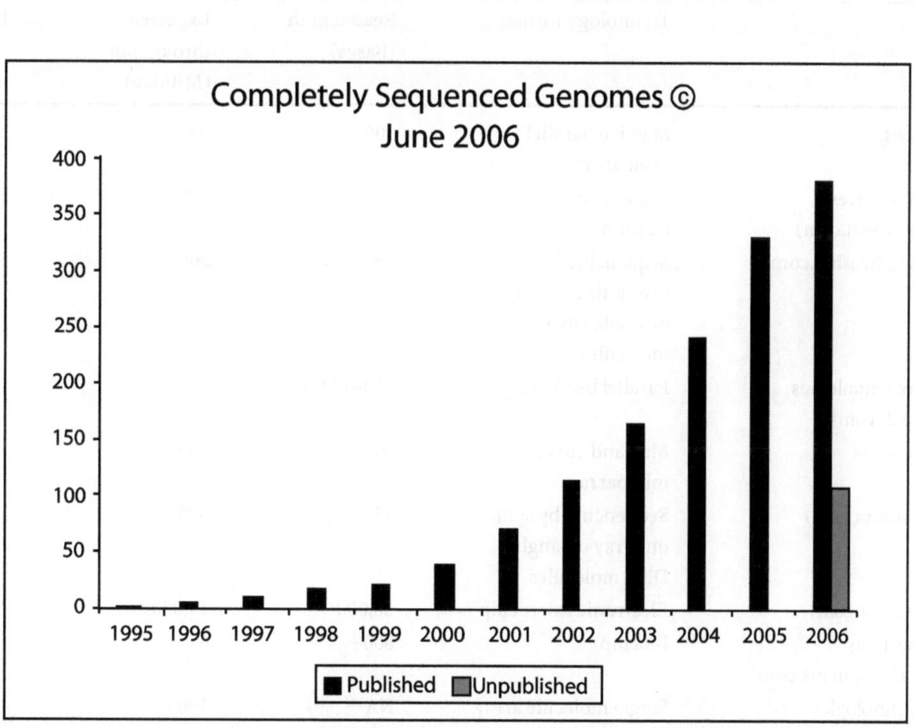

Fig. 2 The increase in the number of sequenced genomes since 1995 (Reproduced with permission from www.genomesonline.org.)

recent strategies and technologies, such as 454 Life Sciences, Agencourt and Solexa, have further driven the cost down and increased the throughput and have hence emerged as alternatives to sequencing genomes using capillary electrophoresis and Sanger sequencing (Marusina 2005) (Table 1). A summary of these technologies is given at the website of Genetic Engineering News (http://www.genengnews.com/sequencing/). These technologies can be grouped into three main categories: Sequencing by Synthesis, Sequencing single DNA molecules and Nanopore sequencing technologies. A brief description of their unique features is given in Table 2. Among these technologies, the Parallel bead array technology is most advanced, with the sequencing machines currently being in the market.

The parallel bead technology (Margulies et al. 2005) was used to rapidly and comprehensively sequence *B. abortus* strain S19 by the Sobral research group at the Virginia Bioinformatics Institute (Crasta et al. 2007, unpublished data). The process uses nebulized DNA fragments and does not require prior cloning. Two runs of sequencing generated over 800,000 sequence reads with an average read length of 100 bp covering over 99.5% of the genome. The gaps, as estimated by aligning the contigs to the closely related genome sequence of the strain 9-941 (Chain et al. 2005; Halling et al. 2005), were mostly less than 300 bp and not more than 1.5 kb. Hence, traditional sequencing using the sequences from the adjacent contigs in PCR-amplification of the genomic DNA and subsequent sequencing efficiently completed the whole-genome sequence.

1.6.4 *Brucella* Genome Sequences

The complete genome sequences of four species/strains of *Brucella* are publicly available (Table 3). All the available genome sequences were published within the last five years and several more (such as *B. abortus* strains S19 and RB51, *B. ovis*) are being sequenced or analysed and are expected to be in the public domain within a year.

The genomes of the members of *Brucella* are strikingly similar (Table 3). Each species within the genus has a genome size of approximately 3.28 Mb. The genome consists of two circular chromosomes,

Table 1 Overview of available rapid genome mapping and sequencing technologies (modified from Service 2006)

Company	Technology format	Read length (Bases)	Expected throughput (MB/day)	In the market
454 Life sciences (www.454.com)	Massive parallel bead array	100	96	Yes
Agencourt Biosciences/ ABI (www.agencourt.com)	Sequencing by ligation	50	200	No
Helicos (www.helicosbio.com)	Sequencing by synthesis on array of single DNA molecules	25	500	Early access program
Microchip Biotechnologies (www.mcbiotech.com)	Parallel bead array	850–1,000	7	No
NimbleGen systems	Map and survey microarrays	30	100	Yes
Solexa (www.solexa.com)	Sequencing by synthesis on array of single DNA molecules	35	500	Early Access Program
LI-COR (www.licor.com)	Electronic microchip	20,000	14,000	No
Network Biosystems (www.networkbiosystems.com)	Biochip	800+	5	No
VisiGen Biotechnologies (www.visigenbio.com)	Single molecule array	NA	1000	No

Table 2 Strategies and methodologies of the rapid sequencing technologies*

Sequencing by synthesis (e.g. 454, Solexa, Agencourt-ABI)	Sequencing single DNA molecules (LI-COR, VisiGen)	Nanopore sequencing technologies
Nebulizing or fragmenting by other means of the genome into 200 to 500-bp fragments*	Isolation of single molecules of DNA	Isolation of single molecules of DNA
Ligation of the adaptors, bead capture or micro-array capture, and clonal amplification of each captured fragment	Synthesizing new strand on single DNA molecule with anchored reagents	Forcing DNA to travel through nanopores
Sequencing of 35–100 b from one end of each clone	Identification of bases on new strand by fluorescence	Stabilisation of this flow in electric fields
Assembly of contigs from a pool of overlapping fragments	Cycling of the above reaction to assemble really long reads on a single strand	Read of bases of the DNA as it emerges from the nanopores by fluorescence or optical density
Close the gaps by directed sequencing or by sequencing of both ends of the template	Close the gaps by additional cloning and sequencing	Close the gaps by additional cloning and sequencing

Extracted with permission from 'Report on Novel Rapid Sequencing Technologies' By BioXcel Corporation, New Haven, CT, 2006

Table 3 Statistics for the available genomes of *Brucella*

Species	Total genome (MB)	Chr. I (Mb)	Chr. II (Mb)	ORFs	Proteins	RNAs	G + C content (%)
Brucella abortus biovar 1 str. 9–941	3.286	2.124	1.162	3,296	2,458	64	57.25
Brucella melitensis 16M	3.295	2.117	1.178	3,197	3,198	66	57.25
Brucella melitensis biovar Abortus 2308	3.278	2.121	1.157	3,350	3,034	68	57.25
Brucella suis 1330	3.315	2.108	1.207	3,388	3,271	64	57.25

the largest, Chromosome I, being on average 2.11 MB. Chromosome II is approximately 1.17 MB. The number of predicted open reading frames (ORFs) range between 3,197 and 3,388 (Chain et al. 2005; Halling et al. 2005). The number of predicted proteins range between 2,458 for *Brucella abortus* 9-941 to 3,271 for *Brucella suis* 1330, with an average number of 2,975 proteins per genome. The predicted RNAs average 65 per genome, with a narrow range of 64–68. The G+C content of all the genomes is the same, being 57.2% for Chromosome I and 57.3% for Chromsome II (DelVecchio et al. 2002a–c; Paulsen et al. 2002; Halling et al. 2005).

The *Brucella* genomes are not only similar in size and content, they also have similar organisation with a high degree of similarity among orthologs, with an average nucleotide or amino acid sequence identity of more than 94% (Chain et al. 2005; Halling et al. 2005). In fact, few fragments could be considered to be truly unique among the three genomes. Fragments previously identified as being in only *B. suis* or *B. melitensis* were now found to be also in *B. abortus*. Considering that the distribution for housekeeping genes responsible for DNA replication, transcription and translation has been well documented for *B. melitensis* (DelVecchio et al. 2002a–c), as well as genes responsible for core metabolism and cell wall synthesis (DelVecchio et al. 2002a–c), one can only assume that due to the strong similarities between *Brucella* genomes, these important regulatory genes will have similar distribution within other *Brucella*.

Comparative genomic analysis also has revealed a remarkable conservation of genome structure and metabolic capabilities between *Brucella spp.* and other members of the α-proteobacteria (Tsolis 2002; Paulsen et al. 2002), and characterization of a type IV secretion system used by *Brucella* to survive within the host phagosome (O'Callaghan et al. 1999; Boschiroli et al. 2002a, b). Cutler et al. (2005) provide an excellent review of the current standing of comparative genomics in *Brucella*.

Brucella spp. are very similar when comparisons of the contents of their individual genomes are made; and yet, they are unique. They have different host preferences, have different levels of pathogenicity within those hosts and have different antibiotic resistances (Halling and Jensen 2006). So despite being similar in a genomic sense, there are other levels at which they differ. Not surprisingly, the proteomes of these organisms are now under intense scrutiny, although at a more preliminary stage than the postgenomic analysis. Two-dimensional gel and mass peptide fingerprinting have been used to compare expression patterns between *Brucella* (Eschenbrenner et al. 2006), and also to identify immunogenic membrane proteins from virulent strains (Khan et al. 2006). Ding et al. (2006) have developed a high efficiency cloning and expression system that combines the synthesis and purification of predicted proteins arrayed in a configuration and platform similar to microarray to examine sero-reactivity of these proteins against serum from immunized animals. The National Institute of Health has funded a five-year (2004–2009) large-scale proteomics initiative to facilitate target identification in *Brucella* and *Brucella*-infected macrophages. The results are expected to elucidate the phagosome composition changes, virulence, plasma membrane changes and MHC peptide presentation. The data will be made available by the Proteomics Resource Center (www.proteomicsresource.org). The rewards of this type of analysis could be valuable, especially in the areas of developing countermeasures against emerging and re-emerging infectious diseases of animals and human.

Unfortunately, *Brucella* has a long history as a bioweapon. It was one of the agents used experimentally by the Japanese before and during World War II (Pappas et al. 2006a, b). In 1952, *B. suis* was the first agent weaponized by the United States (Christopher et al. 2005), and it was also one of the agents developed for offensive purposes by the former Soviet Union (Alibek and Handelman 1999). Supposedly, the Soviet Union developed strains that were resistant to antibiotics and untreatable (Alibek and Handelman 1999), and questions remain about the subsequent where abouts of these resistant strains following the collapse of the Soviet Union. In the unlikely occurrence of a bioterrorism event, it would be important to identify what the organism was before prescribing a specific treatment. Finding an efficient phylogenic grouping technique has become a central goal for investigating the genetic relationship between clinical pathogenic strains and reference strains and for epidemiological surveillance and public health decisions for dealing with certain strains of pathogenic bacteria (Hommais et al. 2005). These methods, which can involve a variety of techniques that range from multilocus enzyme electrophoresis (MLEE), ribotyping, random amplified polymorphic DNA analysis, fluorescent amplified-fragment length polymorphism (FAFLP) analysis, PCR phylotyping using the presence/absence of three genomic DNA fragments, to analysis of variation at mononucleotide repeats in intergenic sequences (Hommais et al. 2005), could be used to reveal any antibiotic-resistance that might be involved. They could also be used to identify a potential source of the infection, which has been of interest in the 2001 intentional release of *Bacillus anthracis*.

Scientists at Texas A&M University have been involved in a two decade long study attempting to define disease resistance to brucellosis in cattle (Westhusin et al. 2007). They recently cloned a bull from a deceased individual that had been documented to be fully resistant to *B. abortus*, *Mycobacterium bovis* and *Salmonella typhimurium*. The cloned bull had similar resistance. A comparative genomic approach could be used to isolate polymorphisms unique to this animal that might be used to create a screening process to identify other animals that were also resistant. It is possible to imagine a further step of selective breeding to generate individuals that are resistant to the disease. Furthermore, these polymorphisms could conceivably lead to the identification of specific proteins involved in disease resistance. Considering the problems and cost associated with keeping herds at a status where they can be certified as *Brucella*-free (England et al. 2004), this type of approach could be very valuable.

1.6.5 Bioinformatics Resources

Advances in genomics and computer technologies are fuelling the development of a database system using digital computation, data, information and networks. One of the major challenges during the post-genomic era is the integration of diverse data sets (Stein 2003). The main goal of bioinformatics a decade ago was to create primary databases from the high-throughput data, mainly from sequencing (Kanehisa and Bork 2003). The current focus is to extend the databases for quantitative data from transcriptomes and proteomes and to provide interoperability between multiple disciplines (Snyder et al. 2007). It is clear that a transdisciplinary approach is inevitable for storing, integrating, displaying the large-data sets based upon distributed computing, information and communication technology (Atkins 2003). The bioinformatics resources developed under such endeavors are expected to assist the scientific community in making use of the large datasets in elucidation of the mechanisms of phenotypic expression. The goal of development bioinformatic resources is, therefore, to provide a scalable, flexible and interoperable technology framework that facilitates discovery of targets, and mechanisms for further validation.

1.6.6 Biological Data Curation in the Age of High-Throughput Technologies

Biological data curation may be defined as the acquisition, storage, analysis, integration and dissemination of biological information. In 1995, the completion of the *H. influenzae* genome sequence marked a watershed in biological research (Fleischmann et al. 1995). Hitherto research in biology was driven primarily by reductionist approaches that attempt to explain natural phenomena by a study of individual components of multicomponent systems (Strange 2005). The development of high-throughput sequencing and other technologies changed the nature of biological experimentation both in terms of the scale and types of data that research laboratories produce. Genomics, transcriptomics, proteomics, metabolomics, lipidomics and phenomics technology platforms have been developed to provide systems-wide measurements for almost all cellular components (Joyce and Palsson 2006). These global, "omics" approaches allow the simultaneous study of different biological elements (genes, RNA, proteins, metabolites), which provide a level of understanding that is impossible to achieve through the reductionist studies.

The generation of high-throughput data presents unique opportunities and challenges for biological data curation. A "sequence-centric" approach to curation of information on an organism typically starts with the annotation of its genomic sequence. In general, the goal of genome annotation is to provide a complete functional description of each gene in an organism and, at the very least, includes the identification of the location and function of each coding sequence. The advent of systems-level biological research makes it possible to provide the functional description of entire networks of cellular components. Specifically, high-throughput data sets provide a powerful complement to the results of traditional reductionist studies that focus on individual molecules, thus allowing for the annotation of biological phenomena at both global and local levels. However, these opportunities are not without challenges; first, the sheer volume of data calls for creative information technology solutions for the storage and display; second, the integration of disparate biological data types, which is crucial to the knowledge discovery process; third, lack of uniform, standardized data representation formats and fourth, the quality of the data with significant amount of noise in the form of technical artifacts, variation between labs and false associations, especially, in proteomics and metabolomics. The gleaning of meaningful information requires both sophisticated data mining techniques as well as intensive manual effort. Curation of omics data will require curators trained in both biology and data mining.

1.6.7 Bioinformatics Resources for *Brucella*

Several research groups are developing bioinformatics resources for the *Brucella* community. Table 4 summarizes some of the major bioinformatics resources, with the curated data and tools available to facilitate research by the *Brucella* research community. The specific goals, objectives and features of four of these resources are described in detail.

1.6.8 The *Brucella* Bioinformatics Portal

The Brucella Bioinformatics Portal (BBP) is a resource specifically developed for the *Brucella* research community (Xiang et al. 2006). The web site allows visualization

Table 4 Publicly available bioinformatics resources for *Brucella* research community

Name	Mission/main features	Link/reference
BBP: Brucella Bioinformatics Portal	• A dedicated resource to *Brucella* community • Query, display of genomic data on Brucella Search of literature on *Brucella* • Community contacts and contact database	http://helab.bioinformatics.med.umich.edu/bbp/; Xiang, Zheng et al. (2006)
CMR: Comprehensive Microbial Resource	• Display information on all the publicly available, complete prokaryotic genomes • Convenience of having all the organisms on a single website • Genome and comparative tools • Searches and downloads for data on genome, gene, evidence and genomic element lists	http://cmr.tigr.org/tigr-scripts/CMR/CmrHomePage.cgi
Pathport: Pathinfo and MiNet Documents	• Federation of best available tools through web services • Development of new tools and visualisation interfaces • Connection to best data sources • Interoperability among tools and across domains • Vetting literature information on Pathosystems and Pathogenesis	http://pathport.vbi.vt.edu/; He et al. (2005)
PATRIC: PathoSystems Resource Integration Center	• A multi-faceted resource and outreach for genomic and associated information to facilitate the discovery of diagnostics, drugs and therapeutics • DNA and protein level curation • Integration of post-genomic data with genome • Targeted annotation to prioritise genomic components for countermeasures	https://patric.vbi.vt.edu/; Snyder et al. (2007)
VLA: Veterinary Laboratory Agency	• Public and animal health through veterinary research, diagnostic and surveillance services • Regional network of 16 veterinary laboratories	http://www.defra.gov.uk/corporate/vla/science/science-bact-bruce.htm
The IGBMC Bioinformatics Platform	• High-throughput platform for comparative and structural genomics • Collect, maintain and deliver a wide range of bioinformatics resources • Develop innovative solutions and participate actively in national and international scientific programs • Software and databases, including analysis tools • Outreach activities	http://bips.u-strasbg.fr/en/Products/Databases/Prokaryote_Genomes.php

and query of the genomic and the associated curation data. BBP adopts several open-source software programs such as GMOD, GBrowse, TextPresso and PubSearch and extends their application for *Brucella* genome annotation. BBP integrates 'PubSearch' of literature search with TextPresso, a NLP-based computational literature mining system into an efficient literature mining and curation system (Limix) for *Brucella*. There are also many interactive graphical interfaces (e.g., MeSH browser and genetic interaction map) for visualization of the data. Probably the most unique feature of this resource is *Brucella* researchers' contact database, and discussion forum, for the *Brucella* research community. BBP also provides links information to all relevant tools, bioinformatics resources.

1.6.9 The Pathogen Portal

The pathogen portal combines information about pathosystems (host, pathogen and their interactions) with powerful analysis and visualization tools to aid in the rapid detection, identification and forensic attribution of high-priority human pathogens, whether causing infectious diseases or potentially used as biological weapons. Also included are some pathogens that affect domestic animals and plants, as bioterrorism against a large domestic food source could also be devastating to the human population affected. Besides the relevant bioinformatics tools and applications for analysis of the omic data, mainly two types of data are available for the *Brucella* community from the web site; the pathogen background information (PathInfo) and Molecular Interaction Network (MINet), which are described below.

MINet documents are developed based on the Molecular Interaction Markup Language (MINetML) to represent the molecular and cellular processes involved in pathogenesis (Wattam et al. unpublished data). Although it focuses on pathogenic interactions, it can be used for any complex interaction. The entire database contains molecular interaction network for 29 different species, and includes a pathway showing the invasion of host cells by *Brucella* (Fig. 3a). What is unique about this

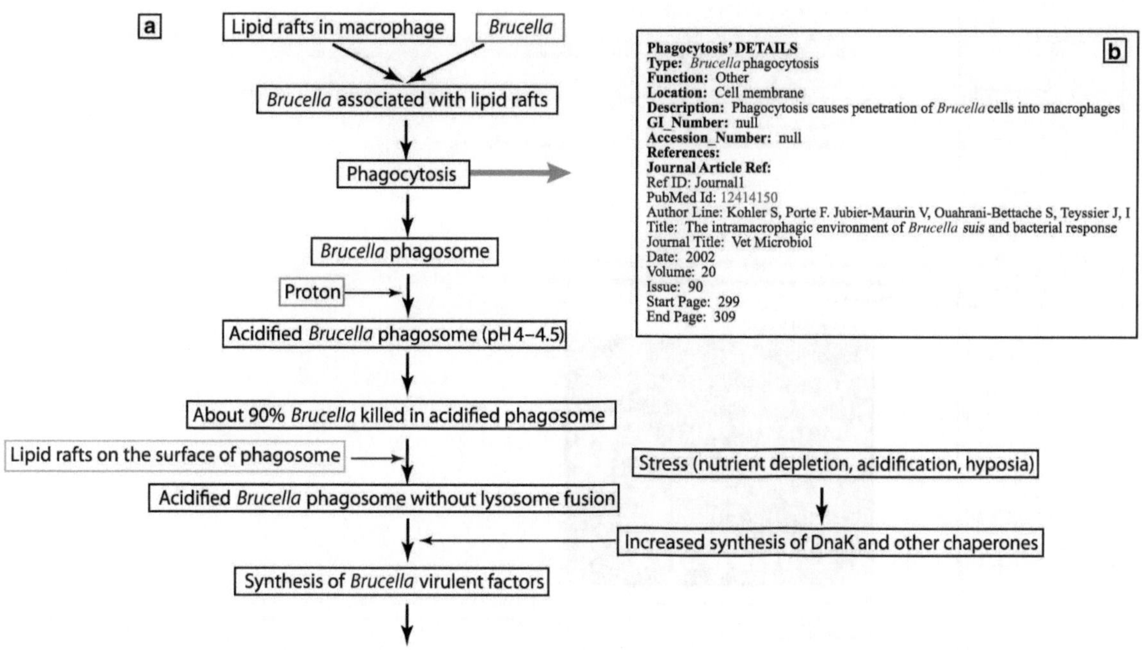

Fig. 3 View of MINet document for *Brucella*. **a** Part of the MINet depicting the interaction between a *Brucella* pathogen and its host cell. In this scenario, the bacterium successfully invades the cell and begins replication. **b** Clicking on the 'Phagocytosis' box displays the literature sources that document the phagocytosis of *Brucella*. Note that PubMed IDs are included with a direct hyperlink to the PubMed database

system is that it gives a visual representation of the interactions between the host cell and the bacterial pathogen, and links each interaction to the literature source (Fig. 3b).

PathInfo documents are based on the Pathogen Information Markup Language (PIML), which is a free, open-source, XML-based format (He et al. 2005). PathInfo provides information of different aspects of the organism, including taxonomy, epidemiology, infection and prevention, as well as diagnostic methods currently used (Fig. 4). The provenance for each section is maintained by directly linking the data to the literature source it is derived from. Currently, the PathInfo database contains background information on 43 different pathogens, including *Brucella*. These documents are available both in ToolBus software and web based documents at the Pathogen portal (http://pathport.vbi.vt.edu/pathinfo/).

1.6.10
The PathoSystems Resource Integration Center

The Pathosystems Resource Integration Center (PATRIC) is one of the eight Bioinformatics Resource Centers (BRCs) established in 2004. The goal of BRCs is to facilitate research on infectious diseases for the development of better therapeutics, vaccines and diagnostics. PATRIC stores, curates and displays all publicly available biological information on eight pathosystems (three bacterial genera and five viral classes) including *Brucella*. The data currently presented at PATRIC is described below.

Nucleotide and Protein Annotation: The whole-genome sequences of the *Brucella* species have been annotated by automated and manual procedures. Automated annotation comprises the identification of coding sequences using the gene prediction

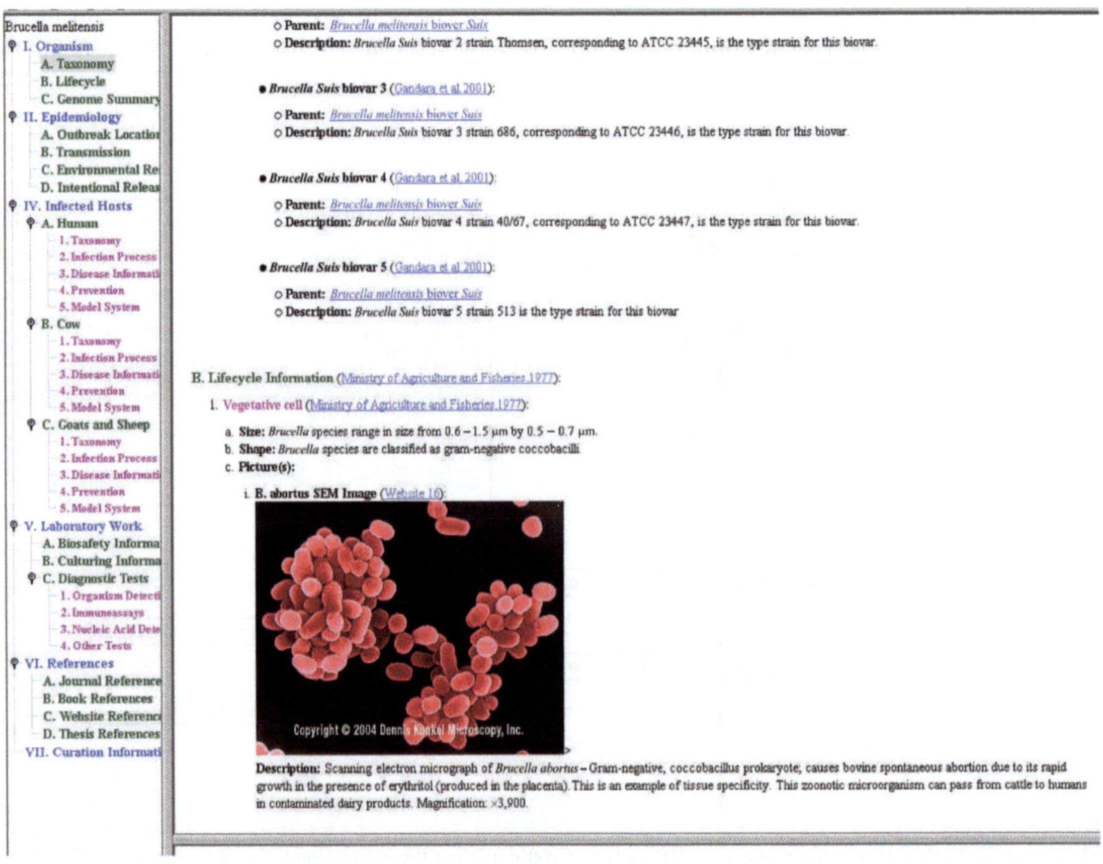

Fig. 4 A portion of the taxonomy section of the PathInfo document for *Brucella melitensis*

programs Glimmer and GeneMark; the correction of start sites using RBSfinder and TICO; and the identification of orthologs using BLASTX against the NCBI non-redundant database (Lukashin and Borodovsky 1998; Delcher et al. 1999; Suzek et al. 2001; Tech et al. 2005). The output of the decision tree is manually inspected by a team of curators. RNA encoding genes are annotated by searching for orthologs in a ribosomal RNA database followed by the manual editing of the coordinates, and the identification of transfer RNAs, transfer-messenger RNAs and regulatory RNAs using tRNAscan-SE and RFAM (Lowe and Eddy 1997; Griffiths-Jones et al. 2003). The coding sequences annotated during the nucleotide level annotation step are then translated and passed through a Protein Annotation Pipeline (Table 5). The annotation of each protein coding gene may be accessed at the PATRIC website. These annotations include coordinates, product names, EC numbers, GO terms, predicted signatures and a list of orthologs as predicted by BLASTP. Also included is the evidence for each annotation.

Metabolic Pathway Annotation: The metabolic map for each bacterium is created from the annotated proteome using the Pathway Tools suite of pathway construction programs (Karp et al. 2002). The pathways database curates data on metabolic and signalling pathways, enzymatic and transport reactions and reaction substrates and products.

Ortholog Groups: For each PathoSystem, the annotated coding sequences are used to create orthologous groups. The orthology assignment procedure has been briefly described by Snyder et al. (2007). There are 3,303 orthologous groups among the four *Brucella* species. For each ortholog group we have created multiple sequence alignments using MUSCLE and phylogenetic trees using PHYLIP (Retief 2000; Edgar 2004).

The PATRIC website provides the user with a query and search tool that retrieves the curated annotations – gene and protein name, feature identifier, GO terms, EC numbers and product description. The user can also query the annotated pathways and perform pathway level comparisons. User-supplied sequences can be analyzed using tools such as BLAST and MUMMER (Kurtz et al. 2004) and Base-by-Base viewer (Brodie et al. 2004).

The Comprehensive Microbial Resource

The Comprehensive Microbial Resource (CMR) curates data on diverse bacteria, including *Brucella* (Peterson et al. 2001). The following *Brucella* genome sequences are curated at CMR: *B. abortus* 9–941; *B. melitensis* 16M; *B. abortus* 2308 and *B. suis* 1330. Most of the data are produced by manual curation of the results of an automated pipeline. Genes are identified by the gene prediction program, Glimmer. The predicted protein coding genes are searched against TIGRFAM and PFAM for functional assignment; tRNAs are identified by tRNAscan; rRNAs are identified manually. The CMR website displays both the legacy annotation

Table 5 The programs and databases that are included in the PATRIC protein annotation pipeline and the type of annotation they produce

Program/database	Features and attributes	Reference
BioPerl modules	Molecular weight; isoelectric point	
Codon Usage script	Codon usage pattern	
Hydropathy script	Hydropathy profile	Stoffer and Volkert (2005)
InterProScan	Domain and motif signatures; functional sites; gene ontology assignments; EC number	Zdobnov and Apweiler (2001)
TIGRFAM	Domains, EC number, gene ontology assignments, TIGR roles	http://www.tigr.org/TIGRFAMs/
BLOCKS	Motif signatures	Henikoff and Henikoff (1996)
MEMSTAT2	Transmembrane domains	McGuffin et al. (2000)
LipoP	Signal peptide sequences (SPaseI cleavage sites)	Juncker et al. (2003)
BLAST	Homologues in the non-redundant NCBI database and the Saier's transporter database	Altschul et al. (1990)
COGnitor	The COG assignment	Tatusov et al. (2000)

as well the TIGR ascribed annotation for each gene. Gene graphic and Region View links provide a graphical display of the gene and its neighborhood. Apart from other nucleotide and protein level annotations, CMR also allows the user to search for PUBMED stored literature for a gene of interest. Orthologs in other species can be search through the 'Genome Region Comparison' tool.

1.6.11 Summary

During the last decade, an explosion in the amount of sequence and other omic data has transformed the biological field from data-poor to data-rich. A paradigm-shift in the way sequencing is done has further increased the speed and efficiency of sequencing. Four genomes are publicly available in *Brucella* and several more are about to be completed. The complete genomes have allowed comparative genomic analysis, which has revealed a remarkable conservation among the genomes despite their unique differences such as host-preference, virulence and pathogenicity. Likewise, the proteomes of these organisms are also under intense scrutiny, although at a more preliminary stage than the post-genomic analysis.

The field of biology is still a subjective and descriptive science. The explosion of data has created some unique challenges such as large technical variation, lack of data standards and poor data quality (Joyce and Palsson 2006). Several bioinformatics resources that are publicly available for the *Brucella* research community are trying to address these challenges. A paradigm-shift, however, is needed in this area to cope up with the data growth, integrate the diverse data types and hence to realize the promise of genomics for practical applications of either understanding the mechanisms of disease progression and host–pathogen interactions or developing countermeasures against natural and induced outbreaks. A transdisciplinary approach is, therefore, essential for developing an infrastructure that will allow interoperability between database systems and development of a systems biology approach and enable the extraction of discernable information for practical applications from the omic data sets.

1.7 Comparative Genomics of *Brucella*

1.7.1 Introduction

Brucellae are gram-negative facultative intracellular coccobacilli, taxonomically placed in the *Rhizobiales* order of the proteobacterial class *Alphaproteobacteria*. The *Alphaproteobacteria* also include obligate intracellular (e.g., *Rickettsia*) and extracellular (e.g., *Agrobacterium*) pathogens as well as symbionts (e.g., Sinorhizobium). *Brucellae* cause brucellosis, a disease of a wide range of mammalian vertebrates, and are responsible for considerable economic losses (Boschiroli et al. 2001). Several species of *Brucella* are identified based on the differences in their pathogenicity and host preference. Among these species, *B. melitensis* (goats), *B. abortus* (cattle), *B. suis* (swine) and *B. canis* (dogs) are also pathogenic to humans, making *Brucella* an important zoonotic pathogen.

The first instance of human infection by *Brucella* was documented by D. Bruce in 1887 (Moreno et al. 2002). The zoonotic infection often occurs through the ingestion of infected food products, direct contact with an infected animal, or inhalation of aerosols. Aerosol transmission is extraordinarily efficient. Infectivity is very high, requiring as few as 10–100 bacteria to establish a successful infection in humans. It is consequently identified as a category B priority agent with potential for use in biological warfare and bio-terrorism and of importance as a potential emerging infectious disease (http://www3.niaid.nih.gov/Biodefense/bandc_priority.htm). Indeed, *Brucella* was the first infectious agent developed for use as a biowarfare agent in the USA in 1954 (Bossi et al. 2004). In recent years, renewed interest in the disease has triggered vigorous research to improve understanding of its basic pathogenic mechanisms as well as host immune response (Boschiroli et al. 2001).

The main goal of this chapter is to describe the evolutionary and phenotypic characteristics of *Brucella* and other related organisms and the comparative genomics approaches to elucidate the subtle differences in genomes that cause major differences, especially, in host specificity, virulence and pathogenicity.

1.7.2 Comparative Analysis of *Brucella* and Related Species

Order *Rhizobiales*

Brucella spp. belong to the order *Rhizobiales*, a diverse group of bacteria composed of 11 families containing soil and aquatic bacteria, plant and animal pathogens, nitrogen-fixing plant symbionts and photosynthetic autotrophs. Like *Brucella*, many have adopted an endopathogenic or endosymbiotic lifestyle. For example, the *Bartonellaceae* contains the causative agents of arthropod-borne diseases such as cat-scratch and trench fevers and are often associated with bacterial endocarditis and angiomatosis. *Bradyrhizobiaceae* and *Rhizobiaceae* contain economically important species known for their ability to live in the cytoplasm of plant cells as well as soil. *Agrobacterium tumefaciens* is an endoparasite responsible for crown gall disease in plants and has become an important tool in agricultural biotechnology as a means of introducing foreign DNA into plant genomes. Other family members, such as *Rhizobium leguminosarum*, induce root hypertrophy, forming characteristic nitrogen-fixing nodules in leguminous plants. Also part of the *Bradyrhizobiaceae, Rhodopseudomonas* are purple non-sulphur phototrophs of interest for their remarkable metabolic repertoire. *R. palustris*, for example, can reduce N_2, evolve H_2 and catabolise a variety of toxic aromatic compounds. *Hyphomicrobiaceae* are morphologically diverse, with many members forming hyphae or prosthecae. *Hyphomicrobium spp.* have been isolated from nearly all soil and water samples tested to date (Hirsch and Conti 1964; Hirsch and Rheinheimer 1968) and have been found to be of use in the removal of nitrate and odorous sulphur compounds from drinking water. The *Methylobacteriaceae*, of which *Methylobacterium* is the major representative, is nearly ubiquitous in the environment, being able to scavenge trace nitrogen, resist desiccation and even moderate levels of gamma irradiation. *Methylobacterium spp.* are chemoorganotrophs and facultative methylotrophs capable of growing on a variety of C_1 substrates such as methanol, formate and formaldehyde. In contrast, the *Methylocystaceae* consists of obligate methanotrophs, utilising only methane and methanol as sole carbon and energy sources. They are also ubiquitous, representing as much as 40% of the bacterial biomass in some aquatic environments.

Phylogenetic analysis of *Rhizobiales*, based on 16S rDNA sequences, indicates a well-defined division of the order into two major groups. The first contains the *Brucellaceae, Bartonellaceae, Phyllobacteriaceae* and *Rhizobiaceae*; the second consists of *Bradyrhizobiaceae, Xanthobacteraceae, Methylocystaceae, Methylobacteriaceae, Beijerinckiaceae* and *Hyphomicrobiaceae*, while *Rhodobiaceae* subtends these two main groups (Lee et al. 2005). The division between the first group and the first two families of the second group is also supported by large-scale, protein-based trees, as illustrated in Fig. 5 (Williams et al. 2005). Sequencing of additional genomes should lead to refined placement of the currently unrepresented families.

The genetic basis for many of the phenotypic characteristics of these organisms has been identified or hypothesized. In *Rhizobium*, genes supporting symbiosis are typically found on a large plasmid, S_i, or a chromosomal region known as the symbiotic island. The *Agrobacterium* plasmid T_i is the carrier of genes responsible for tumorigenesis, whereas the rhizogenic phenotype is imparted by the R_i plasmid. Similarly, the *nod* and *nif* genes required for nodule formation and nitrogen fixation, respectively, in *Rhizobium* and *Sinorhizobium spp.* are found on the megaplasmid pSym. None of these plasmids are required for survival in soil. Parallels also exist between *Sinorhizobium meliloti* and *Brucella abortus* in genes required for intracellular interactions. *BacA* was the first gene found to be required for bacterioid differentiation in *Sinorhizobium spp.*, an early step in the symbiotic pathway. *B. abortus* strains carrying a null mutation in the homologous gene are still capable of infecting mouse macrophages but are unable to sustain chronic infection (LeVier et al. 2000; Ferguson et al. 2005), suggesting an important, conserved role for the gene in intracellular survival in diverse eukaryotic cells. Surprisingly perhaps, virulence genes are also conserved. Components of the *Brucella* and *Bartonella* type IV secretion system, *VirB1–11*, are shared with *A. tumefaciens* (LeVier et al. 2000; Sieira et al. 2004). These species also have virulence-associated protein E in common with *Bradyrhizobium japonicum* (Gottfert et al. 2001; DelVecchio et al. 2002a–c),

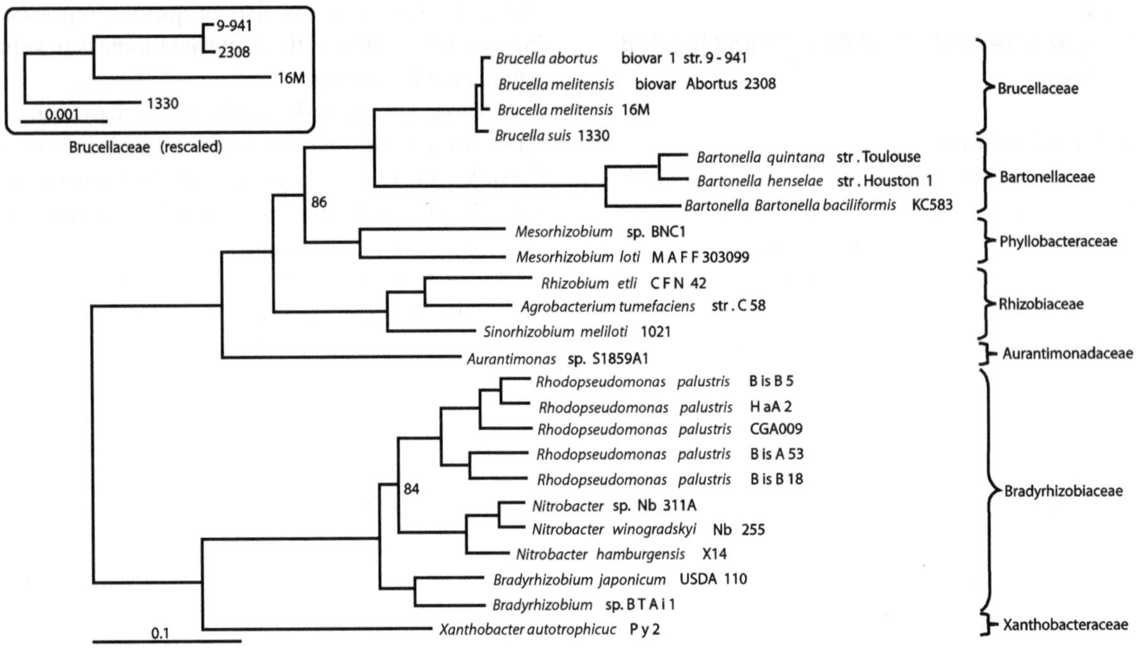

Fig. 5 *Rhizobiales* phylogenetic tree. This tree is based on a masked alignment of 104 selected proteins (33,730 characters) and arose in a MrBayes analysis with the proteins split into five partitions and primed with the neighbour-joining tree (Williams et al. 2007). The identical topology arose in an unpartitioned MrBayes application primed with the rRNA tree, and as the consensus of 50 maximum-likelihood (ML) bootstraps. All nodes had 100% support in the Bayesian analyses and only those indicated (by integers to the right of the node) had <100% ML bootstrap support. The outgroup of 48 alphaproteobacteria that provided the rooting for the order is not shown

a species not generally considered pathogenic to its host. These are some of the well-known genes associated with functional similarities among Rhizobiales. This list will no doubt increase substantially as more genomes are sequenced and that knowledge is in turn fed back into biochemical, morphological and clinical studies.

Family Brucellaceae

The family Brucellaceae comprises the additional genera *Ochrobactrum* and *Mycoplana*, although *Mycoplana* is polyphyletic and species other than *M. dimorpha* and *M. ramosa* are better placed in other families (Garrity et al. 2005). Genome sequences from these additional Brucellaceae genera will provide perspective on the evolution of *Brucella* and its pathogenicity, a perspective not provided by the currently available *Rhizobiales* genomes. The distance that *Brucella spp.* occupy on a genome-derived tree (Fig. 5) is only 1% of the distance to their nearest sister group, *Bartonella*. Selection of strains for genome sequencing can be guided by phylogenetic inferences from 16S rDNA sequences, although this approach can be insufficient when taxa are closely related. Additionally, different tree-building methods can produce markedly different results from the same alignment, and clusters can prove unstable as new sequences are added. Nonetheless, for many taxa, 16S rDNA sequences are the only data available for evaluating phylogenetic relationships.

In contrast to the typical existence of *Brucella* as a facultative intracellular parasite, the other Brucellaceae are found in more diverse ecological situations. *Ochrobactrum* strains can be found in clinical specimens or in environmental samples from soil, sludge, roots and other sources. *M. ramosa* and *M. dimorpha* are soil bacteria. *Ochrobactrum* is diverse, with over 200 16S rDNA sequences currently available. A recent analysis shows that *O. intermedium* and related strains have an especially close relationship to *Brucella*, while other strains of *Ochrobactrum* are increasingly distant (Lebuhn et al. 2006). The analysis of 16S rDNA sequences of *Brucella* and *Ochrobactrum* strains and of *M. ramosa* and *M. dimorpha* with different *Rhizobiales* outgroups (using various alignment, masking and tree-building procedures) revealed that *M. ramosa* appears stably as the closest relative of *Brucella* and *Ochrobactrum* strains, whereas *M. dimorpha* has an unstable position in the phylogenetic trees; in some cases it lies close to *M. ramosa*, but in other cases in can appear within other *Rhizobiales* groups (Williams et al., unpublished data). Genome sequencing of carefully chosen *Ochrobactrum* and *Mycoplana* strains, from soil and other environmental sources, can be expected to provide excellent perspective on the evolution of pathogenicity and other unique aspects of *Brucella*.

Brucella is comprised of several species that infect a wide range of primary mammalian hosts (Lopez-Goni and Moriyon 2004). Eight species of *Brucella* are characterised mainly based on the host-preference and pathogenicity: *B. melitensis* (sheep and goats), *B. abortus* (cattle), *B. suis* (swine), *B. neotomae* (desert rat), *B. ovis* (sheep), *B. canis* (dogs), *B. pinnipidae* (pinnipeds) and *B. cetaceae* (cetaceans). Despite the major differences in host preference, the degree of similarity between these species at the genome sequence level is surprisingly very high (>90%). Therefore, it has been proposed that all the current species be merged into one species (*B. melitensis*) with different strain designations (Lopez-Goni and Moriyon 2004).

Several strains of *Brucella spp.* have been well characterized for virulence or lack thereof. The importance of the attenuated strains in vaccination has been well understood in brucellosis eradication efforts (See Sect. 1.1 for more details). Some of the strains that have been used in vaccination are *B. abortus* S19, a laboratory derived strain used as a live vaccine in cattle since the 1940s (Nicoletti 2002); *B. abortus* 45/20, a smooth strain isolated from pigs (McEwen 1940); *B. melitensis*, Rev. 1 strain with variable attenuation (Banai 2002) and *B. abortus* RB51, a stable rough mutant of *B. abortus* 2308 that is currently being used for vaccination of cattle (Schurig et al. 1991). Availability of the whole-genome sequences of the attenuated strains and their closely related virulent strains provide an excellent opportunity to identify the genome-wide differences and discovery of better targets for clinical studies (see Sect. 1.7.3).

1.7.3 Comparative Genomic Analysis of *Brucella*

The availability of whole-genome sequences of several closely related organisms provides great opportunities to identify and characterise the genes and their regulators associated with distinct properties of organisms such as immune responses, host preference, pathogenicity and virulence. At the time of writing this chapter, four *Brucella* genomes have been sequenced, including *B. melitensis* 16M, *B. suis* 1330, *B. abortus* 9–941 and *B. abortus* 2308, and several more are currently being sequenced (see Sect. 1.6 for details). The four genomes share over 90% sequence identity and the strains are pathogenic to humans in varying degrees of severity; yet the strains have distinct pathogenic properties.

The comparative analyses of *Brucella* genomes have been done with two goals: to determine overall genome structures and to study extensive similarities and gene synteny conservation among the *Brucella spp* (Paulsen et al. 2002; Chain et al. 2005; Halling et al. 2005). Halling et al. (2005) demonstrated that *B. abortus* 9–941, *B. melitensis* and *B. suis* have comparable G+C contents, genome composition and gene numbers. All three genomes are composed of two chromosomes; the first (Chr I) is nearly double the size of the second (Chr II). The results are consistent with those determined in early hybridization studies (Hoyer and McCullough 1968a, b). On the one hand, the similarity and conservation of synteny are especially pronounced in Chr I, with highly similar regions and without rearrangement of homologous backbone sequences (Fig. 6a). Chr II, on the other hand, is more diverse (Fig. 6b). This difference most likely represents distinct evolutionary origins as well as the nature of functional genes in the two chromosomes. Chr I is a classic bacterial circular chromosome (Lobry 1996), while Chr II is likely to have derived from a plasmid since it possesses a cluster of replication genes similar to those from *Agrobacterium* Ti plasmids, and plasmids from other organisms including *Rhizobium spp.* Chr I encodes the majority of the core metabolic machinery for processes such as transcription, translation and protein synthesis, whereas Chr II appear to largely represent auxiliary pathways for speciation and adaptation such as for the utilisation of specific substrates and host–pathogen interactions (Paulsen et al. 2002).

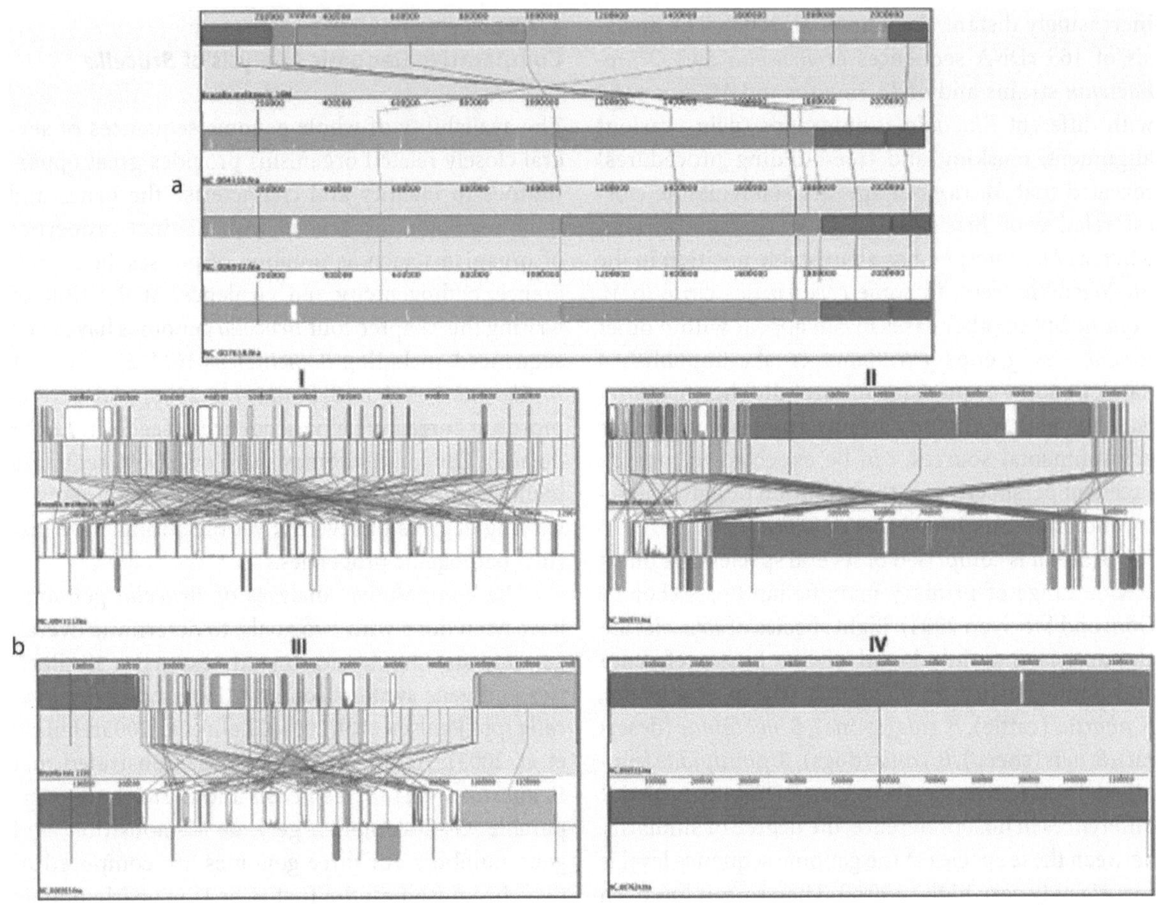

Fig. 6 Genome sequence alignments of four *Brucella* strains. **a** Multiple genome alignment of chromosome I of *B. melitensis* 16M, *B. suis* 1330, *B. abortus* 9-941 and *B. abortus* 2038 ordered from top to bottom. **b** Pair-wise genome alignments chromosome II: **b.I**: *B. melitensis* 16M vs. *B.suis* 1330, **b.II**: *B. melitensis* 16M vs. *B. abortus* 9-941, **b.III**: *B. suis* 1330 vs. *B. abortus* 9-941 and **b.IV**: *B. abortus* 9-941 vs. *B. abortus* 2308. Large solid blocks indicate identical sequences or highly similar sequences, whereas white space indicates sequence variation such as insertions, deletions and inversions (Yu et al. 2007, unpublished data). This and other figures for whole genome alignments in this chapter are generated using MAUVE, a software package for multiple alignment of conserved genomic sequence with rearrangements (Darling et al. 2004)

The genomic sequence differences displayed among the four *Brucella* genomes, especially on Chr II, offer clear clues about the common ancestry among these genomes. *B. melitensis* 16M shows great sequence similarity with *B. abortus* but differs significantly from *B. suis* 1330. The differences are evident from the Mauve sequence alignments between four *Brucella* genomes. There is a large region containing blocks of sequence similarity in the colinear blocks between Chr II of *B. melitensis* 16M and that of *B. abortus* (Fig. 6b.II), whereas no significant similarity blocks exists between *B. melitensis* 16M and *B. suis* 1330 (Fig. 6b.I). This observation supports a hypothesis that *B. melitensis* and *B. abortus* shared a common ancestor and became isolated about 20 million years ago, when radiation of artiodactyls occurred (Moreno et al. 2002). The *B. melitensis–B. abortus* lineage differs significantly from the *B. suis* lineage, which has undergone genetic mutations since it diverged from the most recent common ancestor of all *Brucellae* (Fig. 5). It is interesting to note that *B. suis* displays a much higher level of sequence similarity with *B. abortus* than that with *B. melitensis* (Fig. 6b.II vs. Fig. 6b.I). A 5.222 kb deletion in *B. abortus* 2308 was identified to be the only major difference between the two *B. abortus* genomes, which

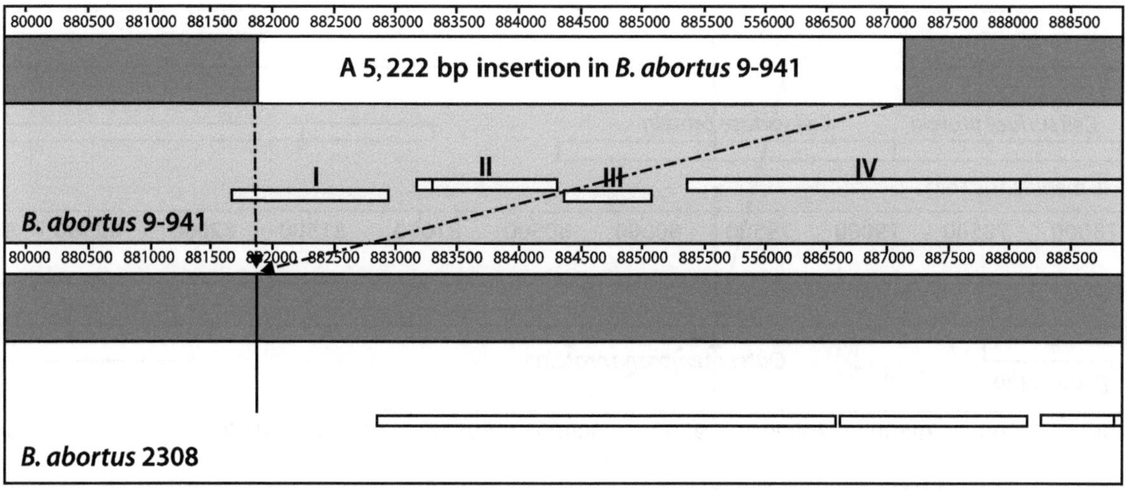

Fig. 7 A 5.222-kbp insertion in *B. abortus* 9-941 relative to *B. abortus* 2308. The DNA fragment encodes four genes: I. sugar ABC transporter, periplasmic sugar-binding protein (BruAb2_0877), II. Hypothetical sugar-binding protein (BruAb2_0879), III. transcriptional activator FtrB (BruAb2_0880) and IV. Nitrite extrusion protein (BruAb2_0881)

has resulted in deletion of four genes (BruAb2_0877, BruAb2_0879, BruAb2_0880 and BruAb2_0881) that encode for three transporter proteins and one transcriptional activator (Fig. 7).

The comparative study also revealed unique gene compositions such as the number of open reading frames (ORFs), the occurrence of fragments, inversions, polymorphic regions and gene inactivations. Differences in the number of ORFs found among the three *Brucella* genomes are mainly because of the differences in annotation of short ORFs. For example, the number of ORFs of less than 100 amino acids annotated (as of 30 January 2007 at https:/patric.vbi.vt.edu) in *B. abortus* 9-941, *B. suis*, and *B. melitensis* 16M are 648, 626 and 550, respectively, while the disparity is even more pronounced when those with less than 50 amino acids are considered (227, 227 and 123, respectively). Genome-specific DNA fragments are also identified by the comparative analysis. Many of these sequences are related to mobile genetic elements as well as functionally and pathogenically important genetic factors. For example, a 2,774 bp fragment found in *B. melitensis* and *B. suis* is missing from *B. abortus*. This fragment encodes a probable surface protein and two partial ORFs with homology to the insertion sequences IS711 and ISBm1. A 25 kb DNA fragment, deleted in *B. abortus*, is especially interesting. Twenty-one putative genes are deleted in their entirety, with two additional genes disrupted at margins. The fragment encodes eight proteins directly involved in polysaccharide biosynthesis with several others involved in transport and related functions. Given the role of outer membrane glycoproteins and exopolysaccharides in the interaction of *Rhizobium* and *Agrobacterium spp.* with their eukaryotic hosts, this region has been hypothesized to affect various *Brucella* phenotypes such as host preference and immune response (Vizcaino et al. 2000, 2001). Another 1,710 bp insertion/deletion is located within a gene that encodes outer membrane protein BruAb1_0072 in *B. abortus* (Fig. 8). The DNA fragment is completely deleted from *B. suis* 1330 and only partially deleted in *B. melitensis* 16M.

Multiple DNA inversions have been discovered including one large segment and two smaller ones in *B. abortus*. The large inversions have been found to have great impact on gene structures within affected regions. An inversion in *B. abortus 9-941* disrupts BRA1003 and BRA0235, two *B. abortus* homologs of *B. suis*, resulting in four pseudogenes (genes that have lost their protein-coding ability). The two homologs encode a putative GAF/GGDEF prokaryotic signalling domain protein and a hypothetical protein, respectively. The first small inversion in this strain, about 2,185 bp and unique to the *B. abortus* 9-941, disrupts BR1062, a proline dipeptidase gene, and the second small inversion, about 2,150 bp and found in *B. melitensis* as well, disrupts the genes encoding a

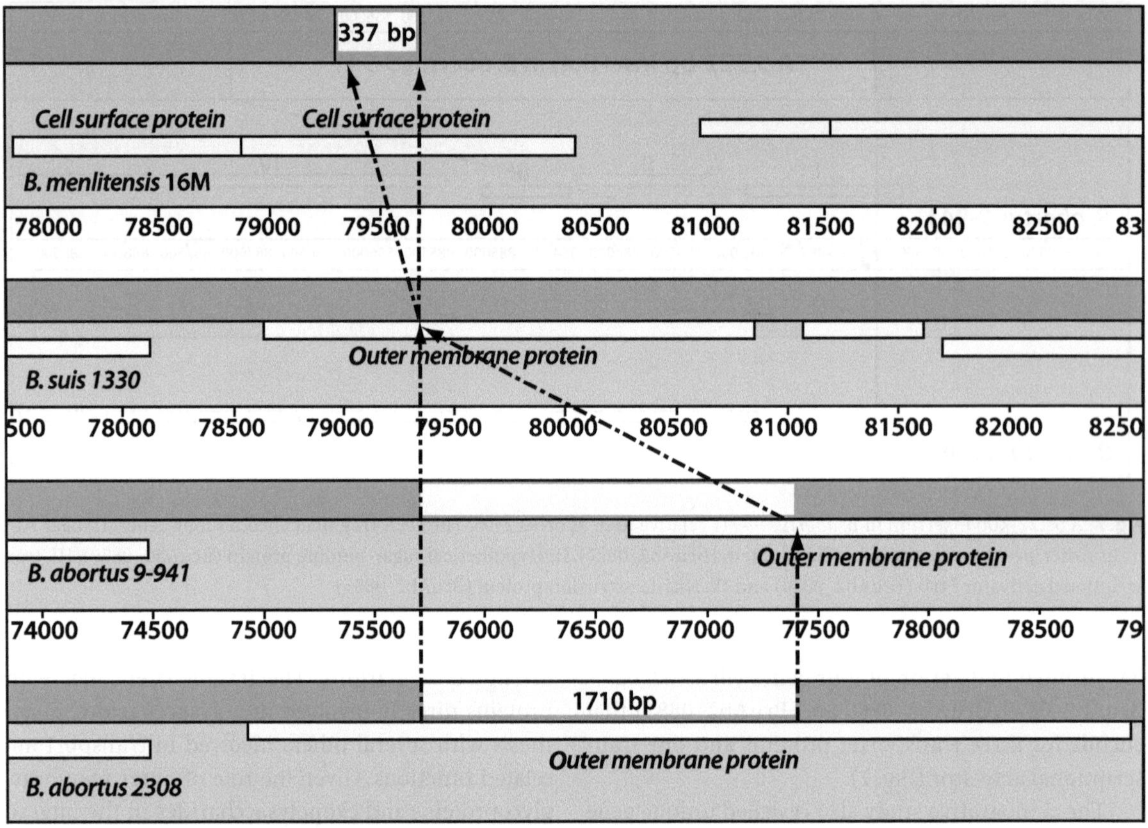

Fig. 8 Distribution of a genome sequence insertion/deletion within a gene that encodes an outer membrane protein (*B. suis* GeneID:1165729) among four sequenced *Brucella* genomes. The genomes are ordered from top to bottom: *B. melitensis* 16M, *B. suis* 1330, *B. abortus* 9–941 and *B. abortus* 2038 (chromosome 1)

putative protein and a protein belonging to the glycosyl transferase family. In addition, several variable ORFs are identified, which encode some of the most important virulence factors of *Brucella*, including an outer membrane protein, a putative bacterial immunoglobulin-like protein, adhesions and an autotransporter and other important virulence factors of *Brucella*. The immunoglobulin-like protein belongs to a group 1 domain common to bacterial surface protein invasins, which is often associated with pathogenicity and host preference (Moreno et al. 2002; Paulsen et al. 2002; Vemulapalli et al. 2006).

A number of pseudogenes have been observed in the *Brucella* genomes: over 200 in *B. abortus*, 152 in *B. melitensis* 16M and 82 in *B. suis* 1330 (Chain et al. 2005). Many of the pseudogenes are species-specific or shared among the *Brucella* species, and are homologous to genes with a broad range of biological functions. Multiple flagellar pseudogenes have been observed: seven in *B. melitensis*, four each in *B. suis* and *B. abortus*. The functions of their normal counterparts vary as well – some are essential for activity (fliG, fliM and motC), some are involved in flagellum assembly (fliI, flgA and flhA) and others are key structural proteins (flgF and flgI). Many pseudogenes are also identified that are associated with those involved in many other biological processes such as nicotinamide synthesis, biotin synthesis, PC biosynthesis, electron transfer and regulatory elements, or to encode autotransporter domain-containing protein systems (Chain et al. 2005).

As the number of genomes is increasing, the need for automated comparative analysis becomes imperative for systematic analysis of the genome features of the closely related organisms such as sequencing errors or real mutations and the sequence variants in speciation and adaptation. With this objective, Yu et al. (2007) developed GenVar, a genome context-based, and sequence variant-oriented analysis pipeline for comparative analysis of closely related bacterial genomes (the software and data is available at https://patric.vbi.vt.edu/downloads/software/GenVar). The pipeline is designed to automatically identify genes that were missed by previous predictions (missed genes) and sequence variants such as genes with disrupted reading frames (split genes) and those with insertions and deletions (Indels). This pipeline has been successfully applied to the four completely sequenced *Brucella* species to identify critical sequence changes behind the differences, allowing the generation of testable hypotheses. In the following sections, a brief description of GenVar is given as well as an explanation of how the pipeline can advance this research in *Brucella* genome analysis.

The GenVar is based on 'Genewisedb', a program to comparatively analyse DNA and protein sequences for the gene structure analysis (Birney et al. 2004). The pipeline is composed of four modules. The first two generate Genewisedb-specific protein database inputs (gwpDB) (Fig. 9.I) and Genewisedb-specific DNA inputs (Fig. 9.II). The gwpDBs constructed for each gene contain only homologous proteins from a limited number of closely related species. Thus, the modules generate small but optimal DNA and protein database inputs to overcome the limitations imposed by using the Genewisedb, which demands

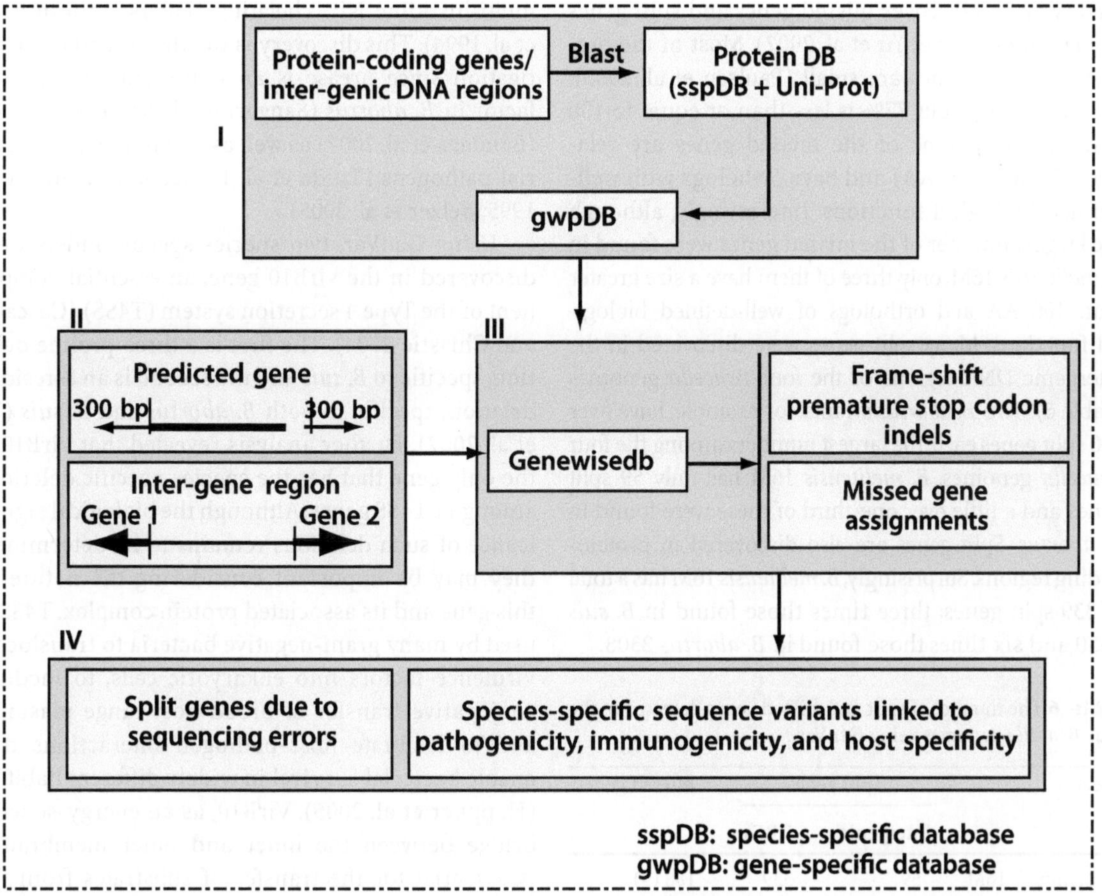

Fig. 9 Data flow in GenVar, a genome-context-based procedure for the identification of missed gene assignments, sequence variants, and for comparative genomic analysis. This procedure consists of four modules, including building Genwisedb-specific protein databases (gvpDB) (I) and Genwisedb-specific DNA inputs (II); detecting sequence variation, and comparatively analysing sequence variants among closely related genomes (III); and species-specific sequence variation and modifications (IV) (Reproduced from Yu et al. 2007)

considerable computational cost, especially on large DNA and protein databases (Birney et al. 2004; Potter et al. 2004). The third module comparatively analyses the two constructed optimal inputs to identify missed genes and sequence variants among closely related species. The missed gene is defined as an intergenic DNA fragment that can fully align with its orthologs from closely related genomes, while the sequence variants include genes with frameshifts, premature stop codons, insertions and deletions (Fig. 9.III). Once the missed and split genes are identified, the fourth module studies the variants within the genome context of closely related species to define species-specific sequence variants and to correlate the variants with the differences among these bacterial species (Fig. 9.IV).

GenVar has been applied to improve annotation quality and to discover missed genes and split genes in *Brucella* (Table 6; Yu et al. 2007). Most of the predicted missed genes are small (Paulsen et al. 2002); an average of about 77% is less than or equal to 100 AA. However, some of the missed genes are relatively large (>100 AA) and have orthologs with well-defined biological functions. Interestingly, although the largest number of the missed genes were found in *B. melitensis* 16M, only three of them have a size greater than 100 AA and orthologs of well-defined biological functions. Many split genes were discovered in the intergenic DNA regions of the four *Brucella* genomes (Table 6). Two *B. abortus* strains, for example, have over 160 split genes each, the largest numbers among the four *Brucella* genomes. *B. melitensis* 16M has only 59 split genes and a little over one-third of these were found in *B. abortus*. Split genes are also discovered in protein-coding regions. Surprisingly, *B. melitensis* 16M has a total of 339 split genes; three times those found in *B. suis* 1330 and six times those found in *B. abortus* 2308.

Table 6 The number of split and missed genes detected in the four *Brucella* genomes using GenVar

Species	Split genes[a]		Missed genes[b]
	Intergenic	ORFs	
B. melitensis 16M	59	339	188 (3)
B. suis 1330	120	111	50 (19)
B. abortus 9–941	165	103	64 (20)
B. abortus 2308	165	52	129 (17)

[a] Premature stop codons and/or frameshifts

[b] The numbers in parenthesis are missed genes that have orthologs with well-defined biological function

Classifications of sequence variants from four *Brucella* genomes revealed many cases of species-specific gene disruptions and some are important for the pathogenesis. For example, species-specific sequence variants are detected in multiple subunits of the flagellar complex (Rodriguez et al. 2005). While the premature stop codons in flagellar biosynthesis genes are common to both *B. melitensis* and *B. abortus*, the frameshifts in flagellar motor switch genes and flagellar biosynthesis genes are specific to *B. abortus*. The selective gene disruption is clearly important to understand the molecular basis of pathogenesis of *Brucella* because of the role of the protein complex (Fretin et al. 2005). A pair of frameshifts in UreE-2 of urease operon 2 was discovered in the *B. abortus* genomes. The frameshifts generated a shortened version of the gene in *B. abortus*, resulting in the loss of multi-histidine Ni^{2+} chelating centers (Sriwanthana et al. 1994). This discovery is worthy of further investigation since urease is an important colonisation factor in *B. abortus* (Sangari et al. 2007) and *B. suis* (Bandara et al. 2007) as well as in a number of bacterial pathogens (Tsuda et al. 1994a, b; Andrutis et al. 1995; Belzer et al. 2005)

Using GenVar, two species-specific indels were discovered in the virB10 gene, an essential component of the Type 4 secretion system (T4SS) (Cascales and Christie 2004). The first is a three-proline deletion, specific to *B. suis* and the second is an 8-residue deletion, specific to both *B. abortus* and *B. suis* (Yu et al. 2007). Further analysis revealed that virB10 is the only gene that has the species-specific deletions among 11 T4SS genes. Although the biological significance of such deletions remains to be determined, they may be important considering the nature of this gene and its associated protein complex. T4SS is used by many gram-negative bacteria to translocate virulence factors into eukaryotic cells, to mediate conjugative transfer of broad-host-range plasmids and to facilitate host–pathogen interactions that enable bacterial survival in widely different habitats (Hoppner et al. 2005). VirB10, as an energy-sensing bridge between the inner and outer membranes, is essential for the transfer of substrates from the inner to the outer membrane (Binns et al. 1995). The specific deletions may control the way that *Brucella* and host interact with each other, contributing to host specificity. Further experiments are needed to validate these hypotheses and to determine the functional roles of the gene in the pathogenesis.

The availability of the whole-genome sequences of the attenuated strains together with their close and virulent relatives allows comprehensive comparative analysis of the genomes for discovery of all the genes and the intergenic regions associated with the attenuation. For example, the analysis of the whole-genome sequence of the *B. abortus* S19, a spontaneously attenuated strain, revealed that 93.9 and 97% of the genes of the strains 9-941 and 2308, respectively, were identical to those of the S19 (Crasta et al., unpublished data). Amongst the non-identical genes only about 50 genes showed consistent differences between the attenuated strain and the two virulent strains. Further analysis and validation of these genes may provide additional clues on the molecular mechanisms of virulence and disease progression.

1.7.4 Summary

The comparative analysis of *Brucella* and the related organisms from the *Rhizobiales* clearly shows high phenotypic diversity of genetically closely related organisms. The organisms vary largely in lifestyle (endosymbiotic, endopathogenic or free-living), pathogenicity (facultative or obligate) and host-preference (plants or animals) of the pathogens and adaptation to various environmental conditions. This provides an excellent opportunity to elucidate the molecular mechanisms through comparative genomic analysis of the rapidly increasing whole-genome sequences. Among eight species of *Brucella*, four are found to infect human and hence are identified as category B priority agents from a biodefense and emerging infectious disease point of view. Comparative genomic analyses of *Brucella* species reveal extensive similarities in genome structure and synteny. Amongst the two chromosomes, chrI is highly similar and represents a classical bacterial chromosome while chrII shows more diversity and is likely to be derived from plasmid. Close observation of the highly similar genomes of *Brucella* using comparative genomic tools such as MAUVE and GenVar reveals genes and intergenic regions that are distinctly different between the four species. Availability and analysis of whole-genome sequences of the virulent and attenuated strains of *Brucella* will provide clues to elucidate the differences in host-preference, pathogenicity and virulence and support discovery of targets to facilitate clinical studies.

1.8 Functional Genomics

1.8.1 Introduction

Comparative genomics reveals the distinguishing features of genomes, and provides a record of the evolutionary changes associated with gene acquisition or loss via recombination, phage lysogeny or gene silencing via sequence variation. These differences may be used to explain phenotypic diversity, including host range and virulence, and to identify species (Bricker 2002; Navarro et al. 2004, 2006; Ratushna et al. 2006). However, functional genomics is necessary to characterize the purpose of each gene and in the process reveal the impact of genomic variation on critical phenotypic parameters. Functional genomics in the *Brucella spp.* is in its infancy. Although characterization of genes associated with survival and virulence began more than 20 years ago, within the last 10 years comparative genomics has fostered a systematic characterization and comparison of genes in multiple genomes of closely related yet distinct genera of organisms (deLorenzo et al. 1990; Sangari and Agüero 1991; Allen et al. 1998; Foulongne et al. 2000; Hong et al. 2000; Lestrate et al. 2000, 2003; Delrue et al. 2001; Endley et al. 2001; Kohler et al. 2002a, b; Kim et al. 2003; Monreal et al. 2003; Dozot et al. 2006). As a result, this section will be heavily weighted toward the discovery phase of functional genomics and will describe approaches to identify and characterize gene function using random mutagenesis, knockout mutant evaluation, transcriptome analysis and proteome analysis. Approaches that analyze higher order structure, including protein folding, protein–protein interactions or defined structural features, are just beginning to be used; in lieu of their description, gene products and mechanisms that are ready for such evaluation will be described.

Once candidate genes have been identified, the rational phase of testing begins in which the contribution of each gene to survival, virulence or other parameters of interest are carefully evaluated.

This involves both the specific inactivation of the gene, so as to not affect expression from neighboring genes and restoration of the virulent phenotype by replacement of the inactivated gene. So far, comparison of mutations in different species has focused on obvious virulence-related functions common to all species. There has been little effort to determine the function or contribution of unique genes or to focus on species-specific sequence variation. One explanation for this oversight is the fact that the animal model systems available are unlikely to reveal subtle differences associated with sequence changes originating from host adaptation. Differences in the mouse model including organ specificity and ultimate clearance are distinct from target species in which the organism affects reproductive tissues or causes abortion, in contrast to the clearance observed in the mouse model. Consequently, future research will require experimentation in the target species, as already recognised for the evaluation of vaccine strains. However, the potential for useful models in laboratory animals cannot be ruled out completely (Kim et al. 2005).

The goal of this section is to describe discovery-based functional genomics approaches to evaluate virulence mechanisms including immune system avoidance, altered intracellular trafficking and growth, and host-adaptation that are common to all *Brucella* species. Our approach is to review the evidence provided to date while proposing new approaches to extend these findings. As such, this commentary is meant to serve as a guide for future experimentation and although it is possible for other researchers to claim similar or even superior insight, we offer ours based on several years of combined experience and interaction. Each section will be subdivided into background and future research sections to present what is known and what in our estimation are the obvious approaches to move forward.

1.8.2 Genomics: Global Approaches

Transposon mutagenesis has proven to be one of the most successful approaches for pathogen gene discovery, but lacks some of the subtleties associated with other approaches. For example, although transposon mutagenesis indicates a role for the gene at a critical stage of the infection process, it does not provide a temporal readout. Depending on the assay employed, the result does not immediately reveal whether the genes identified are involved in invasion, replication or resistance to intracellular or extracellular killing. However, it is possible to modify the protocols to better define these defects. For example, intracellular invasion may be monitored separately from replication based on organism uptake. Furthermore, comparison of survival in the mouse and macrophage models may distinguish between organisms sensitive to extracellular and intracellular killing.

The focus of studies for *Brucella* has been the identification of genes responsible for intracellular survival, since such mutants may persist long enough to stimulate a protective immune response. There are now numerous publications demonstrating that the failure to replicate in cells results in an inability to persist in the host or to cause disease (Allen et al. 1998; Hong et al. 2000; Lestrate et al. 2000, 2003; Delrue et al. 2001, 2004; Kohler et al. 2002a, b, 2003). A great deal of effort has been expended to understand the nature of the intracellular environment and the bacterial mechanisms responsible for enhancing survival within these cells. In contrast, there has been little effort expended to determine the nature of resistance to extracellular killing mechanisms. This may be attributable to the fact that the lipopolysaccharide (LPS) on the cell surface is a major contributor to resistance to complement mediated lysis (Hoffman and Houle 1983, 1995; Corbeil et al. 1988; Fernandez-Prada et al. 2001). Comparison of survival in the mouse model in conjunction with the macrophage model may be used to specifically identify mutants sensitive to extracellular killing. This approach has an advantage over evaluating resistance/sensitivity to complement tested directly in vitro, since it may be expected to include additional factors active in vivo.

In an effort to characterise the numbers of genes contributing to extracellular survival in the host, a bank without evidence of siblings was screened for mutants of reduced fitness. A portion of these mutants ($n = 480$) was screened in both the macrophage and mouse models of infection. Attenuated mutants were identified at a frequency of 3.5 per 100 in the macrophage model and 16 per 100 in the mouse model. The fact that the mutants identified using the mouse model exhibited only a minor reduction in macrophage survival is consistent with the predicted contribution of extracellular killing in the mouse model ($p < 0.005$) (Fig. 10). Interestingly, it was noted that only a handful of the genes identified were involved in LPS biosynthesis while the majority exhibited a smooth phenotype suggesting that other factors control resistance to extracellular killing. Despite its drawbacks, it should be pointed out that macrophage screening provide an efficient approach that is less prone to artifactual readouts, including false negative results observed in the mouse model due to poor tag-hybridization or stochastic loss of mutants in the mixed infection model.

Some work performed with *Brucella* has focused on the use of the mini-Tn5 transposon with a known sequence bias for insertion (Ason and Reznikoff 2004). In an effort to expand on the genetic description of *Brucella* virulence the use of the Himar1 transposable element was evaluated (Chiang and Rubin 2002). Comparison of these two approaches revealed that despite differences in target specificity of the two transposons, the distribution of gene classes identified that are associated with survival was similar

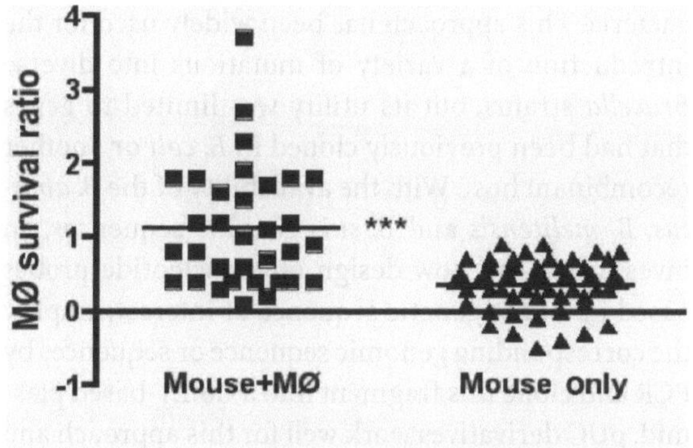

Fig. 10 Comparison of macrophage survival of attenuated *Brucella melitensis* mutants. Mutants were obtained by screening for survival in mice and macrophage in culture and divided into two groups based on identification in the macrophage screen. The replication ratio (CFU_{48h}/CFU_{0h}) for each mutant was determined relative to the parental strain and presented as the $\log_{10}$ of wild-type to mutant. Mutants that were only identified in the mouse model exhibited an average survival ratio that was significantly lower than those mutants identified using both screening methods. The enhanced sensitivity of the mouse model may be explained in part by the contribution of extracellular killing present in the mouse model, but missing from macrophage screening

(Fig. 11). However, nearly two-thirds of the 94 genes were novel identifications consistent with the differences in target specificity (Wu et al. 2006). Among the mutants identified were metabolic loci that may compromise the organism's ability to survive in what has been described as a nutritionally poor environment, that is the macrophage. In support of this hypothesis, all of the enzymes in the purine biosynthetic pathway were identified. Yet despite this, there is a paucity of insertions identified in several metabolic pathways, including 20 amino acid biosynthetic pathways. Of course, one explanation for this result is that the mutant bank does not effectively encompass the entire genome, and this possibility cannot be completely ignored; yet the fact that all the genes in the purine and T4SS pathways were identified suggests that other explanations may be possible. One possibility may be that the organism can acquire some nutrients from the intracellular environment, for example via peptide uptake (Rajashekara et al. 2004). Of course, there are several logistical explanations that may contribute to such discrepancies and final confirmation is best achieved via rational design of knockout mutants to determine the contributions of individual steps of each metabolic pathway to survival.

One alternative to transposon mutagenesis in global screens to identify genes required for survival has been the use of in vitro expression technology (IVET). In *Brucella*, this has been performed by creating libraries of promoter containing DNA fragments upstream of a promoterless gene encoding a reporter gene such as green fluorescent protein (GFP). The constructs were screened for recombinants that were not expressed in vitro, but were active within macrophages (Eskra et al. 2001). Among the genes identified using IVET were several functions that have since been identified in transposon screens, hence proving the validity of the IVET approach. However, the primary limitation associated with IVET and one that is not shared by transposon mutagenesis is the limited ability to detect functions that may be only briefly expressed. Although it may be argued that this may be overcome as a result of the persistence of GFP in the cells, this also has unwanted side effects that may be avoided by the use of improved IVET-style systems. The Cre-Lox (Lambert et al. 2007) and Flp-Frt (Schlake and Bode 1994) systems would permit the identification of new genes or presumably the timing of expression of factors critical for survival.

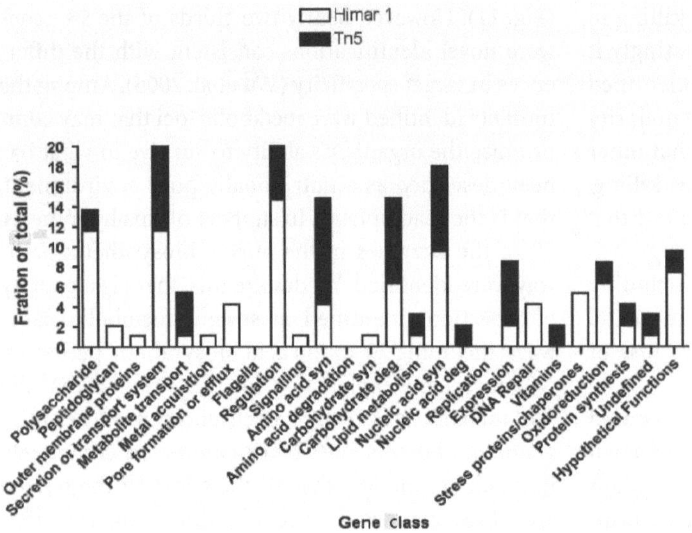

Fig. 11 Comparison of gene class distribution from Himar1 and miniTn5 mutagenesis. Genes were identified by sequencing DNA flanking the transposon insertion using inverse PCR (Hong et al. 2000). Assignments to gene classes were performed as described elsewhere (Wu et al. 2006). Although the number of mutants within each group varied due presumably to transposon insertion specificity, overall the gene classes identified were conserved. This result was not unexpected, since the gene classes necessary for survival are invariable for this organism

However, at this stage there is a surfeit of genes identified that play a role in survival. The use of Cre-Lox in more directed studies such as the identification of T4SS effector functions would be more useful.

1.8.3 Rational Approaches

B. melitensis 16M, *B. suis* 1330 and *B. abortus* 2308 are the strains most widely used in laboratory studies and the virulence properties of both strains in natural and experimental hosts and mammalian cell cultures have been well established. The sequences of the genomes of all three of these bacterial strains are now publicly available (DelVecchio et al. 2002b; Paulsen et al. 2002; Chain et al. 2005) and makes it possible to formulate experimental strategies to directly evaluate the biological functions of individual gene products or subsets of gene products in these bacteria. More specifically, examination of the genes predicted to be present in the *Brucella* genome sequences allows one to develop rational and testable hypotheses regarding the biochemical pathways these bacteria use to carry out their basic physiology and metabolism as well as how they establish and maintain infections in their mammalian hosts.

Gene knockouts. ColE1-based plasmids do not replicate in *Brucella spp.*, and the discovery that these types of plasmids can be introduced by electroporation (Lai et al. 1990) and used for the construction of mutants via gene replacement through homologous recombination (Halling et al. 1991) provided a major breakthrough in the genetic analysis of these bacteria. This approach has been widely used for the introduction of a variety of mutations into diverse *Brucella* strains, but its utility was limited to genes that had been previously cloned in *E. coli* or another recombinant host. With the availability of the *B. abortus*, *B. melitensis* and *B. suis* genome sequences, an investigator can now design oligonucleotide probes based upon the genetic sequence of interest, amplify the corresponding genomic sequence or sequences by PCR and clone this fragment into a ColE1-based plasmid. pUC derivatives work well for this approach and their high copy number makes them easy to manipulate in the laboratory when they are propagated in most of the widely used *E. coli* host strains. The desired mutation (see later) can then be introduced into the *Brucella* gene of interest and the pUC-based plasmid carrying this mutated allele introduced into *Brucella* strains by electroporation (Halling et al. 1991) or conjugal transfer from an *E. coli* host (Tibor et al. 2002). The transformants or transconjugants in which the wild-type gene has been replaced by homologous recombination can then be selected by their antibiotic resistant profiles or other readily detectable phenotypic properties. Recent examples where genome sequence data has been used to formulate experimental strategies aimed at defining the biological functions of specific gene products include studies that have demonstrated the importance of the flagellar (Fretin et al. 2005) and phosphatidylcholine (Comerci et al. 2006; Conde-Alvarez et al. 2006) biosynthesis genes for the virulence of *B. melitensis* 16M and *B. abortus* 2308, respectively. Antibiotic resistance genes are often introduced into the coding regions of cloned *Brucella* genes to create gene disruptions or gene deletions because the antibiotic resistance imparted by the mutated genes serve as effective phenotypic markers for the strains in which they reside. Antibiotic resistance cassettes designed to maintain translational continuity (Menard et al. 1993) can also be used to prevent polar effects when the genes targeted for mutagenesis are components of an operon. Care must be exercised when introducing antibiotic resistance genes into the *Brucella spp.* either on a mutated gene or on the plasmid used for introduction of the mutated gene. The tetracyclines, streptomycin, gentamicin and rifampin are the primary antibiotics used for treatment of human brucellosis (Young 2000), and introduction of genes that confer resistance to any of these antibiotics into *Brucella* is risky and in fact now illegal under NIH/CDC regulations covering research with Select Agents (Leavitt 2005). Kanamycin-, chloramphenicol- and ampicillin-resistance genes have been the most widely used for the construction of *Brucella* mutants by gene replacement because these antibiotics are not used clinically for treatment of brucellosis in humans or animals.

In some instances, that is the construction of mutated strains that have the potential to serve as live vaccine candidates or the introduction of "point" mutations, it is desirable to make *Brucella* mutants that do not carry antibiotic resistance markers. ColE1-based vectors carrying counterselectable markers such as the *sacB* gene have been used for this purpose (Kahl-McDonagh and Ficht 2006). A derivative of the *Brucella* gene of interest carrying a missense or nonsense mutation or an in-frame deletion that removes a significant portion of the coding region can be constructed on the *sacB*-containing plasmid and then this plasmid is introduced into a *Brucella* strain. The first round of selection for the transformants or transconjugants is for the antibiotic resistance conferred by the *sacB* containing vector; in most cases this is ampicillin. This selection ensures that transformants or transconjugants used for further characterization are merodiploid with respect to the gene of interest. Growth in the presence of sucrose can then be used to force resolution of the merodiploid state by homologous recombination. SacB-based counterselection strategies have been used to introduce marked mutations (Ekaza et al. 2000), unmarked deletion mutations (Kahl-McDonagh et al. 2006) and unmarked point mutations (Sieira et al. 2004) into wild-type *Brucella* strains. This approach was used, for instance, to perform site directed mutagenesis of an IHF-binding site in the *virB* promoter region of *B. abortus* 2308 to establish a link between IHF and wild-type *virB* expression in this bacterium (Sieira et al. 2004). SacB-based counterselection has also been used to construct unmarked *Brucella* mutants intended for use in vaccine trials (Kahl-McDonagh et al. 2006; Kahl-McDonagh and Ficht 2006). In addition, this approach has been used to swap one mutation for another in a *B. abortus* strain (Bellaire et al. 2003) and to determine whether or not a particular gene product is essential for *Brucella* viability (Robertson et al. 2000). The use of Flp-Frt based systems in *Brucella* may improve the capacity to make such genetic constructs (Choi and Schweizer 2005).

Genetic complementation. Genetic complementation studies are critical to verify the link between the phenotype exhibited by a mutant and its genotype. To accomplish this, the parental gene is often introduced into a bacterial mutant to test for alleviation of the mutant phenotype. The *Brucella spp.* do not naturally contain plasmids, but certain broad range plasmids will replicate in these bacteria (Rigby and Fraser 1989). The plasmids that have been most widely used for genetic complementation in the *Brucella spp.* have been those derived from the broad host range plasmid pBBR1MCS (Kovach et al. 1994; Elzer et al. 1995). This plasmid replicates in *Brucella* strains with a copy number of approximately 10 copies per genome (Elzer et al. 1994). It has multiple useful unique restriction sites for cloning and variants of this plasmid are available that confer resistance to several different antibiotics (chloramphenicol, kanamycin and ampicillin) that can be used in *Brucella* strains (Kovach et al. 1995). Derivatives of the RK2-based plasmids pGL10 and pMR10 have also been used successfully for genetic complementation of *Brucella* mutants (Gee et al. 2005). The latter plasmids offer the advantage of a low copy number (2–4 copies per genome) and they have an active plasmid partitioning system. pGL10- and pMR10-based plasmids are particularly useful when increased copy number has a negative impact on genetic complementation (such as in the case of genes encoding transcriptional regulators) or when genetic complementation of a mutant is being tested in an experimentally infected animal and prolonged stability of the plasmid in the *Brucella* mutant is critical for an accurate interpretation of the results obtained. In some cases, genetic complementation of mutant phenotype is very difficult or impossible due to gene dosage effects even if a low copy number plasmid is used. In instances such as this, SacB-based counterselection strategies have been used to replace the mutated genomic copy in a *Brucella* mutant with a wild type version of this gene to verify the link between the mutation present in the mutant and its phenotype (LeVier et al. 2000). This approach works well for genes that are transcribed in a monocistronic fashion. It cannot be used, however, to determine whether or not specific mutations have polar effects on downstream genes in an operon.

Functional analysis of Brucella gene products. In addition to serving as databases that allow for the direct evaluation of the biological functions of specific *Brucella* gene products through the construction and phenotypic evaluation of mutants, the *B. melitensis*, *B. suis* and *B. abortus* genome sequences also allow investigators to directly amplify genes of interest from the genomic DNA of *Brucella* strains, clone these genes into expression vectors and produce the corresponding gene products in heterologous hosts such as *E. coli*. Recombinant versions of these proteins can then be purified in large quantities without the biohazards associated with the production of large quantities of *Brucella* strains and the biochemical and immunologic properties of the recombinant proteins examined in laboratory studies. Studies using genome-directed production of recombinant *Brucella* proteins have recently shown that the HmuD is a heme oxygenase (Puri and O'Brian 2006), RibH2 participates in vitamin B2 biosynthesis (Zylberman et al. 2006), Cgh is a bile salt hydrolase (Delpino et al. 2007), PepN is an aminopeptidase (Contreras-Rodriguez et al. 2003; Posadas et al. 2007) and PrpA is a lymphocyte mitogen (Spera et al. 2006). Similar approaches have also been used to produce recombinant versions of PepN and Cgh and these recombinant proteins used to evaluate their capacity to serve as subunit immunogens and/or diagnostic antigens. The information provided in the *B. melitensis*, *B. suis* and *B. abortus* genome sequences has also been exploited to develop DNA vaccine candidates (Yang et al. 2005).

1.8.4 Transcriptomics

DNA microarrays provide a powerful platform for elucidating the transcriptional profile of bacterial pathogens. Transcriptomic studies have facilitated the comprehensive definition of virulence-associated regulons and their regulatory networks for several bacterial pathogens, including *S. typhimurium* (Wilson et al. 2002; Monsieurs et al. 2005; Tamayo et al. 2005), *M. tuberculosis* (Manganelli et al. 2001; Kendall et al. 2004; Hahn et al. 2005; Maciag et al. 2007) and *P. aeruginosa* (Dasgupta et al. 2003; Wagner et al. 2003; Lizewski et al. 2004). However, microarray-based analyses of host–*Brucella* interactions remain, by comparison, in their infancy. To date, microarrays harbouring oligonucleotides that span the entire *Brucella* genome have been employed to identify virulence-associated chromosomal regions in several *Brucella* species (Rajashekara et al. 2004). In addition, microarrays for analysing host cell genes have been employed to obtain transcription profiles of host cell responses to bacterial infection (Eskra et al. 2003; He et al. 2006). Taken together, these studies have provided new insights into the molecular mechanisms mediating *Brucella* pathogenesis, and supported the expectation that this technology, when coupled with emerging computational techniques (Benson and Breitling 2006; Husser et al. 2006; Olson 2006), will contribute to the comprehensive definition of regulons and regulatory networks mediating *Brucella* pathogenesis.

1.8.5 Bacterial Microarrays

Disparate *Brucella* species share highly similar genome sequences (Ugalde 1999; DelVecchio et al. 2002a; Paulsen et al. 2002; Tsolis 2002; Chain et al. 2005). Nevertheless, they display different virulence profiles and host preferences. To better understand the molecular basis of these differences, genomic

DNA extracted from *B. abortus, B. ovis, B. canis, B. suis* and *B. neotomae* was hybridized to a *B. melitensis* 'reference' microarray (Rajashekara et al. 2004). An analysis of the hybridization patterns revealed 217 genes that were present in *B. melitensis*, but absent in other *Brucella* species (Rajashekara et al. 2004). These genes clustered into nine genomic islands, and encoded hypothetical proteins, transporters, transposases and transcriptional regulators that may play a role in defining the host range and virulence of the bacterium. Interestingly, genes believed to participate in bacterial lipopolysaccharide (LPS) biosynthesis were represented in the differential set, consistent with the known role for *B. melitensis* LPS in virulence (Godfroid et al. 1998; Jimenez de Bagues et al. 2004; Cardoso et al. 2006; Tumurkhuu et al. 2006). Although these studies provided an important proof of concept for the use of microarrays in comparative genomic studies involving this pathogen, it is expected that the value of this approach will diminish as the whole-genome sequences of additional *Brucella* species become available, and as purely computational approaches for analyzing genome sequence differences are employed to address comparative genomic questions. On the contrary, recent advances in the selective purification of RNA from *Brucellae* internalized into host cells have created exciting opportunities for new transcriptomic studies to be pursued (Covert et al. 2005), including the possibility of in vivo bacterial transcriptional profiling (Hautefort and Hinton 2000; Snyder et al. 2004; Talaat et al. 2004; Lawson et al. 2006).

1.8.6 Host Cell Microarrays

Transcriptional profiling has been used to define the host response to several bacterial pathogens, including *N. gonorrhoeae* (Binnicker et al. 2003), *P. aeruginosa* (Ichikawa et al. 2000) and *L. pneumophila* (Losick and Isberg 2006). Recently, several groups have applied this approach to study how macrophages respond to *B. abortus* infection in vitro (Eskra et al. 2003; Rajashekara et al. 2005; He et al. 2006). These groups defined over several hundred genes that were differentially transcribed during the early stages (0–12 h) of *Brucella* infection (Eskra et al. 2003; He et al. 2006). The expression of genes encoding pro-inflammatory cytokines and chemokines were upregulated, whereas genes involved in apoptosis, cell cycling and intracellular vesicular trafficking were inhibited. Taken together, the data indicate that *B. abortus* can alter host metabolic and signalling pathways to recruit uninfected macrophages (for future infection) while simultaneously inhibiting apoptosis and intracellular innate immune pathways, thereby permitting intracellular survival of the pathogen. Future studies that employ knockout mouse models or RNA interference mediated gene knockdown approaches shall prove very valuable for correlating the observed changes in gene expression patterns with changes in host cell susceptibility or resistance to *Brucella* infection.

1.8.7 Proteomics

Proteomics seeks to define the entire set of proteins found within an organism at a specific time and under defined environmental conditions (DelVecchio et al. 2002c). Although early proteomics studies of *Brucella* employed one-dimensional SDS-PAGE, immunoblotting and Edman degradation based peptide sequencing to characterize the *Brucella* proteome (Morris 1973; Santos et al. 1984; Gamazo et al. 1989; Brooks-Worrell and Splitter 1992; Lin and Ficht 1995; Teixeira-Gomes et al. 1997; DelVecchio et al. 2002c), current approaches rely upon two-dimensional (2D) gel electrophoresis and peptide mass fingerprinting (PMF). PMF, a powerful technology that combines matrix-assisted laser desorption/ionization (MALDI)-mass spectrometry (MS) with computer-assisted peptide comparison (Appella et al. 2000; Gevaert and Vandekerckhove 2000; Blueggel et al. 2004; Thiede et al. 2005), provides for the high-throughput identification of hundreds of differentially expressed proteins. DelVecchio and colleagues have pioneered the use of this proteomic approach to study *Brucella* physiology (Eschenbrenner et al. 2002, 2006; Mujer et al. 2002; Wagner et al. 2002). For example, when they compared the proteomes of laboratory cultured 16M and Rev 1 strains, several differentially expressed proteins were identified, including proteins regulating iron acquisition, sugar-binding, protein biosynthesis and lipid degradation (Eschenbrenner et al. 2002). The differential expression of Rev 1 iron metabolism pathways was particularly intriguing, and led the authors to hypothesize that misregulation of iron acquisition and utilization pathways may account for the attenuation of virulence displayed by the Rev 1 vaccine strain (Eschenbrenner et al. 2002). In another line of investigation, a comparison of the proteomes of laboratory cultured *B. melitensis* 16M and *B. abortus* 2308 strains revealed significant quantitative and qualitative differences in protein expression patterns (Eschenbrenner et al. 2006). Proteins involved in amino acid binding and transport and Sec-dependent secretion were differentially expressed and led the authors to hypothesize that these pathways may play an important role in defining the host preferences of these two strains. It should be mentioned that because the analyses performed to date have been limited to protein extracts derived from laboratory cultures, these studies have not provided information about the complement of bacterial proteins expressed under physiological conditions. Nevertheless, these important experiments have provided critical baseline data for future in vivo work. Finally, to define the *Brucella* proteins that posses the greatest potential of eliciting a protective immune response (i.e., the bacterial 'immunome'), researchers have combined proteomicapproacheswithimmunoblottingusingantisera from pathogen infected animals or patients (Al Dahouk et al. 2006; Connolly et al. 2006; DelVecchio et al. 2006). To date, these immunoproteomic approaches have succeeded in identifying many candidate proteins; for example, Connolly et al. identified 160 candidate proteins from the *B. abortus* cell envelope, including outer-membrane protein (OMP) 25, OMP31, Omp2b porin and 60 kDa chaperonin GroEL (Connolly et al. 2006). In addition, several additional proteins were proven to possess immunoreactivity with serum derived from patients or infected animals; these included a fumarate reductase flavoprotein subunit, F_0F_1-type ATP synthase alpha subunit and cysteine synthase A (Connolly et al. 2006). Taken together, these immunogenic proteins constitute attractive candidates for developing vaccines against *Brucella* infection, and suggest that immunoproteomics holds significant promise for vaccine or diagnostic developments.

1.8.8 Higher Order (Protein–Protein) Interactions

Investigation of higher order structures has naturally begun with an evaluation of the type IV secretion system (T4SS). This multi-subunit complex is encoded by an operon containing at least 10 genes (VirB1–10) and perhaps more and interacts with several others as demonstrated in studies using bacterial two-hybrid analysis (de Paz et al. 2005). A bacterial two-hybrid approach was also used to explore these interactions and the results suggested that the core components (VirB10 and the T4CP) could be exchanged between *Brucella* and *Agrobacterium*, suggesting that the peripheral proteins may be altered in sequence as a result of adaptation to specific function in different bacteria while the core components are conserved and maintain interacting domains. X-ray crystallography has been used to explore the structure of a number of *Brucella* subcellular components. Most recently, the approach has been used to elaborate on the interactions between proteins in the type IV secretion system that have biological relevance (Paschos et al. 2006). Alternative approaches include the use of histidine tagging and co-immunoprecipitation for co-purification of proteins associated with the histidine tagged gene product.

Earlier studies explored the structural features of a model enzymatic function, lumazine synthase that catalyses the penultimate step in the riboflavin biosynthesis pathway (Ooman et al. 1991; Zylberman et al. 2004). Studies of lumazine synthetase have led to the identification of a promising subunit vaccine candidate (Rosat et al. 2006). X-ray diffraction analysis has been used to study the interaction between antibody and antigen with regard to the immunodominant O-polysaccharide on the cell surface and led to increased knowledge of the differences between M and A antigens and antibody specificity (Rose et al. 1993). NMR has also been used to analyze modifications in a number of virulence factors. First, NMR revealed that a fraction of the cyclic-ß-glucan produced by *Brucella* contained succinyl residues. However, these modifications were shown to have no effect on virulence (Roset et al. 2006). Similar studies have been performed in the analysis of LPS and characterized fatty acid modifications of the lipid A portion affecting the overall toxicity of LPS (Campos et al. 2004).

1.8.9 Conclusions

It is obvious from this review and the distribution of research citations herein that the functional genomics of *Brucella* sp. is in the earliest stages of development. Much emphasis has been placed on the identification of genes encoding functions necessary for virulence. This has led to the development of several live, attenuated vaccine strains and characterization of the intracellular environment in which *Brucella* persists. However, although many genes are now known to be essential for survival their contribution remains undefined. Of the genes identified, the best characterized are those contributing to the biosynthesis of the LPS on the cell surface and those encoding the structural components of the T4SS. Although LPS is thought to be responsible for resistance to extracellular and intracellular killing, it also appears to be important in limiting activation of the invaded cell. Although the T4SS is essential in the regulation of intracellular trafficking and the ultimate outcome of infection, the mechanisms responsible remain undefined. Since the LPS and T4SS are the best studied factors associated with virulence of the *Brucella spp.*, it is clear that much work remains to define and clarify *Brucella* pathogenesis and survival.

1.9 Genome Mapping and Microarray Contributions to Understanding *Brucella* Pathobiology and Host Responses to *Brucella* Infections

1.9.1 Introduction

The overall goal of this section is to provide review and insight into the actual and potential contributions that the genes within the genomes of the *Brucella spp.* play in making these bacteria such successful facultative intracellular pathogens. Particular emphasis will be given to an examination of the studies involving similarities and differences in their genomes within the genus. In addition, emphasis will be placed on studies using microarrays to examine both sides of the host-pathogen interaction.

1.9.2 Biology of a *Brucella* Infection

Vertebrate Infections

The ability of the *Brucella spp.* to infect a wide range of mammalian vertebrates, as described in some of the previous sections, is a strong indication of the evolutionary adaptation this pathogen has undergone. It is perhaps understandable that the brucellae have become such successful intracellular pathogens because they, like other pathogens, have developed mechanisms of avoiding the hosts' immune surveillance and killing systems. At first glance, this may appear paradoxical, given the range of vertebrates' immune systems subverted in the course of establishing a chronic *Brucella* infection. However, when one considers the high degree of similarity both in the qualitative and quantitative genomic content of the four *Brucella* species (*B. melitensis, B. suis* and *B. abortus*) sequenced (Ratushna et al. 2006; Yu et al. 2007, Sect. 1.7), one can surmise that the brucellae have successfully infected a range of mammalian hosts because of the pathogen's adaptation to avoid the shared features of their immune systems.

Anti-Immune Strategies

Finlay and McFadden (2006) have reviewed the various means by which bacteria and viruses evade host immune systems. They classify 12 different strategies by which immune evasion can occur, and these include (1) secretion of modulators or toxins; (2) production of modulators on pathogen surfaces; (3) avoidance of surveillance; (4) antigenic hypervariability; (5) subversion and/or killing of immune cells and phagocytes; (6) blocking induction of acquired immunity; (7) inhibition of complement; (8) inhibition of cytokine production; (9) modulation of apoptosis and autophagy; (10) interference with toll-like receptors; (11) inhibition of antimicrobial peptides and (12) blocking intrinsic cellular pathways.

A variety of published reports during the past 30 years suggest that the *Brucella spp.* exercise or exhibit all or parts of the strategies in the above-listed categories. Many of these infection strategies the *Brucella* utilise have been presented in a special issue of Veterinary Microbiology edited by Halling and Boyle (2002) and a book by Lopez-Goni and Moriyon (2004). A reading of these publications endorses the realization of the need for a closer examination of not only the gene content but of the expression of those genes (i.e. transcriptomics) as a means of understanding the mechanisms by which the brucellae interact with their hosts and cause infection. Thus this section will focus on the use of mutants and microarrays as a means to investigate both the individual and simultaneous expression of genes in *Brucella* and infected hosts and elucidate mechanisms of pathogenesis;

these may in turn suggest novel intervention strategies to the reader.

1.9.3 Genome Mapping for Understanding *Brucella* Pathobiology

Graphic Comparisons of the *Brucella* Genomes

Ratushna et al. (2006) have published a noteworthy comparison of the sequenced genomes of three *Brucella* species: *melitensis, abortus* and *suis* (Fig. 12). They compared the three genomes based on predicted ORFS and found differences confirmed by PCR and RT-PCR: 22 unique genes differentiated *B. suis* from *B. abortus* and *B. melitensis* and only a single unique gene differentiated either *B. abortus* and *B. melitensis* from each other or *B. suis*. They were able to identify a unique set of genes that allows for the differentiation of *all* the known *Brucella spp.* (i.e. biovars) using PCR. In addition, their analysis, aside from validating the conclusion that the genus is highly homogeneous, suggests that there are significant opportunities to explore whether the genomic differences are related to their host preferences and ability to replicate in an intracellular environment.

In support of the need for further measurement of differential gene expression is the in-depth comparative analysis by Yu et al. (see Sect. 1.7, this chapter) of the published genomes of four *Brucella spp.* using a novel informatics software program, Gen-Var. Their analyses reveal that selective gene disruption has occurred in some of genes among the *Brucella spp.* and warrant further testing as to whether these variations are related to the pathogenicity of the different species (e.g., host preference).

Mutations and Affects on Survival and Pathogenesis

Various methods have been described in Sect. 1.8 for preparation and analysis of *Brucella* mutants. *Brucella* gene mutations have resulted in a better understanding of intracellular adaptation and pathogenesis. The *Brucella* Bioinformatics Portal (BBP) (Xiang et al. 2006) has identified 220 mutations found to be associated with survival inside macrophages, HeLa cells or in mice (Table 7). The phenotypes of the *Brucella* mutants confirm the established virulence determinants associated with the *Brucella* type IV secretion system encoded by the *virB* operon (O'Callaghan et al. 1999), the BvrR-BvrS two-component regulatory system encoded by *bvrR* and *bvrS* (Sola-Landa et al. 1998) and O-side chain associated with *Brucella* lipopolysaccharide (Allen et al. 1998; McQuisiton et al. 1999). Transport and metabolism of various metabolites including amino acids, carbohydrates, lipids and inorganic ions appear also to be critical for *Brucella* survival in cell lines and mice (Kohler et al. 2003; Roop et al. 2004). Since the *Brucellae* survive inside phagosomes of phagocytic cells, for example macrophages, bacterial survival associated with these mutated genes suggests that the corresponding metabolites are not accessible to the bacteria inside

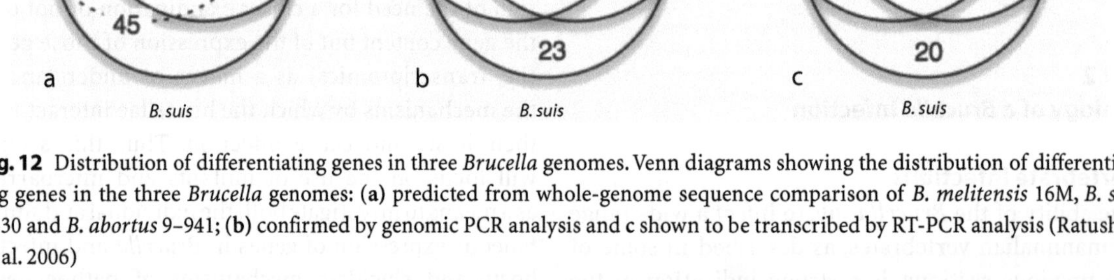

Fig. 12 Distribution of differentiating genes in three *Brucella* genomes. Venn diagrams showing the distribution of differentiating genes in the three *Brucella* genomes: (**a**) predicted from whole-genome sequence comparison of *B. melitensis* 16M, *B. suis* 1330 and *B. abortus* 9–941; (**b**) confirmed by genomic PCR analysis and c shown to be transcribed by RT-PCR analysis (Ratushna et al. 2006)

Table 7 Classification of 220 attenuated *Brucella* genes found from literature search using BBP. The gene names starting with BMEI or BMEII represents locus tags from *B. melitensis* chromosome I or II that currently do not have formal names assigned

COG clusters	Mutated *Brucella* genes affecting intracellular survival
C: Energy production and conversion	8: *caiB, cydB, eryB, fdhA,glpK, narG, norE, pyc*
E: Amino acid transport and metabolism	28: BMEII0626, BMEII0923, *aroC, aspC, cysK, dppA, gcvP, gcvT, glnA, gloA, gltD, glyA, hisC, hisD, hisF, leuA, leuC, livH, lysA, metH, nifS, nikA, pepN, pheA, pheB, serB, thrA, thrC*
F: Nucleotide transport and metabolism	15: *carAB, dut, gpt, hpt, ndrI, purD, purE, purF, purH, purL, purM, purN, pyrB, pyrC, pyrD, rsh*
G: Carbohydrate transport and metabolism	23: BMEII1045, *araG, cbbE, dbsA, eryC, galcD, gluP, gnd, ilvD, malK, manB, mgtA, mosA, mocC, ndvB, pgi, pgm, pmm, rbsK, ugpA, ugpB, xfp, zwf*
H: Coenzyme transport and metabolism	6: BMEI1902, *caiB, cobB, hemH, ilvC, ilvI*
I: Lipid transport and metabolism	6: *aidB, bacA, dxps, pcs, pmtA, uppS*
J: Translation	4: *miaA, pth, rplS, rpsA*
K: Transcription	20: *RpiR, ansC, arsR6, aspB, deoR, gntR, gntR1, gntR10, gntR17, gntR2, gntR4, gntR5, lysR, lysR12, lysR13, lysR18, oxyR, rpoA, rsh, vjbR*
L: Replication, recombination and repair	6: BMEI1229, *alkA, mgps, recA, xseA, uvrA*
M: Cell wall/membrane biogenesis	21: *amiC, galE, gmd, gtrB, lpsA, lpsB, macA, mutM, omp10, omp19, omp25, perA, rfbD, virB1, wbdA, wbkA, wboA, wbkB, wbpL, wbpW, wbpZ*
N: Cell mobility	4: *flgE, flgI, fliC, motB*
O: Posttranslational modification, protein turnover, chaperones	16: BMEI0455, *cydC, cydD, cysY, degP, dnaK, dsbA, dsbB, exsA, glnD, htrA, lon, nrdH, ppiD, tig, trkH*
P: Inorganic ion transport and metabolism	9: BMEII0336, *cysI, fbpA, mgtB, norD, sodC, ssuB, znuA, znuC*
Q: Secondary metabolites biosynthesis, transport and catabolism	2: *dhbC, pncA*
R: General function prediction only	16: BMEI0671, BMEI1258, BMEI1443, BMEI1531, BMEI1859, BMEII0274, BMEII0935, BMEII1037, *bicA, cobW, gltI, gltD, hfq, mosC, rbsC, tldD*
S: Function unknown	3: BMEI1809, BMEII0128, *pncA*
T: Signal transduction mechanisms	15: BMEI1448, *artI, bvrR, bvrS, divK, feuP, feuQ, ftcR, glnL, nodV, ntrC, ntrY, pstP, spotT, vsrB*
U: Intracellular trafficking and secretion	11: *flghA, fliF, virB2- virB6, virB8-virB11*
V: defense mechanisms	4: BMEII0318, *dacF, exsA, norM*
-: Not in COGs	6: BMEI0085, BMEI1339, BMEI1361, BMEI1658, BMEI1844, BMEI1879

the phagosomes and are essential for intracellular growth.

The screening of signature tagged mutants for attenuation identified an attenuated mutant disrupted in *fliF* encoding the MS ring of flagella suggested a possible role for flagella in *Brucella* virulence (Lestrate et al. 2003). This observation led to the discovery of a polar and sheathed flagellum expressed in early log phase growth (Fretin et al. 2005); mutations in *fliF*, *flg*I (P ring monomer) and *fli*C (flagellin) resulted in decreased survivability in mice. FtcR has recently been found to act as a flagellar master regulator in *B. melitensis*, and a *ftc*R mutant has the same decreased virulence phenotype as a *fliF* mutant (Leonard et al. 2007). It is noteworthy that as yet no direct evidence for the synthesis of a flagellum, as opposed to its components, for *Brucella* grown in macrophages or in mice has been produced. It has been suggested that the flagellar system may be used for the export of other as yet unidentified components that help the *Brucella* replicate in an intracellular environment (Fretin et al. 2005).

Brucella pathogenesis relies on interactions between the products of individual *Brucella* genes (e.g., *virB*, *ftc*R). In addition to identifying published *Brucella* gene mutations, BBP can also predict *Brucella* genetic interactions by sampling all accessible *Brucella* publications. Currently, 65 genetic interactions have been identified in the BBP database. Of the 65 genes involved in these interactions, 32 are also found in the attenuated *Brucella* gene mutation list (Table 7). The finding of these predicted gene interactions suggests that a more comprehensive investigation of *Brucella* pathogenesis should be considered, for example by microarrays. The results of this prediction suggest interactions between genes encoding virulence determinants, and a more comprehensive investigation of these interactions using microarrays is likely to uncover new mechanisms for coordinate regulation of virulence genes in response to environmental signals in the host. For example, Fig. 13 demonstrates the interactions among RNA-binding protein *hfq* (host factor 1), *ctrA* (cell cycle transcriptional regulator), *ahpC* (alkyl hydroperoxide reductase C), *sodC* (copper-zinc superoxide dismutase) and 10 other *Brucella* genes related to gene regulation during stationary growth phase. *B. abortus* HF-1 protein encoded by *hfq* contributes to stress resistance during stationary phase and is a major determinant of virulence in mice (Robertson and Roop 1999). Many *Brucella* genes (such as *hem*B, *thi*5, *thi*E and *hde*A) require *hfq*-encoded HF-I for optimal expression during stationary phase (Roop et al. 2003). CtrA is a master response regulator that is essential for viability and is transcriptionally autoregulated. CtrA directly regulates the expression of the *rpo*D, *fts*K, *min*C and *fts*E (Bellefontaine et al. 2002). The *hfq* gene is likely to be negatively regulated by CtrA (Robertson et al. 2000). Bacterial *sod* genes are typically regulated in a growth-phase-dependent manner and their expression is usually maximal during stationary phase. A mutation in *B. abortus hfq* results in greatly reduced *sodC* expression (Roop et al. 2003). Reduced expression of *ahp*C or *sod*C upon entry into stationary phase may increase the killing of the *B. abortus hfq* mutant by host macrophages compared to the parental 2308 strain (Roop et al. 2003). Figure 13 also suggests that CtrA may also regulate *Brucella sodC* expression indirectly by affecting *hfq* expression.

Whole-Genome Microarray Analysis to Understand *Brucella* Biology

Genomic microarrays represent a useful tool for understanding both the relatedness of different *Brucella* strains and the changes in global gene expression patterns that occur upon infection. To date, one study has been published that has employed *Brucella* microarrays to identify genomic islands (Rajashekara et al. 2004) in *B. melitensis* relative to other *Brucella* species. Since the different *Brucella* genomes share a high overall level of similarity, yet the individual *Brucella* have preferences for distinct host populations, it was hypothesized that insertion/deletion events have contributed to biological differences between the *Brucellae*, including host specificity. To study the possible role of insertion/deletion events in the unique biology of the different *Brucella* species, Rajashekara et al. used whole genome microarrays of *B. melitensis* to compare hybridization patterns of *B. suis*, *B. ovis*, *B. abortus*, *B. canis* and *B. neotomae* (Rajashekara et al. 2004). Their survey identified 30 regions present in *B. melitensis* but missing from one

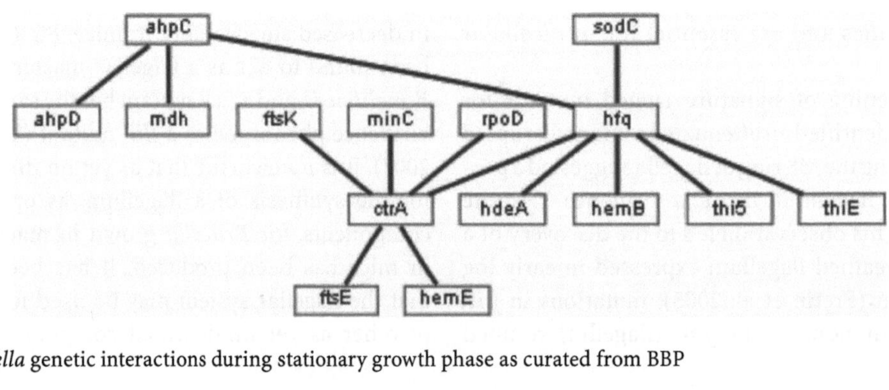

Fig. 13 *Brucella* genetic interactions during stationary growth phase as curated from BBP

or more of the other *Brucella* strains analyzed. These regions range from 58 bp to 44 kb in size; the nine largest, ranging from 5 kb to 44 kb in size, were designated as genetic islands. Most of these islands are absent from multiple isolates of the same *Brucella* species, suggesting that the majority of these deletions are of species-specific characteristics.

An interesting result of this analysis is that the genetic islands identified do not encode adhesins or secreted virulence factors that have been shown to contribute to host specificity in other bacterial species (Moon et al. 1977; Tsolis et al. 1999; Inatsuka et al. 2005). This finding suggests that either (1) adhesins and/or secreted virulence factors may play an important role in adaptation of the *Brucellae* to their different hosts, but are encoded in conserved loci, where they may be differentially expressed or inactivated as the result of point mutations, or (2) host specificity of the *Brucellae* is mediated by mechanisms that are different from those described previously for other bacterial pathogens. The genetic islands differentiating the *Brucella* species do contain a large number of hypothetical genes, suggesting that differences in the biology of the different *Brucella* species may be mediated by functions that are still unknown.

Construction of genomic microarrays that contain all the sequenced *Brucella* species will allow for the characterisation of larger groups of *Brucella* isolates from different species to determine their degree of heterogeneity at the genomic level. Further, studies are underway in several laboratories using microarrays to assay global *Brucella* gene expression profiles during infection of different cells lines and hosts.

1.9.4 Genome Mapping for Understanding Host Response Against *Brucella* Infections

Macrophage Responses to *Brucella* Infections

Various molecular and immunological methods have been described to study the macrophage-*Brucella* interactions (Baldwin and Winter 1994; Roop et al. 2004). Two microarray studies have also been reported showing gene expression profiles in murine macrophages infected with virulent *Brucella* strains (Eskra et al. 2003; He et al. 2006). Of the >6,000 genes on the Affymetrix murine U74A gene microarray chips, Eskra et al. (2003) found over 140 genes that were reproducibly differentially transcribed in RAW264.7 macrophages infected with *B. abortus* for 4 h. He et al. (2006) further analyzed the time course response of J774.A1 macrophages during infection with virulent *B. melitensis* strain 16M using Affymetrix mouse 430 2.0 array containing more than 39,000 genes. Transcription of 243 and 1,053 genes was significantly up- and down-regulated, respectively, between 0 and 4 h. However, compared to uninfected macrophages, only 12 genes were found up- or downregulated at 24 h, and no genes were found differentially regulated at 48 h post infection. Analysis of these gene profiles has provided a more comprehensive picture of the macrophage-*Brucella* interactions.

Immediately after entry into the macrophages, *Brucella* reside in an acidified membrane-bound compartment. Later, *Brucella* gain access to their growth niche known as the replicative phagosome after continual interactions between the *Brucella*-containing compartment and the endoplasmic reticulum of macrophages (Roop et al. 2004). Celli et al. (2003) indicated that the majority of infected *Brucella* stayed in the early phagosome at 4 h post-infection and then most of the *Brucella*-containing vacuoles matured into either degradative or replicative compartments by 24 h post-infection. The replicative phagosomes do not fuse with lysosomes and provide a more hospitable environment than the early-stage acidified compartment for intracellular survival. Gorvel and Moreno (2002) reported that crucial macrophage cell cycle processes such as chromosome condensation of DNA synthesis, mitosis, karyokinesis and cytokinesis were not inhibited while a large number of *Brucella* lived within the replicating niche. In contrast, intracellular *Brucella* replication does not require de novo host protein synthesis, as treatment of Vero cells with cycloheximide does not inhibit *Brucella* replication (Eze et al. 2000). This suggests that the significant *Brucella* growth inside the replicative phagosomes does not interfere with the important host physiological pathways. Furthermore, this correlates with the recent finding that most transcriptional changes in macrophage J774.A1 cells infected with *B. melitensis* strain 16M occur at the early infection stage and all gene expression profiles returned to normal between 24 and 48 h post-infection (He et al. 2006).

A number of studies have indicated that smooth virulent *Brucella* undergo rapid clearance (>90% of *Brucella* killed) within the first 24 h followed by growth of the surviving bacteria after infection of murine J774.A1 or peritoneal macrophages. The mechanism of the 90% reduction at the early stage of infection still remains unclear. In *B. abortus* infected macrophages at 4 h post infection, Eskra et al. (2003) found an increase in the transcription of a number of proinflammatory cytokines and chemokines, such as TNF-α, IL-1β, IL-1α and members of the small inducible cytokine family of proteins. The proinflammatory response may constitute a general recruitment of host antibacterial defenses (Eskra et al. 2003). Active proinflammatory responses were also shown in *B. melitensis* infected macrophages at the early infection stage, as indicated by increased TNF-α, TLR2, IL-1β, IL10ra [IL-10 receptor, alpha] and many other genes (He et al. 2006). Many up-regulated cellular defense factors at the early infection stage may contribute to initial brucellacidal effects such as up-regulated complements and the acid environment in the transferrin-containing endosomes and early phagosomes (He et al. 2006). However, the general host cellular activities were significantly suppressed as indicated by downregulated genes involved in cell growth and/or maintenance, macromolecule and nucleic acid metabolism, biosynthesis and biological process regulation, and intracellular signalling cascade. Macrophages displayed up-regulated response to external stimuli [e.g., Cxcl16, C5ar1, Ncf1 and Fcgr3] but downregulated response to endogenous stimuli [e.g., Gtf2h1, Pole, Rad23b and Msh6], suggesting that at this time infecting *Brucella* had started to escape active macrophage responses to intracellular stress due to infection although inflammatory and other defense responses were still active. The observation of more downregulated than up-regulated gene expression profiles suggests that the most active brucellacidal activity might have happened before 4 h post-inoculation. Analysis of the gene expression profiles at the earlier infection stage may provide more insight into effective brucellacidal mechanism in macrophages.

Several studies indicate that virulent *B. abortus* (Pe and Ficht 2004), *B. suis* (Gross et al. 2000; Dornand et al. 2002) and *B. melitensis* (Fernandez-Prada et al. 2003) inhibit apoptosis in macrophages. Eskra et al. (2003) found a mix of pro- and anti-apoptosis effects based on the gene expression profiles in RAW 264.7 macrophages infected with *B. abortus* strain 2308. He et al. (2006) examined transcription profiles of all apoptosis-related genes in *B. melitensis* strain 16M infected macrophage J774.A1 cells. Although many pro-apoptosis genes were up-regulated, the caspase cascade pathways were not activated. This suggests some upstream component(s) that induces caspase activation was suppressed, for example, release of cytochrome *c* from the mitochondria. It was further demonstrated that 106 mitochondria-associated genes were downregulated while only four mitochondria-associated genes were up-regulated at 4 h post infection (He et al. 2006). It is likely that *B. melitensis* 16M prevents apoptosis in macrophages by suppressing mitochondrial gene expression involved in cytochrome c release, ROS production and mitochondrial membrane permeability, therefore preventing activation of caspase cascades. Prevention of apoptosis in macrophages by *B. melitensis* strain 16M ensures extensive replication after the initial killing stage. This inhibition may contribute to the ability of *Brucella spp.* to persist chronically in the reticuloendothelial system of infected humans and animals.

Mice In Vivo Responses Against *Brucella* Infections

Mouse Studies Before Microarrays Although the mouse does not routinely model some aspects of infection found in natural hosts (e.g., abortion), it has served as an excellent model for persistence of *Brucella* in the tissues of the reticulo-endothelial system. Furthermore, the tools available to study infection in the mouse have enabled mechanistic approaches for identifying specific components of immunity involved in the host response to *Brucella* infection. For this reason, the mouse model has been used extensively for studies of the immune response to both primary *Brucella* infection and to challenge after vaccination. These studies have focused on two types of questions, namely, 'What kinds of host responses are elicited by *Brucella*, and what does this tell us about their pathogenic strategies?' and 'Which host responses correlate with effective immunity to *Brucella* infection?'.

The primary response of mice to infection with *Brucella* is characterised by production of both IFN-γ and IL-10 (Zhan 1993; Fernandez Lago 1996; Murphy 2001). Studies of cytokine production and of antibody isotypes elicited during *Brucella* infection show a Th1 polarisation of the initial response to *Brucella*. The essential role of IFN-γ in limiting *Brucella* infection has been demonstrated, both antibody-mediated depletion of IFN-γ (Zhan and Cheers 1993) and knockout of the genes encoding IFN-γ (Murphy 2001) lead to increased bacterial replication in the tissues, and whereas *Brucella* does not cause morbidity in many strains of inbred mice, in knockout mice that are deficient in IFN-g production, *Brucella* replicates unchecked, leading to lethal morbidity (Murphy 2001).

Transcriptional Responses in Mouse Tissues Determined Using Host Microarrays Host microarrays have been used to generate transcriptional profiles of splenocytes after both primary *Brucella* infection and recall responses to vaccination. These experiments differ from the studies on macrophage responses discussed earlier, in that the gene expression patterns result from interactions that occur between different cell types in tissues in vivo. As such, they also represent the responses of multiple cell types, including macrophages, B and T cells and dendritic cells, to infection.

To identify host responses elicited by infection with virulent *Brucella*, but not attenuated mutants lacking the Type IV secretion system (T4SS), Roux et al. compared early transcriptional responses of mouse splenocytes to infection with *B. abortus*, *B. melitensis* and *B. abortus virB* mutants defective in the T4SS. In this study, transcriptional profiles of whole splenic tissue from *Brucella*-infected mice were compared with those from uninfected mice using Affymetrix mouse 430 2.0 arrays. The largest number of differentially expressed genes belonged to the gene ontology category of inflammation and immunity. At 3 days post infection, both *B. abortus* and *B. melitensis* infection induced expression of genes involved in proinflammatory responses, including IFN-γ, IL-15 and C-C (2 cysteines or Cys-X-Cys)} and C-X-C chemokines. Furthermore, several genes induced by Type I interferons were up-regulated after *Brucella* infection. These included genes involved in antigen processing and presentation, such as the transporter associated with antigen processing (Tap1) and components of the proteasome. In addition, genes encoding diubiquitin, involved in tagging proteins for degradation via the proteasome, and interferon stimulated gene 15 (ISG15), a ubiquitin-like modifier, were induced after infection. This proinflammatory response peaked at three days post-infection and was dependent on the presence of the Type IV secretion system, as *virB* mutants, although present in the same numbers in the spleen, did not induce this response. These findings suggest that the Type IV secretion system elicits a proinflammatory response either directly via secreted effector molecules or indirectly via its effect on intracellular growth. Although *Brucella* elicits an inflammatory response in vivo, this response is likely to be lower than that induced by pathogens that trigger acute inflammation in the spleen, such as *Salmonella*, based on findings that both isolated LPS from *Brucella* and live *Brucella* elicit lower levels of cytokines in macrophages than *Salmonella* LPS or live *Salmonella* (Weiss et al. 2005; Lapaque et al. 2006).

Changes in expression of host genes that are elicited by challenge with *Brucella* antigens may be useful in identifying individuals who have been vaccinated or infected with *Brucella*. Paranavitana et al. (2005a) used microarrays to identify gene expression profiles in splenocytes of mice immunized with the *B. melitensis* vaccine candidate WR201 (*purEK*). Transcriptional profiles of splenocytes from immunized mice stimulated ex vivo with *B. melitensis* extracts were compared with responses of splenocytes from unvaccinated mice (Paranavitana et al. 2005b). This study showed that interferon regulatory factor 1 (IRF-1), suppressor of cytokine signalling 1 (SOCS-1), CD86 and IL-2 receptor are upregulated in vaccinated mice, while interferon regulatory factor 7 (IRF7) and C-C-C chemokine receptor 4 (CXCR4) are downregulated. Although the methodology used in this study was different from the one used in the study by Roux et al. to study the primary response to *Brucella* infection, a common set of proinflammatory genes was identified in both studies, suggesting that induction of these genes may be characteristic of both primary and recall responses to *Brucella* infection (Table 8). Further, 18 genes induced in splenocytes by primary or secondary infection were found to be induced in macrophage-like cell lines infected with *B. abortus* (Eskra et al. 2003) and *B. melitensis* (He et al. 2006) in vitro, suggesting that during infection in vivo, macrophages are one source of these transcripts (Table 8).

Human and Murine Responses Against *Brucella* Infections

In humans, brucellosis is considered a multisystem disease (Sauret and Vilissova 2001), and it induces a

Table 8 Differentially expressed genes identified after primary infection, stimulation of vaccinated splenocytes or macrophage in vitro infection by *Brucella spp.*[a]

Gene	Primary infection	Vaccination	Macrophage cell lines	References
Inflammation and Immunity				
IRF-1	+[b]	+		Paranavitana et al. (2005b); Roux et al. (2007)
IRF-7	+	–[c]		Paranavitana et al. (2005b); Roux et al. (2007)
LRG-47 (Ifi1)	+	+		Paranavitana et al. (2005b); Roux et al. (2007)
SOCS-1	+	+		Paranavitana et al. (2005b); Roux et al. (2007)
SOCS-3	+		+	He et al. (2006); Eskra et al. (2003); Roux et al. (2007)
IFN-γ	+	+		Paranavitana et al. (2005b); Roux et al. (2007)
CCL2 (MCP1)	+	+		Paranavitana et al. (2005a); Roux et al. (2007)
CXCL9 (MIG)	+	+		(Paranavitana et al. 2005a; Roux, et al. 2007)
CXCL10 (IP-10)	+	+	+	Paranavitana et al. (2005a); Eskra et al. (2003); Roux et al. (2007)
CXCL2 (MIP-2)	+	+	+	Eskra et al. (2003); Roux et al. (2007)
CCL4 (MIP-1b)	+		+	Eskra et al. (2003); Roux et al. (2007)
Cox2 (Ptgs2)	+		+	Eskra et al. (2003); He et al. (2006); Roux et al. (2007)
TNF-a	+		+	Eskra et al. (2003); He et al. (2006); Roux et al. (2007)
Serum Amyloid A	+		+	Eskra et al. (2003); Roux et al. (2007)
Fas	+		+	Eskra et al. (2003); Roux et al. (2007)
Irg1	+		+	Eskra et al. (2003); Roux et al. (2007)
Ifit1	+		+	Eskra et al. (2003); Roux et al. (2007)
IL-1 receptor antagonist	+		+	Eskra et al. (2003); He et al. (2006); Roux et al. (2007)
Extracellular matrix				
MMP 13	+		+	He et al. (2006); Roux et al. (2007)
Protocadherin 7	+		+	Eskra et al. (2003); Roux et al. (2007)
Timp2	+		+	He et al. (2006); Roux et al. (2007)
Signal transduction				
Metallothionein-2	+		+	Eskra et al. (2003); Roux et al. (2007)
Transcription regulation				
C/EBP	+		+	Eskra et al. (2003); Roux et al. (2007)
Jun	+		+	Eskra et al. (2003); Roux et al. (2007)
Glutaredoxin 1	+		+	Eskra et al. (2003); Roux et al. (2007)
Cell cycle and cell proliferation				
Schlafen 4			–	He et al. (2006); Roux et al. (2007)

[a] Only genes identified as differentially expressed in more than one experimental condition are shown
[b] +, Induction of expression after infection or stimulation with *Brucella*
[c] –, Repression of expression after infection or stimulation with *Brucella*

wide range of non-specific signs and symptoms (e.g. anorexia, back pain, cephalgia, fatigue, fever, malaise, myalgia, sweats and weight loss). In parallel with these symptoms, brucellosis can generate a plethora of clinical manifestations (anaemia, thrombosis, endocarditis, leucopenia, liver abscess, lymphadenopathy, meningitis, nephritis, optic neuritis, splenomegaly, spondylitis, thrombocytopenia and uveitis). Although there are no vaccines for preventing the disease in humans, there are a number of standard antibiotic regimens for treatment of pregnant women, children and adults (Sauret and Vilissova 2001).

In the mouse model of brucellosis, infection results in a predominantly Th1 cellular immune response characterized by the production of TNF-α, interferon-γ and IL-12 (Cheers 1984; Yingst and Hoover 2003). Even though *Brucella* do not normally infect mice, murine studies both with cell lines and intact animals have yielded numerous insights into how this pathogen evades the immune system. Moreover, it has been a useful model for predicting the efficacy of attenuated vaccine strains for cattle (e.g., *B. abortus* RB51, reviewed in Schurig et al. 2002). However, as pointed out by Dornand et al. (2002), the response of human macrophages to a *B. suis* infection is similar to that of other intracellular pathogens, in that apoptosis of monocytes and macrophages is inhibited. He et al. (2006) have extended this conclusion using a murine microarray to examine the gene expression profiles in macrophages following a *B. melitensis* infection. In contrast, the production of TNF-α is stimulated in murine but not in human macrophage cell lines (Dornand et al. 2002). In attempting to identify the mechanisms by which the *Brucella* interfere with the host's immune system, Oliaro et al. (2005) have provided some compelling evidence that the *Brucella* themselves are producing host associated proteins (e.g., Omps) and soluble factors that modulate the innate immune response (via Vg9Vd2 T-cells) as well as inhibit the production of IL-12 and the Th1 response (via natural killer cells).

To begin to more closely examine the possible host–pathogen interactions on a genomic level, it will be important to employ and refine the use of microarrays to examine the transcriptomics of both the *Brucella* and host cells/tissues early during infection. Such data will allow the generation of hypotheses that can then be tested using *Brucella* mutants as well as macrophage lines and transgenic animals (e.g., knockouts). It should not go unnoticed that the *Brucella spp.* are one of a number of gram negative bacterial pathogens that induce reactive arthritis (Inman 2006), a type of autoimmunity that can lead to significant bone degeneration (Kateeb et al. 1999; Hill and Lillicrap 2003). What are the common antigens or pathogenic mechanisms shared by these intracellular pathogens that lead to reactive arthritis? Certainly in a comparative genomics basis, there is much to be learned from a comprehensive transcriptomics profiling of these various pathogens and their corresponding hosts as to what might be the underlying mechanism(s). Only then do we stand the chance of developing novel therapeutic approaches to help alleviate the serious side effects of brucellosis that affect so many of the world's population.

1.10 Future Directions and Prospects

Despite successful control and eradication programmes in many developed countries, brucellosis is still a major problem throughout the world (Pappas et al. 2005a, b). Reemergence of brucellosis is a serious threat to public health in Albania and the ex-Yugoslavian states and a direct threat to the European Union. It is clear that political intervention is required to control the disease in these areas, and also new tools for diagnosis, epidemiology and vaccination will be needed to provide strong support. Apart from its socioeconomic importance, *Brucella* and its interactions with its host are a fascinating biological subject, which will contribute in a fundamental way to our understanding of bacterial evolution, physiology and host pathogen interactions.

At present, we have access to the sequences of the complete genomes of eight *Brucella* strains, with several more underway. In the context of the over 1,000 genomes sequenced or being sequenced at the time of writing (early 2008), this is a small number. New high-throughput sequencing techniques, with the prediction of the $1,000 genome, promise an ever increasing wealth of genomic sequence data from all bacterial species, including *Brucella*. This mass of data will be of little value unless informatics resources develop proportionally. Databases will be ever increasingly integrated, producing an in silico microbe using the genome, ORFome, proteome, virulome, metaolome and interactome. Another outcome will be the reduction of the number of unique genes found in bacteria, including *Brucella*, and identification of the domains and functions of many other genes now designated as conserved hypothetical genes. Though more comprehensive *Brucella* websites are unlikely to appear in addition to PATRIC and TIGR, websites focusing on specific attributes of bacteria genetics will continue to be refined, focusing on identification and characteristics of signal sequences, promoters, paralogs, orthologs and other functions.

Access to the complete genomes of more *Brucella* strains will impact several fields of study and have both practical and intellectual importance. While species and biovar specific tests are now available, we still need assays to distinguish strains of *Brucella* within a biovar. These assays will be essential tools for molecular epidemiology, increasing our ability to track sources of infection. As more genomic sequences become available, a clearer picture of the phylogenetic relationships between strains, how the genus evolved and where this genus fits in the larger picture of closely related α-proteobacteria will expand (Batut et al. 2004).

One key for understanding the biology of *Brucella* is gene regulation. Expression microarrays are beginning to appear in the literature and with the ever decreasing cost of microarrays and sharing within the research community, improved methods and developing technologies, these reports will likely increase logarithmically over the next five years. Promoters and integrated pathways will be uncovered. Already, whole scale studies of regulator genes has begun, with 88 mutants in the different families of regulators assessed in the mouse virulence model and a comprehensive study of two-component systems are underway (Haine et al. 2005). Extragenic sequences should be investigated more thoroughly to determine their affect on expression, and small non-coding RNAs and riboswitches most certainly play a key role. This work will be aided again by informatics; a compilation of extragenic sequences already appears on the TIGR website.

Our understanding of the genetic basis of *Brucella* virulence is still embryonic. Although major virulence factors such as the VirB T4SS have been identified, none of its translocated effectors are yet known. Comparative structural analysis combined with selective mutagenesis will give a close up view of the whole of specific protein–protein interactions not only within the bacterium but also between the bacterium and the host (Paschos et al. 2006). New virulence screens, diverse model systems and integration of knowledge acquired from genomics will reveal new virulence determinants and finally explain the basis on host preference and specificity.

Understanding virulence and host specificity will allow us to develop a new generation of diagnostic assays and vaccines to successfully control this zoonosis.

Acknowledgements. Part of this work was supported by funding from the Department of Defense (grant no. W911SR-04-C-0045 to B. Sobral), the National Institute of Allergy and Infectious Diseases (contract HHSN26620040035C to B. Sobral) and the National Science Foundation (NSF grant no. OCI-0537461 to O. Crasta).

References

Adams L (2002) The pathology of brucellosis reflects the outcome of the battle between the host genome and the *Brucella* genome. Vet Microbiol 90:553–561

Al Dahouk S, Tomaso H, Prenger-Berninghoff E, Splettstoesser WD, Scholz HC, Neubauer H (2005) Identification of *Brucella* species and biotypes using polymerase chain reaction-restriction fragment length polymorphism (PCR-RFLP). Crit Rev Microbiol 3:191–196

Al Dahouk S, Nockler K, Scholz HC, Tomaso H, Bogumil R, Neubauer H (2006) Immunoproteomic characterization of *Brucella abortus* 1119-3 preparations used for the serodiagnosis of *Brucella* infections. J Immunol Meth 309:34–47

Allardet-Servent A, Bourg G, Ramuz M, Pages M, Bellis M, Roizes G (1988) DNA polymorphism in strains of the genus *Brucella*. J Bacteriol 170:4603–4607

Allardet-Servent A, Carles-Nurit MJ, Bourg G, Michaux S, Ramuz M (1991) Physical map of the *Brucella melitensis* 16M chromosome. J Bacteriol 173:2219–2224

Allen CA, Adams LG, Ficht TA (1998) Transposon-derived *Brucella abortus* rough mutants are attenuated and exhibit reduced intracellular survival. Infect Immun 66: 1008–1016

Alp E, Koc RK, Durak AC, Yildiz O, Aygen B, Sumerkan B, Doganay M (2006) Doxycycline plus streptomycin versus ciprofloxacin plus rifampicin in spinal brucellosis. BMC Infect Dis 6:72

Altenbern RA (1973) Chromosomal mapping of *Brucella abortus*, strain 19. Can J Microbiol 19:109–112

Alton GG, Jones LM, Pietz DE (1975) Laboratory Techniques in Brucellosis. Monograph series. World Health Organization, Geneva, Switzerland

Alton GG, Jones LM, Angus RD, Verger JM (1988) Techniques for the brucellosis laboratory. INRA ISBN 2-7380-0042-8

Altschul SF, Gish W, Miller W, Myers EW, Lipman DJ (1990) Basic local alignment search tool. J Mol Biol 215:403–410

Andrutis KA, Fox JG, Schauer DB, Marini RP, Murphy JC, Yan L, Solnick JV (1995) Inability of an isogenic urease-negative mutant stain of *Helicobacter mustelae* to colonize the ferret stomach. Infect Immun 63:3722–3725

Anonymous (1988) International Committee on Systematic Bacteriology subcommittee on the Taxonomy of *Brucella*. Report of the meeting, 5 Sept 1986, Manchester, England. Int J Syst Bacteriol 38:450–452

Appella E, Arnott D, Sakaguchi K, Wirth PJ (2000) Proteome mapping by two-dimensional polyacrylamide gel electrophoresis in combination with mass spectrometric protein sequence analysis. EXS 88:1–27

Ason B, Reznikoff WS (2004) DNA sequence bias during Tn5 transposition. J Mol Biol 335:1213–1225

Avery OT, MacCleod CM, McCarty M (1944) Studies on the chemical nature of the substance inducing transformation of pneumococcal types. I. Induction of transformation by a desoxyribonucleic acid fraction isolated from *Pneumococcus* type III. J Exp Med 79:137

Baldwin CL, Winter AJ (1994) Macrophages and *Brucella*. Immunol Ser 60:363–380

Banai M (2002) Control of small ruminant brucellosis by use of *Brucella melitensis* rev.1 vaccine: Laboratory aspects and field observations. Vet Microbiol 90:497–519

Bandara B, Contreras A, Contreras-Rodriguez A, Martins AM, Dobrean V, Poff-Reichow S, Rajasekaran P, Sriranganathan N, Schurig GG, Boyle SM (2007) *Brucella suis* urease encoded by ure-1 but not ure-2 is necessary for infection of BALB/c mice by gavage. BMC Microbiol 7:57–70

Bang B (1906) Infectious abortion in cattle. J Comp Pathol 77:191, A202

Bang B (1933) Bernhard Bang. Am J Public Health Nations Health 23:48, A49

Batut J, Andersson SG, O'Callaghan D (2004) The evolution of chronic infection strategies in the[alpha]-proteobacteria. Nat Rev Microbiol 2:933–945

Bellaire BH, Elzer PH, Baldwin CL, Roop RM (2003) Production of the siderophore 2,3-dihydroxybenzoic acid is required for wild-type growth of *Brucella abortus* in the presence of erythritol under low-iron conditions in vitro. Infect Immun 71:2927–2832

Bellefontaine AF, Pierreux CE, Mertens P, Vandenhaute J, Letesson JJ, De Bolle X (2002) Plasticity of a transcriptional regulation network among alpha-proteobacteria is supported by the identification of ctrA targets in *Brucella abortus*. Mol Microbiol 43:945–960

Belzer C, Stoof J, Beckwith CS, Kuipers EJ, Kusters JG, van Vliet AH (2005) Differential regulation of urease activity in *Helicobacter hepaticus* and *Helicobacter pylori*. Microbiology 151:3989–3995

Benson M, Breitling R (2006) Network theory to understand microarray studies of complex diseases. Curr Mol Med 6:695–701

Binnicker MJ, Williams RD, Apicella MA (2003) Infection of human urethral epithelium with *Neisseria gonorrhoeae* elicits an upregulation of host anti-apoptotic factors and protects cells from staurosporine-induced apoptosis. Cell Microbiol 5:549–560

Binns AN, Beaupre CE, Dale EM (1995) Inhibition of *vir*B-mediated transfer of diverse substrates from *Agrobacterium tumefaciens* by the incq plasmid rsf1010. J Bacteriol 177:4890–4899

Birney E, Clamp M, Durbin R (2004) Genewise and genomewise. Genom Res 14:988–995

Blankenship RM, Sanford JP (1975) *Brucella canis*. A cause of undulant fever. Am J Med 59:424–426

Blueggel M, Chamrad D, Meyer HE (2004) Bioinformatics in proteomics. Curr Pharm Biotechnol 5:79–88

Boschiroli ML, Foulongne V, O'Callaghan D (2001) Brucellosis: A worldwide zoonosis. Curr Opin Microbiol 4:58–64

Boschiroli ML, Ouahrani-Bettache S, Foulongne V, Michaux-Charachon S, Bourg G, Allardet-Servent A, Cazevieille C, Lavigne JP, Liautard JP, Ramuz M, O'Callaghan D (2002a) The *Brucella suis vir*B operon is induced intracellularly in macrophages. Proc Natl Acad Sci USA 99:1544–1549

Boschiroli ML, Ouahrani-Bettache S, Foulongne V, Michaux-Charachon S, Bourg G, Allardet-Servent A, Cazevieille C, Lavigne JP, Liautard JP, Ramuz M, O'Callaghan D (2002b) Type IV secretion and *Brucella* virulence. Vet Microbiol 90:341–348

Bossi P, Tegnell A, Baka A, Van Loock F, Hendriks J, Werner A, Maidhof H, Gouvras G, Task Force on Biological and Chemical Agent Threats, Public Health Directorate, European Commission, Luxembourg (2004) Bichat guidelines for the clinical management of brucellosis and bioterrorism-related brucellosis. Euro Surveill 9:E15–E16

Bourg G, O'Callaghan D, Boschiroli ML (2007) The genomic structure of *Brucella* strains isolated from marine mammals gives clues to evolutionary history within the genus. Vet Microbiol 125:375–380

Boussau B, Karlberg EO, Frank AC, Leqault BA, Andersson SG (2004) Computational inference of scenarios for {alpha}-proteobacterial genome evolution. Proc Natl Acad Sci USA 101:9722–9727

Brew SD, Perrett LL, Stack JA, MacMillan AP, Staunton NJ (1999) Human exposure to *Brucella* recovered from a sea mammal. Vet Rec 144:483

Bricker BJ (2002) PCR as a diagnostic tool for brucellosis. Vet Microbiol 90:435–446

Bricker BJ (2004) Molecular diagnostics of animal brucellosis: A review of PCR-based assays and approaches. Lopez-Goni I, Moriyon I (eds) *Brucella*: Molecular and Cellular Biology. Horizon Bioscience, Wymondham, pp 25–51

Bricker BJ, Ewalt DR (2005) Evaluation of the hoof-print assay for typing *Brucella abortus* strains isolated from cattle in the united states: Results with four performance criteria. BMC Microbiol 5:37

Bricker BJ, Halling S (1994) Differentiation of *Brucella* abortus bv. 1, 2, and 4, *Brucella melitensis*, *Brucella ovis*, and *Brucella suis* bv. 1 by PCR. J Clin Microbiol 32:2660–2666

Bricker BJ, Ewalt DR, MacMillan AP, Foster G, Brew S (2000) Molecular characterization of *Brucella* strains isolated from marine mammals. J Clin Microbiol 38:1258–1262

Bricker BJ, Ewalt DR, Halling SM (2003) *Brucella* 'hoof-prints': Strain typing by multi-locus analysis of variable number tandem repeats (VNTRS). BMC Microbiol 3:15

Brodie R, Smith AJ, Roper RL, Tcherepanov V, Upton C (2004) Base-by-base: Single nucleotide-level analysis of whole viral genome alignments. BMC Bioinform 5:96

Brooks-Worrell BM, Splitter GA (1992) Antigens of *Brucella abortus* S19 immunodominant for bovine lymphocytes as identified by one- and two-dimensional cellular immunoblotting. Infect Immun 60:2459–2464

Brown GM (1977) The history of the brucellosis eradication program in the United States. Ann Sclavo 19:20–34

Buck JM (1930) Studies of vaccination during calfhood to prevent bovine infectious abortion. J Agric Res 41:667

Buddle MB (1956) Studies on *Brucella ovis* (n.sp.), a cause of genital disease of sheep in New Zealand and Australia. J Hyg (Lond) 54:351–364

Campos MA, Rosinha GM, Almeida IC, Salqueiro XS, Jarvis BW, Splitter GA, Qureshi N, Bruna-Romero O, Gazzinelli RT, Oliveira SC (2004) Role of toll-like receptor 4 in induction of cell-mediated immunity and resistance to *Brucella abortus* infection in mice. Infect Immun 72:176–186

Capasso L (2002) Bacteria in two-millennia-old cheese, and related epizoonoses in roman populations. J Infect 45: 122–127

Cardoso PG, Macedo GC, Azevedo V, Olicveira SC (2006) *Brucella spp* noncanonical lps: Structure, biosynthesis, and interaction with host immune system. Microb Cell Fact 5:13

Carmichael LE, Bruner DW (1968) Characteristics of a newly-recognized species of *Brucella* responsible for infectious canine abortions. Cornell Vet 48:579–592

Cascales E, Christie PJ (2004) *Agrobacterium vir*B10, an ATP energy sensor required for type IV secretion. Proc Natl Acad Sci USA 101:1722 8–17233

Celli J (2006) Surviving inside a macrophage: The many ways of *Brucella*. Res Microbiol 157:93–98

Celli J, de Chastellier C, Franchini DM, Pizarro-Cerda J, Moreno E, Gorvel JP (2003) *Brucella* evades macrophage killing via *vir*B-dependent sustained interactions with the endoplasmic reticulum. J Exp Med 198:545–556

Chain PS, Comerci DJ, Tolmasky ME, Larimer FW, Malfatti SA, Verqez LM, Aquero F, Land ML, Ugalde RA, Garcia E (2005) Whole-genome analyses of speciation events in pathogenic *Brucellae*. Infect Immun 73:8353–8361

Cheers C (1984) Pathogenesis and cellular immunity in experimental murine brucellosis. Dev Biol Stand 56:237–246

Chiang SL, Rubin EJ (2002) Construction of a mariner-based transposon for epitope-tagging and genomic targeting. Gene 296:179–185

Choi KH, Schweizer HP (2005) An improved method for rapid generation of unmarked *Pseudomonas aeruginosa* deletion mutants. BMC Microbiol 5:30

Christopher GW, Agan MB, Cieslak TJ, Olson PE (2005) History of U.S. Military contributions to the study of bacterial zoonoses. Mil Med 170:39–48

Cloeckaert A (2004) DNA polymorphism and taxonomy of *Brucella* species. Lopez-Goni I, Moriyon I (eds) *Brucella*: molecular and cellular biology. Horizon Bioscience, Wymondham, pp 1–24

Cloeckaert A, Verger JM, Grayon M, Paquet JY, Garin-Bastuji B, Foster G, Godfroid J (2001) Classification of *Brucella spp*. Isolated from marine mammals by DNA polymorphism at the *omp*2 locus. Microb Infect 3:729–738

Cloeckaert A, Grayon M, Grepinet O, Boumedine KS (2003) Classification of *Brucella* strains isolated from marine mammals by infrequent restriction site-pcr and development of specific PCR identification tests. Microb Infect 5:593–602

Comerci DJ, Altabe S, de Mendoza D, Ugalde RA (2006) *Brucella abortus* synthesizes phosphatidylcholine from choline provided by the host. J Bacteriol 188:1929–1934

Conde-Alvarez R, Grillo MJ, Salcedo SP, de Miguel MJ, Fugier E, Gorvel JP, Moriyon I, Iriarte M (2006) Synthesis of phosphatidylcholine, a typical eukaryotic phospholipid, is necessary for full virulence of the intracellular bacterial parasite *Brucella abortus*. Cell Microbiol 8:1322–1335

Connolly JP, Comerci D, Alefantis TG, Walz A, Quan M, Chafin R, Grewal P, Mujer CV, Ugalde RA, DelVecchio VG (2006) Proteomic analysis of *Brucella abortus* cell envelope and identification of immunogenic candidate proteins for vaccine development. Proteomics 6:3767–3780

Contreras-Rodriguez A, Ramirez-Zavala B, Contreras A, Schurig GG, Sriranganathan N, Lopez-Merino A (2003) Purification and characterization of an immunogenic aminopeptidase of *Brucella melitensis*. Infect Immun 71:5238–5244

Cook WE, Williams ES, Thorne ET, Kreeger TJ, Stout G, Bardsley K, Edwards H, Schurig G, Colby LA, Enright F, Elxer PH (2002) *Brucella abortus* strain RB51 vaccination in elk. I. Efficacy of reduced dosage. J Wildl Dis 38:18–26

Corbeil LB, Blau K, Inzana TJ, Nielsen KH, Jaconson RH, Corbeil RR, Winter AJ (1988) Killing of *Brucella abortus* by bovine serum. Infect Immun 56:3251–3261

Corbel MJ (1975) Proposal for minimal standards for descriptions of new species and biotypes of the genus *Brucella*. Int J Syst Bacteriol 25:83–89

Corbel MJ, Banai M (2005) *Genus I.* Brucella *Meyer and Shaw 1920, 173AL*. In Brenner DJ, Krieg NR, Staley JT (eds). Bergey's Manual of Systematic Bacteriology. Vol 2. Springer, pp 370–386

Covert J, Eskra L, Splitter G (2005) Isolation of *Brucella* abortus total RNA from *B. abortus*-infected murine raw macrophages. J Microbiol Meth 60:383–393

Cutler SJ, Whatmore AM, Commander NJ (2005) Brucellosis – New aspects of an old disease. J Appl Microbiol 98:1270–1281

Darling AC, Mau B, Blattner FR, Perna NT (2004) Mauve: Multiple alignment of conserved genomic sequence with rearrangements. Genom Res 14:1394–1403

Dasgupta N, Wolfgang MC, Goodman AL, Arora SK, Jyot J, Lory S, Ramphal R (2003) A four-tiered transcriptional regulatory circuit controls flagellar biogenesis in *Pseudomonas aeruginosa.* Mol Microbiol 50:809–824

De Ley J, Segers P, Lievens A, Denijn M, Vanhoucke M, Gillis M (1987) Ribosomal ribonucleic acid cistron similarities and taxonomic neighborhood of *Brucella* and cdc group vd. Int J Syst Bacteriol 37:35–42

de Paz HD, Sanqari FJ, Bolland S, Garcia-Lobo JM, Dehio C, de la Cruz F, Llosa M (2005) Functional interactions between type IV secretion systems involved in DNA transfer and virulence. Microbiology 151:3505–3516

Delcher AL, Harmon D, Kasif S, White O, Salzberg SL (1999) Improved microbial gene identification with glimmer. Nucl Acids Res 27:4636–4641

deLorenzo V, Jakubzik U, Timmis KN (1990) Mini-tn5 transposon derivatives for insertion mutagenesis, promoter probing, and chromosomal insertion of cloned DNA in gram-negative eubacteria. J Bacteriol 172:6568–6572

Delpino MV, Marchesini MI, Estein SM, Somerci DJ, Cassataro J, Fossati CA, Baldi PC (2007) A bile salt hydrolase of *Brucella abortus* contributes to the establishment of a successful infection through the oral route in mice. Infect Immun 75:299–305

Delrue RM, Martinez-Lorenzo M, Lestrate P, Danese I, Bielarz V, Mertens P, De Bolle X, Tibor A, Gorvel JP, Letesson JJ (2001) Identification of *Brucella spp.* genes involved in intracellular trafficking. Cell Microbiol 3:487–497

Delrue RM, Lestrate P, Tibor A, Letesson JJ, De Bolle X (2004) *Brucella* pathogenesis, genes identified from random large-scale screens. FEMS Microbiol Lett 231:1–12

Delrue RM, Deschamps C, Leonard S, Nijskens C, Danese I, Schaus JM, Bonnot S, Ferooz J, Tibor A, De Bolle X, Letesson JJ (2005) A quorum-sensing regulator controls expression of both the type IV secretion system and the flagellar apparatus of *Brucella melitensis.* Cell Microbiol 7:1151–1161

DelVecchio VG, Wagner MA, Eschenbrenner M, Horn TA, Kraycer JA, Estock F, Elzer P, Mujer CV (2002a) *Brucella* proteomes-a review. Vet Microbiol 90:593–603

DelVecchio VG, Kapatral V, Elzer P, Patra G, Mujer CV (2002b) The genome of *Brucella melitensis.* Vet Microbiol 90:587–592

DelVecchio VG, Kapatral V, Redkar RJ, Patra G, Mujer C, Los T, Ivanova N, Anderson I, Bhattacharyya A, Lykidis A, Reznik G, Jablonski L, Larsen N, D'Souza M, Bernal A, Mazur M, Goltsman E, Selkov E, Elzer PH, Haquis S, O'Callaghan D, Letesson JJ, Haselkorn R, Kyrpides N, Overbeek R (2002c) The genome sequence of the facultative intracellular pathogen *Brucella melitensis.* Proc Natl Acad Sci USA 99:443–448

DelVecchio VG, Alefantis T, Ugalde RA, Comerci D, Marchesini MI, Khan A, Lubitz W, Mujer CV (2006) Identification of protein candidates for developing bacterial ghost vaccines against *Brucella.* Meth Biochem Anal 49:363–377

Deqiu S, Donglou X, Jiming Y (2002) Epidemiology and control of brucellosis in China. Vet Microbiol 90:165–182

Detilleux PG, Deyoe BL, Cheville NF (1991) Effect of endocytic and metabolic inhibitors on the internalization and intracellular growth of *Brucella abortus* in Vero cells. Am J Vet Res 52:1658–1664

Diaz R, Jones LM, Wilson JB (1967) Antigenic relationship of *Brucella* ovis and *Brucella melitensis.* J Bacteriol 93: 1262–1268

Ding XZ, Paulsen IT, Bhattacharjee AK, Nikolich MP, Myers G, Hoover DL (2006) A high efficiency cloning and expression system for proteomic analysis. Proteomics 6:4038–4046

Dornand J, Gross A, Lafont V, Liautard J, Oliaro J, Liautard JP (2002) The innate immune response against *Brucella* in humans. Vet Microbiol 90:383–394

Dozot M, Boigegrain RA, Delrue RM, Hallez R, Ouahrani-Bettache S, Danese I, Letesson JJ, De Bolle X, Kohler S (2006) The stringent response mediator rsh is required for *Brucella* melitensis and *Brucella* suis virulence, and for expression of the type iv secretion system *vir*B. Cell Microbiol 8:1791–1802

Edgar RC (2004) Muscle: Multiple sequence alignment with high accuracy and high throughput. Nucl Acids Res 32:1792–1797

Ekaza E, Guilloteau L, Teyssier J, Liautard JP, Kohler S (2000) Functional analysis of the ClpATPase ClpA of *Brucella suis,* and persistence of a knockout mutant in BALB/c mice. Microbiology 146:1605–1616

Elberg SS, Faunce K (1957) Immunization against *Brucella* infection. VI. Immunity conferred on goats by a nondependent mutant from a streptomycin-dependent mutant strain of *Brucella melitensis.* J Bacteriol 73:211–217

Elzer PH, Phillips RW, Kovach ME, Peterson KM, Roop RM (1994) Characterization and genetic complementation of a *Brucella abortus* high-temperature-requirement a (*htr*A) deletion mutant. Infect Immun 62:4135–4139

Elzer PH, Kovach ME, Phillips RW, Robertson GT, Peterson KM, Roop RM (1995) In vivo and in vitro stability of the broad-host-range cloning vector pBBR1MCS in six *Brucella* species. Plasmid 33:51–57

Endley S, McMurray D, Ficht TA (2001) Interruption of the *cyd*B locus in *Brucella abortus* attenuates intracellular survival and virulence in the mouse model of infection. J Bacteriol 183:2454–2462

England T, Kelly L, Jones RD, MacMillan A, Wooldridge M (2004) A simulation model of brucellosis spread in british cattle under several testing regimes. Prev Vet Med 63:63–73

Eschenbrenner M, Wagner MA, Horn TA, Kraycer JA, Mujer CV, Haguis S, Elzer P, DelVecchio VG (2002) Comparative proteome analysis of *Brucella melitensis* vaccine strain Rev 1 and a virulent strain, 16 M. J Bacteriol 184: 4962–4970

Eschenbrenner M, Horn TA, Wagner MA, Mujer CV, Miller-Scandle TL, DelVecchio VG (2006) Comparative proteome analysis of laboratory grown *Brucella abortus* 2308 and *Brucella melitensis* 16 M. J Proteom Res 5:1731–1740

Eskra L, Canavessi A, Carey M, Splitter G (2001) *Brucella* abortus genes identified following constitutive growth and macrophage infection. Infect Immun 69:7736–7742

Eskra L, Mathison A, Splitter G (2003) Microarray analysis of mRNA levels from RAW264.7 macrophages infected with *Brucella abortus*. Infect Immun 71:1125–1133

Etter RP, Drew ML (2006) Brucellosis in elk of eastern Idaho. J Wildlife Dis 42:271–278

Evans A (1950) Comments on the early history of human brucellosis, Cleghorn G. Observations of the Epidemical Diseases of Minorca (From the Years 1744 to 1749). In: Larson CH, Soule MH (eds) Brucellosis. Waverly Press, Baltimore, MD, pp 1–8

Ewalt DR, Harrington R (1979) Isolation of *Brucella abortus*, strain 19, from cattle. J Am Vet Med Assoc 174:172–173

Ewalt DR, Payeur JB, Martin BM, Cummins DR, Miller WG (1994) Characteristics of a *Brucella* species from a bottlenose dolphin (*Tursiops truncatus*). J Vet Diagn Invest 6:448–452

Eze MO, Yuan L, Crawford RM, Paranavitana CM, Hadfield TL, Bhattacharjee AK, Warren RL, Hoover DL (2000) Effects of opsonization and gamma interferon on growth of *Brucella melitensis* 16M in mouse peritoneal macrophages in vitro. Infect Immun 68:257–263

Farlow J, Smith KL, Wong J, Abrams M, Lytle M, Keim P (2001) *Francisella tularensis* strain typing using multiple-locus, variable-number tandem repeat analysis. J Clin Microbiol 39:3186–3192

Fekete A, Bantle JA, Halling SM, Sanborn MR (1990) Preliminary development of a diagnostic test for *Brucella* using polymerase chain reaction. J Appl Bacteriol 69:216–227

Fekete A, Bantle JA, Halling SM, Stich RW (1992) Amplification fragment length polymorphism in *Brucella* strains by use of polymerase chain reaction with arbitrary primers. J Bacteriol 174:7778–7783

Fernandez-Lago L, et al. (1996). Endogenous gamma interferon and interleukin-10 in *Brucella abortus* 2308 infection in mice. FEMS Immunol Med Microbiol 15:109–114

Fernandez-Prada CM, Kinolich M, Vemulapalli R, Sriranganathan N, Boyle SM, Schurig GG, Hadfield TL, Hoover DL (2001) Deletion of *wbo*A enhances activation of the lectin pathway of complement in *Brucella abortus* and *Brucella melitensis*. Infect Immun 69:4407–4416

Fernandez-Prada CM, Zelazowska EB, Nikolich M, Hadfield TL, Roop RM, Robertson GL, Hoover DL (2003) Interactions between *Brucella melitensis* and human phagocytes: Bacterial surface O-polysaccharide inhibits phagocytosis, bacterial killing, and subsequent host cell apoptosis. Infect Immun 71:2110–2119

Ficht TA, Bearden SW, Sowa BA, Marquis H (1990) Genetic variation at the *omp*2 porin locus of the *Brucellae*: Species-specific markers. Mol Microbiol 4:1135–1142

Fleischmann RD, Adams MD, White O, Clayton RA, Kirkness EF, Kerlavage AR, Bult CJ, Tomb JF, Dougherty BA, Merrick JM (1995) Whole-genome random sequencing and assembly of *Haemophilus influenzae* rd. Science 269:496–512

Finlay BB, McFadden G (2006) Anti-immunology: evasion of the host immune system by bacterial and viral pathogens. Cell 124(4): 767–782

Fosgate GT, Carpenter TE, Chomel BB, Case JT, DeBess EE, Reilly KF (2002) Time-space clustering of human brucellosis, California, 1973–1992. Emerg Infect Dis 8:672–678

Foster G, Jahans KL, Reid RJ, Ross HM (1996) Isolation of *Brucella* species from cetaceans, seals and an otter. Vet Rec 138:583–586

Foulongne V, Bourg G, Cazevieille C, Michaux-Charachon S, O'Callaghan D (2000) Identification of *Brucella* SUIS genes affecting intracellular survival in an *in vitro* human macrophage infection model by signature-tagged transposon mutagenesis. Infect Immun 68:1297–1303

Fretin D, Fauconnier A, Kohler S, Halling S, Leonard S, Nijskens C, Ferooz J, Lestrate P, Delrue RM, Danese I, Vandenhaute J, Tibor A, DeBolle X, Letesson JJ (2005) The sheathed flagellum of *Brucella melitensis* is involved in persistence in a murine model of infection. Cell Microbiol 7:687–698

Gamazo C, Winter AJ, Moriyon I, Riezu-Boj JI, Blasco JM, Diaz R (1989) Comparative analyses of proteins extracted by hot saline or released spontaneously into outer membrane blebs from field strains of *Brucella ovis* and *Brucella melitensis*. Infect Immun 57:1419–1426

Garcia P, Yrivarren JL, Argumans C, Crosby E, Carrillo C, Gotuzzo E (1990) Evaluation of the bone marrow in patients with brucellosis. Clinico-pathological correlation. Enferm Infecc Microbiol Clin 8:19–24

Gee JE, De BK, Levett PN, Whitney AM, Novak RT, Popovic T (2004) Use of 16s rRNA gene sequencing for rapid confirmatory identification of *Brucella* isolates. J Clin Microbiol 42:3649–3654

Gee JM, Valderas MW, Kovach ME, Grippe VK, Robertson GT, Ng WL, Richardson JM, Winkler ME, Roop RM (2005) The *Brucella abortus* Cu/Zn superoxide dismutase is required for optimal resistance to oxidative killing by murine macrophages and wild-type virulence in experimentally infected mice. Infect Immun 73:2873–2880

Gevaert K, Vandekerckhove J (2000) Protein identification methods in proteomics. Electrophoresis 21:1145–1154

Gil A (2000) Zoonosis en los sistemas de produccion animal de las areas urbanas y peiurbanas de america latina. FAO Livestock Information and Policy Branch Livestock Policy Discussion paper

Godfroid F, Taminiau B, Danese I, Denoel P, Tibor A, Weynants V, Cloeckaert A, Godfroid J, Letesson JJ (1998) Identification of the perosamine synthetase gene of *Brucella melitensis* 16M and involvement of lipopolysaccharide O side chain in *Brucella* survival in mice and in macrophages. Infect Immun 66:5485–5493

Godfroid J, Cloeckaert A, Liautard JP, Kohler S, Fretin D, Walravens K, Garin-Bastuji B, Letesson JJ (2005) From the discovery of the Malta fever's agent to the discovery of a marine mammal reservoir, brucellosis has continuously been a re-emerging zoonosis. Vet Res 36:313–326

Gorvel JP, Moreno E (2002) *Brucella* intracellular life: From invasion to intracellular replication. Vet Microbiol 90:281–297

Griffiths-Jones S, Bateman A, Marshall M, Khanna A, Eddy SR (2003) Rfam: An RNA family database. Nucl Acids Res 31:439–441

Gross A, Terraza A, Ouahrani-Bettache S, Liautard JP, Dornand J (2000) In vitro *Brucella suis* infection prevents the programmed cell death of human monocytic cells. Infect Immun 68:342–351

Groussaud P, Shankster SJ, Koylass MS, Whatmore AM (2007) Molecular typing divides marine mammal strains of *Brucella* into at least threegroups with distinct host preferences. J Med Microbiol 56:1512–1518

Guerrero G, Peralta H, Aguilar A, Diaz R, Villalobos MA, Medrano-Soto A, Mora J (2005) Evolutionary, structural and functional relationships revealed by comparative analysis of syntenic genes in *Rhizobiales*. BMC Evol Biol 5:55

Gupta RS (2005) Protein signatures distinctive of alpha proteobacteria and its subgroups and a model for alpha-proteobacterial evolution. Crit Rev Microbiol 31:101–135

Hahn MY, Raman S, Anaya M, Husson RN (2005) The *Mycobacterium tuberculosis* extracytoplasmic-function sigma factor *sig*L regulates polyketide synthases and secreted or membrane proteins and is required for virulence. J Bacteriol 187:7062–7071

Haine V, Sinon A, Van Steen F, Rousseau S, Dozot M, Lestrate P, Lambert C, Letesson JJ, DeBolle X (2005) Systematic targeted mutagenesis of *Brucella melitensis* 16M reveals a major role for GntR regulators in the control of virulence. Infect Immun 73:5578–5586

Halling SM, Bricker BJ (1994) Characterization and occurrence of two repeated palindromic DNA elements of *Brucella spp.*: Bru-rs1 and Bru-rs2. Mol Microbiol 14:681–689

Halling SM, Jensen AE (2006) Intrinsic and selected resistance toantibiotics binding the ribosome: analyses of *Brucella* 23S rrn, L4, L22, EF-Tu1, EF-Tu2, efflux and phylogenetic implications. BMC Microbiol 6:84–99

Halling SM, Zuerner RL (2002) Evidence for lateral transfer to *Brucella*: Characterization of a locus with a Tn-like element (tn2020). Biochem Biophys Acta 1574:109–116

Halling SM, Detilleux PG, Tatum FM, Judge BA, Mayfield JE (1991) Deletion of the *bcsp*31 gene of *Brucella* abortus by replacement. Infect Immun 59:3863–3868

Halling SM, Tatum FM, Bricker BJ (1993) Sequence and characterization of an insertion sequence, IS711, from *Brucella ovis*. Gene 133:123–127

Halling SM, Peterson-Burch BD, Bricker BJ, Zuerner RL, Qing Z, Li LL, Kapur V, Alt DP, Olsen SC (2005) Completion of the genome sequence of *Brucella abortus* and comparison to the highly similar genomes of *Brucella melitensis* and *Brucella suis*. J Bacteriol 187:2715–2726

Hautefort I, Hinton JC (2000) Measurement of bacterial gene expression *in vivo*. Philos Trans R Soc Lond B Biol Sci 355:601–611

He Y, Vines RR, Wattam AR, Abramochkin GV, Dickerman AW, Eckart JD, Sobral BW (2005) Piml: The pathogen information markup language. Bioinformatics 21:116–121

He Y, Reichow S, Ramamoorthy S, Ding X, Lathigra R, Craig JC, Sobral BW, Schurig GG, Sriranganathan N, Boyle SM (2006) *Brucella melitensis* triggers time-dependent modulation of apoptosis and down-regulation of mitochondrion-associated gene expression in mouse macrophages. Infect Immun 74:5035–5046

Henikoff JG, Henikoff S (1996) Blocks database and its applications. Meth Enzymol 266:88–105

Herzberg M, Elberg SS (1955) Immunization against *Brucella* infection. III. Response of mice and guinea pigs to injection of viable and nonviable suspensions of a streptomycin-dependent mutant of *Brucella melitensis*. J Bacteriol 69:432–435

Hirsch P, Conti SF (1964) Biology of budding bacteria. II. Growth and nutrition of *Hyphomicrobium spp*. Arch Mikrobiol 48:358–367

Hirsch P, Rheinheimer G (1968) Biology of budding bacteria. V. Budding bacteria in aquatic habitats: Occurrence, enrichment and isolation. Arch Mikrobiol 62:289–306

Hoffman EM, Houle JJ (1983) Failure of *Brucella abortus* lipopolysaccharide (lps) to activate the alternative pathway of complement. Vet Immunol Immunopathol 5:65–76

Hoffmann EM, Houle JJ (1995) Contradictory roles for antibody and complement in the interaction of *Brucella abortus* with its host. Crit Rev in Microbiol 21:153–163

Holmes BPM, Kiredjian M, Kersters K (1988) *Ochrobactrum anthropi* gen. Nov., sp. Nov. From human clinical specimens and previously known as group vd. Int J Syst Bacteriol 38:406–416

Hommais F, Pereira S, Acquaviva C, Escobar-Paramo P, Denamur E (2005) Single-nucleotide polymorphism phylotyping of *Escherichia coli*. Appl Environ Microbiol 71:4784–4792

Hong PC, Tsolis RM, Ficht TA (2000) Identification of genes required for chronic persistence of *Brucella abortus* in mice. Infect Immun 68:4102–4107

Hoover DL, Fridelander AM (1997) Brucellosis. Medical Aspects of Chemical and Biological Warfare, Office of The Surgeon General, Borden Institute, Walter Reed Army Medical Center: 513–521

Hoppner C, Carle A, Sivanesan D, Hoeppner S, Baron C (2005) The putative lytic transglycosylase *vir*B1 from *Brucella suis* interacts with the type iv secretion system core components *vir*B8, *vir*B9 and *vir*B11. Microbiology 151:3469–3482

Hoyer BH, McCullough NB (1968a) Homologies of deoxyribonucleic acids from *Brucella ovis*, canine abortion organisms, and other *Brucella* species. J Bacteriol 96: 1783–1790

Hoyer BH, McCullough NB (1968b) Polynucleotide homologies of *Brucella* deoxyribonucleic acids. J Bacteriol 95:444–448

Husser CS, Buchhalter JR, Raffo OS, Shabo A, Brown SH, Lee KE, Elkin PL (2006) Standardization of microarray and pharmacogenomics data. Meth Mol Biol 316:111–157

Ichikawa JK, Norris A, Bangera MG, Geiss GK, van't Wout AB, Bumgarner RE, Lory S (2000) Interaction of *Pseudomonas aeruginosa* with epithelial cells: Identification of differentially regulated genes by expression microarray analysis of human cDNAs. Proc Natl Acad Sci USA 97:9659–9664

Inatsuka CS, Julio SM, Cotter PA (2005) *Bordetella* filamentous hemagglutinin plays a critical role in immunomodulation, suggesting a mechanism for host specificity. Proc Natl Acad Sci USA 102:18578–18583

Jahans KL, Foster G, Broughton ES (1997) The characterisation of *Brucella* strains isolated from marine mammals. Vet Microbiol 57:373–382

Jiang X, Baldwin CL (1993) Effects of cytokines on intracellular growth of *Brucella abortus*. Infect Immun 61:124–134

Jimenez de Bagues MP, Terraza A, Gross A, Dornand J (2004) Different responses of macrophages to smooth and rough *Brucella spp.*: Relationship to virulence. Infect Immun 72:2429–2433

Jones SM, Winter AJ (1992) Survival of virulent and attenuated strains of *Brucella abortus* in normal and gamma interferon-activated murine peritoneal macrophages. Infect Immun 60:3011–3014

Joyce AR, Palsson BO (2006) The model organism as a system: Integrating 'omics' data sets. Nat Rev Mol Cell Biol 7: 198–210

Jumas-Bilak E, Maugard C, Michaux-Charachon S, Allardet-Servent A, Perrin A, O'Callaghan D, Ramuz M (1995) Study of the organization of the genomes of *Escherichia coli*, *Brucella melitensis* and *Agrobacterium tumefaciens* by insertion of a unique restriction site. Microbiology 141:2425–2432

Jumas-Bilak E, Michaux-Charachon S, Bourg G, Ramuz M, Allardet-Servent A (1998a) Differences in chromosome number and genome rearrangements in the genus *Brucella*. Mol Microbiol 27:99–106

Jumas-Bilak E, Michaux-Charachon S, Bourg G, Ramuz M, Allardet-Servent A (1998b) Unconventional genomic organization in the alpha subgroup of the proteobacteria. J Bacterioll 80:2749–2755

Juncker AS, Willenbrock H, Von Heijne G, Brunak S, Nielsen H, Krogh A (2003) Prediction of lipoprotein signal peptides in gram-negative bacteria. Protein Sci 12:1652–1662

Kahl-McDonagh MM, Ficht TA (2006) Evaluation of protection afforded by *Brucella* abortus and *Brucella melitensis* unmarked deletion mutants exhibiting different rates of clearance in BALB/c mice. Infect Immun 74:4048–4057

Kahl-McDonagh MM, Elzer PH, Haguis SD, Walker JV, Perry QL, Seabury CM, den Hartigh AB, Tsolis RM, Adams LG, Davis DS, Ficht TA (2006) Evaluation of novel *Brucella melitensis* unmarked deletion mutants for safety and efficacy in the goat model of brucellosis. Vaccine 24:5169–5177

Kanehisa M, Bork P (2003) Bioinformatics in the post-sequence era. Nat Genet 33 (Suppl):305–310

Karp PD, Paley S, Romero P (2002) The pathway tools software. Bioinformatics 18 (Suppl 1):S225–S232

Kaufmann AF, Meltzer MI, Schmid GP (1997) The economic impact of a bioterrorist attack: Are prevention and postattack intervention programs justifiable? Emerg Infect Dis 3:83–94

Keim P, Price LB, Klevytska AM, Smith KL, Schupp JM, Okinaka R, Jackson PJ, Hugh-Jones ME (2000) Multiple-locus variable-number tandem repeat analysis reveals genetic relationships within *Bacillus anthracis*. J Bacteriol 182:2928–2936

Kendall SL, Movahedzadeh F, Rison SC, Wernisch L, Parish T, Duncan K, Betss JC, Stoker NG (2004) The *Mycobacterium tuberculosis* dosRS two-component system is induced by multiple stresses. Tuberculosis (Edinb) 84:247–255

Khan AS, Mujer CV, Alefantis TG, Connolly JP, Mayr UB, Walcher P, Lubitz W, DelVecchio VG (2006) Proteomics and bioinformatics strategies to design countermeasures against infectious threat agents. J Chem Inf Model 46:111–115

Kim S, Watarai M, Kondo Y, Erdenebaatar J, Makino S, Shirahata T (2003) Isolation and characterization of mini-Tn5km2 insertion mutants of *Brucella abortus* deficient in internalization and intracellular growth in HeLa cells. Infect Immun 71:3020–3027

Kim S, Lee DS, Watanabe K, Furuoka H, Suzuki H, Watarai M (2005) Interferon-gamma promotes abortion due to *Brucella* infection in pregnant mice. BMC Microbiol 5:22

Kohler S, Foulongne V, Ouahrani-Bettache S, Bourg G, Teyssier J, Ramuz M, Liautard JP (2002a) The analysis of the intramacrophagic virulome of *Brucella suis* deciphers the environment encountered by the pathogen inside the macrophage host cell. Proc Natl Acad Sci USA 99:15711–15716

Kohler S, Porte F, Jubier-Maurin V, Ouahrani-Bettache S, Teyssier J, Liautard JP (2002b) The intramacrophagic environment of *Brucella suis* and bacterial response. Vet Microbiol 90:299–309

Kohler S, Michaux-Charachon S, Porte F, Ramuz M, Liautard JP (2003) What is the nature of the replicative niche of a stealthy bug named *Brucella*?. Trends Microbiol 11: 215–219

Kortepeter MG, Parker GW (1999) Potential biological weapons threats. Emerg Infect Dis 5:523–527

Kouba V (2003) A method of accelerated eradication of bovine brucellosis in the Czech Republic. Rev Sci Tech 22:1003–1012

Kovach ME, Phillips RW, Elzer PH, Roop RM, Peterson KM (1994) pBBR1MCS: A broad-host-range cloning vector. BioTechniques 16:800–802

Kovach ME, Elzer PH, Hill DS, Robertson GT, Farris MA, Roop RM, Peterson KM (1995) Four new derivatives of the broad-host-range cloning vector pbbr1mcs, carrying different antibiotic-resistance cassettes. Gene 166:175–176

Kurtz S, Phillippy A, Delcher AL, Smoot M, Shumway M, Antoneseu C, Salzberg SL (2004) Versatile and open software for comparing large genomes. Genom Biol 5:R12

Lai F, Schurig GG, Boyle SM (1990) Electroporation of a suicide plasmid bearing a transposon into *Brucella abortus*. Microb Pathog 9:363–368

Lambert JM, Bongers RS, Kleerebezem M (2007) Cre-lox-based system for multiple gene deletions and selectable-marker removal in *Lactobacillus plantarum*. Appl Environ Microbiol 73:1126–1135

Lapaque N, Takeuchi O, Corrales F, Akira S, Moriyon I, Howard JC, Gorvel JP (2006) Differential inductions of TNF-alpha and IGTP, IIGP by structurally diverse classic and non-classic lipopolysaccharides. Cell Microbiol 8:401–413

Lavigne JP, O'Callaghan D, Blanc-Potard AB (2005) Requirement of *mgt*C for *Brucella suis* intramacrophage growth: A potential mechanism shared by *Salmonella enterica* and *Mycobacterium tuberculosis* for adaptation to a low-Mg2+ environment. Infect Immun 73:3160–3163

Lawson JN, Lyons CR, Johnston SA (2006) Expression profiling of *Yersinia pestis* during mouse pulmonary infection. DNA Cell Biol 25:608–616

Le Fleche P, Jacques I, Grayon M, Al Dahouk S, Bouchon P, Denoeud F, Nockler K, Neubauer H, Guilloteau LA, Vergnaud G (2006) Evaluation and selection of tandem repeat loci for a *Brucella mlv*A typing assay. BMC Microbiol 6:9

Leal-Klevezas DS, Martinez-de-la-Vega O, Ramirez-Barba EJ, Osterman B, Martinez-Soriano JP, Simpson J (2005) Genotyping of *Ochrobactrum spp*. by AFLP analysis. J Bacteriol 187:2537–2539

Leavitt MO (2005) Possession, use and transfer of select agents and toxinbs; final rule. In: Services dohah (ed) Federal Register, pp 13294–13325

Lebuhn M, Achouak W, Schloter M, Berge O, Meier H, Barakat M, Hartmann A, Heulin T (2000) Taxonomic characterization of *Ochrobactrum* sp. Isolates from soil samples and wheat roots, and description of *Ochrobactrum tritici* sp. Nov. And *Ochrobactrum grignonense* sp. Int J Syst Evol Microbiol 50:2207–2223

Lebuhn M, Bathe S, Achouak W, Hartmann A, Heulin T, Schloter M (2006) Comparative sequence analysis of the internal transcribed spacer 1 of *Ochrobactrum* species. Syst Appl Microbiol 29:265–275

Lee KB, Liu CT, Anzai Y, Kim H, Aono T, Oyaizu H (2005) The hierarchical system of the 'alphaproteobacteria': Description of *Hyphomonadaceae* fam. Nov., *Xanthobacteraceae* fam. Nov. and *Erythrobacteraceae* fam. Nov. Int J Syst Evol Microbiol 55:1907–1919

Leonard S, Ferooz J, Haine V, Danese I, Fretin D, Tibor A, de Walque S, De Bolle X, Letesson JJ (2007) Ftcr is a new master regulator of the flagellar system of *Brucella melitensis* 16M with homologs in *Rhizobiaceae*. J Bacteriol 189: 131–141

Lestrate P, Delrue RM, Danese I, Didembourg C, Taminiau B, Mertens P, De Bolle X, Tibor A, Tang CM, Letesson JJ (2000) Identification and characterization of *in vivo* attenuated mutants of *Brucella melitensis*. Mol Microbiol 38:543–551

Lestrate P, Dricot A, Delrue RM, Lambert C, Martinelli V, De Bolle X, Letesson JJ, Tibor A (2003) Attenuated signature-tagged mutagenesis mutants of *Brucella melitensis* identified during the acute phase of infection in mice. Infect Immun 71:7053–7060

LeVier K, Phillips RW, Grippe VK, Roop RM, Walker GC (2000) Similar requirements of a plant symbiont and a mammalian pathogen for prolonged intracellular survival. Science 287:2492–2493

Levinson G, Gutman G (1987) Slipped-strand mispairing: A major mechanism for DNA sequence evolution. Mol Biol Evol 4:203–221

Lin J, Ficht TA (1995) Protein synthesis in *Brucella abortus* proteins induced during macrophage infection. Infect Immun 63:1409–1414

Lizewski SE, Schurr JR, Jackson DW, Frisk A, Carterson AJ, Schurr MJ (2004) Identification of *alg*R-regulated genes in *Pseudomonas aeruginosa* by use of microarray analysis. J Bacteriol 186:5672–5684

Lobry JR (1996) Asymmetric substitution patterns in the two DNA strands of bacteria. Mol Biol Evol 13:660–665

Losick VP, Isberg RR (2006) Nf-kappaB translocation prevents host cell death after low-dose challenge by *Legionella pneumophila*. J Exp Med 203:2177–2189

Lowe TM, Eddy SR (1997) TRNASCAN-SE: A program for improved detection of transfer rna genes in genomic sequence. Nucl Acids Res 25:955–964

Lukashin AV, Borodovsky M (1998) Genemark.Hmm: New solutions for gene finding. Nucl Acids Res 26:1107–1115

Luna-Martinez JE, Mejia-Teran C (2002) Brucellosis in Mexico. Vet Microbiol 90:19–30

Maciag A, Dainese E, Rodriguez GM, Milano A, Provvedi R, Pasca MR, Smith I, Palu G, Riccardi G, Manganelli R (2007) Global analysis of *Mycobacterium tuberculosis* zur (*fur*B) regulon. J Bacteriol 189:730–740

Manganelli R, Voskuil M, Schoolnik GK, Smith I (2001) The *Mycobacterium tuberculosis ecf* sigma factor sigmae: Role in global gene expression and survival in macrophages. Mol Microbiol 41:423–437

Margulies M, Egholm M, Altman WE, Attiya S, Bader JS, Bemben LA, Berka J, Braverman MS, Chen YJ, Chen Z, Dewell SB, Du L, Fierro JM, Gomes XV, Godwin BC, He W, Helgessen S, Ho CH, Irzyk GP, Jando SC, Alenquer ML, Jarvie TP, Jirage KB, Kim JB, Knight JR, Lanza JR, Leamon Jh, Lefkowitz SM, Lei M, Li J, Lohman KL, Lu H, Makhijani VB, McDade KE, McKenna MP, Myers EW, Nickerson E, Nobile JR, Plant R, Puc BP, Ronan MT, Roth GT, Sarkis GJ,

Simons JF, Simpson JW, Srinivasan M, Tartaro KR, Tomasz A, Vogt KA, Volkmer GA, Wang SH, Wang Y, Weiner MP, Yu P, Begley RF, Rothberg JM (2005) Genome sequencing in microfabricated high-density picolitre reactors. Nature 437:376–380

Marusina K (2005) Whole genome sequencing in 24 hours. Genet Eng News 25

McEwen A (1940) Experiments on contagious abortion. The immunity of cattle inoculated with vaccines of graded virulence. Vet Rec 52:815

McGuffin LJ, Bryson K, Jones DT (2000) The PSIPRED protein structure prediction server. Bioinformatics 16:404–405

McQuiston JR, Vemulapalli R, Inzana TJ, Schurig GG, Sriranganathan N, Fritzinger D, Hadfield TL, Warren RA, Lindler LE, Snellings N, Hoover D, Halling SM, Boyle SM (1999) Genetic characterization of a Tn5-disrupted glycosyltransferase gene homolog in *Brucella abortus* and its effect on lipopolysaccharide composition and virulence. Infect Immun 67:3830–3835

Menard R, Sansonetti PJ, Parsot C (1993) Nonpolar mutagenesis of the *ipa* genes defines *ipaB, ipaC, and ipa*D as effectors of *Shigella flexneri* entry into epithelial cells. J Bacteriol 175:5899–5906

Mercier E, Jumas-Bilak E, Allardet-Servent A, O'Callaghan D, Ramuz M (1996) Polymorphism in *Brucella* strains detected by studying distribution of two short repetitive DNA elements. J Clin Microbiol 34:1299–1302

Meyer K, Show E (1920) A comparison of the morphological, cultural, and biochemical characteristics of *B. abortus* and *B. melitensis*; studies of the genus *Brucella*. J Infect Dis 27:73–184

Michaux S, Paillisson J, Carles-Nurit MJ, Bourg G, Allardet-Servent A, Ramuz M (1993) Presence of two independent chromosomes in the *Brucella melitensis* 16M genome. J Bacteriol 175:701–705

Michaux-Charachon S, Bourg G, Jumas-Bilak E, Guigue-Talet P, Allardet-Servent A, O'Callaghan D, Ramuz M (1997) Genome structure and phylogeny in the genus *Brucella*. J Bacteriol 179:3244–3249

Michaux-Charachon S, Jumas-Bilak E, Allardet-Servent A, Bourg G, Boschiroli ML, Ramuz M, O'Callaghan D (2002) The *Brucella* genome at the beginning of the post-genomic era. Vet Microbiol 90:581–585

Minas A, Minas M, Stournara A, Tselepidis S (2004) The effects of Rev-1 vaccination of sheep and goats on human brucellosis in Greece. Prev Vet Med 64:41–47

Monreal D, Grillo MJ, Gonzalez D, Marin CM, De Miguel MJ, Lopez-Goni I, Cloeckaert A, Moriyon I (2003) Characterization of *Brucella abortus* O-polysaccharide and core ipopolysaccharide mutants and demonstration that a complete core is required for rough vaccines to be efficient against *Brucella abortus* and *Brucella ovis* in the mouse model. Infect Immun 71:3261–3271

Monsieurs P, De Keersmaecker S, Navarre WW, Bader MW, De Smet F, McClelland M, Fang FC, De Moor B, Vanderleyden J, Marchal K (2005) Comparison of the *pho*PQ regulon in *Escherichia coli* and *Salmonella typhimurium*. J Mol Evol 60:462–474

Moon HW, Nagy B, Isaacson RE, Orskov I (1977) Occurrence of K99 antigen on *Escherichia coli* isolated from pigs and colonization of pig ileum by K99+ enterotoxigenic *E. coli* from calves and pigs. Infect Immun 15:614–620

Moreno E (1997) In search of a bacterial species definition. Rev Biol Trop 45:735–771

Moreno E (1998) Genome evolution within the alpha-proteobacteria: Why do some bacteria not possess plasmids and others exhibit more than one different chromosome? FEMS Microbiol Rev 22:255–275

Moreno E (2002) Brucellosis in Central America. Vet Microbiol 90:31–38

Moreno E, Moriyon I (2002) *Brucella melitensis*: A nasty bug with hidden credentials for virulence. Proc Natl Acad Sci USA 99:1–3

Moreno E, Stackebrandt E, Dorsch M, Wolters J, Busch M, Mayer H (1990) *Brucella abortus* 16s rRNA and lipid a reveal a phylogenetic relationship with members of the alpha-2 subdivision of the class proteobacteria. J Bacteriol 172:3569–3576

Moreno E, Cloeckaert A, Moriyon I (2002) *Brucella* evolution and taxonomy. Vet Microbiol 90:209–227

Moriyon I, Grillo MJ, Monreal D, Gonzalez D, Marin C, Lopez-Goni I, Mainar-Jaime RC, Moreno E, Blasco JM (2004) Rough vaccines in animal brucellosis: Structural and genetic basis and present status. Vet Res 35:1–38

Morris JA (1973) The use of polyacrylamide gel electrophoresis in taxonomy of *Brucella*. J Gen Microbiol 76:231–237

Mujer CV, Wagner MA, Escenbrenner M, Horn T, Kraycer JA, Redkar R, Hagius S, Elzer P, DelVecchio VG (2002) Global analysis of *Brucella melitensis* proteomes. Ann NY Acad Sci 969:97–101

Munford RS, Weaver RE, Patton C, Feeley JC, Feldman RA (1975) Human disease caused by *Brucella canis*. A clinical and epidemiologic study of two cases. JAMA 231:1267–1269

Murphy EA, et al. (2001). Interferon-gamma is crucial for surviving a *Brucella abortus* infection in both resistant C57BL/6 and susceptible BALB/c mice. Immunology 103:511–518

Navarro E, Casao MA, Solera J (2004) Diagnosis of human brucellosis using PCR. Expert Rev Mol Diagn 4:115–123

Navarro E, Segura JC, Castano MJ, Solera J (2006) Use of real-time quantitative polymerase chain reaction to monitor the evolution of *Brucella melitensis* DNA load during therapy and post-therapy follow-up in patients with brucellosis. Clin Infect Dis 42:1266–1273

Nicoletti P (2002) A short history of brucellosis. Vet Microbiol 90:5–9

O'Callaghan D, Cazevieille C, Allardet-Servent A, Boschiroli ML, Bourg G, Foulongne V, Frustos P, Kulakov Y, Ramuz

M (1999) A homologue of the *Agrobacterium tumefaciens virB* and *Bordetella pertussis* ptl type IV secretion systems is essential for intracellular survival of *Brucella suis*. Mol Microbiol 33:1210–1220

Olsen SC, Stoffregen WS (2005) Essential role of vaccines in brucellosis control and eradication programs for livestock. Expert Rev Vaccines 4:915–928

Olson NE (2006) The microarray data analysis process: From raw data to biological significance. NeuroRx 3:373–383

Osterman B (2006) International committee on systematics of prokaryotes. Subcommittee on the taxonomy of *Brucella*. Report of the meeting, 17 September 2003, Pamplona, Spain. Int J Syst Evol Microbiol 56:1173–1175

Oomen RP, Young NM, Bundle DR (1991) Molecular modeling of antibody-antigen complexes between the *Brucella abortus* O-chain polysaccharide and a specific monoclonal antibody. Protein Engg 4:427–433

Ouahrani S, Michaux S, Sri Widada J, Bourg G, Tournebize R, Ramuz M, Liautard JP (1993) Identification and sequence analysis of IS6501, an insertion sequence in *Brucella spp.*: Relationship between genomic structure and the number of IS6501 copies. J Gen Microbiol 139:3265–3273

Ouahrani-Bettache S, Soubrier MP, Liautard JP (1996) IS6501-anchored PCR for the detection and identification of *Brucella* species and strains. J Appl Bacteriol 81: 154–160

Palanduz A, Palanduz S, Guler K, Guler N (2000) Brucellosis in a mother and her young infant: Probably transmission by breast milk. Int J Infect Dis 4:55–56

Pan American Health Organization (1998) Country health profiles. Salud en las Americas 1 and 2

Pappas G, Akritidis N, Bosilkovski M, Tsianos E (2005a) Brucellosis. New Eng J Med 352:2325–2336

Pappas G, Solera J, Akritidis N, Tsianos E (2005b) New approaches to the antibiotic treatment of brucellosis. Int J Antimicrob Agents 26:101–105

Pappas G, Panagopoulou P, Christou L, Akritidis N (2006a) *Brucella* as a biological weapon. Cell Mol Life Sci 63: 2229–2236

Pappas G, Panagopoulou P, Akritidis N, Christou L, Tsianos EV (2006b) The new global map of human brucellosis. Lancet Infect Dis 6:91–99

Paranavitana CM, Zelazowska E, Das R, Izadjoo M, Jett M, Hoover D (2005a) Identification of novel genes in the memory response to *Brucella* infection by cDNA arrays. Mol Cell Probes 19:341–348

Paranavitana C, Zelazowska E, Izadjoo M, Hoover D (2005b) Interferon-gamma associated cytokines and chemokines produced by spleen cells from *Brucella*-immune mice. Cytokine 30:86–92

Paschos A, Patey G, Sivanesan D, Gao C, Bayliss R, Waksman G, O'Callaghan D, Baron C (2006) Dimerization and interactions of *Brucella suis virB8* with *virB4* and *virB* 10 are required for its biological activity. Proc Natl Acad Sci USA 103:7252–7257

Paulsen IT, Seshadri R, Nelson KE, Eisen JA, Heidelberg JF, Read TD, Dodson RJ, Umayam L, Brinkac LM, Beanan MJ, Daugherty SC, Deboy RT, Durkin AS, Kolonay JF, Madupu R, Nelson WC, Ayodeji B, Kraul M, Shetty J, Malek J, Van Aken SE, Riedmuller S, Tettelin H, Gill SR, White O, Salzberg SL, Hoover DL, Lindler LE, Halling SM, Boyle SM, Fraser CM (2002) The *Brucella suis* genome reveals fundamental similarities between animal and plant pathogens and symbionts. Proc Natl Acad Sci USA 99:13148–13153

Pei J, Ficht TA (2004) *Brucella abortus* rough mutants are cytopathic for macrophages in culture. Infect Immunol 72:440–450

Peterson JD, Umayam LA, Dickinson T, Hickey EK, White O (2001) The comprehensive microbial resource. Nucl Acids Res 29:123–125

Posadas DM, Martin FA, Sabio y Garcia JV, Spera JM, Delpino MV, Baldi P, Campos E, Cravero SL, Zorrequieta A (2007) The *tolC* homologue of *Brucella suis* is involved in resistance to antimicrobial compounds and virulence. Infect Immun 75:379–389

Potter SC, Clarke L, Curwen V, Keenan S, Mongin E, Searle SM, Stabenau A, Storey R, Clamp M (2004) The ensemble analysis pipeline. Genom Res 14:934–941

Pourbagher A, Pourbagher MA, Savas L, Turunc T, Demiroglu Z, Erol I, Yalcintas D (2006) Epidemiologic, clinical, and imaging findings in brucellosis patients with osteoarticular involvement. AJR Am J Roentgenol 187:873–880

Puri S, O'Brian MR (2006) The *hmuQ* and *hmuD* genes from *Bradyrhizobium japonicum* encode heme-degrading enzymes. J Bacteriol 188:6476–6482

Ragan VE (2002) The animal and plant health inspection service (aphis) brucellosis eradication program in the United States. Vet Microbiol 90:11–18

Rajashekara G, Glasner JD, Glover DA, Splitter GA (2004) Comparative whole-genome hybridization reveals genomic islands in *Brucella* species. J Bacteriol 186:5040–5051

Rajashekara G, Glasner JD, Krepps M, Splitter GA (2005) Temporal analysis of pathogenic events in virulent and avirulent *Brucella melitensis* infections. Cell Microbiol 7:1459–1473

Ratushna VG, Sturgill DM, Ramamoorthy S, Reichow SA, He Y, Lathigra R, Sriranganathan N, Halling SM, Boyle SM, Gibas CJ (2006) Molecular targets for rapid identification of *Brucella spp.* BMC Microbiol 6:13

Renders N, Licciardello L, IJsseldijk C, Sijmons M, van Alphen L, Verbrugh H, van Belkum A (1999) Variable numbers of tandem repeat loci in genetically homogeneous haemophilus influenzae strains alter during persistent colonisation of cystic fibrosis patients. FEMS Microbiol Lett 173:95–102

Retief JD (2000) Phylogenetic analysis using PHYLIP. Meth Mol Biol 132:243–258

Rhyan JC, Gidlewski T, Ewalt DR, Hennager SG, Lambourne DM, Olsen SC (2001) Seroconversion and abortion in cattle experimentally infected with *Brucella* sp. Isolated from a pacific harbor seal (*Phoca vitulina richardsi*). J Vet Diagn Invest 13:379–382

Rigby CE, Fraser AD (1989) Plasmid transfer and plasmid-mediated genetic exchange in *Brucella abortus*. Can J Vet Res 53:326–330

Robertson GT, Roop RM (1999) The *Brucella abortus* host factor i (hf-i) protein contributes to stress resistance during stationary phase and is a major determinant of virulence in mice. Mol Microbiol 34:690–700

Robertson GT, Reisenauer A, Wright R, Jensen RB, Jensen A, Shapiro L, Roop RM (2000) The *Brucella abortus ccr*M DNA methyltransferase is essential for viability, and its overexpression attenuates intracellular replication in murine macrophages. J Bacteriol 182:3482–3489

Rodriguez JL, Palmer GH, Knowles DP, Brayton KA (2005) Distinctly different *msp*2 pseudogene repertoires in *Anaplasma marginale* strains that are capable of superinfection. Gene 361:127–132

Roop RM, Gee JM, Robertson GT, Richardson JM, Ng WL, Winkler ME (2003) *Brucella* stationary-phase gene expression and virulence. Annu Rev Microbiol 57:57–76

Roop RM, Bellaire BH, Valderas MW, Cardelli JA (2004) Adaptation of the *Brucellae* to their intracellular niche. Mol Microbiol 52:621–630

Rose DR, Przybylska M, To RJ, Kayden CS, Oomen RP, Vorberg E, Young NM, Bundle DR (1993) Crystal structure to 2.45 A resolution of a monoclonal Fab specific for the *Brucella* A cell wall polysaccharide antigen. Protein Sci 2:1106–1113

Roset MS, Ciocchini AE, Ugalde RA, Inon de Iannino N (2006) The *Brucella abortus* cyclic beta-1,2-glucan virulence factor is substituted with o-ester-linked succinyl residues. J Bacteriol 188:5003–5013

Ross HM, Foster G, Reid RJ, Jahans KL, MacMillan AP (1994) *Brucella* species infection in sea-mammals. Vet Rec 134:359

Roth F, Zinsstag J, Orkhon D, Chimed-Ochir G, Hutton G, Cosivi O, Carrin G, Otte J (2003) Human health benefits from livestock vaccination for brucellosis: Case study. Bull World Health Organ 81:876–876

Roux CM, et al. (2007) Brucella requires a functional Type IV secretion system to elicit innate immune responses in mice. Cell Microbiol 9(7):1851–1869

Ruben B, Band JD, Wong P, Colville J (1991) Person-to-person transmission of *Brucella melitensis*. Lancet Infect Dis 337:14–15

Samartino L (2002) Brucellosis in Argentina. Vet Microbiol 90:71–80

Sangari F, Aguero J (1991) Mutagenesis of *Brucella abortus*: Comparative efficiency of three transposon delivery systems. Microb Pathog 11:443–446

Sangari FJ, Seoane A, Rodriguez MC, Aguero J, Garcia Lobo JM (2007) Characterization of the urease operon of *Brucella abortus* and assessment of its role in virulence of the bacterium. Infect Immun 75:774–780

Sanger F, Nicklen S, Coulson AR (1977) DNA sequencing with chain-terminating inhibitors. Proc Natl Acad Sci USA 74:5463–5467

Santos JM, Verstreate D, Perera VY, Winter AJ (1984) Outer membrane proteins from rough strains of four *Brucella* species. Infect Immun 46:188–194

Sauret JM, Vilissova N (2002) Human brucellosis. J Am Board Fam Pract 15(5):401–406. URL: http://www.ncbi.nlm.nih.gov/pubmed/12350062 (PMID: 12350062)

Schurig GG, Roop RM, Bagchi T, Boyle SM, Buhrman D, Sriranganathan N (1991) Biological properties of RB51; a stable rough strain of *Brucella abortus*. Vet Microbiol 28: 171–188

Schlake T, Bode J (1994) Use of mutated FLP recognition target (FRT) sites for the exchange of expression cassettes at defined chromosomal loci. Biochemistry 33:12746–12751.

Service RF (2006) Gene sequencing. The race for the $1000 genome. Science 311:1544–1546

Sieira R, Comerci DJ, Pietrasanta LI, Ugalde RA (2004) Integration host factor is involved in transcriptional regulation of the *Brucella abortus vir*B operon. Mol Microbiol 54:808–822

Simmons GC, Hall WTK (1953) Epididymitis of rams. Aust Vet J 29:33–40

Snyder EE, Kampanya N, Lu J, Nordberg EK, Karur HR, Shukla M, Soneja J, Tian Y, Xue T, Yoo H, Zhang F, Dharmanolla C, Dongre NV, Gillespie JJ, Hamelius J, Hance M, Huntington KI, Jukneliene D, Koziski J, Mackasmiel L, Mane SP, Nguyen V, Purkayastha A, Shallom J, Yu G, Guo Y, Gabbard J, Hix D, Azad AF, Baker SC, Boyle SM, Khudyakov Y, Meng XJ, Rupprecht C, Vinje J, Crasta OR, Czar MJ, Dickerman A, Eckart JD, Kenyon R, Will R, Setubal JC, Sobral BW (2007) Patric: The VBI pathosystems resource integration center. Nucl Acids Res 35:D401–D406

Snyder JA, Haugen BJ, Buckles EL, Lockatell CV, Johnson DE, Donnenberg MS, Welch RA, Mobley HL (2004) Transcriptome of uropathogenic *Escherichia coli* during urinary tract infection. Infect Immun 72:6373–6381

Sohn AH, Probert WS, Glaser CA, Gupta N, Bollen AW, Wong JD, Grace EM, McDonald WC (2003) Human neurobrucellosis with intracerebral granuloma caused by a marine mammal *Brucella spp*. Emerg Infect Dis 9:485–488

Sola-Landa A, Pizarro-Cerda J, Grillo MJ, Moreno E, Moriyon I, Blasco JM, Gorvel JP, Lopez-Goni I (1998) A two-component regulatory system playing a critical role in plant pathogens and endosymbionts is present in *Brucella abortus* and controls cell invasion and virulence. Mol Microbiol 29:125–138

Solera J, Martinez-Alfaro E, Espinosa A, Castillejos MI, Geyo P, Rodriguez-Zapata M (1998) Multivariate model for predicting relapse in human brucellosis. J Infect 36:85–92

Spera JM, Ugalde JE, Mucci J, Comerci DJ, Ugalde RA (2006) A B lymphocyte mitogen is a *Brucella abortus* virulence factor required for persistent infection. Proc Natl Acad Sci USA 103:16514–16519

Sriwanthana B, Island MD, Maneval D, Mobley HL (1994) Single-step purification of *Proteus mirabilis* urease accessory protein UreE, a protein with a naturally occurring histidine tail, by nickel chelate affinity chromatography. J Bacteriol 176:6836–6841

Stein L (2003) Integrating biological databases. Nat Rev (Genetics) 4:337–345

Stevens MG, Olsen SC, Pugh GW, Mayfield JE (1997) Role of immune responses to a *gro*EL heat shock protein in preventing brucellosis in mice vaccinated with *Brucella abortus* strain RB51. Comp Immunol Microbiol Infect Dis 20:147–153

Stoenner HG, Lackman DB (1957) A new species of *Brucella* isolated from the desert wood rat, *Neotoma lepida thomas*. Am J Vet Res 18:947–951

Strange K (2005) The end of "Naive reductionism": Rise of systems biology or renaissance of physiology? Am J Physiol Cell Physiol 288:C968–C974

Suzek BE, Ermolaeva MD, Schreiber M, Salzberg SL (2001) A probabilistic method for identifying start codons in bacterial genomes. Bioinformatics 17:1123–1130

Talaat AM, Lyons R, Howard ST, Johnston SA (2004) The temporal expression profile of *Mycobacterium tuberculosis* infection in mice. Proc Natl Acad Sci USA 101: 4602–4607

Taleski V, Zerva L, Kantardjiev T, Cvetnic Z, Erski-Biljic M, Nikolovski B, Bosnjakovski J, Katalinic-Jankovic V, Panteliadou A, Stojkoski S, Kirandziski T (2002) An overview of the epidemiology and epizootology of brucellosis in selected countries of central and southeast Europe. Vet Microbiol 90:147–155

Tamayo R, Prouty AM, Gunn JS (2005) Identification and functional analysis of *Salmonella enterica* serovar *typhimurium pmr*A-regulated genes. FEMS Immunol Med Microbiol 43:249–258

Tatusov RL, Galperin MY, Natale DA, Koonin EV (2000) The COG database: A tool for genome-scale analysis of protein functions and evolution. Nucl Acids Res 28:33–36

Tcherneva E, Rijpens N, Naydensky C, Herman LM (1996) Repetitive element sequence based polymerase chain reaction for typing of *Brucella* strains. Vet Microbiol 51:169–178

Tcherneva E, Rijpens N, Jersek B, Herman LM (2000) Differentiation of *Brucella* species by random amplified polymorphic DNA analysis. J Appl Microbiol 88:69–80

Tech M, Pfeifer N, Morgenstern B, Meinicke P (2005) Tico: A tool for improving predictions of prokaryotic translation initiation sites. Bioinformatics 21:3568–3569

Teixeira-Gomes AP, Cloeckaert A, Bezard G, Dubray G, Zygmunt MS (1997) Mapping and identification of *Brucella melitensis* proteins by two-dimensional electrophoresis and microsequencing. Electrophoresis 18:156–162

Thiede B, Hohenwarter W, Krah A, Mattow J, Schmid M, Schmidt F, Jungblut PR (2005) Peptide mass fingerprinting. Methods 35:237–247

Tibor A, Wansard V, Bielartz V, Delrue RM, Danese I, Michel P, Walravens K, Godfroid J, Letesson JJ (2002) Effect of *omp*10 or *omp*19 deletion on *Brucella abortus* outer membrane properties and virulence in mice. Infect Immun 70:5540–5546

Tobes R, Ramos JL (2005) Rep code: Defining bacterial identity in extragenic space. Environ Microbiol 7:225–228

Traum J (1914) Report of the Chief of the Bureau of Animal Industry, United States Department of Agriculture, Washington, D.C., p 30

Tsolis RM (2002) Comparative genome analysis of the alpha-proteobacteria: Relationships between plant and animal pathogens and host specificity. Proc Natl Acad Sci USA 99:12503–12505

Tsolis RM, Townsend SM, Miao EA, Miller SI, Ficht TA, Adams LG, Baumler AJ (1999) Identification of a putative *Salmonella typhimurium* host range factor with homology to *ipa*H and *yop*M by signature-tagged mutagenesis. Infect Immunol 67:6385–6393

Tsuda M, Karita M, Morshed MG, Okita K, Nakazawa T (1994a) A urease-negative mutant of *Helicobacter pylori* constructed by allelic exchange mutagenesis lacks the ability to colonize the nude mouse stomach. Infect Immun 62:3586–3589

Tsuda M, Karita M, Mizote T, Morshed MG, Okita K, Nakazawa T (1994b) Essential role of *Helicobacter pylori* urease in gastric colonization: Definite proof using a urease-negative mutant constructed by gene replacement. Eur J Gastroenterol Hepatol 6(Suppl 1):S49–S52

Tumurkhuu G, Koide N, Takahashi K, Hassan F, Islam S, Ito H, Mori I, Yoshida T (2006) Characterization of biological activities of *Brucella melitensis* lipopolysaccharide. Microbiol Immunol 50:421–427

Ugalde RA (1999) Intracellular lifestyle of *Brucella spp*. Common genes with other animal pathogens, plant pathogens, and endosymbionts. Microbes Infect 1:1211–1219

Velasco J, Romero C, Lopez-Goni I, Leiva J, Diaz R, Moriyon I (1998) Evaluation of the relatedness of *Brucella spp*. and *Ochrobactrum anthropi* and description of *Ochrobactrum intermedium* sp. Nov., a new species with a closer relationship to *Brucella spp*. Int J Syst Bacteriol 48:759–768

Vemulapalli TH, Vemulapalli R, Schurig GG, Boyle SM, Sriranganathan N (2006) Role in virulence of a *Brucella abortus* protein exhibiting lectin-like activity. Infect Immunol 74: 183–191

Verger JM, Grimont F, Grimont PAD, Grayon M (1985) *Brucella*, a monospecific genus as shown by deoxyribonucleic acid hybridization. Int J Syst Bacteriol 35:292–295

Verger JM, Grimont F, Grimont PA, Grayon M (1987) Taxonomy of the genus *Brucella*. Ann Inst Pasteur Microbiol 138:235–238

Verger JM, Grayon M, Chaslus-Dancla E, Meurisse M, Lafont JP (1993) Conjugative transfer and *in vitro/in vivo* stability of the broad-host-range incP r751 plasmid in *Brucella spp*. Plasmid 29:142–146

Verger JM, Grayon M, Cloeckaert A, Lefevre M, Ageron E, Grimont F (2000) Classification of *Brucella* strains isolated from marine mammals using DNA-DNA hybridization and ribotyping. Res Microbiol 151:797–799

Vizcaino N, Cloeckaert A, Verger J, Grayon M, Fernandez-Lago L (2000) DNA polymorphism in the genus *Brucella*. Microb Infect 2:1089–1100

Vizcaino N, Cloeckaert A, Zygmunt MS, Fernandez-Lago L (2001) Characterization of a *Brucella* species 25-kilobase DNA fragment deleted from *Brucella abortus* reveals a large gene cluster related to the synthesis of a polysaccharide. Infect Immun 69:6738–6748

Wagner MA, Eschenbrenner M, Horn TA, Kraycer JA, Mujer CV, Haguis S, Elzer P, DelVecchio VG (2002) Global analysis of the *Brucella melitensis* proteome: Identification of proteins expressed in laboratory-grown culture. Proteomics 2:1047–1060

Wagner VE, Bushnell D, Passador L, Brooks AI, Iglewski BH (2003) Microarray analysis of *Pseudomonas aeruginosa* quorum-sensing regulons: Effects of growth phase and environment. J Bacteriol 185:2080–2095

Weiss DS, Takeda K, Akira S, Zychlinsky A, Moreno E (2005) Myd88, but not toll-like receptors 4 and 2, is required for efficient clearance of *Brucella abortus*. Infect Immun 73: 5137–5143

Westhusin ME, Shin T, Templeton JW, Burghardt RC, Adams LG (2007) Rescuing valuable genomes by animal cloning: A case for natural disease resistance in cattle. J Anim Sci 85: 138–142

Whatmore AM, Murphy TJ, Shankster S, Young E, Cutler SJ, Macmillan AP (2005) Use of amplified fragment length polymorphism to identify and type *Brucella* isolates of medical and veterinary interest. J Clin Microbiol 43:761–769

Whatmore AM, Perrett LL, MacMillan AP (2007) Characterisation of the genetic diversity of *Brucella* by multilocus sequencing. BMC Microbiol 7:34

Whatmore AM, Shankster SJ, Perrett LL, Murphy TJ, Brew SD, Thirlwall RE, Cutler SJ, MacMillan AP (2006) Identification and characterization of variable-number tandem-repeat markers for typing of *Brucella spp*. J Clin Microbiol 44:1982–1993

Williams KP, Sobral BW, Dickerman AW (2007) A robust species tree for the alphaproteobacteria.J Bacteriol 189:4578–4586

Wilson JW, Ramamurthy R, Porwollik S, McClelland M, Hammond T, Allen P, Ott CM, Pierson DL, Nickerson CA (2002) Microarray analysis identifies *Salmonella* genes belonging to the low-shear modeled microgravity regulon. Proc Natl Acad Sci USA 99:13807–13812

Wise DJ (1995) Intracellular growth of *Brucella abortus* and *B. melitensis* in murine macrophage-like cell lines and partial characterization of a biologically active extract from *B. abortus* strain RB51. Ph dissertation. Virginia Tech, Blacksburg, VA

World Health Organization (2005) The control of neglected zoonotic diseases; a route to poverty alleviation. Zoonoses and Veterinary Public Health, WHO, Geneva, Switzerland

Wu Q, Pei J, Turse C, Ficht TA (2006) Mariner mutagenesis of *Brucella* melitensis reveals genes with previously uncharacterized roles in virulence and survival. BMC Microbiol 6:102–116

Xiang Z, Zheng W, He Y (2006) Bbp: *Brucella* genome annotation with literature mining and curation. BMC Bioinform 7:347–361

Yagupsky P, Baron EJ (2005) Laboratory exposures to *Brucellae* and implications for bioterrorism. Emerg Infect Dis 11: 1180–1185

Yanagi M, Yamasato K (1993) Phylogenetic analysis of the family *Rhizobiaceae* and related bacteria by sequencing of 16s rRNA gene using PCR and DNA sequencer. FEMS Microbiol Lett 15:115–120

Yang X, Hudson M, Walters N, Bargatze RF, Pascual DW (2005) Selection of protective epitopes for *Brucella melitensis* by DNA vaccination. Infect Immun 73:7297–7303

Yingst S, Hoover DL (2003) T cell immunity to brucellosis. Crit Rev Microbiol 29:313–331

Young EJ (2000) *Brucella* species, pp 2386–2393. In: Mandell GL, Bennett JE, Dolin R (eds) Mandell, Dougles and Bennette's Principles and Practice of Infectious Diseases, 5th ed. Churchill Livingstone, Philadelphia, PA

Yu GX, Boyle SM, Crasta OR (2007) A versatile computational pipeline for bacterial genome annotation improvement and comparative analysis, with *Brucella* as a use case. Nucl Acids Res 35:3953–3962

Zdobnov EM, Apweiler R (2001) Interproscan – an integration platform for the signature-recognition methods in interpro. Bioinformatics 17:847–848

Zhan Y, Cheers C (1993) Endogenous gamma interferon mediates resistance to *Brucella abortus* infection. Infect Immun 61:4899–4901

Zylberman V, Craig PO, Klinke S, Braden BC, Cauerhff A, Goldbaum FA (2004) High order quaternary arrangement confers increased structural stability to *Brucella* sp. lumazine synthase. J Biol Chem 279:8093–8101

Zylberman V, Klinke S, Haase I, Bacher A, Fischer M, Goldbaum FA (2006) Evolution of vitamin B2 biosynthesis: 6,7-Dimethyl-8-ribityllumazine synthases of *Brucella*. J Bacteriol 188:6135–6142

2 *Mycobacterium avium* subspecies *paratuberculosis*

Ling-Ling Li[1], Sushmita Singh[2], John Bannantine[3], Sagarika Kanjilal[1,4], and Vivek Kapur[1,2,5] (✉)

[1] Department of Veterinary and Biomedical Sciences, Penn State, University Park, PA, USA
[2] Biomedical Genomics Center, University of Minnesota, St. Paul, MN, USA
[3] National Animal Disease Center, US Department of Agriculture, Agricultural Research Service, Ames, IA, USA
[4] Department of Pharmacology, Penn State Milton S. Hershey Medical Center, Hershey, PA, USA
[5] Center for Infectious Disease Dynamics, Penn State, University Park, PA, USA, vkapur@psu.edu

2.1 Introduction

2.1.1 The Pathogen and the Disease It Causes

Mycobacterium avium subspecies *paratuberculosis* (*Map*) is the etiological agent of a severe gastroenteritis in ruminants, known as Johne's disease (JD) (Harris and Barletta 2001). The name Johne's disease derives from the name of the German bacteriologist, Heinrich Albert Johne, who, together with Frothingham, demonstrated a connection between cattle enteritis and the presence of acid-fast microorganisms in sections of the intestinal mucosa, in 1895 (Chiodini et al. 1984). However, it was not until 1912 that FW Twort succeeded in cultivating and characterizing the causative agent of this disease, which was, as suspected, a *Mycobacterium*, and which in 1914 was shown to produce enteritis in experimentally infected animals (Kreeger 1991; Cocito et al. 1994).

Bacteria of the genus *Mycobacterium* are gram-positive, acid-fast, rod-shaped bacteria that include a number of important human and animal pathogens. After the initial reports of the disease from Europe, Johne's disease, or "paratuberculosis" as it is also known, was first described as an infection in cattle in North America in 1908, and has been reported from every state since. Prior to 1908, the majority of reports of the disease originated from northern and western Europe (Kreeger 1991). Although systematic global surveys of the incidence of JD are lacking, the disease has been reported from animals on every continent with the exception of Antartica, and from essentially every country in which intensive animal agriculture is practiced (Cocito et al. 1994).

Johne's Disease and Its Pathogenesis

Johne's disease is a debilitating, almost always fatal, chronic granulomatous enteritis that occurs primarily in domestic and wild ruminants. Susceptible animals include cattle, sheep, goats, other ruminants as well as other wild animal species including bison, elk, deer, rabbits and birds. Monogastric domestic and wild animal species can also be infected, albeit infections in these hosts occur less frequently. Wild animals such as deer and rabbits have been implicated as potential disease carriers, and may spread *Map* among other wild and domestic ruminants. This may also serve to limit opportunities to control or eradicate JD from domesticated animal populations (Libke and Walton 1975; Jessup et al. 1981; Chiodini and Van Kruiningen 1983). In addition to causing a well described spectrum of diseases in domestic and wild animals, *Map* has been isolated from humans with Crohn's disease, suggesting a possible involvement of this microorganism in the etiology or pathogenesis of Crohn's disease (Chiodini et al. 1984; Naser et al. 2000).

Natural infections with *Map* are thought to primarily occur via the fecal–oral route. Animals acquire the infection by ingestion of the bacteria mainly via feed, water, or teats that are contaminated with feces from infected animals that contain *Map*. Following the ingestion of *Map*, the bacteria are transported from the intestinal lumen into the intestinal wall via M cells, which overlie the domes of Peyer's patches (Sigurethardottir et al. 2004). It is believed that *Map* fibronectin attachment protein facilitates M-cell

Genome Mapping and Genomics in Animal-Associated Microbes
V. Nene, C. Kole (Eds.)

targeting and invasion through a bridge formed with integrins on the apical surface of intestinal M cells (Secott et al. 2004). A recent study of the small intestinal mucosa of infected goat kids in areas with and without Peyer's patches suggests that *Map* may also gain entry through enterocytes (Sigurdardottir et al. 2005).

After crossing the epithelial lining of the intestine, the bacteria are phagocytozed by subepithelial macrophages, which are thought to be the target cells for *Map* (Valentin-Weigand and Goethe 1999). Results of experimental studies in cattle and sheep reveal that *Map* cells are detectable in intestinal macrophages within a few hours after infection via the oral route (Clarke 1997). Once within the macrophage, like many other pathogenic members of the genus *Mycobacterium*, *Map* are thought to avoid the bactericidal action of the professional phagocytic cells and instead proliferate (Stabel 2000; Coussens 2001). The various mechanisms that enable mycobacteria to survive within phagosomes include inhibition of phagosomal maturation, resistance against antimicrobial molecules, and adaptation to host-induced metabolic constraints (Schnappinger et al. 2003). T-cell-mediated immune responses are essential in determining the outcome of infection and overall severity of disease. It is believed that the ultimate fate of *Map* within macrophages is influenced by the activation state of the phagocytic cell, and that the cytokine interferon-gamma (IFN-γ), which is primarily secreted by the $CD4^+$ T cells, is considered to be an important macrophage activation factor (Sigurethardottir et al. 2004).

In experimentally infected calves and lambs, the first granulomatous lesions have been observed within the interfollicular regions of Peyers patches and mesenteric lymph nodes as early as 3 months (in calves) or one and a half months after infection (in lambs). In both animal species, lesions extend into the intestinal mucosa after several additional months of infection (Clarke 1997). *Map* has also been identified within the mononuclear cell fraction of blood and tissue fluid from infected cattle, suggesting that macrophages may function as vehicles in dissemination of the organisms from infected sites (Valentin-Weigand and Goethe 1999). Generally, it takes several years from the time of initial infection (thought to be in animals <6 months in age) until development of clinical signs (usually animals aged >2 years), although in experimentally infected animals, this process could be faster. For example, oral innoculation of a high dose of *Map* (a total of ~10^9 over 10 doses given weekly) caused progressive infection and clinical disease in Cheviot sheep within nine months after initial exposure (Stewart et al. 2004).

The clinical signs of JD include diarrhea, weight loss, mortality, and decreased milk production in domestic cattle. At necropsy, JD is characterized by a thickened, corrugated appearance of the intestinal mucosa, and microscopically the disease present non-ulcerative granulomatous lesions that contain macrophages packed with acid-fast bacilli (Hamilton et al. 1989).

The pathology associated with JD is thought to result, in part, from a severe immune dysregulation and chronic inflammation that result from *Map* infection. Consistent with this notion, gene expression profiles of ileal tissue samples from cows chronically infected with *Map* when compared to uninfected cows showed significantly greater expression of IL-1α, tumor necrosis factor receptor-associated protein 1, monocyte chemotatic protein-2, b-cadherin, and β1 integrin. Based on these findings, it has been suggested that the overexpression of IL-1α may be partially responsible for many of the clinical signs associated with Johne's disease (Aho et al. 2003).

Economic Importance and Zoonotic Potential of *Map*

As described earlier, JD is a slowly progressing disease that requires an extended period of time for the development of clinical signs. In cattle, animals are thought to become infected with *Map* as calves but often do not manifest clinical signs until 2–5 years of age. Hence, by the time a single infected animal is identified based on presentation of clinical signs, 40% or more of the animals within the herd may already have become infected (Kreeger 1991). Given the relative difficulties associated with identifying preclinical infected animals, it is not surprising therefore that JD can spread rapidly on a farm, and is commonly reported in diary herds throughout the world. The National Animal Health Monitoring System (NAHMS) of the USDA, Animal and Plant Health Inspection Service (APHIS) found that 21.6% of US dairy operations were *Map* positive, resulting in considerable economic losses to the dairy industry (US Department of Agriculture 1977). The costs associated with JD vary widely and range from $200 million to $1.5 billion annually. These figures are most often extrapolated from estimated rates of prevalence, and the computation of financial losses from culling or death of clinically infected cows reduced reproductive efficiency, feed efficiency, and decreased milk production (Stabel 1998; Ott et al. 1999).

A major concern with Johne's disease is the ease with which the bacterium is spread amongst animals within a herd. Subclinical or clinically infected animals shed *Map* in feces and milk, enabling dissemination to susceptible calves, the environment, and, most troublingly, from retail milk sources in Europe and North America (Streeter et al. 1995; Sweeney 1996; Whitlock and Buergelt 1996; Grant et al. 2002). A more recent study of 143 raw milk cheese samples collected at the retail level in Switzerland revealed that 4.2% of the samples tested positive for *Map* based on real-time PCR analysis of a *Map* gene (Stephan et al. 2007). The contamination of retail food by *Map* is of particular concern due to *Map's* implications as a possible cause of Crohn's disease in humans (Greenstein 2003).

2.1.2 Morphology, Taxonomic Position, Life-Cycle, and Host-Range of *Map*

Map is a gram-positive, acid-fast, rod-shaped microorganism of 0.5–1.5 µm in size. On the commonly used HEYM (Herrold's egg yolk medium) agar, *Map* forms either nonpigmented or pigmented rough colonies. The bacterium is very slow growing and fastidious: to grow in the laboratory, most strains require the presence of the siderophore, mycobactin (Lambrecht et al. 1988). Visible colony formation on solid media may take up to four months, despite the presence of various growth supplements and other essential nutrients (Cocito et al. 1994).

Taxonomical Classification of *Map*

The mycobacteria represent a group of high GC content, gram-positive microorganisms comprising more than 100 species. Taxonomical classifications place the genus *Mycobacterium* as a relative of the genera *Corynebacterium*, *Nocardia*, *Rhodococcus*, and *Streptomyces*. Mycobacteria are characterized by their lipid-rich cell wall visualized only by the acid-fast stain (Chacon et al. 2004). The main pathogenic species of mycobacteria include *M. tuberculosis*, *M. bovis*, *M. leprae*, and *M. avium* (Harris and Barletta 2001). Although often grouped with *M. avium* in the *Mycobacterium avium–intracellulare* complex, *M. intracellulare* is recognized as a genetically distinct species by DNA–DNA hybridization and drug susceptibility profiles (Baess 1983; Saito et al. 1989). *M. avium* has been subdivided based on DNA–DNA hybridization tests into three subspecies: *M. avium* subsp. *avium*, *M. avium* subsp. *silvaticum*, and *M. avium* subsp. *paratuberculosis* (Thorel et al. 1990). In vitro doubling time varies between subspecies from 12 to over 20 h, with *Map* being the slowest grower of the three subspecies (Chacon et al. 2004). In addition, *Map* is the only *M. avium* subspecies that depends on the siderophore mycobactin for growth, especially when grown as a primary culture directly from animal tissues (Lambrecht and Collins 1992). More recently, based on differences in IS1245 RFLP, 16-23S rDNA ITS, growth-temperature tolerance, and host range, *M. avium* subsp. *hominissuis* is designated as a new subspecies that is distinguishable from *M. avuim* subsp. *avium* (Mijs et al. 2002).

The Natural Lifecycle and Host-Range of *Map*

Map can survive for long period in the environment (Whittington et al. 2004), but is thought to primarily infect and cause disease in domestic ruminants including cattle, sheep, and goats, as well as many wild ruminants. These include deer, antelopes, mountain goats, bisons, camels, llamas, and even monogastric animals including pigs, rabbits, chickens, and wild birds (for example, *Map* infections have been found in the starling) (Movahedzadeh et al. 2004). *Map* has also been reported from asymptomatic horses and mules, which may prove to be an additional source for dissemination of the organism in the environment (Cocito et al. 1994). Interestingly, *Map* has been isolated from intestinal tissues and breast milk of Crohn's disease patients (Chiodini et al. 1984; Gitnick et al. 1989; Naser et al. 2000). A recent study revealed a significant correlation between the presence of *Map* in intestinal biopsies and active Crohn's disease (Bull et al. 2003a, b). In the laboratory, animals including mice, rats, rabbits, hamsters, gerbils, and guinea pigs can be productively infected with *Map* (Cocito et al. 1994).

2.2 The Genome Sequence of *Map*

The complete genome sequence of K-10, a *Map* isolate from cattle with Johne's disease, has been completed and annotated, and is publicly available and downloadable from GenBank (accession number AE016958) (Li et al. 2005). *Map* K-10, an isolate from a dairy herd in Wisconsin, is a low passage clinical strain isolated by investigators at the USDA National Animal Disease Center in the mid-1970s (Foley-Thomas et al. 1995). This isolate was chosen because it retains virulence for cattle and other natural and experimental hosts, and represents a "typical" clinical isolate of *Map* that is recovered from cattle in the United States (Foley-Thomas et al. 1995). This isolate is also genetically tractable as has been shown by various phage infection and transposon mutagenesis studies (Foley-Thomas et al. 1995; Harris and Barletta 2001).

Overall, the completion of genome sequence of *Map* and comparative genomics analyses have provided a strong foundation for future investigations on the genetics, evolution, natural physiology, and virulence of this important pathogen. Some of the key attributes of the *Map* genome are discussed later.

2.2.1 Characteristics of the *Map* K-10 Genome

Map K-10 has a single circular genome with a sequence of 4,829,781 base pairs, and a G+C content of 69.3% (Table 1; Fig. 1). The G+C content is relatively constant throughout the genome, and the analysis identified only a few genomic regions with lower G+C content, corresponding to prophages or coding RNA sequences (Fig. 1). A single *rrn* operon (16S-23S-5S) was identified in K-10, which is located ~2.75 megabases (Mb) from the putative *oriC* on the opposite strand. This is approximately 1 Mb further from the *oriC* than what is described in other mycobacterial genomes (~1.50 Mb in *Mtb*; Cole et al. 1998) and has been speculated to contribute to the slow growth of *Map* as compared with other mycobacteria.

2.2.2 Repetitive DNA in *Map*

Approximately 1.5% or 72.2 kb of the *Map* K-10 genome is comprised of repetitive DNA, including insertion sequences, multigene families, and duplicated housekeeping genes. The analysis also identified 17 copies of the previously described insertion sequence IS*900* (Green et al. 1989), seven copies of IS*1311* (Whittington et al. 1998), and three copies of IS*Mav2* (Strommenger et al. 2001). A total of 16 additional *Map* insertion sequence elements were identified in the analysis, totaling 19 different insertion sequences with 58 total copies in the K-10 genome. The analysis revealed several insertion sequences with no identifiable homologs in other bacteria: IS_*MAP02* with six copies and *IS_MAP04* with four copies. These newly discovered IS elements are of particular interest for their use as specific potential diagnostic targets due to their absence in other mycobacteria. In addition, within the genome

Table 1 Summary of the complete genome of *Map* K-10 and its comparison with other *Mycobacterium* species

	Map	*Maa*	*Mtb*	*M. bovis*	*M. leprae*	*M. smegmatis*	*M. sp MCS*
Genome size (bp)	4,829,781	5,475,738	4,411,532	4,345,492	3,268,203	6,988,209	5,705,448[a]
G+C content (%)	69.30	68.99	65.61	65.63	57.79	67.40	68.38
Protein coding (%)	91.30	91.94	90.80	90.59	49.50	92.42	92.41
ORFs	4,350	5,312	3,959	3,953	1,604	6,897	5,413[b]
Gene density (bp/gene)	1,112	1,031	1,114	1,099	2,037	1,013	1,054
Average gene length (bp)	1,015	948	1,012	995	1,011	936	982
tRNAs	45	45	45	45	45	47	48
rRNA operon	1	1		1	1	2	2
Plasmid	0	0	0	0	0	0	1

[a] Chromosome only (plasmid not included)

[b] Number of protein-coding ORFs in the chromosome

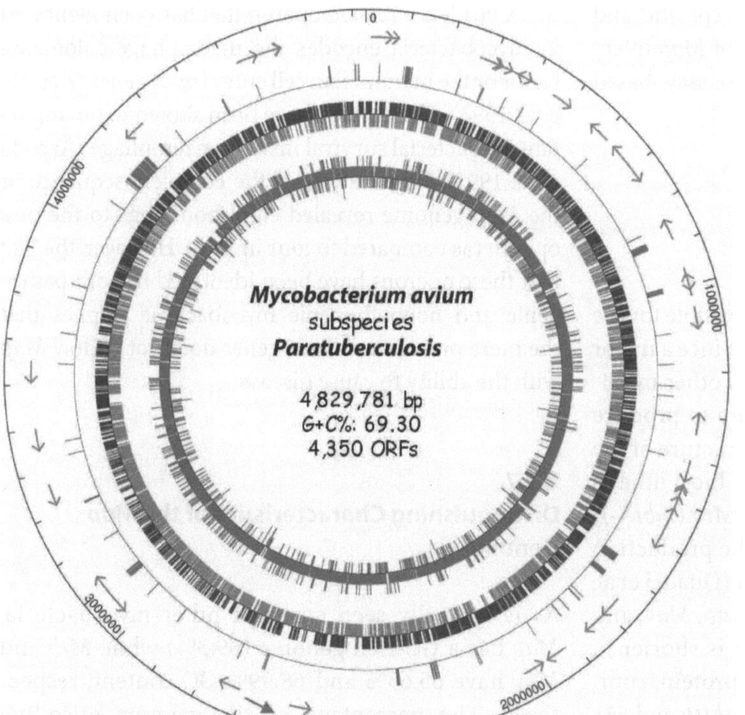

Fig. 1 Circular representation of the *Map* K-10 genome. From inside: Innermost histogram, GC content. Second innermost histograms: location and orientation of ORFs in *Map* K-10, *dark grey* histogram indicates the same direction of transcription as the origin of replication, and *light grey* histogram indicates the opposite direction of transcription as the origin of replication. Dark grey regions on outermost circle indicate unique sequences. Broad-tipped arrow, rRNA operon; fine-tipped arrows, 45 tRNAs outer circle scale. The figure was generated with GENESCENE software (DNAstar, Madison, WI)

of K-10, the analysis also identified 12 homologs to the REP13E12 family, ~1,400 bp insertion sequence that was first described in the *Mtb* genome (Gordon et al. 1999). A large number of simple sequence repeats (SSRs) or variable number tandem repeat (VNTR) sequences were also identified, which have been successfully used as markers for differentiation and subtyping of *Map* strains (Amonsin et al. 2004; Harris et al. 2006; Thibault et al. 2007). These typing methods will be discussed later in this chapter.

2.2.3
Protein Encoding Genes

The K-10 genome contains 4,350 open reading frames (ORFs) with lengths ranging from 114 bp (a ribosomal subunit encoding gene) to 19,155 bp (a peptide synthetase), accounting for 91.5% of the entire genome. A total of 52.5% of the genes are transcribed with the same polarity as that of DNA replication. The analysis showed that a total of 60% of the putative proteins in *Map* had homologs to other microbial proteins with known functions and 40% were hypothetical proteins. The functional redundancy due to gene duplication that was previously observed in *Mtb* (~52% of genes are functionally redundant) exists to an even greater extent in *Map* (Tekaia et al. 1999). For example, a large number ($n \sim 150$) of genes with transcriptional regulatory functions were identified in the *Map* genome. This number is greater than what is found in *Mtb* ($n \sim 100$) and is consistent with the ability of *Map* to survive in a wide range of environmental conditions (Cole et al. 1998).

2.2.4
Unique Regions and Unique Genes

Based on a comparison of nucleotide sequences between Map and all other bacterial genomes, the analysis identified approximately 161 unique sequence regions in the *Map* genome, the longest region being 15.9 kb in length. Within these unique regions, 39 predicted proteins are unique to *Map*, with no identifiable homologs. More importantly, several follow-up studies show that these unique sequences have considerable potential for the development of more specific and sensitive diagnostic assays for detection of *Map* infection with both molecular and immunoassay based approaches (Paustian et al. 2004).

2.2.5 Mycobactin Synthesis

Mycobactin is a siderophore that is responsible for the binding and transport of iron into cells. Since a major phenotypic difference between *Map* and other mycobacteria is the apparent inability *of Map* to produce mycobactin in laboratory culture, the structure of the *Mbt* operon responsible for mycobactin biosynthesis was examined. A cluster of 10 genes in *Mtb* (*mbtA-J*) has been shown to be responsible for the production of mycobactin and the transport of iron (Quadri et al. 1998). The major difference between *Map*, *Mav*, and *Mtb* was in the *mbtA* gene. Gene *mbtA* is shorter in *Map*: encoding a 400 amino acid (aa) protein, compared with a 565 residue polypeptide in *Mtb* and 551 in *Mav*. As a result of this truncation, *MbtA* has only 330 residues in *Map* that match the N-terminal of the EntE domain and lacks more than 200 residues of EntE C-terminal. Since *MbtA* is thought to initiate mycobactin production, the truncation observed in this key gene suggests that the entire cascade leading to mycobactin production may be attenuated in *Map* (Quadri et al. 1998).

2.2.6 Insights into Virulence and Pathogenicity Gleaned from the *Map* Genome

There is a striking paucity in the K-10 genome in the number of the (Pro-Glu/Pro-Pro-Glu rich proteins (PE/PPE) family of proteins that are thought to play an important role in mycobacterial infection from both an antigenic as well as an immunologic standpoint. While these families of proteins comprise 10% of the *Mtb* genome, there were only six PE homologs and 36 PPE homologs in *Map* (comprising 1% of the genome) compared to 99 and 68, respectively, in *Mtb* (Cole et al. 1998). While the exact significance of this observation is unknown, it may suggest a more limited, less variable and different immune response towards *Map* as compared with *Mtb*.

A virulence related operon that has been identified in mycobacteria encodes the macrophage-colonizing factor or the mammalian cell entry (*mce*) genes (Arruda et al. 1993). These genes have been shown to be important for bacterial survival inside a macrophage (Arruda et al. 1993). The analysis of the complete sequence of the *Map* genome revealed eight homologs to the *mce* operon (as compared to four in *Mtb*). However, the fact that these operons have been identified in both pathogenic and nonpathogenic mycobacteria implies that the mere presence of these genes does not endow *Map* with the ability to cause disease.

2.2.7 Distinguishing Characteristics of the *Map* Genome

As is typically seen amongst other mycobacteria, *Map* has a GC-rich genome (69.3%) while *Mtb* and *Mav* have 65.63% and 68.99% GC content, respectively. The percentage of the genome encoding proteins in *Map* (91.5%) is also the highest among the mycobacteria sequenced so far (Li et al. 2005). The *Map* genome contains 4,350 ORFs with lengths ranging from 114 to 19,155 bp, and a total of 52.5% of the genes are transcribed with the same polarity as that of DNA replication, a fraction that is slightly lower than the 59% observed in *Mtb*. A comparison of proteins across mycobacterial genomes using BLASTP indicates presence of several genomic rearrangements in *Map* when compared to *Mtb*. It should be noted that a majority of these unique genes may well exist in mycobateria not yet sequenced and that they encode hypothetical proteins (Marri et al. 2006). A general analysis showed that a total of 60% of the putative proteins in *Map* had homologs to other microbial proteins with known functions and 25% were homologous to hypothetical proteins (Li et al. 2005). A total of 39 predicted proteins are unique to *Map*, with no identifiable homologs in the current databases. Of the predicted proteins, ~75% had homology to those identified in *Mtb* (Cole et al. 1998).

Map has an increased redundancy in genes involved in lipid metabolism as well as transcriptional regulatory genes in remarkable contrast to *M. leprae*, which seems to have evolved by having a minimal set of genes for most pathways with only half its genome

(49.5%) encoding functional proteins (Cole et al. 2001; Li et al. 2005). *Map* has approximately 1.5% of its genome comprised of repetitive sequences, which include a total of 58 insertion elements.

It is believed that insertion sequences preferentially integrate within intergenic regions so as to avoid the disruption of essential genes (Perret et al. 1997). Consistent with this hypothesis, the majority of the IS elements found in *Map* appear to be clustered within intergenic regions. For example, MAP0028c and MAP0029c, MAP0849c and MAP0850c, and MAP2155, MAP2156, and MAP2157 are clustered within 5 kb of each other in noncoding regions of the chromosome. The analysis also shows that insertion sequences in *Map* are absent from the region flanking 32 kb of either side of *oriC*. A similar observation was made for the *Mtb* genome; however, in the case of *Mtb*, this distance is considerably greater at 600 kb (Gordon et al. 1999). It is thought that there may be detrimental effects to chromosomal replication when insertion sequences are located close to the *oriC*, thereby raising the intriguing possibility that the presence of an insertion sequence, (MAP0028c/IS*1311*), 32 kb from the *oriC* in *Map* may contribute to the increased generation interval of *Map* as compared with *Mtb* and other mycobacteria (Gordon et al. 1999).

2.2.8 Genomics-Based Insights into *Map* Metabolism

Similar to other mycobacteria sequenced thus far, the *Map* genome encodes a complete set of enzymes for many metabolic pathways including glycolysis, the pentose phosphate pathway, the TCA cycle, and the glyoxylate cycle. However, there are genes and putative pathways missing in the *Map* genome that have been described in *Mtb*. For example, *Mtb* encodes genes necessary for urease production (*ureABC* and *ureDFG*). Bacterial ureases catalyze the hydrolysis of urea to ammonia and carbon dioxide. The ability to obtain nitrogen from urea is important in the colonization of varying environments, including a host (Burne and Chen 2000). Because *Map* lacks this specific pathway (ureABC and ureDFG), its ability to acquire nitrogen from urease may differ from that of *Mtb* and other urease-encoding bacteria (Li et al. 2005).

Map has a class I fructose-bisphosphate aldolase (*fba*), which is a metal independent form of the enzyme. Other species of mycobacteria have a zinc-dependent form of *fba* (class II). The fact that class II *fba* is present in other actinobacterial genomes and *Map*'s class I *fba* shares 69% identity with a proteobacterial *fba* gene suggests that *Map* acquired this gene by nonorthologous gene displacement, a process in which a protein evolves to fill the functional void created by a missing component of a pathway or key enzyme. Since class I and II *fba* genes are interchangeable from a functional standpoint, it is likely that there are no physiological difference between *Map* and other species (Koonin and Galperin 2003; Marri et al. 2006). *Map* along with *M. bovis* has two isocitrate lyase homologues (*icl* and *aceA*) while *M. leprae* is totally lacking the *icl* gene and *Mtb* has a nonfunctional copy of the *aceA* gene (Cole et al. 1998). Isocitrate lyase is an essential enzyme of the glyoxylate cycle, which is responsible for the growth of the microorganism on acetate and palmitate and therefore key to survival in the microaerophilic conditions inside the host. The existence of both homologues of the isocitrate lyase genes in *Map* may add to its virulence and survival in the host (Marri et al. 2006).

All the genes involved in aerobic respiration are conserved in *Map*, *Mtb*, and *M. bovis*, which is not the case in *M. leprae* where most of these genes are either lost or reduced to pseudogenes. *Map* has phosphoenol pyruvate carboxylase instead of pyruvate carboxylase as in *Mtb*. The PEP carboxylase may give *Map* an additional route to generate ATP through the oxidation of NADH by converting PEP to fumarate and malate. The genes *nirA* and *cysH* for anaerobic respiration, which encode nitrate and phosphate reductases, respectively, are present in duplicate in the *Map* genome flanked by insertion elements. Flanking of the two genes by insertion elements suggests recent duplication of the genes as mediated by the IS elements. Another gene, *narX*, encoding a nitrate reductase is absent from *Map* along with *frdBCD* gene involved in interconversion of fumarate and succinate (Marri et al. 2006). The fumarate reductase complex (frdABCD) functions as an electron transport chain in bacteria under anaerobic conditions and plays a major role in *Mtb* metabolism under starvation conditions (Betts et al. 2002). The absence of the fumarate reductase complex and *narX* gene in

Map suggests that it has an alternate mechanism of survival under anaerobic conditions.

Functional redundancy based on amino acid homology comparisons is particularly high among genes involved in lipid metabolism and oxidoreduction in *Map*. There are as many as 254 predicted genes functioning as oxidoreductases and oxygenases compared to 171 of *Mtb* (Li et al. 2005). As an example, five genes consisting of three essential genes, *fabG2*, *accD4*, *KasB*, and two non-essential genes, *fabG3* and *fabG5*, are present as duplicates in the fatty acid biosynthesis pathway of *Map*. The presence of multiple copies of these genes suggests a possible mechanism to increase virulence through production of unique complex lipids by the combined action of fatty acid synthetases and polyketide synthetases (Marri et al. 2006). Mycobacteria were originally classified as such by the presence of mycolic acids (Minnikin and Goodfellow 1980; Besra et al. 1994). Not only do mycobacteria produce this type of lipid, but these organisms are also known for their ability to produce and utilize a vast array of other lipophilic molecules (Besra et al. 1994). Importantly, the diverse structures that are located primarily on the cell wall are thought to play a role in pathogenesis in many mycobacterial species by their ability to allow entry into host cells or suppress or evade host immune defense mechanisms (Schorey et al. 1997; Rhoades and Ullrich 2000). Increased survival of mycobacteria may also be enabled, in part, by their ability to preferentially utilize fatty acids instead of carbohydrates for basic metabolic needs (Bloch and Segal 1956; Segal and Bloch 1957; Smith 2003). The analysis shows that there are approximately 80 more genes in *Map* ($n \sim 266$) that are predicted to be involved in lipid metabolism than there are in *Mtb*. Although this difference in number of lipid metabolism and biosynthesis related genes is due primarily to genetic redundancy in *Map*, there are some noteworthy differences. For example, *Map* contains a gene (*MAP3194*) encoding hydroxymethylglutaryl-CoA lyase. This is an enzyme that is found in other bacteria as well as in humans, and catalyzes the last step of ketogenesis and leucine catabolism (Ashmarina et al. 1994). The enzyme may play a role in fatty acid biosynthesis by altering what is produced and distributed to the cell membrane (Ashmarina et al. 1994). This difference between *Map* and *Mtb* indicates that there may be differences in lipid metabolism and biosynthesis that may play a role in what is present/absent on the surface, thus, affecting host immune defense mechanisms.

Genes involved in fatty acid degradation (*fad*) while conserved in *Mtb* show significant variation in *Map*. The cholesterol oxidase gene (*choD*) is truncated in *Map* while the *fadH* and *echA*18 genes are altogether absent. The gene *fadD*11 is present in duplicate in Map. Even if *Map* has lost around 19 genes involved in fatty acid degradation that are present in *Mtb*, it has gained about 35 additional ones as a result of duplication of those present in its genome (Marri et al. 2006).

Overall, the availability of the genome sequence of Map has provided key insights on the structure of the genome, as well as metabolic and virulence potential of the organism.

2.3 Population Studies of *Map*

Isolates of *Map* from different clinical sources have relatively few distinguishing phenotypic characteristics. When grown in culture, the major features that phenotypically differentiate *Map* isolates include the growth rate and variation in pigmentation (Stevenson et al. 2002; Motiwala et al. 2006). The two phenotypes are observed on solid culture media in terms of growth rate: in slow-growers, colonies are visible only after 16 weeks of incubation, while in fast-growers, colonies are visible after 6–12 weeks of incubation. The slow-growing and pigmented strains appear to have a host preference for sheep, while relatively fast-growing and non-pigmented strains appear to infect a very broad range of hosts including cattle and goats although interspecies transmission has been reported (Collins et al. 1990; Whittington et al. 2001; Stevenson et al. 2002). Overall, culture characteristics and biochemical assays do not provide sufficient resolution for strain differentiation among *Map* isolates.

Hence, to gain a better understanding of the population structure of *Map*, to identify the origin of source of an infection during outbreaks, as well as to help identify risk factors that influence the transmission of *Map* within and between herds, there has been an increased interest in the application of molecular strain typing methods for the identification of genetic diversity among *Map* isolates over the past two decades. Here, we briefly introduce the typing methods that have been applied for population studies in *Map*.

2.3.1
Multi-Locus Enzyme Electrophoresis (MLEE)

MLEE had long been a standard method to estimate the genetic diversity and structure in natural populations of variety of species of bacteria, and provided an invaluable population genetic framework for bacterial species and extensive data for systematics and useful marker systems for epidemiology before extensive utility of the DNA based methods (Maiden et al. 1998). MLEE were also used to subtype *M. tuberculosis (Mtb)* and the *M. avium* complex (MAC). Based on the electrophoretic differences in several different metabolic ezymes, MLEE can successfully distinguish *Map* from *M. avium* and *M. intracellulare* (Wasem et al. 1991; Yakrus et al. 1992; Feizabadi et al. 1997); however, this technique has not been useful for *Map* strain typing, primarily due to restricted allelic variation and overall difficulty in obtaining sufficient quantities of bacteria for extraction of enzymes.

2.3.2
DNA-Based Tools

In recent years, many DNA-based techniques have been developed or adapted for strain differentiation. These techniques are either based on insertion sequences or single nucleotide polymorphism (SNP), or a combination of both.

Restriction fragment length polymorphism (RFLP) coupled with hybridization to the insertion element, IS*900*, has been widely used for the differentiation of *Map* isolates from animal and human sources (Collins et al. 1990; Francois et al. 1997; Choy et al. 1998; Pavlik et al. 1999). In this procedure, the restriction of *Map* genomic DNA with the endonucleases *Bst*EII, *Pst*I, or both is followed by agarose gel electrophoresis of the DNA fragments. To enhance strain discrimination, probes targeting IS elements (IS*900* is most extensively used) are usually used for DNA hybridization (Motiwala et al. 2006). Using this approach, a large number of *Map* isolates recovered from cattle, sheep, goat, and human in different countries were characterized. While the technique enables a considerably high level of strain discrimination, two distinct groups of *Map* isolates have been identified, one with a broad host range including cattle, goat, and human, while the second group represent isolates primarily recovered from sheep (Collins et al. 1990; Francois et al. 1997; Choy et al. 1998; Pavlik et al. 1999). Interestingly, the genotypes are associated with the phenotype of the isolates: the cattle-type strains (*C* strains) are relatively easy to culture from tissues and feces of infected animals, and, in contrast, sheep strains (*S* strains) are difficult to culture. However, most reports have been unable to identify any distinct nonoverlapping population of *Map* in domestic livestock species, wild ruminants, or humans. The reported range of the index of discrimination for IS*900*-RFLP in different studies has been 0.509–0.599 with *Bst*EII or *Pst*I alone, and 0.697 with both *Bst*EII and *Pst*I (Motiwala et al. 2006). Thus, overall, RFLP shows a relatively low discriminatory index, and moreover, the technique is time consuming and requires relatively large quantities of DNA, which in turn demands large quantities of bacteria that are not always available in a short period of time and have, therefore, limited the application of this approach in routine epidemiological studies of *Map*.

Pulsed-field gel electrophoresis (PFGE) uses an infrequent cutting enzyme that generates high molecular weight fragments. These fragments are separated in a size-dependent manner by gel electrophoresis in alternating electric fields. This technique has not been used as extensively as RFLP for the genotyping of *Map* isolates because it is more time consuming and labor intensive and requires specialized equipment (Levy-Frebault et al. 1989; Coffin et al. 1992; Stevenson et al. 2002). The results of the few PFGE studies that have been conducted for the differentiation of *Map* isolates have confirmed the finding from the RFLP analyses that *Map* isolates can be broadly grouped into sheep and cattle genotypes (de Juan et al. 2005). In a recent study conducted in Spain, 268 isolates were characterized by PFGE using *Sna*BI and *Spe*I endonucleases. A total of 37 distinct multiplex PFGE profiles were identified, including 32 novel profiles. In this investigation, the index of discrimination *Map* using PFGE was found to be 0.693, which is similar to that achieved with RFLP when using both *Bst*EII and *Pst*I. Interestingly, the authors found that cattle in Spain are infected with cattle type strains, while sheep and goats are mainly infected with sheep type strains (Sevilla et al. 2007). In contrast, goats in US are infected mostly by cattle type strains, and may be a reflection of differences in husbandry practices as relates to rearing goats in the two countries.

Multiplex PCR specific for an IS*900* integration loci (MPIL) is another approach that has been developed for strain differentiation of *Map*. Bull et al. (2000) characterized a total of 14 IS*900* loci that are present in the *Map* genome and exploited the sequence information to develop a multiplex PCR typing method. This method is advantageous in comparison to RFLP and PFGE as it is PCR based and requires very small quantities of DNA. In a study performed on *Map* isolates from US sources using MPIL showed that the majority of the *Map* isolates (201 of 210) clustered in one group. In this group, 78% of cattle isolates fell within the same branch (Motiwala et al. 2003). The discriminatory index for *Map* isolates by using MPIL was only 0.456, suggesting that strain diversity achieved by MPIL was not as robust as that achieved by RFLP and PFGE (Motiwala et al. 2003).

Amplified fragment length polymorphism (AFLP) is also a PCR-based fingerprinting technique, which allows high-resolution genotyping for the rapid screening of genetic diversity among bacterial isolates. The AFLP technique is based on the selective PCR amplification of restriction fragments from a total digest of genomic DNA. The technique involves restriction of the DNA and ligation of oligonucleotide adapters, selective amplification of sets of restriction fragments, and gel analysis of the amplified fragments (Vos et al. 1995). An analysis of 104 *Map* isolates from diverse hosts and geographic regions by AFLP showed that 72% of *Map* isolates fell into two major clusters. Interestingly, the AFLP fingerprints of the human *Map* isolates were unique and did not cluster with either cattle or sheep isolates (Motiwala et al. 2003), which was in contrast to previous reports that suggest close genetic concordance between isolates of Map recovered from animal and human hosts (Pavlik et al. 1995; Whittington et al. 2000). The index of discrimination for *Map* isolates in this investigation when using AFLP was only 0.592, lower than that achieved by RFLP and PFGE. However, another study in which 20 *Map* field isolates were subtyped using AFLP revealed 11 genotypes, suggesting an apparent high degree of polymorphism in this sample (O'Shea et al. 2004). However, the relatively small sample size precludes drawing any firm conclusions regarding the discriminatory power of the AFLP approach in differentiating among isolates of Map. Overall, compared with MPIL, AFLP shows a greater resolving power; however, it is an approach that requires considerable technical skill, and is often plagued with repeatability problems. In addition, the results from these analyses are difficult to compare between laboratories, and thus the approach has not been widely used for *Map* strain differentiation (Bensch and Akesson 2005).

IS*1311* PCR and restriction endonuclease analysis (IS*1311* PCR-REA) has been used for *Map* strain differentiation. Southern blot analysis of RFLP patterns using IS*1311* as a probe indicated that there are 7–10 copies of IS*1311* in *Map*. IS*1311* PCR-REA targeted a point mutation in the IS*1311* sequences and enabled distinction of *Map* isolates from sheep and cattle (Whittington et al. 1998). The restriction pattern that was generated on digestion with *HinfI* was due to the presence of a cytosine or thymidine at position 223 (GenBank accession number U16276) in each copy of IS*1311*. IS*1311* PCR-REA studies have further identified a specific mutation in *Map* isolates recovered from bison (Whittington et al. 2001). In *Map* isolates recovered from sheep, there is C at position 223 in all copies of IS*1311*, while in isolates from bison, there is a substitution of C to T at the same site in all copies of IS*1311*. Interestingly, isolates recovered from cattle have both the C and T containing alleles. For instance, the K-10 isolate has seven copies of IS*1311*, of which three copies have the C and four the T allele. This approach has the advantage of allowing rapid differentiation among cattle, sheep, and bison type isolates. However, this method has overall relatively low discriminatory power, and no variation can be detected between *Map* isolates recovered from cattle and therefore has limited utility for molecular epidemiologic investigations.

The availability of the genome sequence of strain K-10 has enabled the identification of repetitive DNA sequences that are present in the *Map* genome. These repetitive sequences are either widely dispersed across the genome or are contiguous to each other. The contiguous repeats can be categorized on the basis of their location, size of repeat unit, and whether or not all repeats are identical (homogeneous or heterogeneous). These repeats are termed "variable number of tandem repeats" (VNTR) that represent a varying number of repeat units per genetic locus, or "simple sequence repeat" (SSR) that are differentiated by the length of the individual repeat unit (van Belkum et al. 1998). The length of the repeat unit ranges from single nucleotide up to 100. The SSR is usually defined as repeat unit containing six or less nucleotides, which usually consists of simple homopolymeric tracts of single nucleotide (mononucleotide repeat) or multimeric tracts (of homogeneous or heterogeneous repeat) such as di or tri-nucleotide repeats. The VNTR is usually defined as longer tandem repeats. The mechanisms underlying repeat variability are also diverse. The variability of the repeats is believed to be caused by slipped strand mispairing, the genetic instability of polynucleotide tracts, especially poly (G-T) (Torres-Cruz and van der Woude 2003). By a combination of inadequate DNA polymerase action and lack of efficient repair during DNA replication, sequence units can either be inserted or deleted (van Belkum et al. 1998). Homologous recombination is likely to be responsible for longer repeat units above 22 nucleotides (Rocha et al. 1999). It is believed that long DNA repeats play a very important role in the strategies of antigenic variation. The percentage of identity between these repeats ranges from 78 to 90% (Rocha et al. 1999).

Methods based on VNTRs have been demonstrated to be of considerable utility for typing of isolates of *M. tuberculosis*. A total of 41 such loci (each with 46–100 nucleotides per unit) were identified based on genome sequence comparison and named as mycobacterial interspersed repetitive units (MIRUs). Of these, 12 loci vary in tandem repeat numbers (Supply et al. 2000). PCR based assays using these 12 loci have been shown to be rapid and reproducible. This method can be automated for large-scale typing projects using high-throughput automation (Supply et al. 2001). VNTRs were also used as a marker for strain typing of *Salmonella enteriac* subsp. *enterica* serovar Typhimurium (Lindstedt et al. 2003) and *Bacillus anthracis* (Kim et al. 2002). A method based on SSR (2–4 nucleotides per unit) has been used for typing stains of *Yersinia pestis* (Adair et al. 2000).

A recent bioinformatic screen of the *Map* K-10 genome in our laboratory identified 185 mono-, di-, and tri-nucleotide tandem repeat sequences dispersed throughout the *Map* genome, of which 78 were perfect repeats with identity of 100%. We also identified 362 longer tandem repeat sequences with length distribution of 6–74 bp, repeat number ranging from 2 to 16, and mutual homology of 67–100% (Li et al. 2005). A brief overview of the SSR and VNTR based typing methods for *Map* is described below.

2.3.3 Multi-Locus Short Sequence Repeats (MLSSR)

The whole-genome sequence of *M. paratuberculosis* strain K-10 was analyzed for SSRs with Tandem Repeat Finder software (TRF version 3.21) (Benson 1999), and 185 mono-, di-, and trinucleotide repeat sequences were identified with 78 perfect repeats. Primers for regions flanking these 78 repeats were designed with Primer 3 software (Rozen and Skaletsky 2000) to yield an average amplification product of 250–400 bases for each locus. The sequences of each SSR locus were aligned, and the numbers of tandem repeats were identified by use of the MegAlign program (DNASTAR, Inc, Madison, WI). Comparative nucleotide sequencing of the 78 loci of six *Map* isolates from different host species and geographic locations identified a subset of 11 polymorphic short sequence repeats (SSRs), with an average of 3.2 alleles per locus. The ORFs or genes flanking each locus of these 11 loci were also identified. Comparative sequencing of these 11 loci was further used to genotype a collection of 33 *Map* isolates representing different multiplex PCR for IS*900* loci (MPIL) or amplified fragment length polymorphism (AFLP) types. The analysis differentiated the 33 *Map* isolates into 20 distinct MLSSR types, consistent with geographic and epidemiologic correlates and with an index of discrimination of 0.96. MLSSR analysis was also clearly able to distinguish between sheep and cattle isolates of *Map* and reproducibly differentiated strains representing the predominant MPIL genotype and AFLP genotypes of *Map* described previously (Motiwala et al. 2003; Amonsin et al. 2004).

The genotypes of *Map* isolates have been investigated in large-scale by using multilocus of SSR in *Map* isolates from broad range of hosts and different geographic regions throughout the US (Ghadiali et al. 2004; Motiwala et al. 2004; Harris et al. 2006). A study conducted on 68 isolates from different geographic locations using 11 loci revealed that four loci (locus 1,

2, 8, and 9) were associated with the highest Simpson's diversity indices (Harris et al. 2006). Locus 1 and 2 are two loci of mononucleotide G repeats, while locus 8 and 9 are trinucleotide repeats. Locus1 G repeat was used to type 33 isolates recovered from different wild species that were also typed by MPIL and enabled differentiation of the *Map* isolates in the common MPIL clade into seven distinct alleles and had an index of 0.75 (Motiwala et al. 2004). Interestingly, a relationship between allele type of locus 1 and host species was demonstrated in isolates recovered from different wild species (Motiwala et al. 2004). Locus 8 also showed some host specificity, the allele of 3-GGT was only found in sheep type isolates, while alleles of 4- or 5-GGT repeats were most found in cattle type isolates. Some loci also revealed one or two base substitutions in some isolates and the majority of the nucleotide substitutions were found in isolates recovered from sheep.

The stability of these four loci was established in two experiments. In the first experiment, three isolates with distinct genotypes were repeatedly inoculated for a total of 10 passages and in the second experiment, 10 colonies from each of the three isolates were used for SSR analysis. Both studies with these three isolates showed that there were no changes in the SSR genotype as a result of in vitro passages or culture, indicating reliability in the utility of SSR analysis in molecular epidemiologic study of *Map* infection (Harris et al. 2006).

A recent study on 94 isolates recovered from cattle, goats, human, and other animal species was typed by two SSR loci (locus1 and 8) and cluster analysis divided the isolates into three distinct clades. Cattle isolates were classified into nine alleles, six of which formed to clade A and three formed to clade C. Sheep isolates were classified to eight alleles, three of which formed a distinct clade B. Clade B contains only sheep isolates, while clade A and clade C contain isolates from broad range of hosts including human (Ghadiali et al. 2004). Simpson's diversity index for this analysis was 0.78, indicating a strain discrimination capability much higher than that of other markers or methods reported (Table 2) (Pavlik et al. 1999; Bull et al. 2000; Motiwala et al. 2003). In another study, 211 *Map* isolates derived from animals throughout the US were differentiated into 61 genotypes by using four SSR loci (Table 2) (locus 1, 2, 8, and 9), with five genotypes accounted for 35.5% of all isolates analyzed (Harris et al. 2006). These findings are in agreement with those of other recent studies in which several genotypes were demonstrated within some herds in Ohio (Motiwala et al. 2005). However, the dominant genotypes in these two studies were different: in the former study dominant genotypes all have 14 Gs, and in the latter study the dominant genotype has seven Gs in locus 1. Although SSR typing method showed much higher discriminatory power, combined other fingerprinting techniques targeting different categories of polymorphisms may be

Table 2 Methods used for genotyping of *Map* isolates

Method	Target region	*D* value	No. of map isolates	References
IS900-RFLP	IS900 (*Bst*EII and *Pst*I)	0.697	1008	(Pavlik et al. 1999)
MPIL	IS900 insertion loci	0.597	247	(Motiwala et al. 2003)
PFGE	*Sna*BI and *Spe*I	0.693	268	(Sevilla et al. 2007)
AFLP	*Mse*I	0.592	86	(Motiwala et al. 2003)
SSR	All 1 to 11 loci	0.967	33	(Amonsin et al. 2004)
SSR	Locus 1 alone	0.751	93	(Motiwala et al. 2004)
SSR	Locus 1 and 8	0.783	94	(Ghadiali et al. 2004)
SSR	Locus 1 alone	0.720	211	(Harris et al. 2006)
SSR	Locus 2 alone	0.800	211	(Harris et al. 2006)
SSR	Locus 8 alone	0.320	211	(Harris et al. 2006)
SSR	Locus 9 alone	0.330	211	(Harris et al. 2006)
SSR	Locus 1, 2, 8, 9	0.961	211	(Harris et al. 2006)
MIRU-VNTR	All 8 loci	0.751	183	(Thibault et al. 2007)
MLVA	All 5 loci	0.316	49	(Overduin et al. 2004)

D value Simpson's diversity index; *IS900-RFLP* IS900-restriction fragment length polymorphism

necessary in *Map* typing because of high degree of genomic homogeneity of *Map* isolates.

In summary, the results of these studies confirmed the utility of the SSR approach as an easy and rapid method based on PCR and sequence analysis that requires only small amount of DNA sample to perform. Moreover, the results are reproducible and have a practical advantage of inter-laboratory comparisons.

2.3.4 Variable Number of Tandem Repeats (VNTR)

VNTR analysis, also termed as multilocus variable number of tandem repeat (MLVA) analysis, has been used for *Map* differentiation from other subspecies of *M. avium* or differentiation within *Map* isolates (Bull et al. 2003a, b; Overduin et al. 2004; Romano et al. 2005; Thibault et al. 2007). Mycobacterial interspersed repetitive units (MIRU) is the VNTR based typing method. Bull et al. (2003a, b) identified MIRU at 18 conserved loci by comparative sequence analysis of *Map* K-10 and *Maa* 104. Six of these loci were found to differ between *Maa* and *Map* in the number of tandem repeat sequences and four sets of primers were designed for locus-specific PCR. Of these four loci, locus 2 and 3 showed different number of tandem repeat sequences among *Map* isolates, which could segregate *Map* isolates into two major groups: most *Map* isolates grouped into common profiles, while two pigmented ovine *Map* strains distinguished by different number of repeats at these loci. PCR at either locus 1 or 4 distinguished *Map* from other subspecies of *M. avium.* The polymorphism at locus 2 and 3 among *Map* isolates were confirmed by other studies (Overduin et al. 2004; Thibault et al. 2007).

Another study performed by Overduin and colleagues investigated 49 *Map* isolates that were previously typed by RFLP (Overduin et al. 2004). Based on in silico comparison of genome sequences of *Maa* 104 and *Map* K-10, 20 VNTR loci were selected for polymorphism screening in 49 *Map* isolates. Five (VNTR1607, 1605, 1658, 3527, and 3249) out of 20 loci were found polymorphic with two or three different number of tandem repeats. VNTR 1658 is the same locus as the locus 3 described by Bull et al. (2003a, b). The VNTR typing of 49 *Map* isolates yielded six different genotypes and DNA sequence analysis of the VNTR loci resulted in two additional alleles. The comparison discriminatory power of these two techniques showed that diversity index for VNTR was lower than that of RFLP (0.316 and 0.448, respectively, Table 2). However, the VNTR typing was able to subdivide the most predominant RFLP type (R01) into six subtypes, providing a promising molecular subtyping approach to study the diversity of *Map* isolates (Overduin et al. 2004).

Romano et al. (2005) selected six VNTR-MIRU loci including three loci (VNTR1605, 3249, 3527) used in the study performed by Overduin et al. (2004) to screen 26 *Map* and 21 *M. avium* subsp. *homonissuis* (*Mah*) isolates. The analysis with the combination of six VNTR-MIRUs identified 15 different alleles in 21 *Mah* isolates; however, all 26 *Map* isolates showed identical VNTR-MIRU phenotype, although five different IS*900*-RFLP profiles were identified with *Pst*I and *Bst*EII. All these 26 *Map* isolates were recovered from cattle and deer from Buenos Aires, Argentina (Romano et al. 2005). This result also showed that diversity index for VNTR was lower than that of RFLP, which is consistent with the result of the study of Overduin et al. (2004).

Recently, Thibault et al. (2007) studied 183 isolates recovered from bovine, caprine, ovine, cervine, leporine, and human origins from 10 different countries by using MIRU-VNTR typing. The isolates were also typed by IS*900*-RFLP and the predominant type comprising 131 isolates was identified as R01. Eight polymorphic loci were identified from a total 34 loci tested and analysis with the combination of eight polymorphic loci identified 21 patterns with a discrimination index of 0.751 (Table 2) (Thibault et al. 2007). One of the eight loci identified in this study, 292, was described as MIRU-locus 2 by Bull et al. (2003a, b). Similar to the results of a recent study (Overduin et al. 2004), 131 isolates of R01 type identified by IS*900*-RFLP was further subdivided into 15 MIRU-VNTR types, suggesting that VNTR typing in combination of other methods would lead to better detection of diversity among *Map* isolates.

In a recent study conducted in our laboratory, several additional novel polymorphic loci were identified by screening 126 VNTR loci with small number of *Map* isolates. Combined with the polymorphic loci identified by other studies, 19 loci were used for typing 37 *Map* isolates recovered from cattle, sheep, goat, human, and bison. These isolates were previously typed by IS*1311*-REA into cattle, sheep, and bison types. Interestingly, different IS*1311*-REA strain types of isolates showed strong association with their VNTR genotypes (data not published) (Li et al, not published). Figure 2 presents a gel image of one of the VNTR PCR loci, VNTR106 (also identified as MIRU-locu2 and 292 in other studies), in which different types of isolates showed different number of tandem repeats.

Taken together, the results of various studies have confirmed the utility of the VNTR approach as an easy, rapid, and inexpensive method based only on PCR. Although VNTR based assays did not show the high discriminatory power as seen with SSR based approaches, it is too early to make this conclusion since there are only limited number of VNTR loci that have been tested so far, and relatively small numbers of *Map* isolates have been analyzed with this method. One advantage of the VNTR based approach is that this technique requires only a simple PCR assay followed by size fractionation and visualization on agarose gels, and hence is more easily applicable for high-throughput epidemiological investigations at a lower cost.

2.4 Concluding Comments

Over the last decade or so, the complete genome sequences of a half dozen or more mycobacterial species have been deciphered. Along with *Mycobacterium avium* subsp. *paratuberculosis* (Li et al. 2005) these include *M. tuberculosis* H37Rv (Cole et al. 1998), *M. leprae* (Cole et al. 2001), *M. tuberculosis* CDC1551 (Camus et al. 2002), *M. bovis* (Garnier et al. 2003), and *M. avium* ssp. *avium* (http://www.tigr.org). Since mycobacterial species constitute a significant group of pathogens causing a variety of diseases such as tuberculosis, leprosy, and Johne's disease in both humans and animals, a comparative study of their genomes would provide valuable insights into their genome structure, physiology, and mechanism of pathogenesis. Considerable new research activity is focused on using microarray-based approaches for comparative genomics studies, as well as for transcriptional profiling of *Map* grown in vitro and in vivo. The availability of genomics

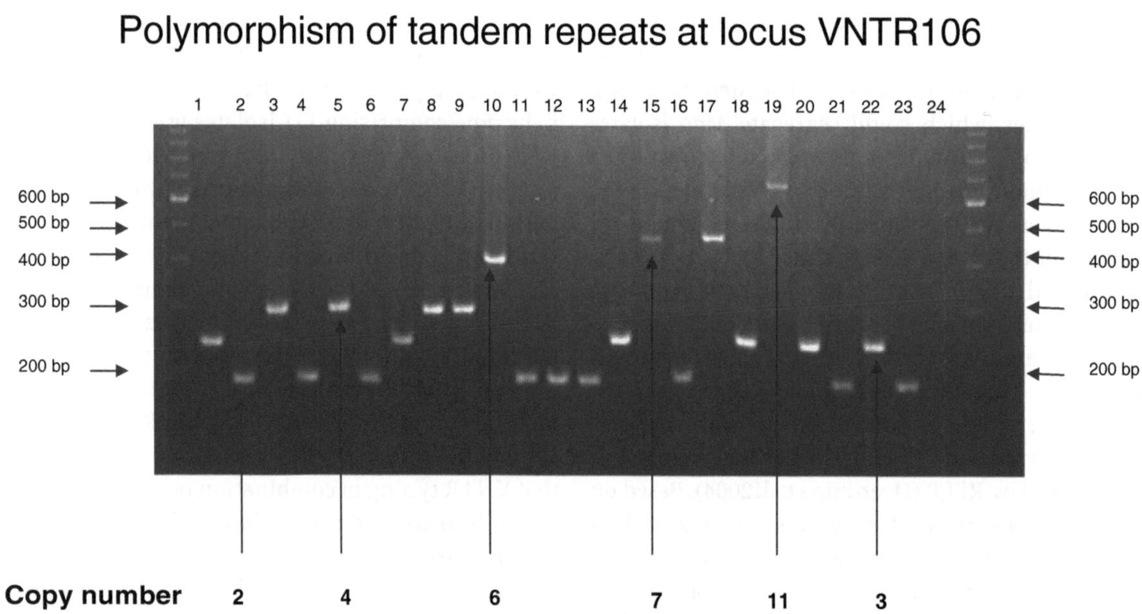

Fig. 2 Amplification products obtained with B, C, and S type isolates for VNTR106. Marker, 100 bp ladder; Lane1, 376_cattle_C; lane2, 333_goat_B; line3, 377_cattle_C; line4, 405_goat_B; line5, 381_cattle_C; line6, 406_goat_B; line7, 708_cattle_C; line8, 727_goat_B; line9, 711_goat_C; line10, JM57_goat_S; line11, B10.3_bison_B; line12, B6.1_bison_B; line13, o20_goat_S; line14, B8.3_bison_C; Line15, 311_goat_S; line16, 7561_bison_B; line17, o18_goat_B; line18, 7288_bison_C; line19, o16_goat_S; Line20, MAP6_human_C; line21, 332_goat_B; line22, 4054_goat_C; line23, 331_goat_B; line24, blank; Marker, 100 bp ladder

and proteomics resources including mutant clone libraries and recombinant proteins through the USDA funded Johne's disease integrated program (www.jdip.org) has enabled comprehensive studies investigating the basic biology of *Map* and the molecular mechanisms with which it causes disease. Finally, the new found information and genomics enabled resources are catalyzing the development of new generations of diagnostic tests and vaccines that will help combat and control the spread of this major animal pathogen.

References

Adair DM, Worsham PL, Hill KK, Klevytska AM, Jackson PJ, Friedlander AM, Keim P (2000) Diversity in a variable-number tandem repeat from *Yersinia pestis*. J Clin Microbiol 38:1516–1519

Aho AD, McNulty AM, Coussens PM (2003) Enhanced expression of interleukin-1alpha and tumor necrosis factor receptor-associated protein 1 in ileal tissues of cattle infected with *Mycobacterium avium* subsp. *paratuberculosis*. Infect Immun 71:6479–6486

Amonsin A, Li L, Zhang Q, Bannantine JP, Motiwala AS, Sreevatsan S, Kapur V (2004) Multilocus short sequence repeat sequencing approach for differentiating among *Mycobacterium avium* subsp. *paratuberculosis* strains. J Clin Microbiol 42:1694–1702

Arruda S, Bomfim G, Knights R, Huima-Byron T, Riley LW (1993) Cloning of an *M. tuberculosis* DNA fragment associated with entry and survival inside cells. Science 261:1454–1457

Ashmarina LI, Rusnak N, Miziorko HM, Mitchell GA (1994) 3-Hydroxy-3-methylglutaryl-CoA lyase is present in mouse and human liver peroxisomes. J Biol Chem 269:31929–31932

Baess I (1983) Deoxyribonucleic acid relationships between different serovars of *Mycobacterium avium*, *Mycobacterium intracellulare* and *Mycobacterium scrofulaceum*. Acta Pathol Microbiol Immunol Scand B 91:201–203

Bensch S, Akesson M (2005) Ten years of AFLP in ecology and evolution: why so few animals? Mol Ecol 14:2899–2914

Benson G (1999) Tandem repeats finder: a program to analyze DNA sequences. Nucl Acids Res 27:573–580

Besra GS, Sievert T, Lee RE, Slayden RA, Brennan PJ, Takayama K (1994) Identification of the apparent carrier in mycolic acid synthesis. Proc Natl Acad Sci USA 91:12735–12739

Betts JC, Betts JC, Lukey PT, Robb LC, McAdam RA, Duncan K (2002) Evaluation of a nutrient starvation model of *Mycobacterium tuberculosis* persistence by gene and protein expression profiling. Mol Microbiol 43:717–731

Bloch H, Segal W (1956) Biochemical differentiation of *Mycobacterium tuberculosis* grown in vivo and in vitro. J Bacteriol 72:132–141

Bull TJ, Hermon-Taylor J, Pavlik I, El-Zaatari F, Tizard M (2000) Characterization of IS*900* loci in *Mycobacterium avium* subsp. *paratuberculosis* and development of multiplex PCR typing. Microbiology 146:2185–2197

Bull TJ, McMinn EJ, Sidi-Boumedine K, Skull A, Durkin D, Neild P, Rhodes G, Pickup R, Hermon-Taylor J (2003a) Detection and verification of *Mycobacterium avium* subsp. *paratuberculosis* in fresh ileocolonic mucosal biopsy specimens from individuals with and without Crohn's disease. J Clin Microbiol 41:2915–2923

Bull TJ, Sidi-Boumedine K, McMinn EJ, Stevenson K, Pickup R, Hermon-Taylor J (2003b) Mycobacterial interspersed repetitive units (MIRU) differentiate *Mycobacterium avium* subspecies *paratuberculosis* from other species of the *Mycobacterium avium* complex. Mol Cell Probes 17:157–164

Burne RA, Chen YY (2000) Bacterial ureases in infectious diseases. Microb Infect 2:533–542

Camus JC, Pryor MJ, Medigue C, Cole ST (2002) Re-annotation of the genome sequence of *Mycobacterium tuberculosis* H37Rv. Microbiology 148:2967–2973

Chacon O, Bermudez LE, Barletta RG (2004) Johne's disease, inflammatory bowel disease, and *Mycobacterium paratuberculosis*. Annu Rev Microbiol 58:329–363

Chiodini RJ, Van Kruiningen HJ (1983) Eastern white-tailed deer as a reservoir of ruminant paratuberculosis. J Am Vet Med Assoc 182:168–169

Chiodini RJ, Van Kruiningen HJ, Merkal RS (1984) Possible role of mycobacteria in inflammatory bowel disease. I. An unclassified *Mycobacterium* species isolated from patients with Crohn's disease. Dig Dis Sci 29:1073–1079

Choy E, Whittington RJ, Marsh I, Marshall J, Campbell MT (1998) A method for purification and characterisation of *Mycobacterium avium* subsp. *paratuberculosis* from the intestinal mucosa of sheep with Johne's disease. Vet Microbiol 64:51–60

Clarke CJ (1997) The pathology and pathogenesis of paratuberculosis in ruminants and other species. J Comp Pathol 116:217–61

Cocito C, Gilot P, Coene M, de Kesel M, Poupart P, Vannuffel P (1994) Paratuberculosis. Clin Microbiol Rev 7:328–345

Coffin JW, Condon C, Compston CA, Potter KN, Lamontagne LR, Shafiq J, Kunimoto DY (1992) Use of restriction fragment length polymorphisms resolved by pulsed-field gel electrophoresis for subspecies identification of mycobacteria in the *Mycobacterium avium* complex and for isolation of DNA probes. J Clin Microbiol 30:1829–1836

Cole ST, Brosch R, Parkhill J, Garnier T, Churcher C, Harris D (1998) Deciphering the biology of *Mycobacterium tuberculosis* from the complete genome sequence. Nature 393:537–544

Cole ST, Eiglmeier K, Parkhill J, James KD, Thomson NR, Wheeler PR, Honore N (2001) Massive gene decay in the leprosy bacillus. Nature 409:1007–1011

Collins DM, Gabric DM, de Lisle GW (1990) Identification of two groups of *Mycobacterium paratuberculosis* strains by restriction endonuclease analysis and DNA hybridization. J Clin Microbiol 28:1591–1606

Coussens PM (2001) *Mycobacterium paratuberculosis* and the bovine immune system. Anim Health Res Rev 2:141–161

de Juan L, Mateos A, Dominguez L, Sharp JM, Stevenson K (2005) Genetic diversity of *Mycobacterium avium* subspecies *paratuberculosis* isolates from goats detected by pulsed-field gel electrophoresis. Vet Microbiol 106:249–257

Feizabadi MM, Robertson ID, Cousins DV, Dawson DJ, Hampson DJ (1997) Use of multilocus enzyme electrophoresis to examine genetic relationships amongst isolates of *Mycobacterium intracellulare* and related species. Microbiology 143(Pt 4):1461–1469

Foley-Thomas EM, Whipple DL, Bermudez LE, Barletta RG (1995) Phage infection, transfection and transformation of *Mycobacterium avium* complex and *Mycobacterium paratuberculosis*. Microbiology 141:1173–1181

Francois B, Krishnamoorthy R, Elion J (1997) Comparative study of *Mycobacterium paratuberculosis* strains isolated from Crohn's disease and Johne's disease using restriction fragment length polymorphism and arbitrarily primed polymerase chain reaction. Epidemiol Infect 118:227–233

Garnier T, Garnier T, Eiglmeier K, Camus JC, Medina N, Mansoor H, Pryor M (2003) The complete genome sequence of Mycobacterium bovis. Proc Natl Acad Sci USA 100:7877–7882

Ghadiali AH, Strother M, Naser SA, Manning EJ, Sreevatsan S (2004) *Mycobacterium avium* subsp. *paratuberculosis* strains isolated from Crohn's disease patients and animal species exhibit similar polymorphic locus patterns. J Clin Microbiol 42:5345–5348

Gitnick G, Collins J, Beaman B, Brooks D, Arthur M, Imaeda T, Palieschesky M (1989) Preliminary report on isolation of mycobacteria from patients with Crohn's disease. Dig Dis Sci 34:925–932

Gordon SV, Heym B, Parkhill J, Barrell B, Cole ST (1999) New insertion sequences and a novel repeated sequence in the genome of Mycobacterium tuberculosis H37Rv. Microbiology 145:881–892

Grant IR, Ball HJ, Rowe MT (2002) Incidence of *Mycobacterium paratuberculosis* in bulk raw and commercially pasteurized cows' milk from approved dairy processing establishments in the United Kingdom. Appl Environ Microbiol 68:2428–2435

Green EP, Tizard ML, Moss MT, Thompson J, Winterbourne DJ, McFadden JJ, Hermon- Taylor J (1989) Sequence and characteristics of IS900, an insertion element identified in a human Crohn's disease isolate of *Mycobacterium paratuberculosis*. Nucl Acids Res 17:9063–9073

Greenstein RJ (2003) Is Crohn's disease caused by a mycobacterium? Comparisons with leprosy, tuberculosis, and Johne's disease. Lancet Infect Dis 3:507–514

Hamilton HL, Follett DM, Siegfried LM, Czuprynski CJ (1989) Intestinal multiplication of *Mycobacterium paratuberculosis* in athymic nude gnotobiotic mice. Infect Immun 57:225–230

Harris NB, Barletta RG (2001) *Mycobacterium avium* subsp. *paratuberculosis* in Veterinary Medicine. Clin Microbiol Rev 14:489–512

Harris NB, Payeur JB, Kapur V, Sreevatsan S (2006) Short-sequence-repeat analysis of *Mycobacterium avium* subsp. *paratuberculosis* and *Mycobacterium avium* subsp. *avium* isolates collected from animals throughout the United States reveals both stability of loci and extensive diversity. J Clin Microbiol 44:2970–2973

Jessup DA, Abbas B, Behymer D (1981) Paratuberculosis in tule elk in California. J Am Vet Med Assoc 179:1252–1254

Kim W, Hong YP, Yoo JH, Lee WB, Choi CS, Chung SI (2002) Genetic relationships of Bacillus anthracis and closely related species based on variable-number tandem repeat analysis and BOX-PCR genomic fingerprinting. FEMS Microbiol Lett 207:21–27

Koonin EV, Galperin MY (2003) Evolution of Central Metabolic Pathways: The Playground of Non-orthologus Gene Displacement. Kluwer, Dordrecht, The Netherlands

Kreeger JM (1991) Ruminant paratuberculosis—a century of progress and frustration. J Vet Diagn Invest 3:373–382

Lambrecht RS, Collins MT (1992) Mycobacterium paratuberculosis. Factors that influence mycobactin dependence. Diagn Microbiol Infect Dis 15:239–246

Lambrecht RS, Carriere JF, Collins MT (1988) A model for analyzing growth kinetics of a slowly growing *Mycobacterium* sp. Appl Environ Microbiol 54:910–916

Levy-Frebault VV, Thorel MF, Varnerot A, Gicquel B (1989) DNA polymorphism in *Mycobacterium paratuberculosis*, "wood pigeon mycobacteria," and related mycobacteria analyzed by field inversion gel electrophoresis. J Clin Microbiol 27:2823–2826

Li L, Bannantine JP, Zhang Q, Amonsin A, May BJ, Alt D, Banerji N, Kanjilal S, Kapur V (2005) The complete genome sequence of *Mycobacterium avium* subspecies *paratuberculosis*. Proc Natl Acad Sci USA 102:12344–12349

Libke KG, Walton AM (1975) Presumptive paratuberculosis in a Virginia white-tailed deer. J Wildl Dis 11:552–553

Lindstedt BA, Heir E, Gjernes E, Kapperud G (2003) DNA fingerprinting of *Salmonella enterica* subsp. enterica serovar typhimurium with emphasis on phage type DT104 based on variable number of tandem repeat loci. J Clin Microbiol 41:1469–1479

Maiden MC, Bygraves JA, Feil E, Morelli G, Russell JE, Urwin R, Zhang Q, Zhou J, Zurth K, Caugant DA, Feavers IM, Achtman M, Spratt BG (1998) Multilocus sequence typing: a portable approach to the identification of clones within populations of pathogenic microorganisms. Proc Natl Acad Sci USA 95:3140–3145

Marri PR, Bannantine JP, Golding GB (2006) Comparative genomics of metabolic pathways in Mycobacterium species: gene duplication, gene decay and lateral gene transfer. FEMS Microbiol Rev 30:906–925

Mijs W, de Haas P, Rossau R, Van der Laan T, Rigouts L, Portaels F, van Soolingen D (2002) Molecular evidence to support a proposal to reserve the designation *Mycobacterium avium* subsp. *avium* for bird-type isolates and '*M. avium* subsp. *hominissuis*' for the human/porcine type of M. avium. Int J Syst Evol Microbiol 52:1505–1518

Minnikin DE, Goodfellow M (1980) Lipid composition in the classification and identification of acid-fast bacteria. Soc Appl Bacteriol Symp Ser 8:189–256

Motiwala AS, Strother M, et al (2003) Molecular epidemiology of *Mycobacterium avium* subsp. *paratuberculosis*: evidence for limited strain diversity, strain sharing, and identification of unique targets for diagnosis. J Clin Microbiol 41:2015–2026

Motiwala AS, Amonsin A, Strother M, Manning EJ, Kapur V, Sreevatsan S (2004) Molecular epidemiology of *Mycobacterium avium* subsp. *paratuberculosis* isolates recovered from wild animal species. J Clin Microbiol 42:1703–12

Motiwala AS, Strother M, Amonsin A, Byrum B, Naser SA, Stabel JR, Shulaw WP, Bannantine JP, Kapur V, Sreevatsan S (2005) Rapid detection and typing of strains of *Mycobacterium avium* subsp. *paratuberculosis* from broth cultures. J Clin Microbiol 43:2111–2117

Motiwala AS, Li L, Kapur V, Sreevatsan S (2006) Current understanding of the genetic diversity of *Mycobacterium avium* subsp. *paratuberculosis*. Microb Infect 8:1406–1418

Movahedzadeh F, Smith DA, Norman RA, Dinadayala P, Murray-Rust J, Russell DG, Kendall SL, Rison SC, McAlister MS, Bancroft GJ, McDonald NQ, Daffe M, Av-Gay Y, Stoker NG (2004) The Mycobacterium tuberculosis ino1 gene is essential for growth and virulence. Mol Microbiol 51:1003–1014

Naser SA, Schwartz D, Shafran I (2000) Isolation of *Mycobacterium avium* subsp *paratuberculosis* from breast milk of Crohn's disease patients. Am J Gastroenterol 95:1094–1095

O'Shea B, Khare S, Bliss K, Klein P, Ficht TA, Adams LG, Rice-Ficht AC (2004) Amplified fragment length polymorphism reveals genomic variability among *Mycobacterium avium* subsp. *paratuberculosis* isolates. J Clin Microbiol 42:3600–3606

Ott SL, Wells SJ, Wagner BA (1999) Herd-level economic losses associated with Johne's disease on US dairy operations. Prev Vet Med 40:179–192

Overduin P, Schouls L, Roholl P, van der Zanden A, Mahmmod N, Herrewegh A, van Soolingen D (2004) Use of multilocus variable-number tandem-repeat analysis for typing *Mycobacterium avium* subsp. *paratuberculosis*. J Clin Microbiol 42:5022–5028

Paustian ML, Amonsin A, Kapur V, Bannantine JP (2004) Characterization of novel coding sequences specific to *Mycobacterium avium* subsp. *paratuberculosis*: implications for diagnosis of Johne's Disease. J Clin Microbiol 42:2675–2681

Pavlik I, Bejckova L, Pavlas M, Rozsypalova Z, Koskova S (1995) Characterization by restriction endonuclease analysis and DNA hybridization using IS900 of bovine, ovine, caprine and human dependent strains of *Mycobacterium paratuberculosis* isolated in various localities. Vet Microbiol 45:311–318

Pavlik I, Horvathova A, Dvorska L, Bartl J, Svastova P, du Maine R, Rychlik I (1999) Standardisation of restriction fragment length polymorphism analysis for *Mycobacterium avium* subspecies *paratuberculosis*. J Microbiol Meth 38:155–167

Perret X, Viprey V, Freiberg C, Broughton WJ (1997) Structure and evolution of NGRRS-1, a complex, repeated element in the genome of Rhizobium sp. strain NGR234. J Bacteriol 179:7488–7496

Quadri LE, Sello J, Keating TA, Weinreb PH, Walsh CT (1998) Identification of a *Mycobacterium tuberculosis* gene cluster encoding the biosynthetic enzymes for assembly of the virulence-conferring siderophore mycobactin. Chem Biol 5:631–645

Rhoades ER, Ullrich HJ (2000) How to establish a lasting relationship with your host: lessons learned from *Mycobacterium* spp. Immunol Cell Biol 78:301–310

Rocha EP, Danchin A, Viari A (1999) Functional and evolutionary roles of long repeats in prokaryotes. Res Microbiol 150:725–733

Romano MI, Amadio A, Bigi F, Klepp L, Etchechoury I, Llana MN, Morsella C, Paolicchi F, Pavlik I, Bartos M, Leao SC, Cataldi A (2005) Further analysis of VNTR and MIRU in the genome of *Mycobacterium avium* complex, and application to molecular epidemiology of isolates from South America. Vet Microbiol 110:221–237

Rozen S, Skaletsky H (2000) Primer3 on the WWW for general users and for biologist programmers. Meth Mol Biol 132:365–386

Saito H, Tomioka H, Sato K, Tasaka H, Tsukamura M, Kuze F, Asano K (1989) Identification and partial characterization of *Mycobacterium avium* and *Mycobacterium intracellulare* by using DNA probes. J Clin Microbiol 27:994–997

Schnappinger D, Ehrt S, Voskuil MI, Liu Y, Mangan JA, Monahan IM, Dolganov G, Efron B, Butcher PD, Nathan C, Schoolnik GK (2003) Transcriptional Adaptation of *Mycobacterium tuberculosis* within macrophages: insights into the phagosomal environment. J Exp Med 198:693–704

Schorey JS, Carroll MC, Brown EJ (1997) A macrophage invasion mechanism of pathogenic mycobacteria. Science 277:1091–1093

Secott TE, Lin TL, Wu CC (2004) *Mycobacterium avium* subsp. *paratuberculosis* fibronectin attachment protein facilitates M-cell targeting and invasion through a fibronectin bridge with host integrins. Infect Immun 72:3724–3732

Segal W, Bloch H (1957) Pathogenic and immunogenic differentiation of *Mycobacterium tuberculosis* grown in vitro and in vivo. Am Rev Tuberc 75:495–500

Sevilla I, Garrido JM, Geijo M, Juste RA (2007) Pulsed-field gel electrophoresis profile homogeneity of *Mycobacterium avium* subsp. *paratuberculosis* isolates from cattle and heterogeneity of those from sheep and goats. BMC Microbiol 7:18

Sigurdardottir OG, Bakke-McKellep AM, Djonne B, Evensen O (2005) *Mycobacterium avium* subsp. *paratuberculosis* enters the small intestinal mucosa of goat kids in areas with and without Peyer's patches as demonstrated with the everted sleeve method. Comp Immunol Microbiol Infect Dis 28:223–230

Sigurethardottir OG, Valheim M, Press CM (2004) Establishment of *Mycobacterium avium* subsp. *paratuberculosis* infection in the intestine of ruminants. Adv Drug Deliv Rev 56:819–834

Smith I (2003) Mycobacterium tuberculosis pathogenesis and molecular determinants of virulence. Clin Microbiol Rev 16:463–496

Stabel JR (1998) Johne's disease: a hidden threat. J Dairy Sci 81:283–288

Stabel JR (2000) Transitions in immune responses to *Mycobacterium paratuberculosis*. Vet Microbiol 77:465–473

Stephan R, Schumacher S, Tasara T, Grant IR (2007) Prevalence of *Mycobacterium avium* subspecies *paratuberculosis* in Swiss raw milk cheeses collected at the retail level. J Dairy Sci 90:3590–3595

Stevenson K, Hughes VM, de Juan L, Inglis NF, Wright F, Sharp JM (2002) Molecular characterization of pigmented and nonpigmented isolates of *Mycobacterium avium* subsp. *paratuberculosis*. J Clin Microbiol 40:1798–1804

Stewart DJ, Vaughan JA, Stiles PL, Noske PJ, Tizard ML, Prowse SJ, Michalski WP, Butler KL, Jones SL (2004) A long-term study in Merino sheep experimentally infected with *Mycobacterium avium* subsp. *paratuberculosis*: clinical disease, faecal culture and immunological studies. Vet Microbiol 104:165–178

Streeter RN, Hoffsis GF, Bech-Nielsen S, Shulaw WP, Rings DM (1995) Isolation of *Mycobacterium paratuberculosis* from colostrum and milk of subclinically infected cows. Am J Vet Res 56:1322–1324

Strommenger B, Stevenson K, Gerlach GF (2001) Isolation and diagnostic potential of ISMav2, a novel insertion sequence-like element from *Mycobacterium avium* subspecies *paratuberculosis*. FEMS Microbiol Lett 196:31–37

Supply P, Mazars E, Lesjean S, Vincent V, Gicquel B, Locht C (2000) Variable human minisatellite-like regions in the Mycobacterium tuberculosis genome. Mol Microbiol 36:762–771

Supply P, Lesjean S, Savine E, Kremer K, van Soolingen D, Locht C (2001) Automated high-throughput genotyping for study of global epidemiology of Mycobacterium tuberculosis based on mycobacterial interspersed repetitive units. J Clin Microbiol 39:3563–3571

Sweeney RW (1996) Transmission of paratuberculosis. Vet Clin North Am Food Anim Pract 12:305–312

Tekaia F, Gordon SV, Garnier T, Brosch R, Barrell BG, Cole ST (1999) Analysis of the proteome of *Mycobacterium tuberculosis* in silico. Tuber Lung Dis 79:329–342

Thibault VC, Grayon M, Boschiroli ML, Hubbans C, Overduin P, Stevenson K, Gutierrez MC, Supply P, Biet F (2007) New variable-number tandem-repeat markers for typing *Mycobacterium avium* subsp. *paratuberculosis* and *M. avium* strains: comparison with IS900 and IS1245 restriction fragment length polymorphism typing. J Clin Microbiol 45:2404–2410

Thorel MF, Krichevsky M, Levy-Frebault VV (1990) Numerical taxonomy of mycobactin-dependent mycobacteria, emended description of *Mycobacterium avium*, and description of *Mycobacterium avium* subsp. *avium* subsp. nov., *Mycobacterium avium* subsp. *paratuberculosis* subsp. *nov.*, and *Mycobacterium avium* subsp. *silvaticum* subsp. *nov.* Int J Syst Bacteriol 40:254–260

Torres-Cruz J, van der Woude MW (2003) Slipped-strand mispairing can function as a phase variation mechanism in *Escherichia coli*. J Bacteriol 185:6990–6994

US Department of Agriculture, A PHIS (1977) Johne's Disease on U.S. Dairy Operation, Fort Collins, Colorado, USA

Valentin-Weigand P, Goethe R (1999) Pathogenesis of *Mycobacterium avium* subspecies *paratuberculosis* infections in ruminants: still more questions than answers. Microbes Infect 1:1121–1127

van Belkum A, Scherer S, van Alphen L, Verbrugh H (1998) Short-sequence DNA repeats in prokaryotic genomes. Microbiol Mol Biol Rev 62:275–293

Vos P, Hogers R, Bleeker M, Reijans M, van de Lee T, Hornes M, Frijters A (1995) AFLP: a new technique for DNA fingerprinting. Nucl Acids Res 23:4407–4414

Wasem CF, McCarthy CM, Murray LW (1991) Multilocus enzyme electrophoresis analysis of the Mycobacterium avium complex and other mycobacteria. J Clin Microbiol 29:264–271

Whitlock RH, Buergelt C (1996) Preclinical and clinical manifestations of paratuberculosis (including pathology). Vet Clin North Am Food Anim Pract 12:345–356

Whittington R, Marsh I, Choy E, Cousins D (1998) Polymorphisms in IS1311, an insertion sequence common to *Mycobacterium avium* and *M. avium* subsp. *paratuberculosis*, can be used to distinguish between and within these species. Mol Cell Probes 12:349–358

Whittington RJ, Hope AF, Marshall DJ, Taragel CA, Marsh I (2000) Molecular epidemiology of *Mycobacterium avium* subsp. *paratuberculosis*: IS900 restriction fragment length polymorphism and IS1311 polymorphism analyses of isolates from animals and a human in Australia. J Clin Microbiol 38:3240–3248

Whittington RJ, Marsh IB, Whitlock RH (2001) Typing of IS 1311 polymorphisms confirms that bison (Bison bison) with paratuberculosis in Montana are infected with a strain of *Mycobacterium avium* subsp. *paratuberculosis* distinct from that occurring in cattle and other domesticated livestock. Mol Cell Probes 15:139–145

Whittington RJ, Marshall DJ, Nicholls PJ, Marshall IB, Reddacliff LA (2004) Survival and dormancy of *Mycobacterium avium* subsp. *paratuberculosis* in the environment. Appl Environ Microbiol 70:2989–3004

Yakrus MA, Reeves MW, Hunter SB (1992) Characterization of isolates of *Mycobacterium avium* serotypes 4 and 8 from patients with AIDS by multilocus enzyme electrophoresis. J Clin Microbiol 30:1474–1478

3 *Anaplasma*

Kelly A. Brayton(✉), Michael J. Dark, and Guy H. Palmer

Programs in Genomics and Vector-borne Diseases, Department of Veterinary Microbiology and Pathology, Washington State University, Pullman, Washington 99164-7040, USA, kbrayton@vetmed.wsu.edu

3.1 Introduction

Anaplasma marginale is the most globally prevalent vector-borne pathogen of livestock, with endemic regions in all the six permanently inhabited continents. Its discovery occurred during what could be called the first "golden era" of pathogen identification – the years spanning the start of the twentieth century. Yet, as genomic studies have revealed, *A. marginale* reflects millions of years of evolution, with adaptation to both its tick vector and to the ungulates that serve as the mammalian reservoir host. This long evolution, resulting in a pathogen that is able to persist for long periods of time in its mammalian host until episodic transmission can occur, represents not only one of the most fascinating models of persistent bacterial infection, but also presents daunting challenges in disease control. However, recent studies, made possible by complete genome sequence data, have both illuminated basic mechanisms and identified heretofore unsuspected points for pathogen control.

3.2 History

A. marginale was first identified as a distinct pathogen by Sir Arnold Theiler in South Africa (Theiler 1910). Cattle imported from England, to improve productivity of both meat and milk, were highly susceptible to both tick infestation and *A. marginale* infection when compared with native cattle. This susceptibility, resulting from both intrinsic genetic differences in the breed and lack of prior exposure, resulted in disease outbreaks with severe morbidity and mortality. Importantly, cattle dying of acute disease, termed anaplasmosis, have high levels of cell-associated bacteremia, microscopically identifiable using any of the several Romanowsky type stains. While there is little doubt that the improvement of microscopy and development of new cellular staining methods in the latter part of the nineteenth century were critical to the discovery of *A. marginale*, perhaps more important was the relatively new concept that arthropods, especially ticks, could serve as vectors of microbial pathogens. The observations of Smith and Kilborne, published in 1893, established that *Boophilus* ticks (now reorganized into the genus *Rhipicephalus*) could transmit the causative agent of Texas Fever (Smith and Kilborne 1893), the protozoan *Babesia bovis*, and created an intellectual framework leading to the discovery of additional vector-borne pathogens. Nonetheless, Theiler was alone in his ability to identify *A. marginale* as a unique disease agent; reanalysis of the illustrations in Smith and Kilborne's original monograph, published more than 20 years before the work of Theiler, clearly shows the presence of *A. marginale* in infected Texas cattle, an observation mistakenly interpreted as one of the developmental stages of *Babesia* (Smith and Kilborne 1893; Dikmans 1933).

Theiler named his unique organism *Anaplasma*, for its apparent lack of a microscopically definable internal structure (Theiler 1910). He described a small cluster of vacuole-bound organisms within a mature erythrocyte that appeared as a single basophilic (due to the bacterial DNA) coccus within the acidophilic erythrocytic cytoplasm (Fig. 1a). Only decades later, with the advent of transmission electron microscopy, it was revealed that each "coccus" was actually a membrane-bound vacuole containing a small colony of 2–8 individual bacteria (Fig. 1b) (de Robertis and Epstein 1951; Ristic 1968). Theiler continued

Genome Mapping and Genomics in Animal-Associated Microbes
V. Nene, C. Kole (Eds.)

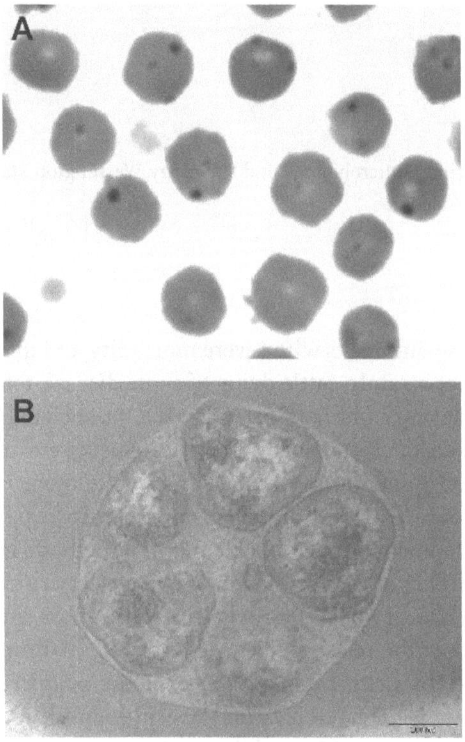

Fig. 1 *Anaplasma marginale* in erythrocytes. **a** Giemsa stained blood smear at ×100 magnification. **b** Electron photomicrograph at ×100,000 magnification of a vacuole containing five *A. marginale* organisms. Image kindly provided by Susan Noh

his investigations to demonstrate that *A. marginale* was not contagious but relied on tick transmission. Perhaps most impressive was his isolation of a less virulent *Anaplasma* in 1911, which he designated *A. marginale* subtype *centrale*, based on a subtle difference in intracellular location within infected erythrocytes, and its prompt deployment as a live vaccine (Theiler 1911, 1912). He demonstrated that deliberately inoculating imported, highly susceptible cattle and allowing them to survive a period of usually mild disease provided protection against subsequent severe morbidity and death after exposure to virulent *A. marginale*. Remarkably, this milder *A. marginale* subspecies *centrale* remains the only consistently effective vaccine in use today. The genetic relationship between *A. marginale* and this subspecies *centrale* will be discussed in Sect. 4 on taxonomy.

The reanalyzed monographs of Smith and Kilbourne clearly show that *A. marginale* was already present in the western hemisphere in 1893 (Dikmans 1933). In both Africa and North America, this detection was associated with the severe disease and high cell-associated bacteremia levels associated with infection of highly susceptible cattle breeds. The microscopic detection of intraerythrocytic *Anaplasma* spp. in a diverse number of wild ruminant species and the definitive identification in several of these hosts is consistent, with this organism having a long history of infection in wild ruminants (Kuttler 1984; de la Fuente et al. 2003b; Scoles et al. 2006). Although a limited number of wild ruminants have been examined in detail, studies in several North American deer and elk species indicate that the host–pathogen interaction is well established only with minimal diseases (Christensen et al. 1958, 1960; Osebold et al. 1959; Kreier and Ristic 1963; Christensen and McNeal 1967; Howarth et al. 1969; Renshaw et al. 1977; Smith et al. 1982; Zaugg 1988; Kuttler and Zaugg 1988; Zaugg et al. 1996). Studies in cattle breeds native to the tropical endemic regions similarly support this relationship (Kuttler et al. 1988; Guglielmone 1995; Bock et al. 1999). Thus, *A. marginale* appears to be an infection of wild ruminants, including those wild cattle later domesticated and bred within the tropical regions, and "emerged" only when highly susceptible cattle were imported into a region of a well established cycle of infection and transmission. However, the details of this natural history are vague; although there is almost no question that *A. marginale* was imported into Australia via infected cattle, the degree to which it was introduced into other regions, such as South America, southern and eastern Asia, and southern Europe, due to cattle movement rather than transmission to cattle from existing endemic foci in wild ruminants is unknown. Recently, de la Fuente and colleagues have been investigating this issue using wild ruminants in southern Europe and with phylogenetic analyses using strains collected worldwide (de la Fuente et al. 2004b, 2005c, 2007). However, at this point, genetic analysis of strains has not been able to resolve clear patterns of strain movement and relationships between infections in domestic and wild ruminants – questions that broader genomic approaches may well shed light upon. While transmission between wild and domestic animals may have a role in the spread of the infection or contribute to *A. marginale* gene flow, it is also clear that *A. marginale* can be maintained and continually transmitted solely among domestic cattle without the need for a wild animal reservoir.

3.3 Impact on Animal Health

A. marginale stands out among the major vector-borne livestock pathogens (*Anaplasma*, *Babesia*, *Ehrlichia*, *Theileria*, and *Trypanosoma*) due to its transmissibility by multiple vectors (Dikmans 1950; Ewing 1981; Minjauw 2001), which results in a much larger global footprint in terms of endemicity – over 80% of the world's cattle population are within the areas of *A. marginale* transmission (FAO 1994). However, *A. marginale* is most prevalent within the tropical regions and thus is frequently coendemic with one or more of these other major pathogens (FAO 1994; Minjauw 2001). This coendemicity makes it difficult to assign economic cost to individual pathogens, as diagnosis is more frequently at the disease level, that is, because anemia and fever due to one or more of these pathogens and methods focused on tick control can simultaneously impact multiple pathogens. Globally, tick-borne diseases, which include all of the aforementioned pathogens with the exception of the tsetse fly-transmitted *Trypanosoma* spp., have an estimated economic cost in the tens of billions of dollars (de Castro 1997; Minjauw 2001; Jongejan and Uilenberg 2004). However, bearing in mind that this includes loss due to morbidity and mortality (especially in regions where definitive diagnosis is rarely sought), costs of preventive measures (principally tick control and in some regions vaccination) and treatments, loss of productivity (weight gain, milk yield, animal traction), reproductive cost due to spontaneous abortion, loss of market (export to disease- free countries), and loss of opportunity to upgrade genetic stock due to higher disease susceptibility (de Castro 1997; Jongejan and Uilenberg 2004), any estimate may well be off by several orders of magnitude. More refined estimates may be possible by examining specific regions/countries where only anaplasmosis and babesiosis are present, including Latin America from northern Argentina to Mexico, and Australia (McCosker 1979; Losos 1986; Brown 1997). Notably, the United States (excluding the Commonwealth of Puerto Rico) is free of bovine babesiosis and thus costs directly attributed to *A. marginale* infection should be calculable. However, there has been no recent analysis on a national scale – the oft quoted US $300 million (McCallon 1973) should be viewed with some skepticism, as this is based on a quite dated study and the cost varies dramatically by region, according to vector tick prevalence, and type of livestock management. A 2006 study by the Canadian Food Inspection Agency estimated the production costs of endemic anaplasmosis, if it were to become established in Canada, at US $10–30 million per year (Lord et al. 2006).

Apart from the total economic costs, tick-borne diseases, including *A. marginale*, have a magnified impact on the economic security and well-being in the poorest countries. Roughly 70% of the world's poor are dependent on livestock, not only as a source of meat and milk, but also as labor for fieldwork and transport (Perry and Sones 2007). Furthermore, herd sizes are frequently small, and as a result, morbidity and mortality in even a few animals can have devastating effects that ripple through poor families, resulting in loss of food security and opportunities for economic and educational progression. Consequently, development of improved control methods, principally thorough vaccination, is a high priority for these diseases – a goal that may be accelerated through genomic studies.

3.4 Taxonomy

The taxonomy of *A. marginale* (Fig. 2) has been a long, strange trip – having been variously classified as a virus and a protozoan parasite. As noted earlier *A. marginale* represented the first rickettsial pathogen to be identified; nonetheless, it was not correctly placed within the Order Rickettsiales, until the seventh edition of Bergey's manual in 1957. The taxonomy was further confused by the introduction of additional genus or species names to describe *A. marginale* strains that appeared to have different morphological features within infected erythrocytes (e.g., *Anaplasma caudatum*, *Paranaplasma caudate*, and *Paranaplasma discoides*) (Kreier and Ristic 1963; Ristic and Krier 1984; Smith et al. 1989), features that are now known to be host proteins associated with the vacuole. Fortunately, the nomenclature of *A. marginale* and related bacteria has been unified and solidified by the recent comprehensive analysis of Dumler (Dumler et al. 2001). This analysis is a leading

- Bacteria (superkingdom)
  - Proteobacteria (phylum)
    - Alpha-proteobacteria (class)
      - Rickettsiales (order)
        - Anaplasmataceae (Family)
          - *Anaplasma* (Genus)
            - *A. marginale* (type species)
              - *A. marginale ss centrale*
            - *A. bovis*
            - *A. ovis*
            - *A. phagocytophilum*
            - *A. platys*

Fig. 2 Current taxonomic classification of *Anaplasma* species

example of how genetic approaches can effectively unify taxonomy and facilitate comparative approaches to better understand pathogen biology. At the genus level within the Family Anaplasmataceae, *Anaplasma* and *Ehrlichia* are genetically most closely related (Fig. 3). This genetic relatedness is reflected in shared biology: bacteria in these two genera infect and persist in animal reservoir hosts and are transmitted intrastadially or transstadially by ixodid ticks (Kocan 1986; Kocan and Bezuidenhout 1987; Prozesky and Du Plessis 1987; Palmer et al. 2000; Paddock and Childs 2003). Within the tick, the development is highly conserved, initial colonization occurs within the midgut epithelium (Fig. 4a), followed by invasion and a second round of replication in the salivary gland acini prior to transmission via saliva during tick feeding (Kocan 1986; Kocan and Bezuidenhout 1987; Prozesky and Du Plessis 1987) (Fig. 4b). As midgut epithelial cells are phagocytic, mechanisms that allowed survival in the tick were likely critical in later adaptation to the mammalian host, where some but not all of the *Anaplasma* and *Ehrlichia* spp. infect and survive in phagocytic cells (Dumler et al. 2001) (Table 1). The separation of these two lineages is estimated to have occurred some 390 million years ago (Ohashi et al. 2001), resulting in unique genetic repertoires and biological mechanisms for survival in the mammalian host.

A. marginale is the type species for the genus *Anaplasma* (Dumler et al. 2001). The other formally recognized species include *A. bovis*, *A. ovis*, *A. platys*, and *A. phagocytophilum* (Fig. 2) (Dumler et al. 2001). Within the mammalian host, these pathogens invade and replicate in mature hematopoietic cells (Table 1). Although previous taxonomic distinction had incorporated the specific mammalian cell type targeted by each pathogen, genomic and biologic studies do not support the use of this criterion, as data is most consistent with adaptation to the mammalian host, and a specific cell type being a relatively late event in the evolution of *Anaplasma* spp. Therefore, it is not surprising that, for example, *A. marginale* and *A. phagocytophuilum* have very similar gene content and are highly syntenic at the genomic level, although the former infects erythrocytes and the latter neutrophils (Lohr et al. 2004; Brayton et al. 2005; Dunning Hotopp et al. 2006). Although *A. phagocytophilum* has mechanisms required for survival in the normally bactericidal environment of the neutrophil (Rikihisa 2003, 2006; Carlyon and Fikrig 2006; Lee and Goodman 2006), many of these may also be present in *A. marginale* due to the common requirement for survival and replication in the phagocytic midgut epithelial cells of the tick midgut. In contrast to *A. marginale* and *A. phagocytophilum* (Brayton et al. 2005; Dunning Hotopp et al. 2006), there is little genetic or genomic data available for *A. bovis*, *A. ovis*, and *A. platys* due to their lesser importance as animal pathogens. However, when available, this data may prove quite informative in better understanding the evolutionary biology within the genus *Anaplasma*.

As described in Sect. 2 on the history of *Anaplasma*, Theiler reported the isolation of a less virulent type of *A. marginale*, which he designated *A. marginale*

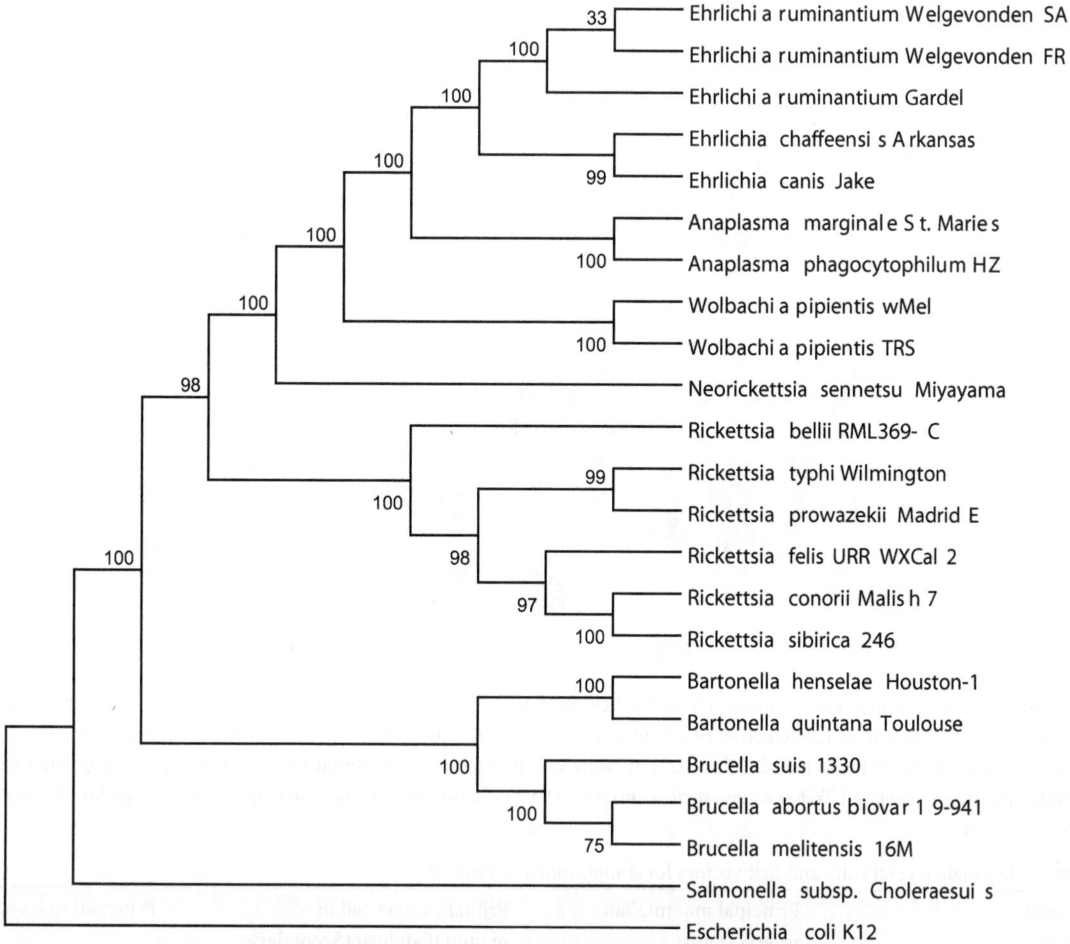

Fig. 3 Maximum parsimony tree of the Rickettsiales. Alignments of concatenated amino acid sequences for groEL, groES, atpA, recA were used to infer the relationships for the species listed using Mega 3.1 (Kumar et al. 2004). The numbers indicate bootstrap percentages based on 1,000 repetitions

subtype *centrale* and developed for use as a vaccine (Theiler 1912). This organism is commonly referred to as *A. centrale* and is often erroneously assumed to have been formally accepted as a separate *Anaplasma* species. However, its relationship to *A. marginale senso stricto* remains unclear and, consequently, the official taxonomic classification defaults to its original designation as a type or subspecies of *A. marginale* (Theiler 1911, 1912; Dumler et al. 2001). Whole genome sequencing of the original *A. marginale* ss *centrale* is currently underway, and comparisons at this level may help to resolve this taxonomic question. A wide distribution of *A. marginale* strains, including those isolated in Africa, will be needed for comparison with *A. marginale* ss *centrale*, originally isolated in South Africa. Very few isolations of *A. marginale* ss *centrale* have been reported; however, two reports find *A. centrale* in Japan (Inokuma et al. 2001; Kawahara et al. 2006). These Japanese isolates should also be compared with the South African isolate to understand the genotypic variation within this (sub) species. These studies, combined with studies examining vector competence and mammalian host range, should begin to clarify whether there is a broad *A. marginale senso lato* that includes different genotypes and phenotypes or whether there is actually speciation within what is now considered a single species, *A. marginale*.

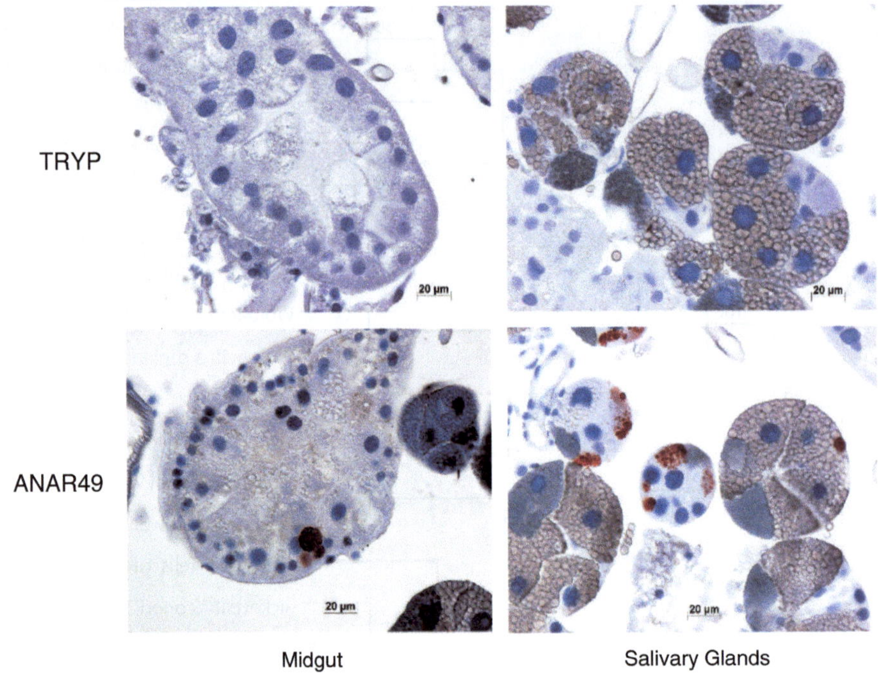

Fig. 4 Immunohistochemistry of *A. marginale* in the tick. Sections of *Dermacentor andersoni* ticks fed on calves infected with the St. Maries strain were used for detection of colonies (*red color*) by immunohistochemistry with *A. marginale*-specific monoclonal antibody ANAR49 (*lower panels*). The sections were counterstained with hematoxylin. An isotype control monoclonal antibody against an unrelated *Trypanosoma brucei* antigen (TRYP) did not elicit a reaction (*top panels*). Image kindly provided by Massaro Ueti

Table 1 Mammalian reservoirs and tick vectors for *Anaplasma* and *Ehrlichia*

Pathogen	Principal mammalian reservoir host	Primary target cell in mammalian host (Secondary sites)	Principal tick vector
Anaplasma marginale	Cattle, wild ruminants	Erythrocytes (endothelial cells)	*Dermacentor, Rhipicephalus*
A. marginale ss centrale	Cattle, wild ruminants	Erythrocytes	*Rhipicephalus(?)*
A. bovis	Cattle	Monocytes	Unknown
A. ovis	Sheep, goats, wild ruminants	Erythrocytes	*Dermacentor, Rhipicephalus*
A. phagocytophilum	Sheep, horses, wild ruminants, mice	Neutrophils (endothelial cells)	*Ixodes*
A. platys	Dogs	Platelets	Unknown
Ehrlichia canis	Dogs, wild canids	Monocytes, macrophages	*Rhipicephalus*
E. chaffeensis	Deer (dogs)	Monocytes, macrophages	*Amblyomma*
E. ewingii	Dog, wild canids	Neutrophils	*Amblyomma*
E. muris	Rodents	Monocytes, macrophages	*Haemaphysalis*
E. ruminantium	Sheep, goats, cattle, wild ruminants	Endothelial cells (neutrophils)	*Amblyomma*

3.5 Strains

3.5.1 Genotype

The *msp1α* gene has been used as a genotypic marker for *A. marginale* strains (Allred et al. 1990). This single copy gene contains a series of repeats of 86–89 bp near the 5′ end that vary both in sequence and in number. This utilitarian marker is stable during infection in a given animal, and is not found in other *Anaplasma* species (Palmer et al. 2001). The repeat types have been designated first alphabetically, then using some greek letters, and finally using numbers (Allred et al. 1990; Palmer et al. 2001; Lew et al. 2002; Shkap et al. 2002; de la Fuente et al. 2003b, 2004b, 2005e, 2007; Mtshali et al. 2007). Table 2 gives an example of a few of the repeat sequences. As of 2007, 79 different repeat sequences had been detected, and this number continues to grow with continued typing studies (de la Fuente et al. 2007).

Little investigation has been done to correlate *msp1α* genotype with genome diversity; however, preliminary studies indicate that, within a herd, strains with the same or similar *msp1α* genotype (i.e., 5B and 6B) are closely related (Rodriguez et al. 2005) and thus that *msp1α* genotype can change over time. The frequency of *msp1α* genotype change is unknown, but it is not a rapid event as detected for *msp2*, with many variants occurring within a single infection. A previous study has shown that *msp1α* genotype was maintained for a period of 1 year in several animals (Palmer et al. 2001). The genotypic changes in the related strains would be accomplished most easily by expansion or reduction of the number of *msp1α* repeat units present in the gene.

3.5.2 Strain Diversity

There is considerable genetic and biologic heterogeneity among strains within *A. marginale* (Carson et al. 1970; Eriks et al. 1994; Brown et al. 1998; Torioni de Echaide et al. 1998; Rurangirwa et al. 2000; Palmer et al. 2001; Futse et al. 2003; Barigye et al. 2004; de la Fuente et al. 2004b, 2005a, b, e, 2007; Lohr et al. 2004; Palmer et al. 2004; Stich et al. 2004; Rodriguez et al. 2005; Ocampo Espinoza et al. 2006; Mtshali et al. 2007). As noted earlier, virulence was one of the differences among strains to be identified first and has been exploited for use as live vaccines. This history notwithstanding, the genetic basis for virulence remains completely unknown. This, in part, reflects that deliberate in vitro or in vivo passage for attenuation has not been consistently effective (Ristic et al. 1968; Welter and Woods 1968; Kuttler 1969, 1972; Kuttler and Zaraza 1969; Lora and Koechlin 1969; Welter and Ristic 1969; Carson et al. 1970, 1977; Osorno et al. 1975; Ristic and Carson 1977; Vizcaino et al. 1978; Corrier et al. 1980; Henry et al. 1983; Kuttler and Zaugg 1988; Pipano 1995; Shkap and Pipano 2000; Bock and de Vos 2001; Melendez et al. 2003) and thus searching for genetic change upon a common strain genetic background has not been possible. Current approaches are using comparison of whole-genome sequences among virulent and less virulent or avirulent strains. Whether these comparisons will identify key regions of differences associated with virulence, as was notably done in genome comparisons between *Mycobacterium tuberculosis* and BCG, remains to be seen (Brosch et al. 2000, 2001, 2002; Pym et al. 2002).

A second focus of investigation among *A. marginale* strains is the variation in tick transmissibility. Although the level of *A. marginale* bacteremia has influence on the efficiency of feeding ticks to acquire infection (Eriks et al. 1993), differences among pathogen strains appear to represent the more significant transmission determinant (Smith et al. 1986; Wickwire et al. 1987; Scoles et al. 2006; Ueti et al. 2007) (Table 3). The St. Maries strain is a prototypically efficiently transmitted strain: essentially 100% of feeding *Dermancetor andersoni* adult male ticks acquire infection followed by efficient replication, initially in the midgut and then in the salivary gland, resulting in levels of $\leq 10^5$ *A. marginale* in the salivary gland at the time of transmission (Eriks et al. 1994; Futse et al. 2003; Scoles et al. 2005). In contrast, several *A. marginale* strains have been isolated and shown to be very inefficient in transmission. Among these, the Florida strain of *A. marginale* has been perhaps the best studied. The Florida strain either does not efficiently invade the midgut epithelium and/or fails to replicate efficiently within the midgut, a phenotype that is conserved in at least four different tick species (Scoles et al. 2007). The genetic basis underlying this pathogen strain phenotype is not

Table 2 *msp1α* repeat sequences

Repeat form	Sequence	Reference
A	DDSSSASGQQQESSVSSQSE-ASTSSQLG	Allred et al. (1990)
B	ADSSSAGGQQQESSVSSQSDQASTSSQLG	Allred et al. (1990)
C	ADSSSAGGQQQESSVSSQSGQASTSSQLG	Allred et al. (1990)
D	ADSSSASGQQQESSVSSQSE-ASTSSQLGG	Allred et al. (1990)
E	ADSSSASGQQQESSVSSQSE-ASTSSQLG	Allred et al. (1990)
F	TDSSSASGQQQESSVSSQSGQASTSSQLG	Palmer et al. (2001)
G	DDSSSASGQQQESSVSSQSGQASTSSQSG	Palmer et al. (2001)
H	TDSSSASGQQQESSVSSQSGQASTSSQSG	Palmer et al. (2001)
I	DDSSSASGQQQESSVSSQSGQASTSSQLG	Palmer et al. (2001)
J	ADSSLAGGQQQESSVSSQSDQASTSSQLG	Palmer et al. (2001)
K	ADGSSAGGQQQESSVSSQSDQASTSSQLG	de la Fuente et al. (2003b)
L	AGSSSADGQQQESSVSSQSDQASTSSQLG	de la Fuente et al. (2003b)
M	ADSSSASGQQQESSVSSQSGQASTSSQLG	de la Fuente et al. (2003b)
N	TDSSSASGQQQESSVSSQSDQASTSSQLG	de la Fuente et al. (2003b)
O	-----SAGGQQQESSVSSQSDQASTSSQLG	de la Fuente et al. (2003b)
P	TDSSSASGQQQESSGSSQSGQASHSAQSG	de la Fuente et al. (2003b)
Q	ADSSSASGQQQESSVSSQSDQASTSSQLG	de la Fuente et al. (2003b)
R	ADSSSAGGQQHESSVSSQSDQASTSSQWG	de la Fuente et al. (2003b)
S	ADGSSAGGQQQESSVSSQSDQASTSSQLG	de la Fuente et al. (2003b)
T	AGSSSAGGQQQESSVSSQSDQASTSSQLG	de la Fuente et al. (2003b)
U	DDSSSASGQQQESSVSSQSDQASTSSQLG	de la Fuente et al. (2003b)
V	ADSSSAGGQQ-ESSVSSQSDQASTSSQLG	de la Fuente et al. (2003b)
W	TDSSSASGQQQESSVSSQSGQASTSSSRG	de la Fuente et al. (2003b)
α	ADSSSASG------VLSQSGQASTSSSLG	de la Fuente et al. (2004a)
ß	TDSSSAGDQQQGSGVSSQSGQASTSSQLG	de la Fuente et al. (2004a)
Γ	TDSSSASGQQQESSVSSQSD-ASTSSQLG	de la Fuente et al. (2004a)
π	ADSSSAGGQQQESSVSSQSGQASTSSQFG	de la Fuente et al. (2004a)
Σ	ADSSSAGGQQQESSVSSQSE-ASTSSQLG	de la Fuente et al. (2004a)
σ	ADSSSAGGQQQESIVSSQSDHASTSSQLG	de la Fuente et al. (2004a)
μ	ADSSSASGLQQESSVSSQSGQASTSSQLG	de la Fuente et al. (2004a)
τ	TDSSSASGQQQESSVLSPSGQASTSSQLG	de la Fuente et al. (2004a)
Φ	TDSSSASGQQQESSVSSQSE-ASTSSQLG	de la Fuente et al. (2004a)
Aus. type 1	ADGSSAGDQQQESSVSSQSG-ASTSSQSG	Lew et al. (2002)
Is. 1	SGSSSASGQQQESSVLSQGGQASTSSQLG	Shkap et al. (2002)
Is. 3	ADSSSASGQQQESSVLSQSGQASTSSQLG	Shkap et al. (2002)
Is. 4	TDSSSASGQQQESSVLSQSGQASTSSQLG	Shkap et al. (2002)
It 5	ADSSSASGQQQESSVSSQSD-ASTSSQLG	de la Fuente et al. (2005b)
It 6	ADSSSASGQQQESSVSSHSE-ASTSSQLG	de la Fuente et al. (2005b)
It 7	TDSSSASGQQQESSVSSHSE-ASTSSQLG	de la Fuente et al. (2005b)

Table 3 Strain-specific transmission phenotypes of *Anaplasma marginale*

Strain	St. Maries	Mississippi	ss centrale
Bacteremia during feeding[1]	1.0×10^8 ml^{-1}	3.2×10^8 ml^{-1}	1.3×10^8 ml^{-1}
Presence in midgut lumen[2]	100%	100%[3]	100%
Colonization of midgut epithelium[2]	100%	2%	54%
Level in midgut epithelium	$10^{6.8\ (\pm 0.45)}$	NQ[4]	$10^{5.4\ (\pm 0.67)}$
Colonization of salivary gland[2]	100%	0%	71%
Level in salivary gland	$10^{7.7\ (\pm 0.45)}$	NQ[3]	$10^{7.4\ (\pm 0.80)}$
Transmission[5]	Positive	Negative	Negative

[1] Bactermia during tick acquisition feeding
[2] Percentage of *A. marginale* positive (number *msp5* PCR positive/total examined)
[3] The level in the midgut lumen was $10^{3.7\ (\pm 0.59)}$
[4] *NQ* Nonquantifiable, below the minimum linear range detectable level of 5×10^1 bacteria
[5] Transmission was verified by sampling for 100 days; all negative animals were negative at all timepoints by *msp5* PCR and MSP5 CI-ELISA

well understood. Ability to bind tick cells in vitro has been associated with differences in the amino acid repeat structure in the N-terminus of major surface protein-1a (MSP1a) and this has been proposed to underlie the failure of the Florida strain to initially invade (de la Fuente et al. 2001). However, recent data show that these MSP1a 28 or 29 aa N-terminal repeats are not required for invasion and replication in the *Dermacentor andersoni* midgut epithelium and that the MSP1a N-terminal repeat structure is not predictive of either midgut invasion or efficient transmission (Ueti et al. 2007). In addition to these initial steps in the midgut, there are also pathogen strain differences at the level of the salivary gland (Ueti et al. 2007). Understanding the mechanisms underlying these differences would allow better prediction of transmission patterns, which can vary greatly both temporally and geographically. The highly transmissible St. Maries strain was the first *A. marginale* genome sequenced (Brayton et al. 2005) (described in detail in Sect. 7), and the poorly tick-transmissible Florida strain was then sequenced to provide a first genome-wide compilation of differences associated with transmission. Although there is a high degree of synteny and conservation of gene content between these two strains, there is a high degree of sequence diversity including 2,021 insertion/deletions (indels) and 7,588 single nucleotide polymorphisms (SNPs). These differences form a database, which can then be compared with the other highly and poorly transmissible strains using either whole genome or targeted approaches. The goal is to reduce the number of candidate genes to a manageable number for genetic knock-out and complementation studies for definitive assignment of pathogen genes required for efficient transmission.

3.6 Population Studies

While a number of studies have confirmed the presence of *A. marginale* for diagnostic purposes using *msp1α* (Lew et al. 2002; Shkap et al. 2002; de la Fuente et al. 2003b, 2004a, 2005d), only three studies have been conducted on a herd scale. The first study looked at *msp1α* genotypes in a herd within an endemic region (eastern Oregon, USA), where *D. andersoni* is the biological vector (Palmer et al. 2001). This study demonstrated that the *msp1α* genotype was stable within individual animals over time, and within the herd there was a family of related *msp1α* genotypes, for example, AFAFIFFH, AFAFIFH, etc. Each infected animal contained a single genotype of *A. marginale* – this and further studies led to the concept of exclusion of infection for this organism (de la Fuente et al. 2002, 2003a).

The second study detected three families of unrelated *msp1α* genotypes in a herd in Kansas, corresponding to the B family (2B, 3B, 4B, 5B, and 6B), the DE family (5D, 2DE, 5DE, 6DE, and 9DE), and EMΦ (which was monomorphic) (Palmer et al. 2004). Interestingly, this study was the first to detect strain superinfection. Within the herd, 7% of individual animals were infected with two distinctly different *msp1α* genotypes. In each case, one of the genotypes present was the EMΦ type, while the second genotype belonged to one of the other two families (B or DE). This study led to the hypothesis that, within a herd, the *msp1α* genotype could serve as a marker for variation of the *msp2* gene, a much more complicated phenomenon discussed in further detail in Sect. 3.7.5.1. In fact, the *msp2* gene repertoire of the EMΦ strain is totally unique when compared with B and DE family strains, and therefore *msp2* is implicated as the factor that governs a strain's ability to superinfect (Rodriguez et al. 2005; Futse et al. 2008). More recent data suggest that superinfection may be common if the genetic distance between strains is sufficient: a molecular and serological study demonstrated that 71% of cattle vaccinated by infection with *A. marginale ss centrale* then exposed in the field to wild type *A. marginale* were positive for both organisms (Molad et al. 2006).

The third study in South Africa sampled 24 commercial and communal farms in the Free State Province (Mtshali et al. 2007). The *msp1α* sequence was amplified from 125 of 215 cattle tested. As this is the first report of African *msp1α* genotypes, it is interesting to note that repeat sequences previously detected in the North American studies were detected in these samples (i.e., F, H, Q), as well as the previously unreported repeat sequences (Palmer et al. 2001; de la Fuente et al. 2003b).

3.7 Genome Sequence and Resources

3.7.1 Genome Architecture and General Features

The St. Maries strain genome, completed in 2005, provided the first complete genome sequence for the genus *Anaplasma* (Brayton et al. 2005). The St. Maries strain was chosen for sequencing, as it represented a virulent strain that was known to be efficiently transmitted by both *Rhipicephalus* (*Boophilus*) *microplus* and *Dermacentor andersoni* ticks (Eriks et al. 1994; Futse et al. 2003). Because of the obligate intracellular nature of the pathogen, a bacterial artificial chromosome (BAC)-based, clone-by-clone sequencing strategy was adopted to obtain the sequence, as this afforded the ability to screen for clones of interest and to remove clones of bovine origin from the project. This sequencing strategy also had the benefit of separating long repeats into separate assembly projects, thus simplifying the assembly of the complete genome sequence. The BAC library (Brayton et al. 2001) contains 50-fold redundancy of clones and serves as a resource for population analysis of gene sequences. The BAC tiling path chosen to generate the complete genome sequence is displayed in Fig. 5.

Table 4 shows the general features of the genome. The completed genome contains 1,197,687 bp, and has a G + C content of 49.8%, somewhat higher than is typical for other obligate intracellular Rickettsiales (31%) (Andersson et al. 1998; Ogata et al. 2001; Wu et al. 2004; Brayton et al. 2005; Collins et al. 2005). The *A. marginale* genome contains an unusually high number of open reading frames (ORFs), for example, there are 2,417 ORFs over 100 bases in length that contain a start codon. In contrast, the *Ehrlichia ruminantium* Welgevonden strain genome has only 1,028 such ORFs (Collins et al. 2005). Many of the *A. marginale* ORFs substantially overlap either on the same or the opposite strand, and do not have homologs in the public databases. This large number of ORFs represents a challenge for annotation, as the most likely coding candidates must be identified from this large

Table 4 General features of the *A. marginale* genome

Genome size	1,197,687 bp
G + C	49%
Protein coding (%)	86%
Protein coding genes	949
Functional assignment	567
Conserved family assignment	107
Conserved hypothetical	126
Hypothetical	151
Functional pseudogenes	16
Split domain ORFS	8
Gene density	0.79
Mean gene length	1,077
Ribosomal RNAs	3
Transfer RNAs	37

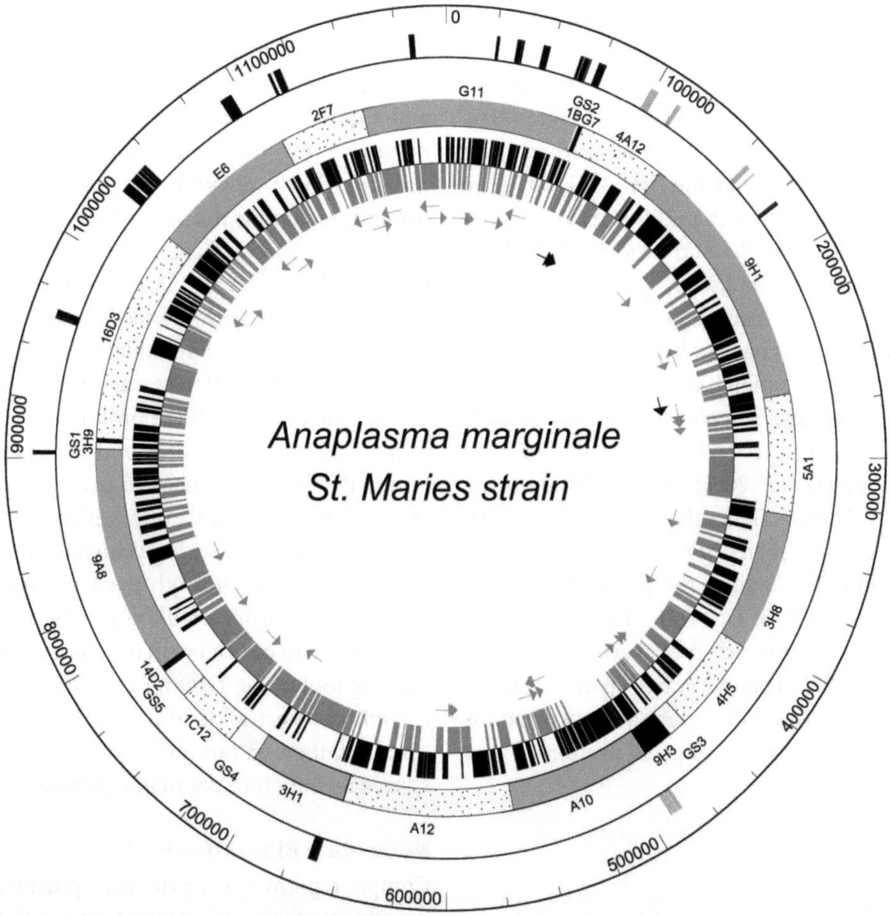

Fig. 5 Genome map of *A. marginale* (St Maries strain). The innermost and second circles show the position and orientation of rRNA *(black arrows)* and tRNA *(gray arrows)* genes. The third and fourth circles show the positions of the predicted CDSs in the reverse *(gray)* and forward *(black)* orientations. The fifth circle shows the positions of the BACs (full BACs are *gray* or *stippled*; partial BACs are *black*) and gap-spanning PCR fragments *(white)* that were sequenced. The sixth circle shows the positions of the msp2 *(black)* and msp1 *(gray)* superfamily genes. The outermost circle gives the scale in bases

selection. The reported annotation contains 1,005 genes, including 949 predicted coding sequences (CDSs), 37 tRNA genes, three rRNA genes, and 16 functional pseudogenes (Fig. 5). The 949 CDSs include four genes for which the introduction of a stop codon has resulted in two separate open reading frames. These four genes, *murC*, *aspS*, *mutL*, and *aatA*, may reflect the reductive evolution process described for the Rickettsiales (Andersson et al. 1998), with the recent acquisition of a single base change that results in premature truncation of the gene. The 37 tRNA genes are representative of all 20 amino acids. The three rRNA genes are organized as a single 16S gene separated by ~163 kb from a ribosomal operon containing the 23S and 5S genes. This split operon structure and the presence of a single set of rRNA genes seems to be typical for the Rickettsiales (Rurangirwa et al. 2002).

In *Escherichia coli*, 89% of the genes are predicted to start with ATG, while only 9% and 2% are predicted to start with GTG or TTG (Blattner et al. 1997). The alternative start codons, as well as the ATG start codon, are not recognized equally, with GTG start codon initiation efficiency about half that of the ATG codon and TTG start codon initiation efficiency half again that of the GTG start codon (Lewin 1990). The *A. marginale* genome has a high proportion of alternative start codons used in the predicted genes. Just over half (493) of the 949 CDSs are predicted to start with ATG codons, while 270 and 186 CDSs start with GTG and TTG codons, respectively. Often there is redundancy in the start codons, with an alternative start codon being positioned closely to an ATG start codon. For example, the gene encoding MSP1a has always been annotated as beginning with the first ATG, but gene prediction algorithms predict a GTG start codon just upstream to be used as the start codon (GTG TGT GTT ATG...), and the multicopy gene, *orfX*, starts with GTG GTG ATG.... The usage of these predicted alternative start codons has not been investigated.

The origin of replication has not been determined experimentally for any *Anaplasma* nor any closely related species (Wu et al. 2004; Brayton et al. 2005; Collins et al. 2005; Foster et al. 2005; Dunning Hotopp et al. 2006; Frutos et al. 2006). Genes that are often clustered near the origin (*dnaA*, *gyrA*, *gyrB*, *rpmH*, *dnaN*) were found distributed throughout the genome; however, there was a clear shift in GC skew (Fig. 6) that corresponded with a shift in octomer skew (Lobry 1996; Salzberg et al. 1998). Therefore, base pair 1 was set arbitrarily in one of the shift regions and corresponds to the start of AM1359, a hypothetical protein.

3.7.2 Metabolism

Central Metabolic Pathways

Metabolic pathways were reconstructed from 283 proteins annotated with EC numbers (Fig. 7). Similar metabolic capacities are present amongst the Anaplasmataceae (Wu et al. 2004; Brayton et al. 2005; Collins et al. 2005; Foster et al. 2005; Dunning Hotopp et al. 2006; Frutos et al. 2006). Glycolysis does not appear to be functional, as a sugar transport system and a glucokinase are not present, and the enzymes necessary for the Entner–Doudoroff pathway are missing. Most enzymes for gluconeogenesis are present, including the steps from pyruvate through to fructose-6-phosphate. All enzymes necessary for the tricarboxylic acid (TCA) cycle are present. The gluconeogenesis pathway feeds into the aminosugar metabolism and the nonoxidative pentose phosphate pathway for which all necessary enzymes are present. The organism may utilize proline as a primary carbon source as several proline transporters were found. Proline can be used to make glutamate, which can feed into the TCA cycle, which can then lead to gluconeogenesis, peptidoglycan biosynthesis, or fatty acid biosynthesis through pyruvate and acetyl-CoA (Fig. 7). *A. phagocytophilum* p44/MSP2 has recently been shown to have a porin function capable of transporting glutamine (Huang et al. 2007). If MSP2, the *A. marginale* ortholog of p44/MSP2, serves this same function, then glutamine can also serve as a carbon source feeding into the major pathways.

Nucleotide Biosynthesis

Complete pathways for de novo purine and pyrimidine biosynthesis are present, as are found in other Anaplasmataceae (Wu et al. 2004; Collins et al. 2005; Foster et al. 2005; Dunning Hotopp et al. 2006; Frutos et al. 2006). This is in contrast to the sequenced Rickettsiaceae (Andersson et al. 1998; Ogata et al. 2001, 2005, 2006; McLeod et al. 2004), which do not

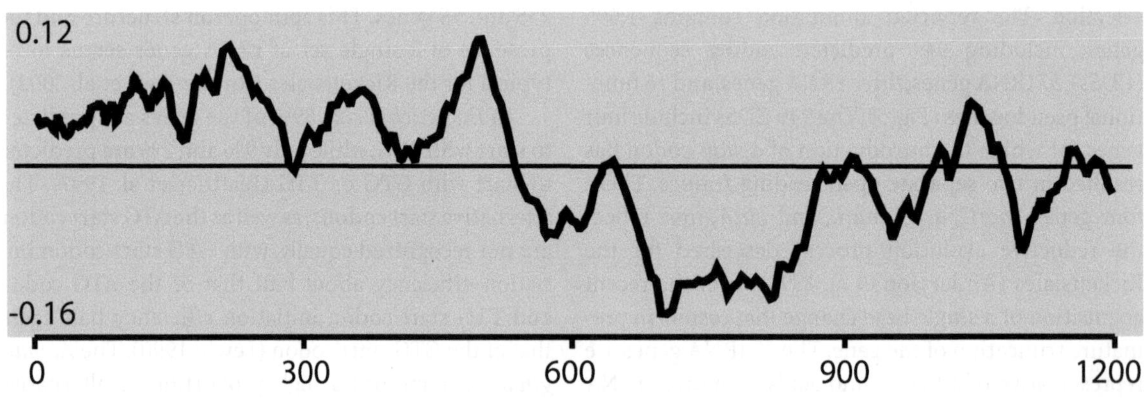

Fig. 6 GC skew of *A. marginale*. The GC skew is calculated as (G − C/G + C). Skew plot was generated using Artemis and a 50,000 bp window

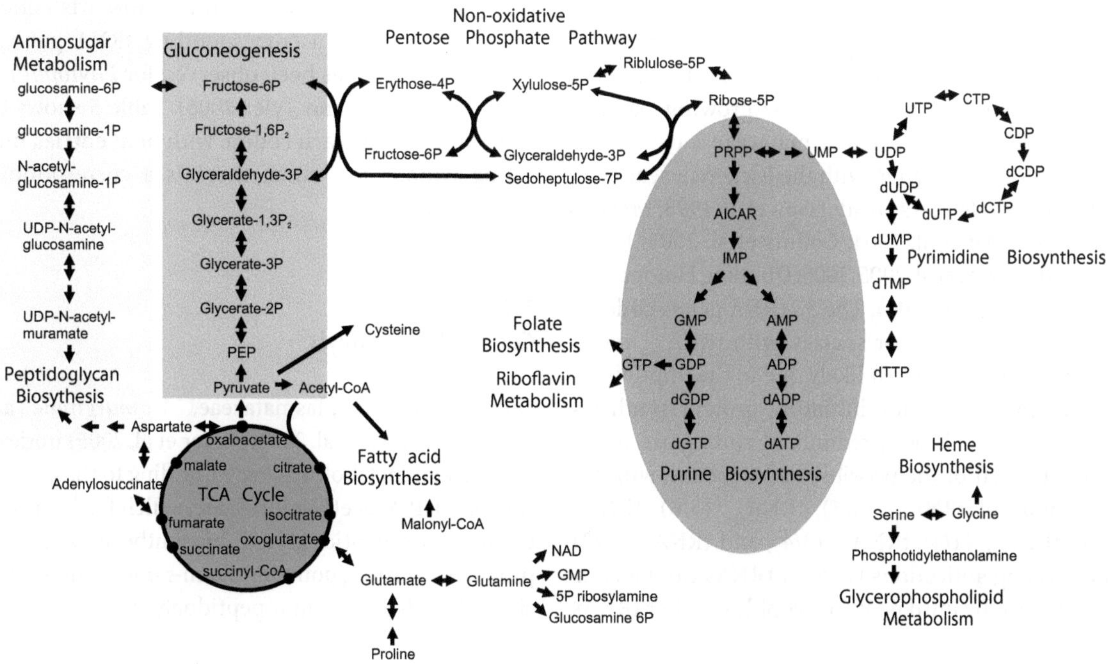

Fig. 7 Major metabolic pathways for *A. marginale*. Pathways depicting aminosugar metabolism, gluconeogenesis, the TCA cycle, the nonoxidative pentose phosphate pathway, purine, and pyrimidine biosynthesis are shown. Pathways for peptidoglycan biosynthesis, fatty acid biosynthesis, folate biosynthesis, riboflavin metabolism, heme biosynthesis, and glycerophospholipid metabolism are indicated as being present

synthesize nucleosides but can convert nucleoside monophosphates into all the necessary nucleotides. It has been suggested (Collins et al. 2005) that the Rickettsiaceae, which grow free in the cytoplasm of the host (Dumler et al. 2001), have easy access to host nucleoside monophosphates and have dispensed with energy intensive de novo synthesis of nucleotides. In contrast, the Anaplasmataceae replicate within an intracellular vacuole (Dumler et al. 2001) and have retained their ability to synthesize nucleotides.

Amino Acid Biosynthesis and Translational Machinery

Rickettsiales typically have a very limited capacity to synthesize amino acids, and *A. marginale* is no exception. There are no complete pathways for amino acid biosynthesis, and enzymes for the terminal biosynthetic step were detected for only nine amino acids, including serine, glycine, tyrosine, cysteine, phenylalanine, proline, aspartate, glutamine, and glutamate. Presumably, many of the amino acids are imported; however, only a transporter for proline was definitively identified, and a putative assignment was made for an alanine transporter. The importation of proline leads to the production of glutamate, glutamine, aspartate, and cysteine.

Most tRNA synthetases are present, with the exception of those for asparagine and glutamine, the two most commonly missing aminoacyl tRNA sythetases from bacteria and archaea (Tumbula et al. 2000). These two aminoacyl tRNAs can be made through transamidations of mischarged Glu-$tRNA_{Gln}$ and Asp-$tRNA_{Asn}$. A heterotrimeric amidotransferase encoded by the *gatA*, *gatB*, and *gatC* genes has been shown to catalyze both these conversions in *Chlamydia trachomatis* (Raczniak et al. 2001). Two of the genes (*gatA* and *gatB*) had been reported in the *A. marginale* genome when originally annotated (Brayton et al. 2005); however, with additional sequences being deposited in the databases, it appears that AM1030, originally annotated as a hypothetical protein, could be *gatC*. These genes are not maintained in the operon structure found in *C. trachomatis* (Stephens et al. 1998). Transamidations using this heterotrimer would provide the complete set of aminoacyl tRNAs for *A. marginale*.

The genome contains 37 tRNA genes, as identified by tRNAscan-SE (Lowe and Eddy 1997), representing a full complement of amino acids. This number of tRNA genes is typical of slow growing bacteria (Rocha 2004). Other Anaplasmataceae have similar numbers of tRNA genes, with the Rickettsiaseae typically having a few less (Andersson et al. 1998; McLeod et al. 2004; Wu et al. 2004; Collins et al. 2005; Foster et al. 2005; Ogata et al. 2005, 2006; Dunning Hotopp et al. 2006; Frutos et al. 2006). The 37 tRNA genes code for 35 different isoacceptor species, with tRNA$_{Met}$ represented three times, most likely due to the importance of this amino acid for initiating protein synthesis. Most amino acids with redundant coding are represented by half of the possible codon boxes, with the exceptions of tRNA$_{Arg}$ (4/6), tRNA$_{Leu}$ (5/6), tRNA$_{Ile}$ (1/3), tRNA$_{Ala}$ (1/4), tRNA$_{Gly}$ (3/4), and tRNA$_{Lys}$ (2/2). The missing anticodons types of tRNAs can be compensated for by third position wobble of other tRNAs. There appears to be no codon bias towards codons for which there is a corresponding tRNA gene, in contrast to what has been observed for *Phytophthora* species (Tripathy and Tyler 2006). Table 5 shows the percent usage for each codon, with bold entries indicating codons for which there is a corresponding tRNA gene.

3.7.3 Cell Wall Biogenesis

Unlike most Anaplasmataceae, *A. marginale* and *W. pipientis* (Wu et al. 2004; Foster et al. 2005) undergo aminosugar metabolism (Fig. 7), leading to the production of UDP-*N*-acetyl-muramate, which feeds into the pathway for peptidoglycan biosynthesis. Aspartate provides a starting point for diami-nopimelate synthesis, which also feeds into peptidoglycan biosynthesis.

Table 5 Codon usage

Amino acid	Codon and percent usage											
Arg	AGG*	30.2	AGA	19.8	CGC	19.6	CGT	13.2	CGG	11.9	CGA	5.3
Leu	CTG	25.8	TTG	22.2	CTT	17.1	CTC	12.9	CTA	12.8	TTA	9.2
Ser	AGC	24.2	TCT	20.6	TCC	16.4	AGT	15.6	TCA	12.6	TCG	10.6
Ala	GCA	28.1	GCG	26.1	GCT	23.4	GCC	22.4				
Gly	GGC	29.0	GGG	27.3	GGT	26.3	GGA	17.4				
Pro	CCG	25.9	CCC	25.7	CCA	24.4	CCT	24.0				
Thr	ACC	29.1	ACT	25.6	ACA	25.3	ACG	20.0				
Val	GTG	35.5	GTT	27.8	GTA	22.0	GTC	14.7				
Ile	ATA	44.2	ATT	32.5	ATC	23.3						
Asn	AAC	52.7	AAT	47.3								
Asp	GAT	56.9	GAC	43.1								
Cys	TGC	57.0	TGT	43.0								
Gln	CAG	57.9	CAA	42.1								
Glu	GAG	54.2	GAA	45.8								
His	CAC	57.2	CAT	42.8								
Lys	AAG	55.2	AAA	44.8								
Phe	TTT	62.2	TTC	37.8								
Tyr	TAC	53.0	TAT	47.0								
Met	ATG	100										
Trp	TGG	100										

* Underline indicates codons for which there is a corresponding tRNA gene

This diaminopimelate producing pathway is also used for lysine biosynthesis in many organisms; however, the necessary enzyme (lysA) for the production of lysine is missing. Most of the enzymes necessary for peptidoglycan synthesis are present, as in *R. prowazekii* (Andersson et al. 1998). *A. marginale* and *W. pipientis* appear to be exceptions among the Anaplasmataceae in being able to produce a peptidoglycan layer, although the lack of the enzymes for lipid A precludes the ability to synthesize LPS. Several Anaplasmataceae (but not *A. marginale*) have been observed to be fragile (Ristic and Krier 1984; Lin and Rikihisa 2003; Collins et al. 2005), and it is thought that the lack of peptidoglycan and LPS are accountable for this. Dunning Hotopp et al. (2006) suggest that organisms infecting the immune cells of a vertebrate host (such as *A. phagocytophilm* and *E. chaffeensis*) would have a selective advantage through the loss of peptidoglycan, as this binds to Toll-like receptor 2 and activates leukocytes.

3.7.4 Transporters

Cytoplasmic Membrane Transporters

A. marginale is typical of obligate intracellular bacteria that have been shown to possess a limited transporter repertoire in spite of the massive metabolite flux between the organism and its host (Ren and Paulsen 2005). Fifty-six proteins are predicted to comprise 32 transporters (Table 6) using TransportDB, a relational database for predicting the cytoplasmic membrane transport complement (Ren et al. 2007). The transporter profile is quite similar to that of the closely related Anaplasmataceae, with *W. pipientis* having a few more transporters (Table 6). There are 11 ATP-dependent transporters. Since *A. marginale* is predicted not to synthesize lipid A, it is surprising to find a putative exporter for lipid A (*msbA2*). Also included in the ATP-dependent category is the signal recognition particle protein. The genome encodes 20 secondary transporters, with an expansion in the Major Facilitator Superfamily (MFS), including four proline/betaine transporters. The *sec* pathway for the secretion of polypeptides appears to be intact; however, the *tat* transport system contains only *tatC*, with *tatA*, *B*, and *E* being missing. There is a single "unclassified" transporter for Mg^{2+}.

The Type IV Secretion System

In addition to the cytoplasmic membrane transporters, a type IV secretion system (T4SS) has been identified. The best studied T4SS is found in *Agrobacterium tumefaciens*, which contains 11 *vir* genes in a single operon, which assemble into a conjugal transfer system that exports DNA from the Ti plasmid into the host cell, resulting in tumorigenesis (Li et al. 1998, 1999). The archetypical T4SS is shown in Fig. 8. With the generation of genome sequences for hundreds of bacteria, the identification of a T4SS has become more commonplace. The T4SS has a variety of functions, but for pathogenic bacteria, it is thought to be the means for injecting virulence factors into the host cell (Backert and Meyer 2006). All Rickettsiales sequenced to date have a T4SS, although they

Table 6 Transporters

	AM St.M	AP HZ	EC Ark	ER W1	WP wMel
Genome size(Mb)	1.2	1.5	1.2	1.52	1.5
Total transporter proteins	32	31	28	30	40
No. of transporters per Mb genome	26.67	20.67	23.33	19.74	26.67
ATP-Dependent	11 (34%)	9 (29%)	9 (32%)	9 (30%)	12 (30%)
The ATP-binding Cassette (ABC) Superfamily	8	8	8	8	11
The H+- or Na+-translocating F-type, V-type, and A-type ATPase (F-ATPase) Superfamily	1	1	1	1	1
The type II (General) secretory pathway (IISP) family	2	0	0	0	0

(Continued)

Table 6 (Continued)

	AM St.M	AP HZ	EC Ark	ER W1	WP wMel
Secondary transporter	20 (63%)	21 (68%)	18 (64%)	20 (67%)	27 (68%)
The auxin efflux carrier (AEC) family	1	1	1	1	1
The alanine or glycine:Cation symporter (AGCS) family	1	1	2	2	3
The amino acid–polyamine–organocation (APC) Family	0	0	0	0	1
The cation diffusion facilitator (CDF) Family	1	1	1	1	1
The monovalent cation:Proton antiporter-2 (CPA2) family	1	1	1	1	1
The monovalent cation (K+ or Na+): Proton antiporter-3 (CPA3) Family	1	1	1	1	1
The dicarboxylate/amino acid:Cation (Na+ or H+) Symporter (DAACS) family	2	3	1	1	2
The drug/metabolite transporter (DMT) superfamily	0	0	0	0	3
The major facilitator superfamily (MFS)	6	5	5	7	8
The multidrug/oligosaccharidyl-lipid/ polysaccharide (MOP) flippase superfamily	1	1	0	0	0
The NhaC Na+:H+ antiporter (NhaC) family	0	0	0	0	2
The cytochrome oxidase biogenesis (Oxa1) family	1	2	1	1	1
The resistance-nodulation-cell division (RND) superfamily	2	2	2	2	2
The twin arginine targeting (Tat) family	1	1	1	1	1
The tripartite ATP-independent periplasmic transporter (TRAP-T) family	1	1	1	1	0
The K+ transporter (Trk) family	1	1	1	1	0
Unclassified	1 (3%)	1 (3%)	1 (4%)	1 (3%)	1 (3%)
The Mg2+ transporter-E (MgtE) family	1	1	1	1	1

have fewer *vir* genes than *A. tumefaciens* and are usually arranged as two operons (Andersson et al. 1998; McLeod et al. 2004; Wu et al. 2004; Brayton et al. 2005; Collins et al. 2005; Foster et al. 2005; Ogata et al. 2005, 2006; Dunning Hotopp et al. 2006; Frutos et al. 2006). One operon contains *virB3*, *virB4*, and *virB6*, followed by three or four *virB6*-like proteins (*R. bellii virB3* is separated from the rest of the locus by ~8 kb), while the second operon contains *virB8*, *virB9*, *virB10*, *virB11*, and *virD4*. All Rickettsiales sequenced to date contain a putative *virB2* and *virB3*, although these are not annotated in all cases, and appear to be missing the genes for *virB1*, *virB5*, and *virB7* that are found in *A. tumefaciens*. In *A. marginale*, there are additional T4SS copies of *virB8* and *virB4* that occur singly at separate loci in the genome.

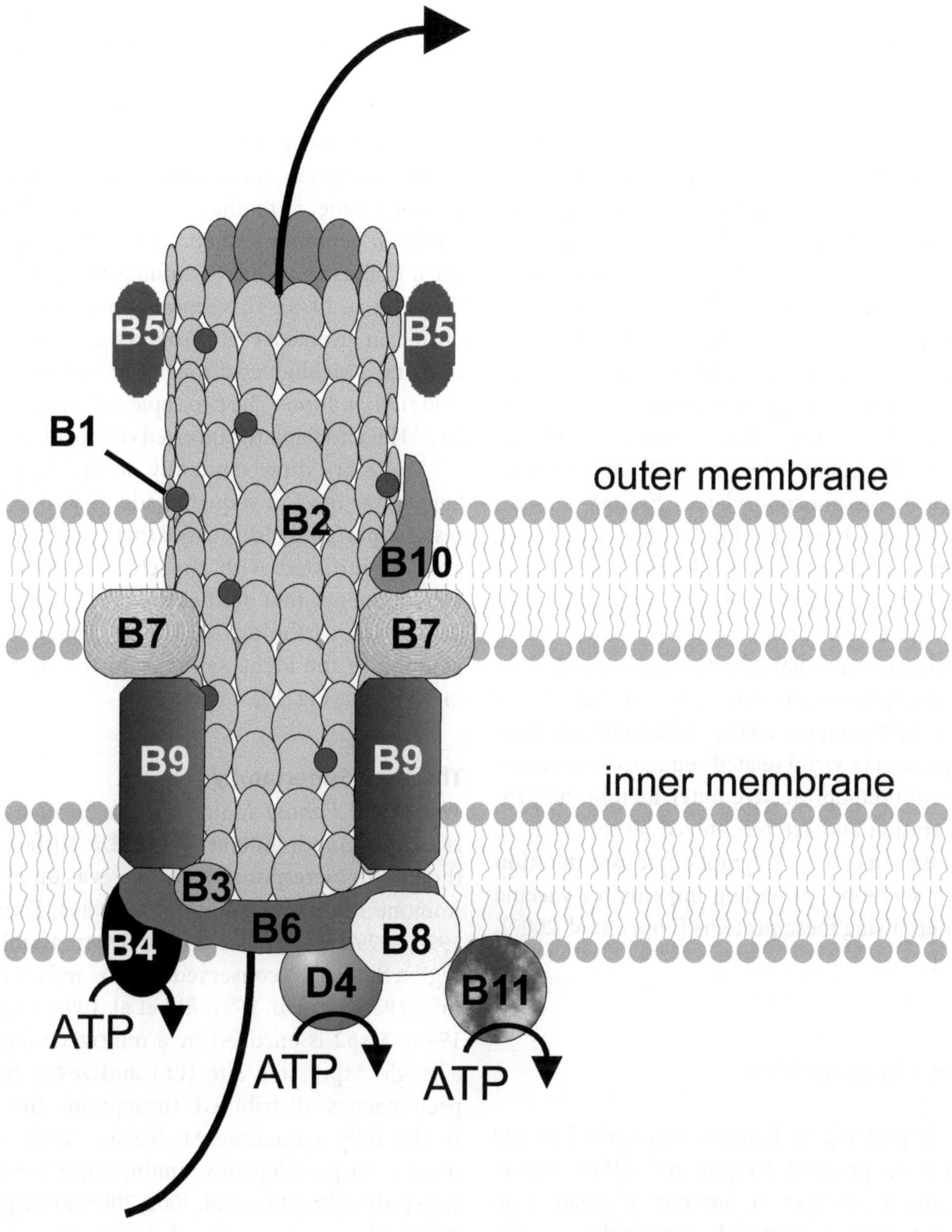

Fig. 8 Diagram of a generic Type Four Secretory System. The components found in the *Agrobacterium tumefaciens* T4SS are shown in their relative position in the membrane. The arrow indicates the direction of transport. Only *virB2, 3, 4, 6, 8-11*, and *virD4* have been identified in *A. marginale*. Adapted from KEGG (Ogata et al. 1999)

Few effector molecules transported by the T4SS are known for organisms closely related to *A. marginale*. Seven putative effector proteins (Beps) were recently identified for *Bartonella henselae* (Schulein et al. 2005). The Beps appear to be responsible for (1) a massive rearrangement of the actin cytoskeleton to allow for uptake of the bacteria, (2) stimulation of a proinflammatory response by activation of NF-κB, and (3) creating an environment where the host cell is refractory to apoptotic stimuli. Three of these proteins, BepD, BepE, and BepF, carry short repeated peptide sequences with presumed tyrosine phosphorylation motifs, and BepD has been shown to be phosphorylated upon entry into the host cell. The

B. henselae Beps were shown to have bipartite targeting sequences for the T4SS, including a region at the N-terminus and a 142 aa region adjacent to an unconserved, positively charged tail; however, this motif has not been found in other T4SS effector molecules (Schulein et al. 2005). The T4SS effector proteins from *A. tumefaciens* have the consensus sequence R-X_7-R-X-R-X-R-X-X_n, where K can substitute for R (Vergunst et al. 2005); however, only one protein in *A. phagocytophilum* has been found to have a similar sequence to this motif, with other Anaplasmataceae being completely devoid of this sequence (Dunning Hotopp et al. 2006). More recently, *A. phagocytophilum* AnkA (Caturegli et al. 2000) was shown to be transported by the T4SS into the host cell, where it was phosphorylated through interaction with the Abl-1 tyrosine kinase (Lin et al. 2007). Knockdown of AnkA or Abl-1 inhibited infection of host cells with *A. phagocytophilum*, suggesting that the Abl-1 signal transduction pathway is important in facilitating intracellular infection (Lin et al. 2007).

Recently, it has been demonstrated that the *A. marginale* T4SS proteins virB9, virB10, and conjugal transfer protein (a virB9 related sequence) are members of an immunoprotective outer membrane protein preparation and are recognized by serum IgG2 and stimulate memory T lymphocyte proliferation and IFN-γ secretion, providing impetus for vaccine development using these proteins (Lopez et al. 2007).

3.7.5 Paralogous Gene Families

The two largest repeat families correspond to the major surface protein2 (*msp2*) and MSP1 superfamilies, discussed later. A number of small gene families are present in the *A. marginale* genome. There are two small families of putative cell surface proteins containing two or three members and a family of four exported protein genes. The *Anaplasma* appendage associated protein gene (*aaap*) (Stich et al. 2004) involved in actin filament formation has two paralogs immediately upstream in the St. Maries strain. Transporter proteins account for several gene families, with the largest being the ATP binding cassette proteins (8) and the major facilitator superfamily proteins (6) (See Table 6 for a complete list of transporter families). Twelve families of paralogous genes contain two to four members and range from different enzymes containing shared domains to undefined products. There are no families of insertion sequences present in the genome.

Transcription terminator *rho*, normally a single copy gene, is duplicated in the *Anaplasma* and *Ehrlichia* genomes sequenced to date (Brayton et al. 2005; Collins et al. 2005; Dunning Hotopp et al. 2006; Frutos et al. 2006). This repeat appears to have arisen during an inversion event around the origin of replication, and is highlighted through comparison with *Ehrlichia ruminantium*. The two copies of *rho* are separated by 333 kb and flank the inverted element (Fig. 9).

There are three tandemly occurring genes that have a low level of sequence identity to *orfX* of the *msp2* superfamily (discussed further); however, they are not initiated with the characteristic "MLLK" start sequence that defines the *orfX* paralogs. These genes have orthologs in the *A. phagocytophilum* genome in the locus syntenic to the *A. marginale msp3* locus (Fig. 10).

The *Msp2* Superfamily

The* Msp2 *Family Major surface protein (Msp) 2 is an immunodominant 38–44 kDa surface protein thought to be responsible for the evasion of the host immune response (Palmer et al. 1986a; French et al. 1998, 1999). Msp2 has a central hypervariable (HVR) region flanked by conserved amino- and carboxy-termini (Palmer et al. 1994; Eid et al. 1996; French et al. 1998). Msp2 is encoded by a multigene family with a single expression site (ES) and seven functional pseudogenes distributed throughout the genome in the fully sequenced St. Maries strain, with the number of pseudogenes ranging from 5–9 in different strains (Brayton et al. 2002, 2005; Rodriguez et al. 2005). The *msp2* functional pseudogenes, truncated at both the 5′ and the 3′ ends, are not positioned near a promoter and cannot make a functional protein from their present location; however, they have been shown to recombine into the expression site in whole or in part to generate a new functional variant of Msp2 (Brayton et al. 2001, 2002). The combinatorial diversity afforded by segmental gene conversion allows for several thousand of Msp2 variants, which can account for lifelong evasion of the host immune response (Brayton et al. 2002).

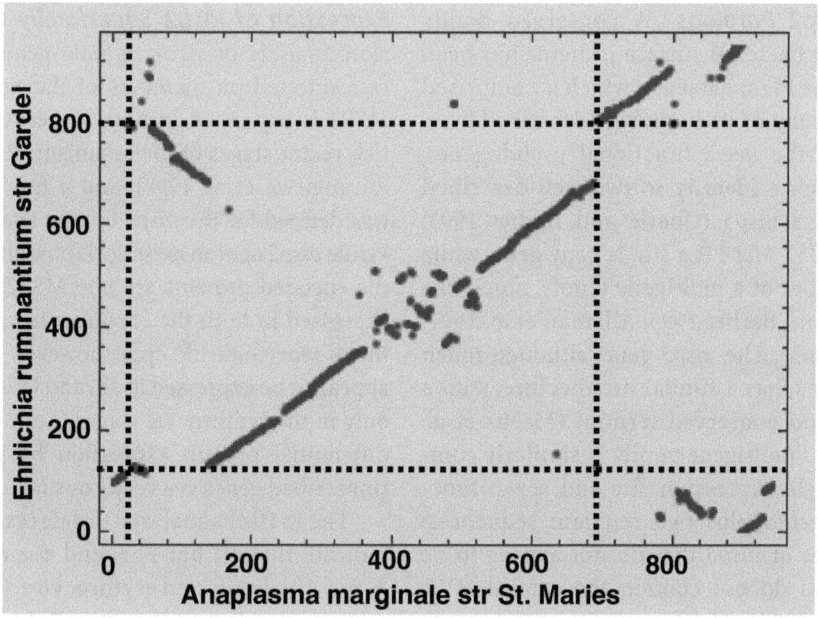

Fig. 9 Duplicated *rho* genes in *A. marginale* and *E. ruminantium*. GenePlot (http://www.ncbi.nlm.nih.gov/sutils/geneplot.cgi?tax1=234826&tax2=302409) showing pairwise genome comparisons of protein homologs. The positions of the duplicated *rho* genes for each genome are shown with a *dashed line*

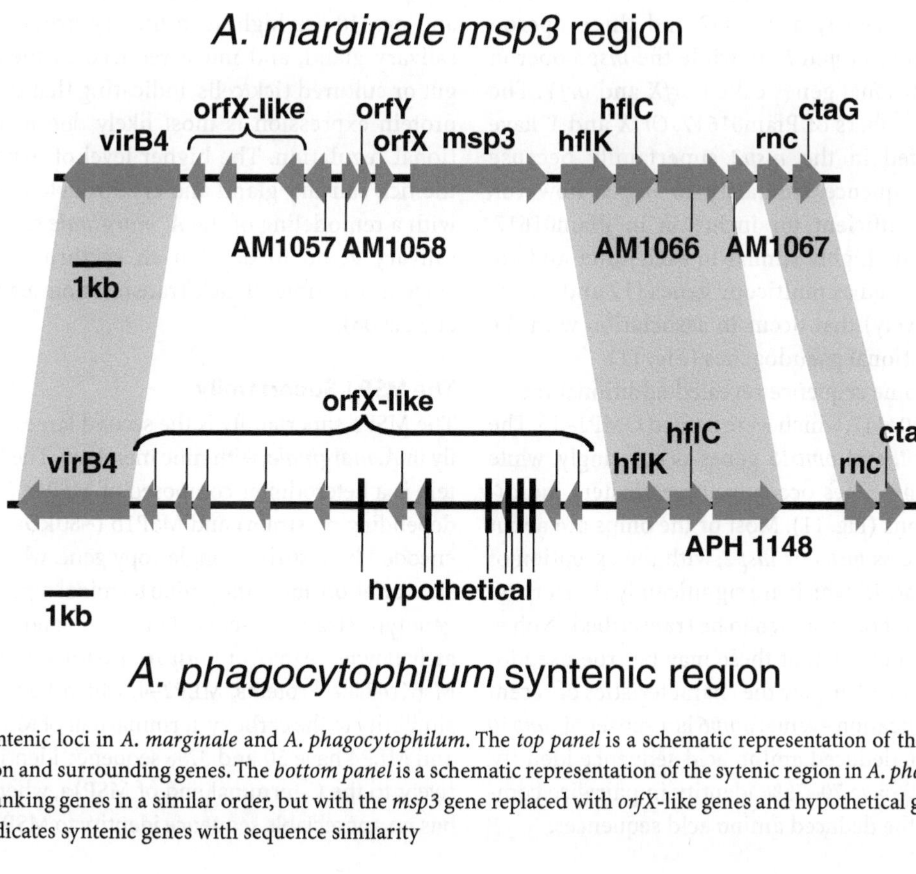

Fig. 10 Syntenic loci in *A. marginale* and *A. phagocytophilum*. The *top panel* is a schematic representation of the *A. marginale msp3* operon and surrounding genes. The *bottom panel* is a schematic representation of the sytenic region in *A. phagocytophilum*, showing flanking genes in a similar order, but with the *msp3* gene replaced with *orfX*-like genes and hypothetical genes. The gray shading indicates syntenic genes with sequence similarity

Additional* Msp2 *Paralogs A conserved family (Pfam01617) of bacterial surface proteins has been assembled in the Pfam database, which is comprised of MSP2 and many of its homologs described here. In addition to the *msp2* functional pseudogenes, *msp2* has sequence identity to two well-described genes: *msp4* and *msp3* (Oberle and Barbet 1993; Meeus et al. 2003). *Msp4* is a single copy gene, while *msp3* is a member of a multigene family, much like *msp2* (Oberle and Barbet 1993; Alleman et al. 1997; Meeus et al. 2003). The *msp3* gene, although much larger than *msp2*, has a similar architecture, with a central HVR and conserved termini (Meeus et al. 2003). The *msp3* multigene family is similarly composed of a single expression site and seven functional pseudogenes plus two remnant sequences: short sequences of *msp3* that do not appear to be transcribed and do not contain the central HVR (Brayton et al. 2005). The functional pseudogenes for *msp2* and *msp3* are distributed throughout the genome and are often (4/7) closely positioned in a tail-to-tail arrangement in the pseudogene complex (Brayton et al. 2001). Both *msp2* and *msp3* are transcribed as part of an operon: the *msp2* operon contains four genes, one *msp2* and three operon associated genes (*opag1–3*), while the *msp3* operon has two additional genes called *orfX* and *orfY*. The *opags* are members of Pfam01617. *OrfX* and *Y* have been included in the *msp2* superfamily because they have sequence similarity to *msp3*; however, this is not sufficient for inclusion in Pfam01617. *OrfX* and *Y* are highly similar to each other and are each represented as multicopy genes (12 and 7 copies, respectively) that occur in association with the *msp2/3* functional pseudogenes (Fig. 11).

The genome sequence revealed additional members of Pfam01617, which were named OMP1–15. The *omp1*, *omp14*, and *omp15* genes occur singly, while the remaining genes occur in three clusters indicative of operons (Fig. 11). Most of the omps are about the same size as *msp4* or *msp2*, with the exception of *omp2, 3, 6,* and *15*, which are significantly shorter and which have not been shown to be transcribed (Noh et al. 2006), suggesting that these may be true pseudogenes. *Omps 6–10* display the characteristics of recent tandem duplication events: *omp6* is a subset of *omp10* and has 99% deduced amino acid sequence identity, and *omp 7–9* have 70–75% identity in pairwise comparisons of the deduced amino acid sequences.

Expression of* Msp2 *Superfamily Genes Expression analysis of *A. marginale* genes has primarily concentrated on members of the *msp2* superfamily. MSP2 is expressed in both the erythrocyte and the tick vector stages of the organism (Palmer et al. 1994; Rurangirwa et al. 1999), and a functional promoter was defined for the *msp2* operon (Barbet et al. 2005). While *msp2* operon message is ubiquitously expressed, the encoded proteins are not: MSP2 and OpAG2 are expressed in both the erythrocyte and tick stages of the *A. marginale* life cycle; however, OpAG1 does not appear to be expressed at all, and OpAG3 is expressed only in the erythrocyte stages (Lohr et al. 2002). This differential protein expression by polycistronically transcribed genes is a very unusual finding.

The 14 OMPs analyzed to date (excluding OMP15) indicate that all but 2, 3, and 6 are transcribed in *A. marginale* infected erythrocytes, tick midgets, and salivary glands, and that *omps7–10* are polycistronically transcribed (Noh et al. 2006). Protein expression of OMP1, 4, 7, 8, 9, and 11 was demonstrated in erythrocytes and shown to be markedly reduced when *A. marginale* is cultured in tick cells. Quantitative PCR analysis indicates the transcript level for these genes and *omp10* are highest in the erythrocyte and tick salivary gland, and much reduced in the tick midgut or cultured tick cells, indicating that the reduced protein expression is most likely due to transcriptional regulation. The higher level of expression in the tick salivary gland and erythrocyte is consistent with a remodeling of the *A. marginale* surface in the salivary gland to develop an erythrocyte-infective stage at the time of tick transmission feeding (Noh et al. 2006).

The MSP1 Superfamily

The MSP1 superfamily is the second largest gene family in *A. marginale*, with nine members. The MSP1 protein is a heterodimer composed of MSP1a (60–80 kDa, depending on strain) and MSP1b (~80 kDa). MSP1a is encoded by *msp1α*, a single copy gene, which exhibits size variation due to the amino terminal repeats used for genotype strains (discussed in Sects. 5 and 6). Immediately downstream from *msp1α* are three CDSs termed Msp1α-Like Proteins, MLP2–4, which have structural similarity to the carboxy-terminal half of MSP1a. MLP2 and MLP4 have 30 and 37% sequence identity, respectively, to the C-terminal end of MSP1a, whereas MLP3 has no appreciable sequence identity to MSP1a. MSP1b

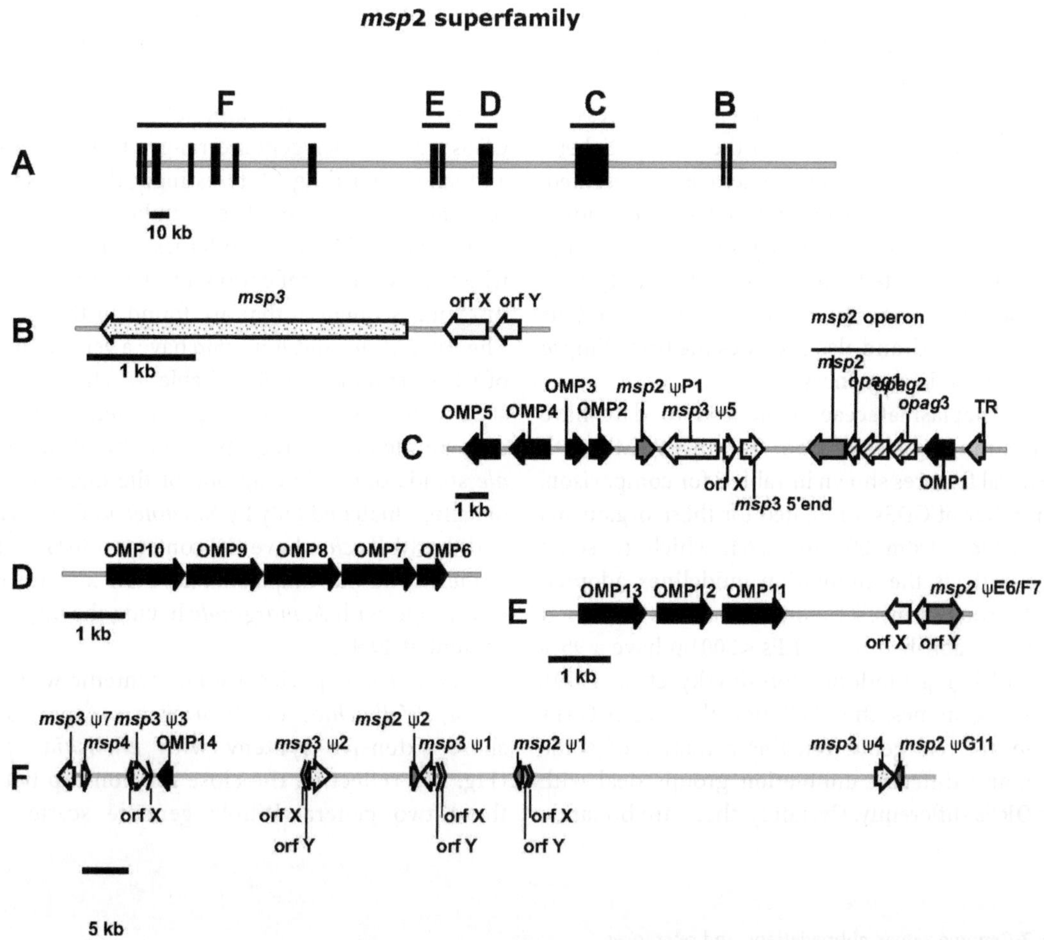

Fig. 11 *A. marginale msp2* superfamily schematic. **a** The distribution of superfamily members in a 360 kb region of the genome. The loci identified in *A* are shown in more detail in **b–f. b** The *msp3* operon containing *orfX*, *orfY*, and *msp3*. **c** The locus shown contains the *msp2* operon, several OMPs, and pseudogenes as well as *orfX* and a putative transcriptional regulator (TR) that is found in the syntenic region in *E. ruminantium*, *E. chaffensis*, and *A. phagocytophilum*. **d** A putative operon of five OMP genes. **e** Another putative operon of OMP genes that is situated near a *msp2* pseudogene. The 95 kb region depicted in (f) illustrates the positional relationship of the *msp2/3* pseudogenes with *orfX/Y*. OMP14 and *msp4* also occur in this region of the genome

is encoded by a five-member multigene family designated as *msp1β-1*, *msp1β-2*, and *msp1βpg1–3*, with "pg" designating a partial gene. Transcript has been detected for both full-length genes and also for *msp1βpg3*; however, only MSP1b-1 was definitively identified by mass spectrometry in the MSP1 complex (Macmillan et al. 2006). The presence of the partial genes may reflect the reductive evolution process, or they may serve to provide variation for MSP1b. MSP1b varies considerably between strains: St Maries *msp1β-1* is 95% identical, whereas St Maries *msp1β-2* is 77% identical to the corresponding gene in the Florida strain. As in MSP1a, variation in MSP1b must be a rather infrequent event, as it was not detected during the course of infection in a donor calf, nor in the subsequent infection in a recipient calf after tick transmission (Macmillan et al. 2006).

3.8 Comparative Genomics

To date, 15 Rickettsiales have annotated genomes available in the public databases, with several others in various stages of completion (Andersson et al. 1998; Ogata et al. 2000, 2001, 2005, 2006; McLeod et al. 2004; Wu et al. 2004; Brayton et al. 2005; Collins et al. 2005; Foster et al. 2005; Dunning Hotopp et al. 2006; Frutos et al. 2006). Table 7 lists the completed genomes and their relevant abbreviations, accession numbers, and references. Within the family Rickettsiaceae, only one of the two genera is represented, with five *Rickettsia* species sequenced – these genome sequences represent both subtypes in this genus: the typhus and the spotted fever groups. The family Anaplasmataceae has had representatives of all the four genera sequenced, and also provides the first glimpse of multiple strain sequences.

The Anaplasmataceae sequenced to date have genomes ranging in size from 0.86 to 1.52 Mb, with the general features shown in Table 8 for comparison. The number of CDSs annotated for these organisms varies widely from 805 to 1,264, which to some degree, reflects the annotation guidelines adopted by each group of investigators. In *Bacillus subtilis*, it has been established that ORFs <300 bp have a 95% chance of being random (Borodovsky et al. 1999). For other genomes, the likelihood that small ORFs <300 bp are real genes remains a matter of some debate, and different annotation groups deal with these ORFs differently. Certainly, there are bonafide genes that are quite small, like the ribosomal protein genes; however, many of the small ORFs are annotated as hypothetical proteins. If these small ORFs are removed from the annotation, the number of CDSs >300 bp per genome ranges from 674 to 967, a much tighter range. Interestingly, the two smallest genomes have the smallest numbers of CDSs, and *W. pipientis* wMel has the largest number of CDSs, which may be a reflection of the prophages and insertion sequences that are found in this genome. The *Anaplasma* and *Ehrlichia* have a very tight range of CDSs, from 812 to 877 (Table 8). The annotation style obviously affects other parameters, such as percent coding and average gene length, yet *A. marginale* stands out as having one of the highest coding densities, matched only by *N. sennetsu*. The *Ehrlichia* and the *Wolbachia* have GC contents ~30%, while *N. sennetsu* and the *Anaplasma* have GC contents in the 40% range, with *A. marginale* having the highest GC content at 49.9%.

Anaplasma species are not syntenic with *Rickettsia*, *Wolbachia*, or *Neorickettsia* species, but share extensive synteny with *Ehrlichia* species (Fig. 12), reflecting the close relationship between these two genera. Whole genome scatter plots

Table 7 Genome names, abbreviations, and references

Genus	Species	Strain	Abbreviation	Accession number	Reference
Anaplasma	*marginale*	St Maries	AMStM	CP000030	Brayton et al. (2005)
Anaplasma	*phagocytophilum*	HZ	APHZ	CP000235	Dunning Hottop et al. (2006)
Ehrlichia	*ruminantium*	Welgevonden	ERW1	CR767821	Collins et al. (2005)
Ehrlichia	*ruminantium*	Welgevonden	ERW2	CR925678	Frutos et al. (2006)
Ehrlichia	*ruminantium*	Gardel	ERG	CR925677	Frutos et al. (2006)
Ehrlichia	*chaffeensis*	Arkansas	ECArk	CP000236	Dunning Hottop et al. (2006)
Ehrlichia	*canis*	Jake	ECJake	CP000107	Copeland (unpublished)
Wolbachia	*pipientis*	wMel	WPwMel	AE017196	Wu et al. (2004)
Wolbachia	*pipientis*	TRS	WPTRS	AE017196	Foster et al. (2005)
Neorickettsia	*sennetsu*	Miyayama	NSMi	CP000237	Dunning Hottop et al. (2006)
Rickettsia	*bellii*	RML369-C	RB	CP000087	Ogata et al. (2006)
Rickettsia	*conorii*	Malish 7	RC	AE006914	Ogata et al. (2000, 2001)
Rickettsia	*felis*	URRWXCal2	RF	CP000053	Ogata et al. (2005)
Rickettsia	*prowazekii*	Madrid E	RP	AJ235269	Andersson et al. (1998)
Rickettsia	*typhi*	Wilmington	RT	AE017197	McLeod et al. (2004)

Table 8 Genome comparisons

	AMStM	APHZ	ERW1	ERW2	ERG	ECArk	ECJake	WPwMel	WolTRS	NSMi
Size (bp)	1,197,687	1,471,282	1,516,355	1,512,977	1,499,920	1,176,248	1,315,030	1,267,782	1,080,084	859,006
CDS annotated	949	1264	888	958	950	1104	925	1,195	805	932
CDS <300 bp	72	422	76	107	101	281	67	228	131	245
CDS >300 bp	877	842	812	851	849	823	858	967	674	687
tRNA	37	37	36	36	36	37	36	34	34	33
rRNA	3	3	3	3	3	3	3	3	3	3
GC	49.9	41.6	27.5	27.5	27.5	30.1	29.0	35.4	34.2	41.4
Coding %	85.4	68.2	62.0	63.2	63.9	79.6	72.1	80.2	67.0	87.7
Gene length	1,078	794	1,059	1000	1009	847	1025	850	899	808

See Table 7 for abbreviations of species names

between closely related genera typically reveal an X-shaped pattern called an X-alignment, indicative of a symmetrical inversion around the origin of replication and/or terminus (Eisen et al. 2000). This can be most clearly seen when comparing *E. chaffeensis* with *E. canis* or *E. ruminantium* (Fig. 12). Interestingly, *E. ruminantium* and *E. canis* have almost perfect synteny, with no inversions. When comparing X-alignments using *A. marginale*, there is more scatter indicating multiple inversion events around the origin and terminus. These appear to be mediated by repeat sequences in the genomes, often *msp2* pseudogenes, and as mentioned earlier, the duplicated *rho* gene has also been involved in one of these large-scale inversions. The scatter plots reveal a 14 kb region that is syntenic amongst all Rickettsiales, and likely amongst most bacteria (data not shown), that contains the ribosomal protein genes (circled area in Fig. 12).

Comparison of the *A. marginale* deduced proteome with that of the other 14 available Rickettsiales generated lists of common CDSs using a cutoff e-value of 1e-5. Nearly half (424) of the annotated *A. marginale* CDSs are common among all Rickettsiales (Table 9). While most of these common proteins perform housekeeping functions, 81 are annotated as unnamed proteins with conserved homologs in the databases, often with an entry in the Pfam database (Bateman et al. 2004). Another 85 proteins are common to the sequenced Anaplasmataceae, but not found in the Rickettsiaceae (Table 9). These include genes involved in gluconeogenesis, purine, pyrimidine, and flavin biosynthesis, the methylerythritol phosphate pathway, and 28 unnamed genes of unknown function. Comparison of *A. marginale* with *A. phagocytophilum* and *Ehrlichia* reveals an additional 47 conserved CDSs (Table 9). This list includes many of the *msp2* superfamily OMPs, but not MSP2 itself, as this protein is also found in the *Wolbachia* (annotated as surface protein WSP), but not in *Neorickettsia*. There are just 23 CDSs uniquely shared between *A. marginale* and *A. phagocytophilum*, including three additional *msp2* superfamily protein genes, with the remainder being annotated as conserved hypothetical proteins (Table 9).

There are 107 genes unique to *A. marginale*, including those encoding surface proteins such as MSP1 (Palmer et al. 1986b, 2001; Barbet and Allred 1991; Viseshakul et al. 2000), OpAG1 (Lohr et al. 2002), and MSP1a-like proteins (MLPs) 2–4 (Brayton et al. 2005). Genes encoding the secreted protein *Anaplasma* appendage associated protein (aaap) (Stich et al. 2004) and two AAAP-like proteins (annotated as AM879 and AM880) are also unique to *A. marginale*. The current analysis conflicts with previous reports (Dunning Hotopp et al. 2006) and shows that *A. marginale* is unique amongst the Rickettsiales in its ability to synthesize pantothenate, or vitamin B5, a necessary precursor for Coenzyme A synthesis (i.e., the only *Rickettsiales* with panB, panC, and IlvC; Table 9). The Anaplasmataceae have varying abilities to synthesize Coenzyme A, with the *Ehrlichia* retaining all the enzymatic capability to synthesize Coenzyme A from pantothenate, and *N. sennetsu*, *A. phagocytophilum*, and the *Wolbachia* only able to catalyze the last two steps of the Coenzyme A pathway. The remaining 89

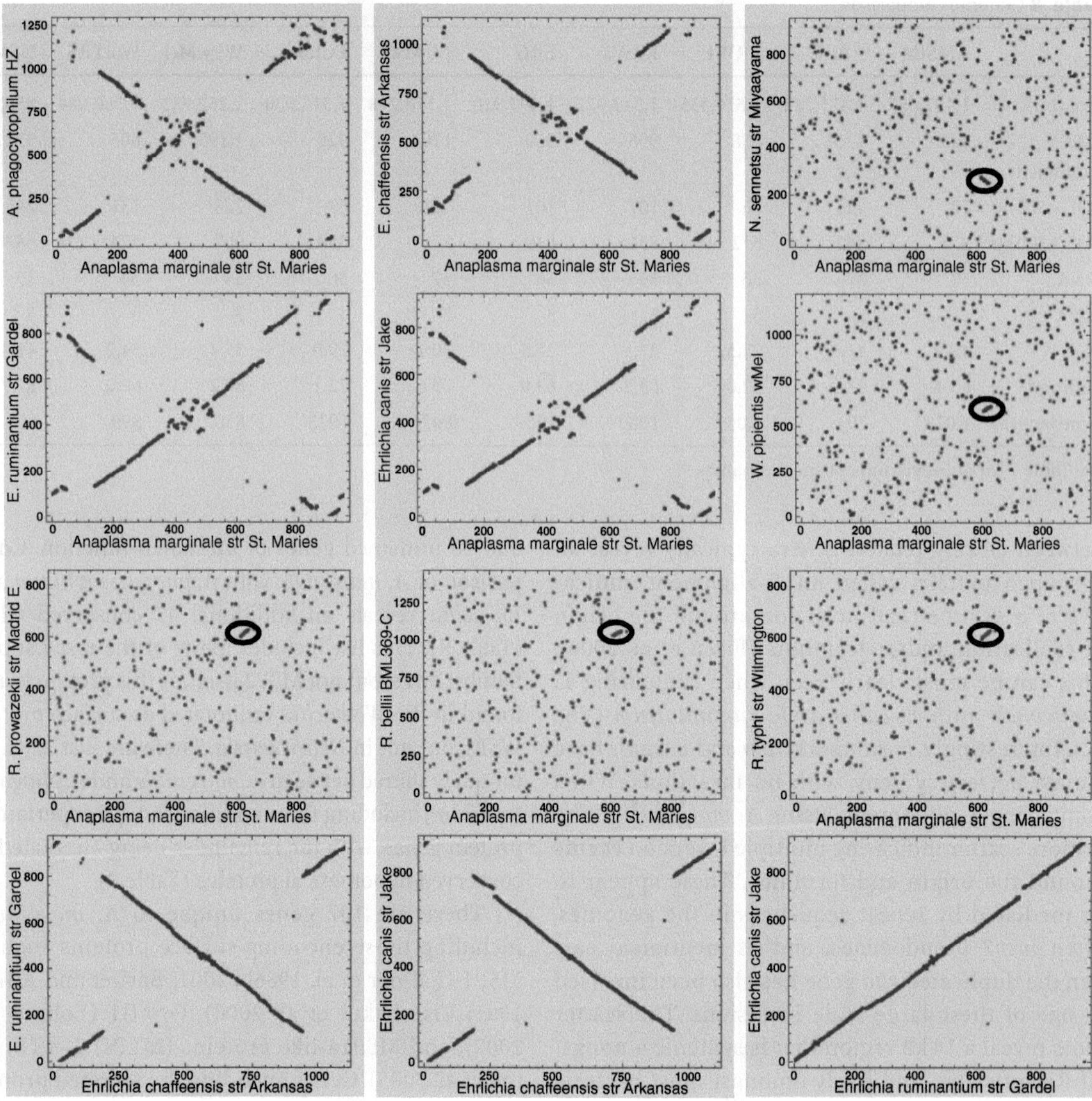

Fig. 12 Whole genome alignments. GenePlot (www.ncbi.nlm.nih.gov/sutils/geneplot.cgi?) dot plot alignments showing pairwise genome comparisons of protein homologs. The ovals highlight a syntenic segment containing ribosomal protein genes

genes unique to *A. marginale* are, by definition, hypothetical proteins.

3.9 Future Scope

The availability of the genome sequence has opened several avenues for research on *A. marginale*, including the ability to map mass spectrometry data back to the genome, thus identifying potential vaccine candidates (Lopez et al. 2007), comparative genomic approaches to dissect phenotypic traits, such as virulence and transmissibility, and has provided the tools to understand genetic diversity in this species. Acquisition of additional strain sequences will be helpful in determining the nature of strain variability, allowing a more precise definition of strain and species for these organisms. Additionally, determination of strain variation will aid in determining which of the uncharacterized conserved hypothetical and unique hypothetical genes in Anaplasmataceae may be important in organism dissemination and infection

Table 9 Lists of CDS

A. CDS common to *Rickettsiales*

aatA	acnA	acpP	acpS	adk	adx1	alaS	aprE	argS	asd	aspS
atpA	atpB	atpD	atpG	atpH	barA	bcr2	birA	carB	ccmA	ccmc
ccmE	ccmF	cdsA	clpA	clpB	clpP	clpX	coaE	coq7	cycM	cysS
czcR	dcd	def	dhkA	dksA	dnaA	dnaB	dnaE	dnaG	dnaJ	dnaK
dnaN	dnaZ	dut	efp	engA	era	fab1	fabB	fabD	fabG	fabH
fabZ	fdxA	ffh	fmt	folD	ftsA	ftsH	ftsK	ftsY	ftsZ	fumC
fusA	gatA	gatB	gcp	gidA	glnA	glnA	gltA	gltX	gltX2	glyA
glyQ	glyS	gmk	gor	gpsA	greA	groEL	groES	grpE	grxC1	gyrA
gyrB2	hemA	hemB	hemC	hemE	hemF	hemH	hemK	hflC	hflK	hflX
hisS	hscA	hscB	hslU	hslV	htpG	icd	ileS	infA	infB	infC
iscS	ispA	ispB	kefB	ksgA	lepA	lepB	leuS	lgt	ligA	lipA
lipB	lon	lpdA	lspA	lysS	maeB	maf	map	mdh	metS	mfd
mgtE	miaA	mkl	mpp	mraW	mreB	msbA2	msp5	mutL	mutS	ndk
nuoD	nuoE	nuoF	nuoG	nifU	nrdA	nrdB	nth	ntrX	ntrY	nuoA
nuoB	nuoC1	nusB	nusG	oma87	pccB	nuoH	nuoI	nuoJ	nuoL	nuoL2
nuoL3	nuoM	nuoN1	nusA	pheS	pheT	phnL	pkcI	pcnB	pdhA	pdhC
pepA	perM	petA	petB	petC	pgsA	proP1	proS	psd	pssA	plsC
pnp	polA	pdhB	ppa	ppdK/ ppsA	prfA	prfB	priA	recF	recG	recJ
recR	pstB	pth	purC	pycA	pyrG	pyrH	radA	rbfA	recA	rnhB
rpiB	rplA	rplB	rho (1)	rho (2)	rimM	rluC	rnc	rnd	rnd	rne
rnhA	rplP	rplQ	rplR	rplS	rplC	rplD	rplE	rplF	rplI	rplK
rplM	rplN	rplO	rpmH	rpmI	rpmJ	rpoA	rplT	rplU	rplV	rplX
rplY	rpmA	rpmB	rpmF	rpmG	rpsF	rpsG	rpsH	rpsI	rpoB	rpoC
rpoD	rpoH	rpsA	rpsB	rpsC	rpsD	rpsE	rpsS	rpsT	rrf	ruvA
rpsJ	rpsK	rpsL	rpsM	rpsN	rpsO	rpsP	rpsQ	rpsR	secY	serS
sfhB	smpB	ruvB	ruvC	sdhA	sdhB	sdhC	secA	secB	secD	secF
tdpX1	tgt	thdF	thrS	sodB	ssb	ssnA	sucA	sucB	sucC	sucD
suhB	tatC	truA	trxA	trxB	trxB2	thyl	tig	tmk	tolC	topA
trbG	trmD	trmU	trpS	ubiG	ubiH	uvrD	valS	tsf	tuf	tuf
typA	tyrS	ubiA	ubiB	ubiE	ubiG	xerD	xthA2	yajC	yidC	virB10
virB11	virB4	virB4	virB6	virB8	virB9	virD4	xerC	znuC	AM026	AM042
AM052	AM074	AM092	AM106	AM127	AM116	AM139	AM154	AM202	AM242	AM245
AM268	AM275	AM296	AM315	AM316	AM320	AM343	AM392	AM401	AM414	AM415
AM419	AM420	AM435	AM436	AM447	AM465	AM475	AM477	AM486	AM487	AM492
AM504	AM516	AM555	AM561	AM562	AM656	AM657	AM680	AM685	AM694	AM709
AM757	AM758	AM810	AM811	AM812	AM832	AM833	AM841	AM846	AM858	AM915
AM927	AM929	AM947	AM980	AM1010	AM1027	AM1066	AM1079	AM1080	AM1088	AM1093
AM1094	AM1108	AM1146	AM1168	AM1183	AM1212	AM1237	AM1264	AM1276	AM1303	AM1319
AM1350	AM1355									

(Continued)

Table 9 (Continued)

B. CDS common to Anaplasmataceae										
argD	atpE	bioA	bolA	carA	coaD	dxr	eno	fbaB	fpbA	gapA
gcpE	glmS	glpX	gshB	guaA	guaB	hflK	ispD	ispE	ispF	lytB
pgk	plsX	pmbA	pstA	pstC	pstS	purA	purD	purE	purF	purH
purK	purL	purM	purQ	pyrB	pyrC	pyrD	pyrE	pyrF	qor	ribA
ribA	ribD	ribF	ribH	rpe	rpmE	rpoZ	surE	tktA	tpiA	yibO
tldD	AM007	AM093	AM119	AM160	AM198	AM237	AM240	AM299	AM342	AM441
AM445	AM480	AM499	AM500	AM508	AM660	AM664	AM753	AM829	AM830	AM836
AM882	AM957	AM967	AM1186	AM1298	AM1337	AM1349				
C. CDS common to *Anaplasma* spp and *Ehrlichia* spp										
cinA	dfp	dsbD	folB	omp10	omp12	omp13	omp4	omp5	omp7	omp8
omp9	opag3	orf Y	pep1	thiC	thiO	trkH	AM002	AM070	AM156	AM216
AM219	AM271	AM324	AM387	AM410	AM501	AM578	AM592	AM821	AM881	AM920
AM936	AM977	AM1046	AM1048	AM1050	AM1051	AM1179	AM1191	AM1285		
D. CDS common to *A. marginale* and *A. phagocytophilum*										
omp3	omp6	omp15	AM136	AM347	AM551	AM589	AM613	AM616	AM628	AM629
AM649	AM675	AM689	AM788	AM921	AM950	AM952	AM968	AM1030	AM1073	AM1078
AM1226										
E. CDS unique to *A. marginale*										
aaap	ana29	ilvC	mlp2	mlp3	mlp4	msp1a	msp1B pg1	msp1B pg2	msp1B pg3	msp1B-1
msp1B-2	omp2	opag1	panB	panC	AM009	AM020	AM027	AM122	AM185	AM247
AM265	AM283	AM291	AM306	AM345	AM346	AM348	AM351	AM354	AM356	AM357
AM359	AM360	AM366	AM368	AM371	AM378	AM380	AM382	AM395	AM413	AM470
AM489	AM495	AM529	AM537	AM541	AM543	AM545	AM546	AM547	AM550	AM582
AM594	AM595	AM603	AM612	AM615	AM634	AM670	AM671	AM673	AM676	AM682
AM683	AM688	AM712	AM760	AM773	AM776	AM819	AM879	AM880	AM955	AM959
AM974	AM997	AM1006	AM1021	AM1041	AM1057	AM1058	AM1077	AM1111	AM1121	AM1135
AM1141	AM1161	AM1165	AM1189	AM1193	AM1216	AM1225	AM1232	AM1238	AM1239	AM1240
AM1241	AM1242	AM1320	AM1324	AM1341	AM1352	AM1354	AM1359			

of host species. Finally, mapping polymorphisms between strains will aid in epidemiological studies (via SNP fingerprinting) and in examination of the evolution of the Anaplasmataceae as a family.

References

Alleman AR, Palmer GH, McGuire TC, McElwain TF, Perryman LE, Barbet AF (1997) *Anaplasma marginale* major surface protein 3 is encoded by a polymorphic, multigene family. Infect Immun 65:156–163

Allred DR, McGuire TC, Palmer GH, Leib SR, Harkins TM, McElwain TF, Barbet AF (1990) Molecular basis for surface antigen size polymorphisms and conservation of a neutralization-sensitive epitope in *Anaplasma marginale*. Proc Natl Acad Sci USA 87:3220–3224

Andersson SG, Zomorodipour A, Andersson JO, Sicheritz-Ponten T, Alsmark UC, Podowski RM, Naslund AK, Eriksson AS, Winkler HH, Kurland CG (1998) The genome sequence of *Rickettsia prowazekii* and the origin of mitochondria. Nature 396:133–140

Backert S, Meyer TF (2006) Type IV secretion systems and their effectors in bacterial pathogenesis. Curr Opin Microbiol 9:207–217

Barbet AF, Allred DR (1991) The msp1 beta multigene family of *Anaplasma marginale*: nucleotide sequence analysis of an expressed copy. Infect Immun 59:971–976

Barbet AF, Agnes JT, Moreland AL, Lundgren AM, Alleman AR, Noh SM, Brayton KA, Munderloh UG, Palmer GH (2005) Identification of functional promoters in the *msp2* expression loci of *Anaplasma marginale* and *Anaplasma phagocytophilum*. Gene 353:89–97

Barigye R, Garcia-Ortiz MA, Rojas Ramirez EE, Rodriguez SD (2004) Identification of IgG2-specific antigens in Mexican *Anaplasma marginale* strains. Ann NY Acad Sci 1026:84–94

Bateman A, Coin L, Durbin R, Finn RD, Hollich V, Griffiths-Jones S, Khanna A, Marshall M, Moxon S, Sonnhammer EL, Studholme DJ, Yeats C, Eddy SR (2004) The Pfam protein families database. Nucl Acids Res 32(Database issue): D138–D141

Blattner FR, Plunkett G IIIrd, Bloch CA, Perna NT, Burland V, Riley M, Collado-Vides J, Glasner JD, Rode CK, Mayhew GF, Gregor J, Davis NW, Kirkpatrick HA, Goeden MA, Rose DJ, Mau B, Shao Y (1997) The complete genome sequence of *Escherichia coli* K-12. Science 277:1453–1474

Bock RE, de Vos AJ (2001) Immunity following use of Australian tick fever vaccine: a review of the evidence. Aust Vet J 79:832–839

Bock RE, Kingston TG, De Vos AJ (1999) Effect of breed of cattle on innate resistance to infection with *Anaplasma marginale* transmitted by *Boophilus microplus*. Aust Vet J 77:748–751

Borodovsky M, Hayes WS, Lukashin AV (1999) Statistical predictions of coding regions in prokaryotic genomes by using inhomogeneous markov models. In: Charlebois RL (ed) Organization of the Prokaryotic Genome. ASM Press, Washington DC, pp 12

Brayton KA, Knowles DP, McGuire TC, Palmer GH (2001) Efficient use of a small genome to generate antigenic diversity in tick-borne ehrlichial pathogens. Proc Natl Acad Sci USA 98:4130–4135

Brayton KA, Palmer GH, Lundgren A, Yi J, Barbet AF (2002) Antigenic variation of *Anaplasma marginale msp2* occurs by combinatorial gene conversion. Mol Microbiol 43:1151–1159

Brayton KA, Kappmeyer LS, Herndon DR, Dark MJ, Tibbals DL, Palmer GH, McGuire TC, Knowles DP Jr (2005) Complete genome sequencing of *Anaplasma marginale* reveals that the surface is skewed to two superfamilies of outer membrane proteins. Proc Natl Acad Sci USA 102:844–849

Brosch R, Gordon SV, Pym A, Eiglmeier K, Garnier T, Cole ST (2000) Comparative genomics of the *mycobacteria*. Int J Med Microbiol 290:143–152

Brosch R, Pym AS, Gordon SV, Cole ST (2001) The evolution of mycobacterial pathogenicity: clues from comparative genomics. Trends Microbiol 9:452–458

Brosch R, Gordon SV, Marmiesse M, Brodin P, Buchrieser C, Eiglmeier K, Garnier T, Gutierrez C, Hewinson G, Kremer K, Parsons LM, Pym AS, Samper S, van Soolingen D, Cole ST (2002) A new evolutionary scenario for the *Mycobacterium tuberculosis* complex. Proc Natl Acad Sci USA 99:3684–3689

Brown DCG (1997) Dynamics and impact of tick-borne disease of cattle. Trop Animal Health Prod 29:1S–3S

Brown WC, Zhu D, Shkap V, McGuire TC, Blouin EF, Kocan KM, Palmer GH (1998) The repertoire of *Anaplasma marginale* antigens recognized by CD4(+) T-lymphocyte clones from protectively immunized cattle is diverse and includes major surface protein 2 (MSP-2) and MSP-3. Infect Immun 66:5414–5422

Carlyon JA, Fikrig E (2006) Mechanisms of evasion of neutrophil killing by *Anaplasma phagocytophilum*. Curr Opin Hematol 13:28–33

Carson CA, Adams LG, Todorovic RA (1970) An antigenic and serologic comparison of two virulent strains and an attenuated strain of *Anaplasma marginale*. Am J Vet Res 31:1071–1078

Carson CA, Sells DM, Ristic M (1977) Cell-mediated immune response to virulent and attenuated *Anaplasma marginale* administered to cattle in live and inactivated forms. Am J Vet Res 38:173–179

Caturegli P, Asanovich KM, Walls JJ, Bakken JS, Madigan JE, Popov VL, Dumler JS (2000) ankA: an *Ehrlichia phagocytophila* group gene encoding a cytoplasmic protein antigen with ankyrin repeats. Infect Immun 68:5277–5283

Christensen JF, McNeal DW (1967) *Anaplasma marginale* infection in deer in the Sierra Nevada foothill area of California. Am J Vet Res 28:599–601

Christensen JF, Osebold JW, Rosen MN (1958) Infection and antibody response in deer experimentally infected with *Anaplasma marginale* from bovine carriers. J Am Vet Med Assoc 132:289–292

Christensen JF, Osebold JW, Harrold JB, Rosen MN (1960) Persistence of latent *Anaplasma marginale* infection in deer. J Am Vet Med Assoc 136:426–427

Collins NE, Liebenberg J, de Villiers EP, Brayton KA, Louw E, Pretorius A, Faber FE, van Heerden H, Josemans A, van Kleef M, Steyn HC, van Strijp MF, Zweygarth E, Jongejan F, Maillard JC, Berthier D, Botha M, Joubert F, Corton CH, Thomson NR, Allsopp MT, Allsopp BA (2005) The genome of the heartwater agent *Ehrlichia ruminantium* contains multiple tandem repeats of actively variable copy number. Proc Natl Acad Sci USA 102:838–843

Corrier DE, Vizcaino O, Carson CA, Ristic M, Kuttler KL, Trevino GS, Lee AJ (1980) Comparison of three methods of immunization against bovine anaplasmosis: an examination of postvaccinal effects. Am J Vet Res 41: 1062–1065

de Castro JJ (1997) Sustainable tick and tickborne disease control in livestock improvement in developing countries. Vet Parasitol 71:77–97 de la Fuente J, Garcia-Garcia JC, Blouin EF, McEwen BR, Clawson D, Kocan KM (2001) Major surface protein 1a effects tick infection and transmission of *Anaplasma marginale*. Int J Parasitol 31:1705–1714

de la Fuente J, Garcia-Garcia JC, Blouin EF, Saliki JT, Kocan KM (2002) Infection of tick cells and bovine erythrocytes with one genotype of the intracellular ehrlichia *Anaplasma marginale* excludes infection with other genotypes. Clin Diagn Lab Immunol 9:658–668

de la Fuente J, Blouin EF, Kocan KM (2003a) Infection exclusion of the rickettsial pathogen *Anaplasma marginale* in the tick vector *Dermacentor variabilis*. Clin Diagn Lab Immunol 10:182–184

de la Fuente J, Golsteyn Thomas EJ, van den Bussche RA, Hamilton RG, Tanaka EE, Druhan SE, Kocan KM (2003b) Characterization of *Anaplasma marginale* isolated from North American bison. Appl Environ Microbiol 69:5001–5005

de la Fuente J, Passos LM, Van Den Bussche RA, Ribeiro MF, Facury-Filho EJ, Kocan KM (2004a) Genetic diversity and molecular phylogeny of *Anaplasma marginale* isolates from Minas Gerais, Brazil. Vet Parasitol 121:307–316

de la Fuente J, Vicente J, Hofle U, Ruiz-Fons F, Fernandez De Mera IG, Van Den Bussche RA, Kocan KM, Gortazar C (2004b) *Anaplasma* infection in free-ranging Iberian red deer in the region of Castilla-La Mancha, Spain. Vet Microbiol 100:163–173

de la Fuente J, Lew A, Lutz H, Meli ML, Hofmann-Lehmann R, Shkap V, Molad T, Mangold AJ, Almazan C, Naranjo V, Gortazar C, Torina A, Caracappa S, Garcia-Perez AL, Barral M, Oporto B, Ceci L, Carelli G, Blouin EF, Kocan KM (2005a) Genetic diversity of *Anaplasma* species major surface proteins and implications for anaplasmosis serodiagnosis and vaccine development. Anim Health Res Rev 6:75–89

de la Fuente J, Massung RF, Wong SJ, Chu FK, Lutz H, Meli M, von Loewenich FD, Grzeszczuk A, Torina A, Caracappa S, Mangold AJ, Naranjo V, Stuen S, Kocan KM (2005b) Sequence analysis of the *msp4* gene of *Anaplasma phagocytophilum* strains. J Clin Microbiol 43:1309–1317

de la Fuente J, Naranjo V, Ruiz-Fons F, Hofle U, Fernandez De Mera IG, Villanua D, Almazan C, Torina A, Caracappa S, Kocan KM, Gortazar C (2005c) Potential vertebrate reservoir hosts and invertebrate vectors of *Anaplasma marginale* and *A*. phagocytophilum in central Spain. Vector Borne Zoonotic Dis 5:390–401

de la Fuente J, Torina A, Caracappa S, Tumino G, Furla R, Almazan C, Kocan KM (2005d) Serologic and molecular characterization of *Anaplasma* species infection in farm animals and ticks from Sicily. Vet Parasitol 133:357–362

de la Fuente J, Torina A, Naranjo V, Caracappa S, Vicente J, Mangold AJ, Vicari D, Alongi A, Scimeca S, Kocan KM (2005e) Genetic diversity of *Anaplasma marginale* strains from cattle farms in the province of Palermo, Sicily. J Vet Med B Infect Dis Vet Public Health 52:226–229

de la Fuente J, Ruybal P, Mtshali MS, Naranjo V, Shuqing L, Mangold AJ, Rodriguez SD, Jimenez R, Vicente J, Moretta R, Torina A, Almazan C, Mbati PM, de Echaide ST, Farber M, Rosario-Cruz R, Gortazar C, Kocan KM (2007) Analysis of world strains of *Anaplasma marginale* using major surface protein 1a repeat sequences. Vet Microbiol 119:382–390

de Robertis E, Epstein B (1951) Electron microscope study of anaplasmosis in bovine red blood cells. Proc Soc Exp Biol Med 77:254–258

Dikmans G (1933) Anaplasmosis II. A short review and a preliminary demonstration of its identity in Louisiana. J Am Vet Med Assoc 82:741–748

Dikmans G (1950) The transmission of anaplasmosis. Am J Vet Res 11:5–16

Dumler JS, Barbet AF, Bekker CP, Dasch GA, Palmer GH, Ray SC, Rikihisa Y, Rurangirwa FR (2001) Reorganization of genera in the families *Rickettsiaceae* and Anaplasmataceae in the order *Rickettsiales*: unification of some species of *Ehrlichia* with *Anaplasma*, *Cowdria* with *Ehrlichia* and *Ehrlichia* with *Neorickettsia*, descriptions of six new species combinations and designation of *Ehrlichia equi* and 'HGE agent' as subjective synonyms of *Ehrlichia phagocytophila*. Int J Syst Evol Microbiol 51:2145–2165

Dunning Hotopp JC, Lin M, Madupu R, Crabtree J, Angiuoli SV, Eisen J, Seshadri R, Ren Q, Wu M, Utterback TR, Smith S, Lewis M, Khouri H, Zhang C, Niu H, Lin Q, Ohashi N, Zhi N, Nelson W, Brinkac LM, Dodson RJ, Rosovitz MJ, Sundaram J, Daugherty SC, Davidsen T, Durkin AS, Gwinn M, Haft DH, Selengut JD, Sullivan SA, Zafar N, Zhou L, Benahmed F, Forberger H, Halpin R, Mulligan S, Robinson J, White O, Rikihisa Y, Tettelin H (2006) Comparative genomics of emerging human ehrlichiosis agents. PLoS Genet 2:e21

Eid G, French DM, Lundgren AM, Barbet AF, McElwain TF, Palmer GH (1996) Expression of major surface protein 2 antigenic variants during acute *Anaplasma marginale* rickettsemia. Infect Immun 64:836–841

Eisen JA, Heidelberg JF, White O, Salzberg SL (2000) Evidence for symmetric chromosomal inversions around the replication origin in bacteria. Genome Biol 1:RESEARCH0011

Eriks IS, Stiller D, Palmer GH (1993) Impact of persistent *Anaplasma marginale* rickettsemia on tick infection and transmission. J Clin Microbiol 31:2091–2096

Eriks IS, Stiller D, Goff WL, Panton M, Parish SM, McElwain TF, Palmer GH (1994) Molecular and biological characterization of a newly isolated *Anaplasma marginale* strain. J Vet Diagn Invest 6:435–441

Ewing SA (1981) Transmission of *Anaplasma marginale* by arthropods. In: Jones RJH (ed) 7th Natl Anaplasmosis Conf. Mississippi State Univ Press, Mississippi State Univ, USA, pp 395–423

FAO (1994) Tick and Tickborne Diseases of Sheep and Goats. In: Report of the FAO First Expert Consultation. FAO, Rome, pp 1–14

Foster J, Ganatra M, Kamal I, Ware J, Makarova K, Ivanova N, Bhattacharyya A, Kapatral V, Kumar S, Posfai J, Vincze T, Ingram J, Moran L, Lapidus A, Omelchenko M, Kyrpides N, Ghedin E, Wang S, Goltsman E, Joukov V, Ostrovskaya O, Tsukerman K, Mazur M, Comb D, Koonin E, Slatko B (2005) The *Wolbachia* genome of *Brugia malayi*: endosymbiont evolution within a human pathogenic nematode. PLoS Biol 3:e121

French DM, McElwain TF, McGuire TC, Palmer GH (1998) Expression of *Anaplasma marginale* major surface protein 2 variants during persistent cyclic rickettsemia. Infect Immun 66:1200–1207

French DM, Brown WC, Palmer GH (1999) Emergence of *Anaplasma marginale* antigenic variants during persistent rickettsemia. Infect Immun 67:5834–5840

Frutos R, Viari A, Ferraz C, Morgat A, Eychenie S, Kandassamy Y, Chantal I, Bensaid A, Coissac E, Vachiery N, Demaille J, Martinez D (2006) Comparative genomic analysis of three strains of *Ehrlichia ruminantium* reveals an active process of genome size plasticity. J Bacteriol 188:2533–2542

Futse JE, Ueti MW, Knowles DP Jr, Palmer GH (2003) Transmission of *Anaplasma marginale* by *Boophilus microplus*: retention of vector competence in the absence of vector-pathogen interaction. J Clin Microbiol 41:3829–3834

Futse JE, Brayton KA, Dark MJ, Knowles DP, Jr., Palmer GH (2008) Superinfection as a driver of genomic diversification in antigenically variant pathogens. Proc Natl Acad Sci USA 105:2123–2127

Guglielmone AA (1995) Epidemiology of babesiosis and anaplasmosis in South and Central America. Vet Parasitol 57:109–119

Henry ET, Norman BB, Fly DE, Wichmann RW, York SM (1983) Effects and use of a modified live *Anaplasma marginale* vaccine in beef heifers in California. J Am Vet Med Assoc 183:66–69

Howarth JA, Roby TO, Amerault TE, McNeal DW (1969) Prevalence of *Anaplasma marginale* infection in California deer as measured by calf inoculation and serologic techniques. Proc Annu Meet US Anim Health Assoc 73:136–147

Huang H, Wang X, Kikuchi T, Kumagai Y, Rikihisa Y (2007) Porin activity of *Anaplasma phagocytophilum* outer membrane fraction and purified P44. J Bacteriol 189:1998–2006

Inokuma H, Terada Y, Kamio T, Raoult D, Brouqui P (2001) Analysis of the 16S rRNA gene sequence of *Anaplasma centrale* and its phylogenetic relatedness to other ehrlichiae. Clin Diagn Lab Immunol 8:241–244

Jongejan F, Uilenberg G (2004) The global importance of ticks. Parasitology 129(Suppl):S3–S14

Kawahara M, Rikihisa Y, Lin Q, Isogai E, Tahara K, Itagaki A, Hiramitsu Y, Tajima T (2006) Novel genetic variants of *Anaplasma phagocytophilum*, *Anaplasma bovis*, *Anaplasma centrale*, and a novel *Ehrlichia* sp. in wild deer and ticks on two major islands in Japan. Appl Environ Microbiol 72:1102–1109

Kocan KM (1986) Development of *Anaplasma marginale* in ixodid ticks: coordinated development of a rickettsial organism and its tick host. In: Sauer JR, Hair JA (eds) Morphology, Physiology, and Behavioral Ecology of Ticks. Ellis Horwood, Chichester, UK, pp 472–505

Kocan KM, Bezuidenhout JD (1987) Morphology and development of *Cowdria ruminantium* in *Amblyomma* ticks. Onderstepoort J Vet Res 54:177–182

Kreier JP, Ristic M (1963) Anaplasmosis. XII. The growth and survival in deer and sheep of the parasites present in the blood of calves infected with the Oregon strain of *Anaplasma marginale*. Am J Vet Res 24:697–702

Kumar S, Tamura K, Nei M (2004) MEGA3: Integrated software for molecular evolutionary genetics analysis and sequence alignment. Brief Bioinform 5:150–163

Kuttler KL (1969) Serial passage of an attenuated *Anaplasma marginale* in splenectomized calves. Proc Annu Meet US Anim Health Assoc 73:135

Kuttler KL (1972) Comparative response to premunization using attenuated *Anaplasma marginale* virulent *A. marginale* and *A. centrale* in different age groups. Trop Anim Health Prod 4:197–203

Kuttler KL (1984) *Anaplasma* infections in wild and domestic ruminants: a review. J Wildl Dis 20:12–20

Kuttler KL, Zaraza H (1969) Premunization with an attenuated *Anaplasma marginale*. Proc Annu Meet US Anim Health Assoc 73:104–112

Kuttler KL, Zaugg JL (1988) Characteristics of an attenuated *Anaplasma marginale* of deer origin as an anaplasmosis vaccine. Trop Anim Health Prod 20:85–91

Kuttler KL, Clifford DJ, Touray BN (1988) Prevalence of anaplasmosis and babesiosis in N'Dama cattle of the Gambia. Trop Anim Health Prod 20:37–41

Lee HC, Goodman JL (2006) *Anaplasma phagocytophilum* causes global induction of antiapoptosis in human neutrophils. Genomics 88:496–503

Lew AE, Bock RE, Minchin CM, Masaka S (2002) A msp1alpha polymerase chain reaction assay for specific detection and differentiation of *Anaplasma marginale* isolates. Vet Microbiol 86:325–335

Lewin B (1990) Genes. Oxford University Press, Oxford, UK

Li PL, Everhart DM, Farrand SK (1998) Genetic and sequence analysis of the pTiC58 *trb* locus, encoding a mating-pair formation system related to members of the type IV secretion family. J Bacteriol 180:6164–6172

Li PL, Hwang I, Miyagi H, True H, Farrand SK (1999) Essential components of the Ti plasmid *trb* system, a type IV macromolecular transporter. J Bacteriol 181:5033–5041

Lin M, Rikihisa Y (2003) *Ehrlichia chaffeensis* and *Anaplasma phagocytophilum* lack genes for lipid A biosynthesis and incorporate cholesterol for their survival. Infect Immun 71:5324–5331

Lin M, den Dulk-Ras A, Hooykaas PJJ, Rikihisa Y (2007) *Anaplasma phagocytophilum* AnkA secreted by type IV secretion system is tyrosine phosphorylated by Abl-1 to facilitate infection. Cell Microbiol 9:2644–2657

Lobry JR (1996) Asymmetric substitution patterns in the two DNA strands of bacteria. Mol Biol Evol 13:660–665

Lohr CV, Brayton KA, Shkap V, Molad T, Barbet AF, Brown WC, Palmer GH (2002) Expression of *Anaplasma marginale* major surface protein 2 operon-associated proteins during mammalian and arthropod infection. Infect Immun 70:6005–6012

Lohr CV, Brayton KA, Barbet AF, Palmer GH (2004) Characterization of the *Anaplasma marginale msp2* locus and its synteny with the omp1/p30 loci of *Ehrlichia chaffeensis* and *E. canis*. Gene 325:115–121

Lopez JE, Palmer GH, Brayton KA, Dark MJ, Leach SE, Brown WC (2007) Immunogenicity of *Anaplasma marginale* type IV secretion system proteins in a protective outer membrane vaccine. Infect Immun 75:2333–2342

Lora CA, Koechlin A (1969) An attenuated *Anaplasma marginale* vaccine in Peru. Am J Vet Res 30:1993–1998

Lord F, Geale D, Belaissaoui S, Coronel C (2006) An overview of anaplasmosis and assessing the risks for Canada. Canadian Food Inspection Agency, Canada

Losos GJ (1986) Babesiosis. In: Losos GJ (ed) Infectious Diseases of Domestic Animals. Churchill-Livingstone, New York, pp 41–46

Lowe TM, Eddy SR (1997) tRNAscan-SE: a program for improved detection of transfer RNA genes in genomic sequence. Nucl Acids Res 25:955–964

Macmillan H, Brayton KA, Palmer GH, McGuire TC, Munske G, Siems WF, Brown WC (2006) Analysis of the *Anaplasma marginale* major surface protein 1 complex protein composition by tandem mass spectrometry. J Bacteriol 188:4983–4991

McCallon BR (1973) Prevalence and economic aspects of anaplasmosis. In: Jones EW (ed) Proc Sixth Natl Anaplasmsois Conf. Heritage Press, Stillwater, OK, pp 1–3

McCosker PJ (1979) Global aspects of the management and control of ticks of veterinary importance. In: Rodriguez J (ed) Recent Advances in Acarology. Academic Press, New York, pp 45–53

McLeod MP, Qin X, Karpathy SE, Gioia J, Highlander SK, Fox GE, McNeill TZ, Jiang H, Muzny D, Jacob LS, Hawes AC, Sodergren E, Gill R, Hume J, Morgan M, Fan G, Amin AG, Gibbs RA, Hong C, Yu XJ, Walker DH, Weinstock GM (2004) Complete genome sequence of *Rickettsia typhi* and comparison with sequences of other *rickettsiae*. J Bacteriol 186:5842–5855

Meeus PF, Brayton KA, Palmer GH, Barbet AF (2003) Conservation of a gene conversion mechanism in two distantly related paralogues of *Anaplasma marginale*. Mol Microbiol 47:633–643

Melendez RD, Toro Benitez M, Niccita G, Moreno J, Puzzar S, Morales J (2003) Humoral immune response and hematologic evaluation of pregnant Jersey cows after vaccination with *Anaplasma centrale*. Vet Microbiol 94:335–339

Minjauw Ba MA (2001) Epidemiology and economics of tick-borne diseases and its effects on the livelihoods of the poor in east and southern Africa and India, DFID Report. DFID, London, pp 1–73

Molad T, Mazuz ML, Fleiderovitz L, Fish L, Savitsky I, Krigel Y, Leibovitz B, Molloy J, Jongejan F, Shkap V (2006) Molecular and serological detection of *A. centrale*- and *A. marginale*-infected cattle grazing within an endemic area. Vet Microbiol 113:55–62

Mtshali MS, de la Fuente J, Ruybal P, Kocan KM, Vicente J, Mbati PA, Shkap V, Blouin EF, Mohale NE, Moloi TP, Spickett AM, Latif AA (2007) Prevalence and genetic diversity of *Anaplasma marginale* strains in cattle in South Africa. Zoonoses Public Health 54:23–30

Noh SM, Brayton KA, Knowles DP, Agnes JT, Dark MJ, Brown WC, Baszler TV, Palmer GH (2006) Differential expression and sequence conservation of the *Anaplasma marginale msp2* gene superfamily outer membrane proteins. Infect Immun 74:3471–3479

Oberle SM, Barbet AF (1993) Derivation of the complete *msp4* gene sequence of *Anaplasma marginale* without cloning. Gene 136:291–294

Ocampo Espinoza V, Vazquez JE, Aguilar MD, Ortiz MA, Alarcon GJ, Rodriguez SD (2006) *Anaplasma marginale*: lack of cross-protection between strains that share MSP1a variable region and MSP4. Vet Microbiol 114:34–40

Ogata H, Goto S, Sato K, Fujibuchi W, Bono H, Kanehisa M (1999) KEGG: Kyoto encyclopedia of genes and genomes. Nucl Acids Res 27:29–34

Ogata H, Audic S, Barbe V, Artiguenave F, Fournier PE, Raoult D, Claverie JM (2000) Selfish DNA in protein-coding genes of *Rickettsia*. Science 290:347–350

Ogata H, Audic S, Renesto-Audiffren P, Fournier PE, Barbe V, Samson D, Roux V, Cossart P, Weissenbach J, Claverie JM, Raoult D (2001) Mechanisms of evolution in *Rickettsia conorii* and *R. prowazekii*. Science 293:2093–2098

Ogata H, Renesto P, Audic S, Robert C, Blanc G, Fournier PE, Parinello H, Claverie JM, Raoult D (2005) The genome sequence of *Rickettsia felis* identifies the first putative conjugative plasmid in an obligate intracellular parasite. PLoS Biol 3:e248

Ogata H, La Scola B, Audic S, Renesto P, Blanc G, Robert C, Fournier PE, Claverie JM, Raoult D (2006) Genome sequence of *Rickettsia bellii* illuminates the role of amoebae in gene exchanges between intracellular pathogens. PLoS Genet 2:e76

Ohashi N, Rikihisa Y, Unver A (2001) Analysis of transcriptionally active gene clusters of major outer membrane protein multigene family in *Ehrlichia canis* and *E. chaffeensis*. Infect Immun 69:2083–2091

Osebold JW, Christensen JF, Longhurst WM, Rosen MN (1959) Latent *Anaplasma marginale* infection in wild deer demonstrated by calf inoculation. Cornell Vet 49:97–115

Osorno MB, Solana PM, Perez JM, Lopez TR (1975) Study of an attenuated *Anaplasma marginale* vaccine in Mexico natural challenge of immunity in an enzootic area. Am J Vet Res 36:631–633

Paddock CD, Childs JE (2003) Ehrlichia chaffeensis: a prototypical emerging pathogen. Clin Microbiol Rev 16:37–64

Palmer GH, Barbet AF, Davis WC, McGuire TC (1986a) Immunization with an isolate-common surface protein protects cattle against anaplasmosis. Science 231:1299–1302

Palmer GH, Barbet AF, Kuttler KL, McGuire TC (1986b) Detection of an *Anaplasma marginale* common surface protein present in all stages of infection. J Clin Microbiol 23: 1078–1083

Palmer GH, Eid G, Barbet AF, McGuire TC, McElwain TF (1994) The immunoprotective *Anaplasma marginale* major surface protein 2 is encoded by a polymorphic multigene family. Infect Immun 62:3808–3816

Palmer GH, Brown WC, Rurangirwa FR (2000) Antigenic variation in the persistence and transmission of the ehrlichia *Anaplasma marginale*. Microbes Infect 2: 167–176

Palmer GH, Rurangirwa FR, McElwain TF (2001) Strain composition of the ehrlichia *Anaplasma marginale* within persistently infected cattle, a mammalian reservoir for tick transmission. J Clin Microbiol 39:631–635

Palmer GH, Knowles DP, Jr., Rodriguez JL, Gnad DP, Hollis LC, Marston T, Brayton KA (2004) Stochastic transmission of multiple genotypically distinct *Anaplasma marginale* strains in a herd with high prevalence of *Anaplasma* infection. J Clin Microbiol 42:5381–5384

Perry B, Sones K (2007) Science for development. Poverty reduction through animal health. Science 315:333–334

Pipano E (1995) Live vaccines against hemoparasitic diseases in livestock. Vet Parasitol 57:213–231

Prozesky L, Du Plessis JL (1987) Heartwater. The development and life cycle of *Cowdria ruminantium* in the vertebrate host, ticks and cultured endothelial cells. Onderstepoort J Vet Res 54:193–196

Pym AS, Brodin P, Brosch R, Huerre M, Cole ST (2002) Loss of RD1 contributed to the attenuation of the live tuberculosis vaccines *Mycobacterium bovis* BCG and *Mycobacterium microti*. Mol Microbiol 46:709–717

Raczniak G, Becker HD, Min B, Soll D (2001) A single amidotransferase forms asparaginyl-tRNA and glutaminyl-tRNA in *Chlamydia trachomatis*. J Biol Chem 276: 45862–45867

Ren Q, Paulsen IT (2005) Comparative analyses of fundamental differences in membrane transport capabilities in prokaryotes and eukaryotes. PLoS Comput Biol 1:e27

Ren Q, Chen K, Paulsen IT (2007) TransportDB: a comprehensive database resource for cytoplasmic membrane transport systems and outer membrane channels. Nucl Acids Res 35: D274–279

Renshaw HW, Vaughn HW, Magonigle RA, Davis WC, Stauber EH, Frank FW (1977) Evaluation of free-roaming mule deer as carriers of anaplasmosis in an area of Idaho where bovine anaplasmosis is enzootic. J Am Vet Med Assoc 170:334–339

Rikihisa Y (2003) Mechanisms to create a safe haven by members of the family Anaplasmataceae. Ann NY Acad Sci 990:548–555

Rikihisa Y (2006) *Ehrlichia* subversion of host innate responses. Curr Opin Microbiol 9:95–101

Ristic M (1968) Anaplasmosis. In: Weinman D, Ristic M (ed) Infectious Blood Diseases of Man and Animals, Vol 2. Academic Press, New York, pp 473–542

Ristic M, Carson CA (1977) Methods of immunoprophylaxis against bovine anaplasmosis with emphasis on use of the attenuated *Anaplasma marginale* vaccine. Adv Exp Med Biol 93:151–188

Ristic M, Krier JP (1984) Genus I *Anaplasma* Theiler 1970, 7AL. In: Krieg, NR, Holt, JG (eds) Bergey's Manual of Systematic Bacteriology, Vol 1. Williams and Wilkins, Baltimore, pp 720–722

Ristic M, Sibinovic S, Welter CJ (1968) An attenuated *Anaplasma marginale* vaccine. Proc Annu Meet US Anim Health Assoc 72:56–69

Rocha EP (2004) Codon usage bias from tRNA's point of view: redundancy, specialization, and efficient decoding for translation optimization. Genom Res 14:2279–2286

Rodriguez JL, Palmer GH, Knowles DP, Jr., Brayton KA (2005) Distinctly different *msp2* pseudogene repertoires in *Anaplasma marginale* strains that are capable of superinfection. Gene

Rurangirwa FR, Stiller D, French DM, Palmer GH (1999) Restriction of major surface protein 2 (MSP2) variants during tick transmission of the *ehrlichia Anaplasma marginale*. Proc Natl Acad Sci USA 96:3171–3176

Rurangirwa FR, Stiller D, Palmer GH (2000) Strain diversity in major surface protein 2 expression during tick transmission of *Anaplasma marginale*. Infect Immun 68:3023–3027

Rurangirwa FR, Brayton KA, McGuire TC, Knowles DP, Palmer GH (2002) Conservation of the unique rickettsial rRNA gene arrangement in *Anaplasma*. Int J Syst Evol Microbiol 52:1405–1409

Salzberg SL, Salzberg AJ, Kerlavage AR, Tomb JF (1998) Skewed oligomers and origins of replication. Gene 217:57–67

Schulein R, Guye P, Rhomberg TA, Schmid MC, Schroder G, Vergunst AC, Carena I, Dehio C (2005) A bipartite signal mediates the transfer of type IV secretion substrates of *Bartonella henselae* into human cells. Proc Natl Acad Sci USA 102:856–861

Scoles GA, Broce AB, Lysyk TJ, Palmer GH (2005) Relative efficiency of biological transmission of *Anaplasma marginale* (*Rickettsiales:* Anaplasmataceae) by *Dermacentor andersoni* (*Acari: Ixodidae*) compared with mechanical transmission by Stomoxys calcitrans (*Diptera: Muscidae*). J Med Entomol 42:668–675

Scoles GA, McElwain TF, Rurangirwa FR, Knowles DP, Lysyk TJ (2006) A Canadian bison isolate of *Anaplasma marginale* (*Rickettsiales:* Anaplasmataceae) is not transmissible by *Dermacentor andersoni* (*Acari: Ixodidae*), whereas ticks from two Canadian *D. andersoni* populations are competent vectors of a U.S. strain. J Med Entomol 43:971–975

Scoles GA, Ueti MW, Noh SM, Knowles DP, Palmer GH (2007) Conservation of transmission phenotype of *Anaplasma marginale* (*Rickettsiales:* Anaplasmataceae) strains among *Dermacentor* and *Rhipicephalus* ticks (*Acari: Ixodidae*). J Med Entomol 44:484–491

Shkap V, Pipano E (2000) Culture-derived parasites in vaccination of cattle against tick-borne diseases. Ann N Y Acad Sci 916:154–171

Shkap V, Molad T, Fish L, Palmer GH (2002) Detection of the *Anaplasma centrale* vaccine strain and specific differentiation from *Anaplasma marginale* in vaccinated and infected cattle. Parasitol Res 88:546–552

Smith RD, Woolf A, Hungerford LL, Sundberg JP (1982) Serologic evidence of *Anaplasma marginale* infection in Illinois white-tailed deer. J Am Vet Med Assoc 181:1254–1256

Smith RD, Levy MG, Kuhlenschmidt MS, Adams JH, Rzechula DL, Hardt TA, Kocan KM (1986) Isolate of *Anaplasma marginale* not transmitted by ticks. Am J Vet Res 47:127–129

Smith RD, Hungerford LL, Armstrong CT (1989) Epidemiologic investigation and control of an epizootic of anaplasmosis in cattle in winter. J Am Vet Med Assoc 195:476–480

Smith T, Kilborne FL (1893) Investigations into the nature, causation, and prevention of Texas or southern cattle fever, Vol. Bulletin 1 Bulletin of the Bureau of Animal Industries, United States Department of Agriculture, USA

Stephens RS, Kalman S, Lammel C, Fan J, Marathe R, Aravind L, Mitchell W, Olinger L, Tatusov RL, Zhao Q, Koonin EV, Davis RW (1998) Genome sequence of an obligate intracellular pathogen of humans: *Chlamydia trachomatis.* Science 282:754–759

Stich RW, Olah GA, Brayton KA, Brown WC, Fechheimer M, Green-Church K, Jittapalapong S, Kocan KM, McGuire TC, Rurangirwa FR, Palmer GH (2004) Identification of a novel *Anaplasma marginale* appendage-associated protein that localizes with actin filaments during intraerythrocytic infection. Infect Immun 72:7257–7264

Theiler A (1910) *Anaplasma marginale* (gen. spec. nov.). The marginale points in the blood of cattle suffering from a specific disease. In: Theiler A (ed) Report of the Government Veterinary Bacteriologist, 1908–1909. Transvaal, South Africa, pp 7–64

Theiler A (1911) Further investigations into anaplasmosis of South African cattle. In: Theiler A (ed.) First Report of the Director of Veterinary Research, Union of South Africa. Johannesburg, South Africa., pp 7–46

Theiler A (1912) Gallsickness of imported cattle and the protective inoculation against this disease. Agric J Union South Afr 3:7–46

Torioni de Echaide S, Knowles DP, McGuire TC, Palmer GH, Suarez CE, McElwain TF (1998) Detection of cattle naturally infected with *Anaplasma marginale* in a region of endemicity by nested PCR and a competitive enzyme-linked immunosorbent assay using recombinant major surface protein 5. J Clin Microbiol 36:777–782

Tripathy S, Tyler BM (2006) The repertoire of transfer RNA genes is tuned to codon usage bias in the genomes of *Phytophthora sojae* and *Phytophthora ramorum.* Mol Plant Microb Interact 19:1322–1328

Tumbula DL, Becker HD, Chang WZ, Soll D (2000) Domain-specific recruitment of amide amino acids for protein synthesis. Nature 407:106–110

Ueti MW, Reagan JO Jr, Knowles DP Jr, Scoles GA, Shkap V, Palmer GH (2007) Identification of midgut and salivary glands as specific and distinct barriers to efficient tick-borne transmission of *Anaplasma marginale.* Infect Immun 75:2959–2964

Vergunst AC, van Lier MC, den Dulk-Ras A, Stuve TA, Ouwehand A, Hooykaas PJ (2005) Positive charge is an important feature of the C-terminal transport signal of the VirB/D4-translocated proteins of *Agrobacterium.* Proc Natl Acad Sci USA 102:832–837

Viseshakul N, Kamper S, Bowie MV, Barbet AF (2000) Sequence and expression analysis of a surface antigen gene family of the rickettsia *Anaplasma marginale.* Gene 253:45–53

Vizcaino O, Carson CA, Lee AJ, Ristic M (1978) Efficacy of attenuated *Anaplasma marginale* vaccine under laboratory and field conditions in Colombia. Am J Vet Res 39:229–233

Welter CJ, Woods RD (1968) Preliminary evaluation of an attenuated *Anaplasma marginale* vaccine in cattle. Vet Med Small Anim Clin 63:798–802

Welter CJ, Ristic M (1969) Laboratory and field trials with an attenuated *Anaplasma marginale* vaccine. Proc Annu Meet US Anim Health Assoc 73:122–130

Wickwire KB, Kocan KM, Barron SJ, Ewing SA, Smith RD, Hair JA (1987) Infectivity of three *Anaplasma marginale* isolates for *Dermacentor andersoni.* Am J Vet Res 48:96–99

Wu M, Sun LV, Vamathevan J, Riegler M, Deboy R, Brownlie JC, McGraw EA, Martin W, Esser C, Ahmadinejad N, Wiegand C, Madupu R, Beanan MJ, Brinkac LM, Daugherty SC, Durkin AS, Kolonay JF, Nelson WC, Mohamoud Y, Lee P, Berry K, Young MB, Utterback T, Weidman J, Nierman WC, Paulsen IT, Nelson KE, Tettelin H, O'Neill SL, Eisen JA (2004) Phylogenomics of the reproductive parasite *Wolbachia pipientis* wMel: a streamlined genome overrun by mobile genetic elements. PLoS Biol 2:E69

Zaugg JL (1988) Experimental anaplasmosis in mule deer: persistence of infection of *Anaplasma marginale* and susceptibility to *A. ovis.* J Wildl Dis 24:120–126

Zaugg JL, Goff WL, Foreyt W, Hunter DL (1996) Susceptibility of elk (*Cervus elaphus*) to experimental infection with *Anaplasma marginale* and *A. ovis.* J Wildl Dis 32:62–66

4 *Ehrlichia*

Basil A. Allsopp[1](✉) and Jere W. McBride[2]

[1] Department of Veterinary Tropical Diseases, Faculty of Veterinary Science, University of Pretoria, Private Bag X04, Onderstepoort 0110, South Africa, Basil.Allsopp@up.ac.za

[2] Departments of Pathology and Microbiology and Immunology, Center for Biodefense and Emerging Infectious Diseases, Sealy Center for Vaccine Development, Institute for Human Infections and Immunity, University of Texas Medical Branch, Galveston, Texas, USA

4.1 Introduction

Ehrlichia species are gram negative obligately intracellular tick-borne alpha-proteobacteria belonging to the order Rickettsiales. There are five named species: *E. canis*, *E. chaffeensis*, *E. ewingii*, *E. muris*, and *E. ruminantium*, all of which naturally infect a variety of vertebrate hosts and two (*E. chaffeensis* and *E. ewingii*) are considered zoonotic pathogens (Dumler et al. 2001). In recent years, however, many other potential *Ehrlichia* species have been detected in animals and ticks by using molecular techniques, but remain largely uncharacterized (Parola et al. 2003; Teglas et al. 2005; Kawahara et al. 2006; Loftis et al. 2006; Shpynov et al. 2006; Allsopp et al. 2007). This review will be limited to three genetically well characterized species, *E. canis*, *E. chaffeensis*, and *E. ruminantium*, responsible for the most important veterinary and human ehrlichioses. The genomes of these three organisms have been completely sequenced, and in the case of *E. ruminantium* three different strains have been completely sequenced and comparatively analyzed.

4.1.1 History of *Ehrlichia*

The first known experimental investigations into a disease caused by an *Ehrlichia* species were conducted in South Africa in 1898, when Dixon (Dixon 1898) and Edington (Edington 1898) demonstrated that heartwater could be induced by blood passage from infected to susceptible animals. At the time, the importance of haemotophagous arthropods as vectors of human and animal disease was a hot research topic, and two years later Lounsbury published his finding that the bont tick (*Amblyomma hebraeum*) was the vector of heartwater in South Africa (Lounsbury 1900). The causative organism itself had not been demonstrated and was initially thought to be a virus (Spreull 1904). Another quarter of a century elapsed before the Canadian rickettsiologist, Edmund Vincent Cowdry, demonstrated that the infectious agent in the tissues of infected animals and ticks was a rickettsial organism, which he named *Rickettsia ruminantium*, now known as *Ehrlichia ruminantium* (Cowdry 1925a, b).

Following the identification of *E. ruminantium*, a rickettsia-like agent, *Rickettsia canis* (*E. canis*), was identified in 1935 in the monocytes of tick-infested Algerian dogs (Donatien and Lestoquard 1935). More than a half century passed after the discovery of *E. canis* before the first human infection with an *Ehrlichia* species was reported and *E. chaffeensis* was identified as an emerging zoonotic pathogen responsible for two human infections in 1991 (Anderson et al. 1991).

4.1.2 Taxonomic Position

E. canis and *E. ruminantium* were originally classified as species of *Rickettsia* (see Sect. 4.2.2) since both agents were discovered before the genus *Ehrlichia* was established. In 1937, Moshkovski created *Ehrlichia* as a subgenus of *Rickettsia*, in an attempt to segregate organisms with tropism for monocytes, and the type species *E. canis* was designated (Moshkovski 1937). However, the subgenus was erroneously proposed

Genome Mapping and Genomics in Animal-Associated Microbes
V. Nene, C. Kole (Eds.)

and subsequently (1945) *Ehrlichia* was elevated to full genus status (Moshkovski 1945). *R. ruminantium* was not designated as a member of the genus *Ehrlichia*, and became the sole species in the newly created genus *Cowdria*, based primarily on the fact that the two organisms exhibited tropism for different cell types (monocytes and endothelial cells) (Moshkovski 1947). *E. chaffeensis* was discovered in 1991 (see Sect. 4.2.2) and was immediately assigned to its present classification.

The advent of molecular phylogenetics has completely revolutionized systematics, especially bacterial systematics, in the last decade. This has resulted in the reclassification of numerous organisms in four genera (*Anaplasma, Neorickettsia, Ehrlichia*, and *Cowdria*) within the order Rickettsiales. The most commonly used gene for bacterial classification is the small subunit ribosomal RNA gene (*rrs*; 16S rRNA), and the Ribosomal Database Project (RDP) now contains 489,840 16S rRNA aligned gene sequences (RDP Release 9.59, 5 March 2008, http://rdp.cme.msu.edu/) (Cole et al. 2007). As the 16S genes of more organisms of the family Rickettsiaceae were sequenced during the 1990s, it became evident that organisms classified in the genus *Ehrlichia* did not constitute a monophyletic group. Organisms classified as *Anaplasma*, *Cowdria*, and *Neorickettsia* were found to constitute three distinct clades, each of which included various organisms genetically consistent with *Ehrlichia* species. A reorganization of the order Rickettsiales was undertaken, based on 16S rRNA and *gro*ESL gene sequences (Dumler et al. 2001), which resulted in the order being split into two families, Rickettsiaceae and Anaplasmataceae, and reclassification of the members of the genera in those families. The genus *Ehrlichia* is now included in the family Anaplasmataceae, along with the genera *Anaplasma*, *Neorickettsia*, and *Wolbachia*. The type specimen of *E. ruminantium* is the Welgevonden strain (Du Plessis 1985), which was obtained from an *Amblyomma hebraeum* tick collected in the north eastern Transvaal in South Africa, the same geographical area as the originally identified causative agent of heartwater (Cowdry 1925a). The type specimen of *E. canis* is the Oklahoma strain obtained from a naturally infected dog in Oklahoma, USA (Dawson et al. 1991a, b). The type specimen of *E. chaffeensis* is the Arkansas strain obtained from a naturally infected human patient from Fort Chaffee, Arkansas, USA (Dawson et al. 1991a, b).

4.1.3 Ultrastructure

Ehrlichia species are gram negative bacteria, which stain purplish-blue with Romanovsky type stains (Fig. 1) (Cowdry 1925a, b; Rikihisa 1991). Common to all members of the family Anaplasmataceae, *Ehrlichia* species reside in an intracellular vacuole (morula) bounded by a membrane derived from the eukaryotic host cell membrane (Dumler et al. 2001). The individual bacteria are coccoid to pleomorphic and vary in size from small (0.4 µm), medium (0.7 µm), large (1 µm), and occasionally very large (≤2 µm) (Pienaar 1970; Popov et al. 1995, 1998). Two distinct morphological forms can

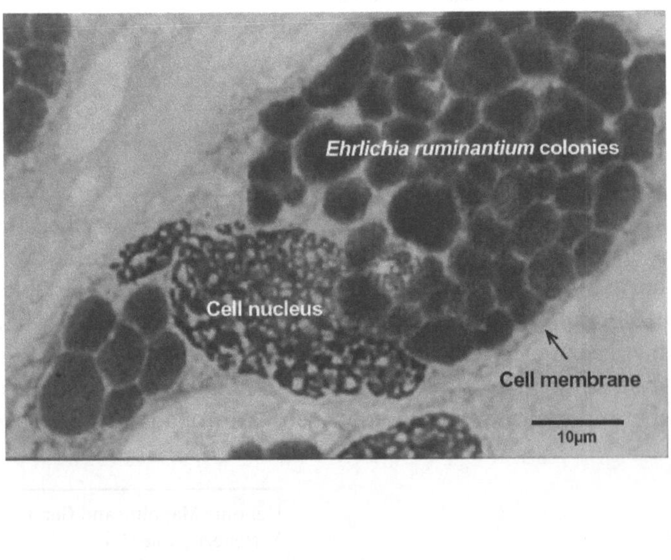

Fig. 1 Light micrograph of dense-cored elementary bodies of *Ehrlichia ruminantium* growing in tissue culture in bovine fibroblastoid cells. The cells were allowed to dry, fixed with methanol, and stained with eosin–methylene blue (Kyro-Quick stain; Kyron Laboratories, Benrose, South Africa). Bar = 10 µm. Reproduced with permission from Dr. Erich Zweygarth, Onderstepoort Veterinary Institute, Private Bag X05, Onderstepoort 0110, South Africa

be distinguished, dense-cored and reticulate cells (Fig. 2), which both have a characteristic gram negative envelope including a cytoplasmic membrane and an outer membrane (cell wall) separated by a periplasmic space. In contrast to free-living gram negative bacteria, however, ehrlichial cell walls do not contain peptidoglycan. The dense-cored cells are usually small, coccoid, and contain an electron-dense nucleoid that occupies most of the cytoplasm. By analogy to Chlamydiae these cells are also often referred to, particularly in the early literature, as elementary bodies. Reticulate cells are larger, have a dispersed filamentous nucleoid, and their cytoplasm contains ribosomes. Sometimes organisms of intermediate size and electron density are referred to as intermediate bodies (Pienaar 1970; Prozesky 1987a; Popov et al. 1995, 1998). Reticulate cells multiply by binary fission and intermediate bodies can also be observed dividing. It is suggested that dense-cored cells are the infectious form of the organism (Prozesky et al. 1986; Jongejan et al. 1991a–c; Zhang et al. 2007). The morphology of *E. ruminantium* in experimentally infected ticks appears to be similar to that in mammalian cells, consisting of both dense-cored and reticulate forms (Kocan and Bezuidenhout 1987).

4.1.4 Life-Cycle

All *Ehrlichia* species are transmitted by nymphal or adult ticks, which become infected as larvae or nymphs, respectively, by ingesting a blood meal from a persistently infected vertebrate host. *Ehrlichia* species are maintained transstadially, primarily in *Amblyomma* ticks, unlike *Rickettsia* species, which are maintained transovarially. Studies with *E. ruminantium* demonstrate the appearance of intracellular colonies in tick gut cells 15 days after the ticks have fed to repletion (Kocan and Bezuidenhout 1987). After moulting, infected ticks can transmit infectious elementary bodies after feeding for 38 h (nymphs) or 75 h (adults) (Bezuidenhout 1987). During the period between attachment and blood meal acquisition, *E. ruminantium* colonies appear for the first time in the salivary gland (Kocan and Bezuidenhout 1987). It has been suggested that the organisms remain largely dormant when the ticks are inactive, but become activated when feeding begins, possibly in response to chemical signals from host blood (Katavolos et al. 1998). In the case of *E. ruminantium* warming of infected ticks to 37°C without allowing them to feed does not result in salivary gland proliferation of the ehrlichial organisms nor do the ticks become infective (Bezuidenhout 1987). The *E. ruminantium* organisms delivered from infected ticks are much more virulent than elementary bodies delivered artificially by an infected blood needle challenge (Collins et al. 2003). The basis for the enhanced virulence is not yet understood, but it may be related to the regulation of tick and/or pathogen genes prior to transmission, which then enhance their survival once they have been inoculated into the mammalian host.

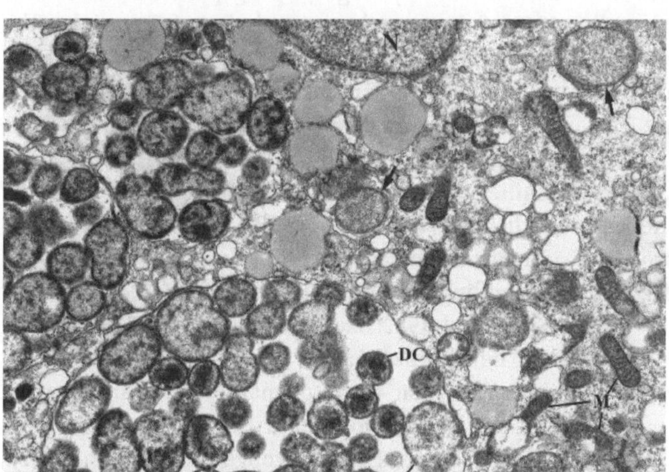

Fig. 2 Electron photomicrograph of multiple *E. canis* morulae containing both smaller dense-cored (DC) and larger reticulate cell (RC) morphological forms of the organism. Some morulae contain a single *E. canis* organism (*arrows*). Host cell nucleus (N) and mitochondria (M) are also seen adjacent to the morulae. Bar = 1 μm. Reproduced with permission from Dr. V. L. Popov, University of Texas Medical Branch, Galveston, Texas, USA

4.1.5 Entry and Development in Host Cells

Morphological and molecular features of the development of *Ehrlichia* have been described in both mammalian (Du Plessis 1982; Zhang et al. 2007) and invertebrate host cells (Kocan et al. 1987a). Transmission electron microscope studies of in vitro cultures in mammalian endothelial cells infected with *E. ruminantium* have shown that the infectious dense-cored cells transform into reticulate forms 2–4 days post-infection, and then undergo binary fission for about 2 days. Intermediate bodies are seen 4–5 days post-infection and finally large numbers of dense-cored elementary bodies are seen 5–6 days after infection, at which time the host cells rupture spontaneously (Fig. 1) (Jongejan et al. 1991a–c). In the mammalian host, *E. ruminantium* initially replicates in reticulo-endothelial cells (monocytes/macrophages) in lymph nodes, and rupture of these cells releases dense-cored elementary bodies, which are distributed in the bloodstream and subsequently infect endothelial cells (Du Plessis 1970; Prozesky and Du Plessis 1987).

The molecular mechanisms involved in the host cell entry process of *E. chaffeensis* are well characterized. The initial interaction involves binding of dense-cored ehrlichial organisms to host cell receptors E- and L-selectin, triggering receptor mediated endocytosis via a clathrin-independent process (Barnewall et al. 1997; Zhang et al. 2003, 2007) involving caveolae and glycosylphosphatidylinositol-anchored proteins (Lin and Rikihisa 2003b). The vacuoles in which the organisms reside have characteristics of early endosomes and are inhibited from fusing with lysosomes by a two component regulatory system (Barnewall et al. 1997; Kumagai et al. 2006). After entry the dense-cored ehrlichial cells transform into reticulate cells that divide by binary fission and then mature into dense-cored cells that are released from the infected host cell (Zhang et al. 2007). Each morphological type of *Ehrlichia* (dense-cored, intermediate, and reticulate) differentially expresses specific surface proteins (Popov et al. 2000; Doyle et al. 2006; Zhang et al. 2007).

4.1.6 Epidemiology

E. ruminantium is transmitted by ticks of the genus *Amblyomma* (Table 1) and is the etiological agent of heartwater, or cowdriosis, in cattle, sheep, goats, and some wild ruminants. Ten *Amblyomma* species capable of transmitting the organism occur in Africa, but the most important vectors are *A. variegatum* and *A. hebraeum* (Bezuidenhout 1987). The distribution of the disease coincides with that of the vector species and encompasses the whole of sub-Saharan Africa and its offshore islands; it also includes the French Antillean islands of Guadeloupe and Antigua, where infected *Amblyomma variegatum* ticks were introduced, possibly as early as the eighteenth century (Maillard and Maillard 1998). The native American tick *A. maculatum* is also known to be an effective vector of *E. ruminantium*, having a vector potency similar to that of *A. variegatum* (Mahan et al. 2000). African wild ruminants are probably the original reservoir of the disease (Neitz 1967) and 15 species have been shown to be susceptible to infection, either naturally or experimentally (Table 1). The disease can also be maintained in a domestic stock population in the absence of wild ruminants, a situation which occurs in Madagascar, Guadeloupe, and São Tomé (Uilenberg 1983). Ten species of non-African ruminants are also known to be susceptible (Table 1), including the very widely distributed American white-tailed deer (*Odocoileus virginianus*).

After its initial discovery in 1935, *E. canis* remained largely unnoticed until it was reported in the United States in 1963 (Ewing 1963). It was subsequently identified as the aetiological agent responsible for outbreaks of a epizootic highly fatal idopathic hemorrhagic syndrome in military working dogs in Vietnam in late 1968, and also of a disease (canine tropical pancytopenia) reported in 1963 among British military dogs in Singapore (Huxsoll et al. 1969). Subsequently, transmission of *E. canis* by *Rhipicephalus sanguineus* ticks (the brown dog tick) was experimentally demonstrated (Groves et al. 1975). The disease caused by *E. canis* is now known as canine monocytic ehrlichiosis (CME), which is distinct from canine granulocytic ehrlichiosis (CGE) caused by *Ehrlichia ewingii*. *E. canis* has a worldwide distribution, mainly in tropical and subtropical regions, corresponding with the distribution of *R. sanguineus* (Keefe et al. 1982; Skotarczak 2003).

The distribution of *E. chaffeensis* in the USA coincides with that of its primary vector, *Amblyomma americanum* (the lone star tick), which is distributed from west central Texas throughout the southeastern, south central, and mid-Atlantic states (Childs and Paddock 2003; Paddock and Childs 2003). The white-tailed deer (*Odocoileus virginianus*) is the primary

Table 1 The susceptibility of various wild ruminants to *Ehrlichia ruminantium* infection. Adapted from Oberem and Bezuidenhout 1987

Common name	Scientific name	Susceptibility		
		Clinical	Subclinical	Refractory
African hosts				
Giraffe	*Giraffa camelopardalis*		+	+
Black wildebeest	*Connochaetes gnu*	+	+	+
Blue wildebeest	*Connochaetes taurinus*			+
Red hartebeest	*Alcelaphus buselaphus*			+
Blesbok	*Damaliscus dorcas phillipsi*	+	+	
Duiker	*Cephalophus sp.*			+
Springbok	*Antidorcas marsupialis*	+		
Scimitar-horned oryx	*Oryx dammah*			+
African buffalo	*Syncerus caffer*		+	
Bushbuck	*Tragelaphus scriptus*			+
Eland	*Taurotragus oryx*	+	+	
Non-African hosts				
White-tailed deer	*Odocoileus virginianus*	+		
Fallow deer	*Cervus dama*	+		
Timor deer	*Cervus timorensis*	+		
Water buffalo	*Bubalus bubalis*	+		
Barbary sheep	*Ammotragus lervia*	+		
Himalayan tahr	*Hemitragus jemlahicus*	+		
Nilgai	*Boselaphus tragocamelus*	+		
Blackbuck	*Antilope cervicapra*	+		
Mouflon	*Ovis orientalis*	+		

Note: Blank cells indicate lack of information for the respective category

reservoir, and canids may act as an important secondary natural reservoir (Paddock and Childs 2003). *E. chaffeensis* has also been detected in *Dermacentor variabilis* and *Ixodes pacificus* ticks (Walker et al. 2004). The recent emergence of *E. chaffeensis* as a human health problem is apparently the result of increases in reservoir host and vector populations, increases of susceptible human populations, and increasing human habitation in rural tick-infested areas (Paddock and Childs 2003).

4.1.7 Diseases

The first reference to a disease that may have been caused by a species of *Ehrlichia* was made by the Voortrekker pioneer, Louis Trichardt, in 1838. While trekking through the northern Transvaal of South Africa, many of Trichardt's sheep died 3 weeks after they had suffered massive tick infestation, and it was later speculated that the disease was heartwater (Neitz 1947). *E. ruminantium* infects wild and domestic ruminants, which may suffer from peracute, acute, subacute, or clinically inapparent forms of the disease that manifest with a wide range of clinical signs. In the naturally acquired disease the morbidity and mortality rates are influenced by the species, breed, and age of the animals affected, the virulence of the *E. ruminantium* stock, and sometimes the season (Uilenberg 1983). Infected animals that show clinical symptoms usually die unless promptly treated with antibiotics (Van Amstel and Oberem 1987; Van de Pypekamp and Prozesky 1987).

The peracute form of heartwater is most likely to affect exotic *Bos taurus* breeds of cattle and goats such as Angoras, and peracutely affected animals often collapse suddenly and die after a few paroxysmal convulsions. Clinical signs of illness are frequently not observed prior to death except in some cases where fever may be the only sign of illness (Neitz 1968; Uilenberg 1983). Acute heartwater, the most common form of the disease, is characterized by a fever (≥40°C), which usually persists for 3–6 days. The body temperature usually falls below normal shortly before death and the majority of animals manifest terminal nervous signs. The subacute form of heartwater is characterized by a fever, which may remain high for 10 days or longer, but the clinical signs are less severe than in the acute form. Animals may die suddenly or gradually recover within a few days. Finally, the clinically inapparent form of the disease (carrier state) is common in animals that have been treated or have recovered naturally from infection (Andrew and Norval 1989; Camus 1992; Peter et al. 2002). These animals appear to be healthy, and yet remain chronic carriers, which are infective to ticks for months or years, a situation which poses serious problems for long term control of the disease (Yunker 1996).

E. canis primarily infects monocytes and macrophages in dogs, causing the potentially fatal hemorrhagic disease known as canine monocytic ehrlichiosis (CME). The disease affects all breeds of dogs, but German shepherds appear to be more susceptible to infection and exhibit increased disease severity and mortalities (Huxsoll et al. 1969, 1970; Van Heerden 1982). The disease has three characterized stages: acute, subclinical, and chronic (Buhles et al. 1974; Van Heerden 1982). Usually 10–14 days after transmission of *E. canis* from an infected tick, the acute phase develops, characterized by depression, fever, pallor of mucous membranes (anemia), lethargy, anorexia and weight loss, and hematological abnormalities, including thrombocytopenia, anemia, and lymphopenia (Buhles et al. 1974; Reardon and Pierce 1981). The acute phase usually resolves spontaneously 2–4 weeks after onset of symptoms and then progresses to the subclinical phase in which the dogs become carriers and organisms are sequestered in the spleen. During this phase, clinical symptoms are mild or absent but hematological abnormalities persist. In experimental infections, more than 60% of infected dogs enter the chronic stage 40–80 days after infection, but longer subclincal phases have been reported (Nims et al. 1971). The chronic stage is characterized by more severe thrombocytopenia, leukopenia, hypergammaglobulinaemia, and hemorrhage. Death occurs due to severe internal hemorrhage or secondary bacterial infections, and perivascular plasma cell infiltrates in organs is a common histiological feature (Nims et al. 1971; Buhles et al. 1974; Codner et al. 1992).

E. chaffeensis is the etiological agent of an emerging zoonosis, human monocytic ehrlichiosis (HME), and also causes mild to severe disease in dogs (Dawson and Ewing 1992; Breitschwerdt et al. 1998). HME is present as an undifferentiated flu-like illness, with most patients reporting fever, headache, malaise, dizziness, chills, and myalgia. A large proportion (40–60%) of patients with HME requires hospitalization, and meningoencephalitis, renal failure, and adult respiratory distress syndrome become life-threatening manifestations in severe cases. Such cases are most often described in older patients (>60 years old) and immunocompromised individuals. The case fatality rate is approximately 3%, owing to the difficulty of making accurate clinical diagnoses, and because of the absence of antibodies in early infections at a time when therapeutic decisions are critical (McQuiston et al. 1999; Patel and Byrd 1999).

4.1.8 Economic Importance

While *E. canis* and *E. chaffeensis* are certainly of social and public health importance, they do not constitute a significant economic burden. *E. ruminantium*, on the contrary, poses a serious economic burden in countries where it is endemic, and has the potential for similar consequences in countries to which it may be introduced. Within the pan African distribution of the disease (Sect. 4.1.6) it has been estimated that 150,000,000 domestic ruminants are at risk of infection (Minjauw and McLeod 2003). The disease also prevents the introduction of high-producing animals into an endemic area to upgrade local stock, because the introduced animals are frequently susceptible to the rapidly fatal peracute form of the disease (Sect. 4.1.7). Heartwater also causes serious losses when animals are moved from heartwater-free to heartwater-infected areas (Simpson et al. 1987). Heartwater losses are regarded as just another factor to be endured in the endemic areas of Africa, and so presumptive diagnoses are usually not confirmed and the real economic cost is therefore difficult to quantify. Estimates have been made, however, which indicate that the losses can be very high. One survey in Zimbabwe put the cost at US$ 5.6 million year (Mukhebi et al. 1999), and another report estimated annual costs for Angola, Botswana, Malawi, Mozambique, South

Africa, Swaziland, Tanzania, and Zambia to a total of US$ 44.7 million (Minjauw and McLeod 2003). Whatever the actual figures may be, the economic importance of heartwater in Africa is comparable to that of East Coast fever, trypanosomosis, rinderpest, and dermatophilosis (Provost and Bezuidenhout 1987). The serious costs of endemic heartwater for livestock owners explains the constant worry over the threat that the disease could spread from the Antillean islands to the American mainland (Deem 1998). Large areas of north, central, and south America are climatically suitable for *A. variegatum*, and moreover *A. maculatum* and the white-tailed deer already constitute a viable native sylvatic tick-host pair for the maintenance of *E. ruminantium* (Sect. 4.1.6). The establishment of endemic heartwater in the Americas will remain a potent economic threat until a safe and effective vaccine becomes available.

4.1.9 *Ehrlichia* as Zoonotic Pathogens

Ehrlichia chaffeensis is recognized as an important emerging zoonotic agent in the United States (Sect. 4.1.6), and by far the majority of cases have occurred in states where lone star ticks and white-tailed deer are abundant. Other animals that are hosts of *A. americanum* may be reservoirs of *E. chaffeensis*, including wild canids, domestic ruminants, and birds (Paddock and Childs 2003). Reported cases of HME are seasonal, with the peak incidence in May through July, and increased risk is associated with recreational and occupational activities that involve human interactions in areas inhabited by *A. americanum* ticks and white-tailed deer (Paddock and Childs 2003). The annual incidence based on passive surveillance is at least 3.2 cases per 100,000 population (McQuiston et al. 1999; Walker et al. 2004). However, substantially higher (10–100 times) incidence has been reported in active surveillance studies in HME endemic regions such as southeastern Missouri (Olano et al. 2003b).

E. ruminantium is considered to be solely a ruminant pathogen, but in 2001 *E. ruminantium* DNA (the 16S rRNA gene) was detected by PCR in a dog suffering from "atypical" canine ehrlichiosis in South Africa (Allsopp and Allsopp 2001). An identical *E. ruminantium* 16S rRNA gene sequence and pCS20 and *gltA* sequences identical to known *E. ruminantium* genotypes were later detected in serum from a woman who had died 3 weeks after the death of her pet dog from "biliary fever" (Allsopp et al. 2005). Subsequently, two cases of fatal encephalitis of unknown origin in children also yielded 16S rRNA, pCS20, and *gltA* sequences typical of *E. ruminantium* (Allsopp et al. 2005). Although ehrlichial organisms have not been isolated in any of these cases, it appears that *E. ruminantium* has infected humans and dogs and this could signal the emergence of *E. ruminantium* as a zoonotic agent in heartwater-endemic areas of Africa. Because of the rapid course of the disease, it has been suggested that children in these areas who present with a clinical picture of encephalitis of unknown origin should be treated with doxycycline (Louw et al. 2005).

E. canis is globally distributed and there is increasing evidence of human infections with this agent. The first reported case involved a healthy asymptomatic adult male in Venezuela, who was in close contact with dogs carrying the organism (Perez et al. 1996). More recently, *E. canis* DNA was detected in six humans with clinical signs consistent with HME, in Venezuela (Perez et al. 2006). Despite these observations, human infections with *E. canis* do not appear to occur with high frequency.

4.2 Genomics

The term genomics is currently often used to define the study of complete genomes, although the term was first coined to mean "a marriage of molecular and cell biology with classical genetics fostered by computational science" (McKusick and Ruddle 1987). In this section we use the broader definition of the term. Molecular characterization of *Ehrlichia* species was for a long time precluded because the organisms were difficult to cultivate in vitro, a fact that currently prevents full characterization of *E. ewingii*. However, most ehrlichial agents have been cultivated, including *E. canis*, the first to be cultured (Nyindo et al. 1971), *E. ruminantium* (Bezuidenhout et al. 1985), and most recently *E. chaffeensis* (Dawson et al. 1991a, b). These advances were rapidly followed by the cloning and sequencing of genes for classification and phylogeny, molecular diagnostics, the characterization of surface antigens, and the recent completion of genome sequences for all three species.

4.2.1 Genetic Variability of *Ehrlichia*

Genetic Variability of *E. ruminantium*

Because a reliable vaccine for heartwater is currently unavailable, a substantial amount of research has been devoted towards developing a broadly efficacious vaccine. Understanding the genetic, and ultimately the antigenic, variability among strains of the organism is obviously of vital importance for vaccine development. Surprisingly, for much of the twentieth century, *E. ruminantium* was considered to be a relatively homogeneous organism, but the genomics revolution has recently changed this perception. Genetic characterization has revealed that *E. ruminantium* is, in fact, a relatively diverse organism. This finding is also supported by observations in nature, suggesting a diverse host range that includes canines and humans (Sect. 4.1.8). In the following section, we consider the well characterized genetic features that may provide insight into the most important pathogenic characteristics of the organism and differences between strains.

E. ruminantium Ribosomal RNA Genotypes

The 16S rRNA gene has been very widely used as a taxonomic and phylogenetic tool for the classification of bacteria (Olsen and Woese 1993) and we noted in Sect. 4.1.2 how the gene has been used to redefine the genus *Ehrlichia*. This was the first gene of *E. ruminantium* to be sequenced, and it was sequenced concurrently by two different laboratories using different stocks of the organism. One group used the Crystal Springs isolate, a classical heartwater-producing stock from southern Africa (Dame et al. 1992), while the other group used an isolate from Senegal, West Africa (Van Vliet et al. 1992). The two 16S rRNA sequences were divergent and this prompted wider surveys to examine strain diversity using this gene (Allsopp et al. 1997; Allsopp and Allsopp 2001). Currently eight distinct 16S rRNA genotypes of *E. ruminantium* have been identified, one of which (Pretoria North) has not been isolated in tissue culture (Allsopp et al. 2007) (Table 2).

Because of the biological differences between these eight genotypes and the widespread use of the 16S rRNA gene for bacterial classification, it is important to note the criteria by which they are all considered to be members of the species *E. ruminantium*. The definition of what constitutes a bacterial species has not been completely defined, but some empirical guidelines have been generally accepted in the past, such as a whole-genome DNA–DNA hybridization level of ≥70% (Wayne et al. 1987) or a 16S gene sequence identity of ≥97% (Stackebrandt and Goebel 1994). A recent comparison of wholly sequenced bacterial genomes has shown that a 70% DNA–DNA reassociation level corresponds, on average, to 93–94% average nucleotide identity (ANI) and to a 99% 16S gene sequence identity (Konstantinidis and Tiedje 2005). These authors have suggested that the criteria for defining a prokaryotic species should be put at 94–99% ANI, corresponding to 99.0–99.9% 16S identity. Each of the eight *E. ruminantium* 16S genes, which we are discussing, has a sequence identity of >99.4% with respect to the others, and they are therefore all *E. ruminantium* by the latest stringent criteria. A maximum likelihood tree of these sequences was constructed using the PHYML program (Guindon and Gascuel 2003), together with orthologs from six other *Ehrlichia spp.* and outgroup rooted using *Anaplasma marginale* (Fig. 3). The *E. ruminantium* genotypes form a tight cluster, well distinguished from all the other *Ehrlichia* species with the exception of the enigmatic *Ehrlichia* species from Panola Mountain, Georgia, USA (Loftis et al. 2006). The 16S gene sequence of this organism has an identity >99.2% with respect to all eight *E. ruminantium* genes, and so there appears to be justification for considering that it, too, should be identified as a strain of *E. ruminantium*. We return to this when discussing the biological differences between *E. ruminantium* strains in the following sections.

E. ruminantium Infectivity and Pathogenicity

Infectivity and pathogenicity are not solely the properties of the *E. ruminantium* genotype involved, but also depend on the tick vector as noted in Sect. 4.1.4. This presents a major practical challenge to vaccine development research and it will be discussed in more detail in Sect. 4.2.5. This section explores

Table 2 Data for eight different 16S rRNA genotypes of *E. ruminantium*

Genotype	Origin		Pathogenicity			Remarks	Reference
	Geographical	Biological	Cattle	Sheep goats	Mice		
Ball3	South Africa	Bovine	+	+	–	S. African blood "vaccine" isolate	Haig 1952
Gardel	Guadeloupe, French West Indies	*A. variegatum*	+	+	0	Exceptionally virulent in Dutch goats.	Uilenberg et al. 1985
Kiswani	Kenya	Bovine	+	+	ND		Kocan et al. 1987b
Mara 87/7	South Africa	*A. hebraeum*	+	+	+	Widespread classical heartwater-producing genotype in S. Africa	Du Plessis et al. 1989
Omatjenne	Namibia	*H. truncatum*	–	+/–	–	From *Amblyomma*-free area, unlikely to be transmitted by *H. truncatum*	Du Plessis 1990
Pretoria North	South Africa	Dog	ND	ND	ND	From a dog suffering "atypical ehrlichiosis." Not available in tissue culture	Allsopp and Allsopp 2001
Senegal	Senegal	Bovine	+	+	+/–		Jongejan et al. 1988
Welgevonden	South Africa	*A. hebraeum*	+	+	+	Widespread classical heartwater-producing genotype in S. Africa	Du Plessis 1985

\+ Pathogenic; +/– Mildly pathogenic; – Nonpathogenic; 0 Noninfective; ND Not done

what we know about the genetic variations between strains with differing infectivity and pathogenicity. Assessing infectivity and pathogenicity prior to the mid-1990s was very difficult, since there were no methods available to genotype *E. ruminantium* stocks. Hence it was not possible to confirm that the different stocks contained individual genotypes and this may have resulted in inconsistent results and conclusions among various studies. The ability to infect mice was the earliest method used to assess variability between stocks, and using this method three different types of infectivity and pathogenicity were recognized: there were pathogenic stocks that killed the host with varying degrees of virulence; there were stocks that were infective but not pathogenic; and there were noninfective stocks that failed to establish any infection. The pathogenicities of the different *E. ruminantium* genotypes are illustrated in Table 2.

The Welgevonden and Mara 87/7 stocks produce the classical peracute and acute heartwater clinical disease in ruminants, as described in Sect. 4.1.7, but not all strains of the organism produce similar manifestations. The Omatjenne strain (Table 1) appears to be the first apathogenic strain of *E. ruminantium* to be isolated. It was obtained from a single *Hyalomma truncatum* tick taken off a healthy cow on a farm in a heartwater- and *Amblyomma*-free area of Namibia

Fig. 3 Maximum likelihood phylogenetic tree inferred from an alignment of small subunit ribosomal RNA gene sequences. The taxa are eight *E. ruminantium* genotypes and six other species of *Ehrlichia* with *Anaplasma marginale* as the outgroup. The scale bar indicates two nucleotide substitutions per 1,000 bases

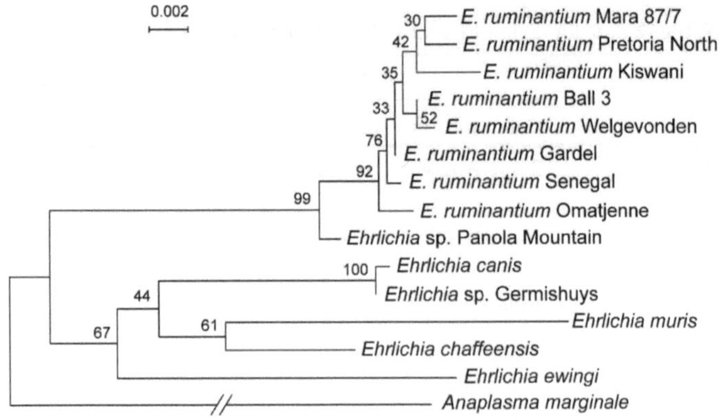

(Du Plessis 1990). Despite being healthy, 81% percent of the cattle on this farm tested seropositive for heartwater using an immunofluorescent antibody test (Du Plessis and Malan 1987). The Omatjenne 16S genotype was later detected, by PCR and probing, in 70 healthy boergoats in a heartwater-free area of South Africa (Allsopp et al. 1997), which suggests that strains carrying this 16S genotype may be apathogenic to small ruminants as well as to cattle. Despite the fact that the original isolation of the Omatjenne strain was from a *Hyalomma truncatum* tick, this species may not be the primary vector of the organism, since the larvae and nymphs feed exclusively on scrub hares (*Lepus saxatilis*) and rodents, while only the adults feed on cattle (Allsopp et al. 2007).

Other apparently nonpathogenic cases of infection are well documented. Examination of healthy goats in a heartwater-free area of South Africa has shown, by PCR amplification, that many animals have very low levels of infection, with 16S genotypes identical to those of both nonpathogenic and virulent stocks of *E. ruminantium* (Allsopp et al. 2007). In the study area, the known tick vectors *A. variegatum* and *A. hebraeum* are not found, and although the "tortoise tick" *A. marmoreum* does occur, the ticks responsible for transmitting the nonpathogenic organisms have not been identified. Also of considerable interest is the *Ehrlichia* species transmitted by *A. americanum* ticks from Panola Mountain State Park in Georgia, USA, described in Sect. 4.1.2. The Panola Mountain *Ehrlichia* species was identified in *A. americanum* ticks and was determined to be related to, but distinguishable from, other *E. ruminantium* genotypes by PCR amplification and DNA sequencing of multiple genes (16S rRNA, *glt*A, *map*1, *map*1-1, and *map*2) (Loftis et al. 2006). Whether this is a recent introduction into the USA, or a long-standing but previously unknown infection, is not clear, but the organism did not produce clinical heartwater in a goat on which infected *A. americanum* ticks were fed. Furthermore, it does seem likely that if this *Ehrlichia* species caused clinical disease, it would have been previously noted by local veterinarians. It is interesting that *A. americanum* ticks are not capable of transmitting some heartwater-producing strains of *E. ruminantium* of African and Caribbean origin (Uilenberg 1982; Uilenberg et al. 1985), indicating that much more work needs to be done on the pathogenicity and transmissibility of the Panola Mountain *Ehrlichia* sp.

The Ball 3 stock needs mention here since it is used as an infection and treatment heartwater "vaccine" in South Africa because of its atypical pathogenicity in ruminants (Van der Merwe 1987). This stock produces an early temperature rise days before the animal becomes seriously sick, allowing antibiotics to be administered during the febrile response. Conversely, the highly virulent Welgevonden stock may cause death within hours of a rapid temperature rise and is therefore not suitable for use in infection and treatment. It is unfortunate, however, that the Ball 3 vaccine confers only limited protection against virulent field challenge with classical heartwater-causing stocks like Welgevonden (Du Plessis et al. 1989).

Several studies have shown that there are considerable sequence polymorphisms of various *E. ruminantium* genes both within and between 16S genotypes (Allsopp et al. 2001, 2003; Van Heerden et al. 2004b). The conclusions seem to be as follows: many different *E. ruminantium* genotypes are in circulation; they have a range of pathogenicities and infectivities; these characteristics do not correlate with the 16S genotype; this gene cannot therefore be used on its own as a marker for virulence. It is in fact unlikely that any single marker gene will be found to be suitable for virulence characterization. Although the genes responsible for virulence have not been identified, they are probably scattered throughout the genome, and since extensive recombination occurs naturally between different genotypes of *E. ruminantium* (Allsopp and Allsopp 2007), different combinations of virulence genes will arise in newly generated strains. A single marker gene is unlikely to be found to be a valid indicator of differently assorted genes contributing to the overall virulence of new genotypes. A further discussion of what is known about recombination in *E. ruminantium* will be found in the section on replication, repair, and recombination in Sect. 4.2.3.

Emergence of Novel *E. ruminantium* Phenotypes

Canine ehrlichiosis is common in South Africa and it is normally diagnosed on the basis of clinical symptoms and blood smear examination, but it has long been noted that dogs often show clinical symptoms suggestive of ehrlichiosis, but morulae cannot be visualized in blood smears. Such cases are often tested using a PCR assay specific for *E. canis* (Oklahoma) (McBride et al. 1996) and most of them give a negative result. Many of these cases, however, test positive for *E. ruminantium* by the pCS20 assay (Mahan et al. 1992; Van Heerden et al. 2004b) and in one instance sequences of other *E. ruminantium*-specific genes were obtained (Allsopp and Allsopp 2001). Although the presence of *E. ruminantium* DNA was confirmed in these cases, unequivocal evidence has not been established linking the illness with *E. ruminantium*.

The detection of *E. ruminantium* DNA sequences in three human serum samples has been mentioned in Sect. 4.1.8. These patients were unrelated individuals who were not obviously immunocompromised, the pCS20 assay was positive, and 16S, pCS20, and *glt*A sequences identical to known *E. ruminantium* genotypes were obtained (Allsopp et al. 2005). Two of the cases were children who died with lesions strongly suggestive of *E. ruminantium* infection, and brain tissue samples from these cases were subjected to immunohistochemical staining for *Rickettsia* species and were found to be negative (Louw et al. 2005). The presence of *E. ruminantium* in these patients was confirmed by multiple detection methods, which suggests that infection by *E. ruminantium* could have been responsible for the deaths. Based on this evidence, it seems that some strains of *E. ruminantium* may cause fatal infections in humans.

Genetic Variability of *E. canis*

E. canis is globally distributed and evidence of infection has been documented in Africa, India, Asia, Southeast Asia, South America, North America, Europe, Caribbean islands, and Middle Eastern countries. The isolates best characterized at the molecular level originated from the United States, where *E. canis* shows very little genetic diversity in 16S rRNA genes, and also in genes encoding immunoreactive proteins, such as major outer membrane proteins (p28/p30) and tandem repeat containing proteins, gp140 and gp36 (McBride et al. 1999, 2000; Yu et al. 2007, 2000a, b; Doyle et al. 2005a). A recent report of molecular and antigenic differences between *E. canis* from ticks, humans, and dogs in Venezuela revealed that the 16S rRNA gene was highly conserved (99%) with *E. canis* Oklahoma strain (Perez et al. 2006). Antigenic profiles obtained using Western immunoblotting also demonstrated a high degree of similarity between *E. canis* from Venezuela and the United States. Most recently, the gene encoding a tandem repeat containing major immunoreactive protein, designated gp36, was examined in numerous isolates from the US, Brazil, and Cameroon. Interestingly, the gp36 was highly conserved, demonstrating only four amino acid changes in the Brazilian and Cameroonian *E. canis*. However, the number of repeats varied in all *E. canis* strains examined, and thus, the strains could be distinguished based on the number of repeats in the gene (Doyle et al. 2005a). More comprehensive antigenic and molecular analysis of geographically separated strains has been performed between the United States and Israeli isolates (Jake and 611 strains), in which the four genes encoding four major immunoreactive proteins (gp19, gp36, gp140, and gp200) have been completely sequenced and compared (McBride et al., unpublished). Substantial diversity was observed in all genes except gp19, which only exhibited a single amino acid difference. Conversely, gp36 exhibited significant diversity including two important amino acid substitutions in the antibody epitope containing the repeat region. Because of these substitutions, the gp36 protein from the US strain did not react with sera from Israeli dogs infected with *E. canis*. Diversity in the gp200 and gp140 genes was also noted, demonstrating that diversity does exist in geographically separated strains of *E. canis*. Divergence in the gp36 gene indicates that it is the best gene target for genotyping studies, but other genes (gp140 and gp200) may provide equally useful information.

Genetic Variability of *E. chaffeensis*

Over 20 *E. chaffeensis* isolates from infected patients in the US have been cultivated in vitro and the existence of molecular divergence among *E. chaffeensis* strains is now well recognized. Early studies that focussed on genes encoding the major outer membrane protein family demonstrated that *E. chaffeensis* could be separated genetically into distinct genetic groups (Yu et al. 1999a, b). This finding was further supported by a subsequent study that similarly found three distinct genetic groups (I, II, and III) based on DNA sequences at six gene loci within the 22-gene p28 multigene locus (Cheng et al. 2003). Following early reports based on p28 genes, several well characterized tandem repeat containing genes, including, gp47, gp120, and VLPT (variable length PCR target), have been sequenced from US strains and differences in the numbers of repeat units and the sequences of repeats has been noted in these genes from among the different strains (Sumner et al. 1999; Doyle et al. 2005a). The gp120 and VLPT genes have been used most frequently for molecular distinction between *E. chaffeensis* isolates (Sumner et al. 1999; Standaert et al. 2000; Yabsley et al. 2003). The sequence of gp120 is conserved between isolates, but the number of tandem repeat units varies from two to five, with three units being most commonly (52.5%) observed in human isolates. VLPT has variations in three nucleotide positions and variations in the number of tandem repeats (3–6). The four (62.7%) and five (34.3%) repeat variants are most commonly detected in deer, while three (27.7%) and four (55.5%) repeat variants predominate in humans (Sumner et al. 1999; Yabsley et al. 2003). The most recently characterized tandem repeat gene, gp47, in *E. chaffeensis* Arkansas strain has seven nearly identical 19 amino acid (ASVSEGDAVVNAVSQETPA) tandem repeats, while other strains (Jax, St. Vincent, and Vanderbilt) have a distinct 33 amino acid (EGNASEPVVSQEAAPVS-ESGDAANPV-SSSENAS) tandem repeat and fewer repeat units (3.4–4.5) (Doyle et al. 2005a).

4.2.2 Genome Sequencing

All *Ehrlichia* species are obligately intracellular, fragile, and difficult to grow in tissue culture, and these circumstances delayed the initiation of genome sequencing of the organisms until the bacterial genome sequencing era was well established. It was also difficult initially to prepare useful quantities of host cell free DNA, a difficulty compounded by the fact that free *Ehrlichia* cells have a density similar to that of mammalian mitochondria. The problem was first partially solved for *E. ruminantium* by the use of an immunoaffinity chromatographic method to purify the organisms, although total yields were low (Brayton et al. 1997) and eventually better differential centrifugation procedures were developed (De Villiers et al. 1998). Representative genomic libraries suitable for genome sequencing were also difficult to produce, and in some cases *Ehrlichia* DNA was found to be unstable in some favored cloning vectors (Brayton et al. 1999). In 1995, the publication of the first complete bacterial genome, *Haemophilus influenzae*, introduced shotgun genome sequence assembly as the preferred method for bacterial genomes (Fleischmann et al. 1995). This eliminated the need to construct maps of small genomes before attempting to sequence them and allowed the use of libraries with smaller inserts that were easier to produce, thereby greatly accelerating the rate at which bacterial genomes could be sequenced. The benefits were rapidly applied to *Ehrlichia* genome sequencing projects and all three of the species under consideration were completely sequenced by 2006.

E. ruminantium Genome

E. ruminantium was the first *Ehrlichia* species to be fully sequenced. The Welgevonden stock was used, this being the type specimen of the species, and because of the widespread cross-immunity which this stock stimulates. Random clones from two small insert libraries were sequenced using Sanger chain-termination chemistry and the assembly generated 72 contigs (Collins et al. 2005). It proved to be difficult to obtain random sequences to fill the gaps and the contigs were ordered using a physical map (De Villiers et al. 2000), which had been under construction some time before use of the shotgun assembly technique became commonplace. The contigs were joined by generating and sequencing PCR products and the final assembly contained 25,648 reads with an average length of 569 bp, giving 9.6-fold coverage of the genome. The genome 1,516,355 bp in length has a low G+C content (27.5%), a feature which is common for endosymbionts and intracellular pathogens. There are 888 protein coding sequences of average length 1,059 bp, 41 stable RNA species, and 32 pseudogenes, hence only 62% of the genome is predicted to be coding sequence. The features of the genome are summarized in Table 3, and a complete list of all CDSs is given in Table 4.

Subsequently two other stocks of *E. ruminantium* were sequenced (Frutos et al. 2006): the Gardel strain, isolated on Guadeloupe (Table 2), and a sample of the Welgevonden strain, which was transferred to Guadeloupe from South Africa in 1985 and then maintained for 18 years in tissue culture. These two genomes are both somewhat smaller than that of the type strain, and Gardel is a slightly unusual genotype of *E. ruminantium* owing to its geographical isolation from other strains (Allsopp and Allsopp 2007). The Gardel strain has the smallest genome (1,499,920 bp), while the relocated Welgevonden strain is intermediate in size between the other two (1,512,977 bp). A more detailed comparison of these three genomes will be given in Sect. 4.2.4. In subsequent sections, reference to the *E. ruminantium* genome will mean the genome of the type strain (Welgevonden), unless otherwise indicated, and the predicted number of 888 protein coding sequences for this genome will be used as the basis for numerical estimates of the percentages of different types of genes.

Table 3 General features of the genome of *E. ruminantium* (Welgevonden type strain). Adapted from Collins et al. 2005

Size	1,516,355 bp
G+C content	27.5%
Total number of CDSs	920
Average length	1,032 bp
Probable pseudogenes	32 (3.5%)
Average length	276 bp
Predicted protein coding sequences	888 (96.5%)
Average length	1,059 bp
% Predicted protein coding sequence	62.0%
Stable RNAs	
Number of ribosomal RNAs	3,362
Number of transfer RNAs	
Number of other RNAs (tmRNA, rnpB)	
Tandem repeats	82,172 bp (5.4% of genome)
Dispersed repeats (direct and inverted)	43,976 bp (2.9% of genome)

E. canis Genome

The genome of *E. canis* (Jake strain) was sequenced by the Joint Genome Institute (USA) using a combination of random small insert (3 and 8 kbp) and fosmid (40 kbp) libraries. The complete genome sequence was determined using 40,000 reads providing ~20-fold coverage. The circular genome was 1,315,030 bp in length, smaller than *E. ruminantium* (1.5 Mbp) and larger than *E. chaffeensis* (1.2 Mbp). The *E. canis* genome has a slightly higher G+C content (29%) than the other two species, and it encodes 967 proteins, 40 RNA species, 17 pseudogenes, with a substantial amount of non-coding sequence (27%) in the genome. A unique bias in coding for the polar amino acids serine and threonine in proteins associated with host–pathogen interactions was noted in *E. canis* (Mavromatis et al. 2006).

E. chaffeensis Genome

The *E. chaffeensis* (Arkansas strain) genome was sequenced using a whole genome shotgun approach at the Institute for Genomic Research (TIGR) from a total of 57,000 sequencing reads. The genome is 1,176,248 bp in length, smaller than either *E. ruminantium* or *E. canis*. *E. chaffeensis* has a larger number

Table 4 Functional classification of *E. ruminantium* protein-coding genes. Adapted from Collins et al. 2005

Energy metabolism (56)		
ATP-synthase complex (8)		
Erum0820	*atpA*	ATP synthase alpha chain
Erum8360	*atpB*	ATP synthase A subunit
Erum4580	*atpC*	ATP synthase epsilon chain
Erum4590	*atpD*	ATP synthase beta chain
Erum8370	*atpE*	ATP synthase C subunit
Erum8380	*atpF*	Probable ATP synthase B subunit
Erum3990	*atpG*	ATP synthase gamma chain
Erum0830	*atpH*	Probable ATP synthase delta chain
Electron transport (34)		
Erum7740	*coax*	Probable cytochrome c oxidase subunit I
Erum7730	*coxB*	Probable cytochrome c oxidase subunit II
Erum0170	*coxC*	Cytochrome c oxidase subunit III
Erum0240	*fdxA*	Ferredoxin
Erum4200	*fdxB*	Ferredoxin, 2FE-2S
Erum3100	*nuoA*	Probable NADH-quinone oxidoreductase chain A
Erum3090	*nuoB*	NADH-quinone oxidoreductase chain B
Erum3070	*nuoC*	Probable NADH-quinone oxidoreductase chain C
Erum4420	*nuoD*	NADH-quinone oxidoreductase chain D
Erum4430	*nuoE*	NADH-quinone oxidoreductase chain E
Erum4810	*nuoF*	NADH-quinone oxidoreductase chain F
Erum4270	*nuoG*	NADH-quinone oxidoreductase chain G
Erum4280	*nuoH*	NADH-quinone oxidoreductase chain H
Erum3710	*nuoI*	NADH-quinone oxidoreductase chain I
Erum4800	*nuoJ*	NADH-quinone oxidoreductase chain J
Erum4790	*nuoK*	NADH-quinone oxidoreductase chain K
Erum4780	*nuoL*	NADH-quinone oxidoreductase chain L
Erum4770	*nuoM*	NADH-quinone oxidoreductase chain M
Erum4760	*nuoN*	NADH-quinone oxidoreductase chain N
Erum5040	*petA*	Ubiquinol-cytochrome c reductase iron-sulphur subunit
Erum5030	*petB*	Cytochrome b
Erum5020	*petC*	Cytochrome c1 precursor
Erum6260	*Qor*	Probable quinone oxidoreductase
Erum6810	*sdhA*	Succinate dehydrogenase flavoprotein subunit
Erum6800	*sdhB*	Succinate dehydrogenase iron-sulfur subunit
Erum1890	*sdhC*	Probable succinate dehydrogenase cytochrome b-556 subunit
Erum1891	*sdhD*	Probable succinate dehydrogenase cytochrome b small subunit
Erum0430		Possible NADH-ubiquinone oxidoreductase subunit
Erum1240		Probable NADH-quinone oxidoreductase subunit
Erum1570		Probable cytochrome b561
Erum5440		Probable NADH-quinone oxidoreductase subunit
Erum6700		Probable NADH-quinone oxidoreductase subunit
Erum6720		Probable c-type cytochrome
Erum7570		Probable NADH-ubiquinone oxidoreductase
Pyruvate dehydrogenase and TCA cycle (14)		
Erum7920	*acnA*	Aconitate hydratase
Erum6330	*fumC*	Fumarate hydratase class II
Erum0750	*gltA*	Citrate synthase
Erum8530	*Icd*	Isocitrate dehydrogenase [NADP]
Erum4090	*Mdh*	Malate dehydrogenase
Erum7520	*pdhA*	Pyruvate dehydrogenase E1 component, alpha subunit
Erum0980	*pdhB*	Probable pyruvate dehydrogenase E1 component, beta subunit

Table 4 (Continued)

Erum0670	*pdhC*	Dihydrolipoamide acetyltransferase, E2 component of pyruvate dehydrogenase complex
Erum2650	*sucA*	2-Oxoglutarate dehydrogenase E1 component
Erum8200	*sucB*	Dihydrolipoamide succinyltransferase, E2 component of 2-oxoglutarate dehydrogenase complex
Erum1520	*sucC*	Succinyl-CoA synthetase, beta subunit
Erum1510	*sucD*	Succinyl-CoA synthetase, alpha subunit
Erum1420		Probable dihydrolipoamide dehydrogenase, E3 component of pyruvate, or 2-oxoglutarate dehydrogenase complex
Erum5130		Probable dihydrolipoamide dehydrogenase, E3 component of pyruvate or 2-oxoglutarate dehydrogenase complex
Central intermediary metabolism (24)		
Erum4840	*Eno*	Enolase
Erum0650	*fbaB*	Probable fructose-bisphosphate aldolase class I
Erum0010	*gapB*	NAD(P)-dependent glyceraldehyde 3-phosphate dehydrogenase
Erum6470	*glpX*	Fructose-1,6-bisphosphatase class II GlpX
Erum5150	*gpmI*	2,3-Bisphosphoglycerate-independent phosphoglycerate mutase
Erum1200	*maeB*	NADP-dependent malic enzyme
Erum0070	*metK*	S-adenosylmethionine synthetase
Erum8570	*ndk*	Nucleoside diphosphate kinase
Erum0360	*pgk*	Phosphoglycerate kinase
Erum7840	*ppa*	Inorganic pyrophosphatase
Erum6690	*ppdK*	Pyruvate phosphate dikinase
Erum7490	*ppnK*	Probable inorganic polyphosphate/ATP-NAD kinase
Erum7240	*pyr*	Uridylate kinase
Erum0560	*rpe*	Ribulose-phosphate 3-epimerase
Erum4100	*rpiB*	Ribose 5-phosphate isomerase B
Erum4570	*Tal*	Probable transaldolase
Erum5600	*tkt*	Transketolase
Erum4040	*tpiA*	Triose phosphate isomerase
Erum0890		Probable aminomethyl transferase
Erum1560		Probable 2-nitropropane dioxygenase
Erum2530		Probable glutathione S-transferase
Erum3230		Possible NAD-glutamate dehydrogenase
Erum4020		Probable pyridine nucleotide-oxidoreductase
Erum4160		Probable NifU-like protein
Purine and pyrimidine metabolism (29)		
Deoxyribonucleotide metabolism (3)		
Erum5190	*Dut*	Probable deoxyuridine 5′-triphosphate nucleotidohydrolase
Erum5650	*nrdA*	Probable ribonucleoside-diphosphate reductase alpha chain
Erum3270	*nrdB*	Probable ribonucleoside-diphosphate reductase beta chain
Purine ribonucleotide biosynthesis (17)		
Erum5880	*adk*	Adenylate kinase
Erum6740	*gmk*	Guanylate kinase
Erum0740	*guaA*	GMP synthase [glutamine-hydrolyzing]
Erum7500	*guaB*	Inosine-5′-monophosphate dehydrogenase
Erum7900	*prsA*	Ribose-phosphate pyrophosphokinase
Erum5630	*purA*	Adenylosuccinate synthetase
Erum2460	*purB*	Adenylosuccinate lyase
Erum7000	*purC*	Phosphoribosylaminoimidazole-succinocarboxamide synthase

(continued)

Table 4 (Continued)

Erum7770	*purD*	Phosphoribosylamine – glycine ligase
Erum1060	*pure*	Phosphoribosylaminoimidazole carboxylase catalytic subunit
Erum0900	*purF*	Glutamine phosphoribosylpyrophosphate amidotransferase
Erum8290	*purH*	Bifunctional purine biosynthesis protein PurH
Erum7940	*purK*	Phosphoribosylaminoimidazole carboxylase ATPase subunit
Erum6510	*purl*	Probable phosphoribosylformylglycinamidine synthase II
Erum6580	*purM*	Phosphoribosylformylglycinamidine cyclo-ligase
Erum6370	*purN*	Phosphoribosylglycinamide formyltransferase
Erum6450	*purQ*	Possible phosphoribosylformylglycinamidine synthase I
Pyrimidine ribonucleotide biosynthesis (9)		
Erum6110	*cmk*	Probable kinase
Erum6990	*dcd*	Probable deoxycytidine triphosphate deaminase
Erum4250	*pyrB*	Aspartate carbamoyltransferase
Erum6350	*pyrC*	Dihydroorotase
Erum1810	*pyrD*	Dihydroorotate dehydrogenase
Erum8490	*pyre*	Probable phosphoribosyltransferase
Erum3040	*pyrF*	Orotidine 5'-phosphate decarboxylase
Erum1160	*pyrG*	CTP synthase
Erum7460	*tmk*	Probable thymidylate kinase
Fatty acid metabolism (12)		
Erum3430	*acpS*	Probable holo-[acyl-carrier-protein] synthase
Erum5320	*bccA*	Probable acetyl-/propionyl-coenzyme A carboxylase alpha chain
Erum7470	*fabD*	Probable malonyl CoA-acyl carrier protein transacylase
Erum2150	*fabF*	3-Oxoacyl-[acyl-carrier-protein] synthase II
Erum3840	*fabG*	3-Oxoacyl-[acyl carrier protein] reductase
Erum5720	*fabH*	3-Oxoacyl-[acyl-carrier-protein] synthase III
Erum2860	*fabI*	Enoyl-[acyl-carrier-protein] reductase [NADH]
Erum8280	*fabZ*	(3R)-hydroxymyristoyl-[acyl carrier protein] dehydratase
Erum2840	*matA*	Probable malonyl-CoA decarboxylase
Erum0550	*plsC*	Probable 1-acyl-sn-glycerol-3-phosphate acyltransferase
Erum5730	*plsX*	Fatty acid/phospholipid synthesis protein
Erum7220		Probable cytidylyltransferase
Macromolecule synthesis and modification (19)		
Erum3060	*ccmE*	Cytochrome c-type biogenesis protein CcmE
Erum7750	*ctaB*	Probable protoheme IX farnesyltransferase
Erum8080	*ctaG*	Cytochrome c oxidase assembly protein
Erum0880	*ccmF*	Cytochrome c-type biogenesis protein CcmF
Erum2210	*dsbB*	Disulphide bond formation protein B
Erum6910	*dsbE*	Probable thiol:disulphide interchange protein
Erum6600	*gpsA*	Glycerol-3-phosphate dehydrogenase [NAD(P)+]
Erum8440	*Lgt*	Prolipoprotein diacylglyceryl transferase
Erum6360	*lipB*	Lipoate-protein ligase B
Erum1220	*Lnt*	Probable apolipoprotein N-acyltransferase
Erum8120	*lspA*	Lipoprotein signal peptidase
Erum3370	*mdmC*	Probable O-methyltransferase
Erum1980	*pgpA*	Probable phosphatidylglycerophosphatase A
Erum8300	*pgsA*	Probable CDP-diacylglycerol–glycerol-3-phosphate 3-phosphatidyltransferase

Table 4 (Continued)

Erum3160	*pssA*	Probable CDP-diacylglycerol – serine O-phosphatidyltransferase
Erum3170	*Psd*	Probable phosphatidylserine decarboxylase proenzyme
Erum3720	*sipF*	Prokaryotic type I signal peptidase
Erum4211		Possible cytochrome c-type biogenesis protein
Erum7040		Probable cytochrome c oxidase assembly protein
Amino acid metabolism (26)		
Erum3490	*aatA*	Aspartate aminotransferase A
Erum4480	*argB*	Acetylglutamate kinase
Erum7830	*argC*	*N*-acetyl-gamma-glutamyl-phosphate reductase
Erum2110	*argD*	Acetylornithine/succinyldiaminopimelate aminotransferase
Erum0510	*argF*	Ornithine carbamoyltransferase
Erum3770	*argG*	Argininosuccinate synthase
Erum1830	*argH*	Argininosuccinate lyase
Erum3800	*argJ*	Arginine biosynthesis bifunctional protein ArgJ
Erum0060	*Asd*	Aspartate-semialdehyde dehydrogenase
Erum1880	*aroE*	3-Phosphoshikimate 1-carboxyvinyltransferase
Erum5170	*carA*	Carbamoyl-phosphate synthase small chain
Erum6310	*carB*	Carbamoyl-phosphate synthase, large subunit
Erum2670	*dapA*	Dihydrodipicolinate synthase
Erum5770	*dapB*	Dihydrodipicolinate reductase
Erum0390	*dapD*	2,3,4,5-Tetrahydropyridine-2,6-dicarboxylate *N*-succinyltransferase
Erum0940	*dapE*	Probable succinyl-diaminopimelate desuccinylase
Erum0340	*dapF*	Diaminopimelate epimerase
Erum0610	*glnA*	Glutamine synthetase
Erum6840	*glyA*	Serine hydroxymethyltransferase
Erum4150	*iscS*	Cysteine desulfurase
Erum5340	*lysA*	Probable diaminopimelate decarboxylase
Erum4460	*pccB*	Propionyl-CoA carboxylase beta chain
Erum0030	*proC*	Pyrroline-5-carboxylate reductase
Erum3850	*putA*	Proline dehydrogenase/delta-1-pyrroline-5-carboxylate dehydrogenase
Erum1480		Possible truncated glutamine synthetase
Erum7720		Probable aspartate kinase
Biosynthesis of co-factors (61)		
Biotin biosynthesis (5)		
Erum3870	*bioA*	Adenosylmethionine-8-amino-7-oxononanoate aminotransferase
Erum6500	*bioB*	Biotin synthase
Erum0220	*bioC*	Possible biotin synthesis protein BioC
Erum1740	*bioF*	Probable 8-amino-7-oxononanoate synthase
Erum2520		Probable biotin–[acetyl-CoA-carboxylase] synthetase
Folic acid (7)		
Erum4080	*folB*	Possible dihydroneopterin aldolase
Erum3680	*folC*	Probable folylpolyglutamate synthase/dihydrofolate synthase
Erum6730	*fold*	Methylenetetrahydrofolate dehydrogenase/methenyltetrahydrofolate cyclohydrolase
Erum4000	*folE*	GTP cyclohydrolase I
Erum6520	*folK*	Probable 2-amino-4-hydroxy-6-hydroxymethyldihydropteridine pyrophosphokinase
Erum6280	*folP1*	Probable dihydropteroate synthase 1
Erum6290	*folP2*	Probable dihydropteroate synthase 2
Heme and porphyrins (7)		
Erum0630	*hemA*	5-Aminolevulinic acid synthase

(continued)

Table 4 (Continued)

Erum2720	*hemB*	Delta-aminolevulinic acid dehydratase
Erum3690	*hemC*	Porphobilinogen deaminase
Erum5380	*hemD*	Probable uroporphyrinogen-III synthase
Erum0180	*hemE*	Uroporphyrinogen decarboxylase
Erum4550	*hemF*	Coproporphyrinogen III oxidase
Erum6180	*hemH*	Ferrochelatase
Menaquinone and ubiquinones (13)		
Erum4750	*dxr*	1-Deoxy-D-xylulose 5-phosphate reductoisomerase
Erum5660	*ispA*	Probable geranyltranstransferase
Erum0600	*ispB*	Octaprenyl-diphosphate synthase
Erum1030	*ispD*	Probable 2-C-methyl-D-erythritol 4-phosphate cytidylyltransferase
Erum3340	*ispE*	Probable 4-diphosphocytidyl-2-C-methyl-D-erythritol kinase
Erum1020	*ispF*	2-C-methyl-D-erythritol 2,4-cyclodiphosphate synthase
Erum4730	*ispG*	Probable 1-hydroxy-2-methyl-2-(E)-butenyl 4-diphosphate synthase
Erum5180	*ispH*	4-Hydroxy-3-methylbut-2-enyl diphosphate reductase
Erum5790	*ubiA*	4-Hydroxybenzoate octaprenyltransferase
Erum2600	*ubiB*	Probable ubiquinone biosynthesis protein UbiB
Erum7700	*ubiE*	Ubiquinone/menaquinone biosynthesis methyltransferase UbiE
Erum0080	*ubiF*	Probable 2-octaprenyl-3-methyl-6-methoxy-1,4-benzoquinol hydroxylase
Erum4110	*ubiG*	Probable 3-demethylubiquinone-9 3-methyltransferase
Riboflavin (6)		
Erum0800	*ribB*	3,4-Dihydroxy-2-butanone 4-phosphate synthase
Erum7390	*ribE*	Probable riboflavin synthase, alpha subunit
Erum1140	*ribD*	Riboflavin biosynthesis protein RibD
Erum8130	*ribF*	Riboflavin kinase/FAD synthetase
Erum3130	*ribH*	Probable 6,7-dimethyl-8-ribityllumazine synthase
Erum0310		Probable riboflavin biosynthesis protein
Thiamine (8)		
Erum2970	*thiC*	Thiamine biosynthesis protein ThiC
Erum1910	*thiD*	Probable phosphomethylpyrimidine kinase
Erum2060	*thiE*	Probable thiamine-phosphate pyrophosphorylase
Erum8480	*thiF*	Probable adenylyltransferase ThiF
Erum7630	*thiG*	Thiazole biosynthesis protein
Erum4980	*thiL*	Probable thiamine-monophosphate kinase
Erum5680	*thiO*	Probable thiamine biosynthesis oxidoreductase
Erum7640		Thiamin S protein
Other (15)		
Erum2160	*acpP*	Acyl carrier protein
Erum2960	*coaE*	Probable dephospho-CoA kinase
Erum3460	*coaD*	Probable phosphopantetheine adenylyltransferase
Erum8140	*grxC*	Probable glutaredoxin 3
Erum0770	*gshA*	Possible gamma-glutamylcysteine synthetase
Erum6640	*gshB*	Glutathione synthetase
Erum5290	*lipA*	Lipoic acid synthetase
Erum0230	*nadA*	Quinolinate synthetase A
Erum0140	*nadC*	Nicotinate-nucleotide pyrophosphorylase [carboxylating]
Erum2910	*nadD*	Probable nicotinate-nucleotide adenylyltransferase
Erum2710	*nadE*	Probable glutamine-dependent NAD(+) synthetase

Table 4 (Continued)

Erum1850	*pdxH*	Pyridoxamine 5′-phosphatate oxidase	Erum5360	*priA*	Primosomal protein N'
Erum2920	*pdxJ*	Pyridoxal phosphate biosynthetic protein PdxJ	Erum6900	*radA*	DNA repair protein RadA
Erum7540	*trxA*	Thioredoxin 1	Erum6440	*radC*	DNA repair protein RadC
Erum3470	*trxB*	Thioredoxin reductase	Erum8500	*recA*	RecA protein (Recombinase A)
Information transfer (173)			Erum6250	*recB*	Probable exodeoxyribonuclease V beta chain
DNA replication, repair, recombination and degradation (48)			Erum0520	*recF*	Probable DNA replication and repair protein RecF
Erum0410	*dfp*	Probable DNA/pantothenate metabolism flavoprotein	Erum0420	*recG*	ATP-dependent DNA helicase RecG
Erum2870	*dnaA*	Chromosomal replication initiator protein DnaA	Erum8550	*recJ*	Probable single-stranded-DNA-specific exonuclease RecJ
Erum5710	*dnaB*	Replicative DNA helicase	Erum4920	*recO*	Possible DNA repair protein RecO
Erum1870	*dnaE*	DNA polymerase III, alpha subunit	Erum2570	*recR*	Probable recombination protein RecR
Erum3310	*dnaG*	Probable DNA primase	Erum4520	*rmuC*	DNA recombination protein RmuC
Erum7880	*dnaN*	DNA polymerase III, beta subunit	Erum6760	*ruvA*	Probable junction DNA helicase RuvA
Erum4990	*dnaQ*	DNA polymerase III, epsilon subunit	Erum6770	*ruvB*	Holliday junction DNA helicase RuvB
Erum0040	*dnaZ*	Probable DNA polymerase III, gamma subunit	Erum0160	*ruvC*	Crossover junction endodeoxyribonuclease RuvC
Erum3810	*exoA*	Probable exodeoxyribonuclease	Erum2140	*smf*	DNA processing protein chain A
Erum2420	*gyrA*	DNA gyrase subunit A	Erum2830	*ssb*	Single-strand DNA binding protein
Erum4260	*gyrB*	DNA gyrase subunit B	Erum3400	*topA*	DNA topoisomerase I
Erum2940	*holB*	DNA III, delta' subunit	Erum3110	*uvrA*	uvrABC system protein A
Erum2930	*hupB*	DNA-binding protein HU-beta	Erum2390	*uvrD*	DNA helicase II
Erum1080	*ihfA*	Probable integration host factor alpha subunit	Erum0370	*xseA*	Exodeoxyribonuclease VII
Erum6140	*ihfB*	Possible integration host factor beta subunit	Erum7560	*xseB*	Probable exodeoxyribonuclease VII small subunit
Erum6940	*ligA*	NAD-dependent DNA ligase	Erum0530		Possible uracil DNA glycosylase
Erum7290	*Mfd*	Transcription-repair coupling factor	Erum1180		Probable integrase/recombinase XerD or XerC
Erum2130	*mutL*	DNA mismatch repair protein MutL	Erum5640		Possible Holliday junction resolvase
Erum4330	*mutM*	Formamidopyrimidine-DNA glycosylase	Erum6590		Probable integrase/recombinase XerD or XerC
Erum2700	*mutS*	DNA mismatch repair protein MutS	Erum7170		Probable methylpurine-DNA glycosylase
Erum2430	*nth*	Endonuclease III			
Erum0490	*polA*	DNA polymerase I			

(continued)

Table 4 (Continued)

Degradation of RNA (6)		
Erum3540	*pnp*	Polyribonucleotide nucleotidyltransferase
Erum8070	*Rnc*	Ribonuclease III
Erum7260	*rnhA*	Ribonuclease HI
Erum1760	*rnhB*	Ribonuclease HII
Erum5800	*rnpA*	Probable ribonuclease P protein component
Erum5510		Probable ribonuclease
RNA synthesis and modification (12)		
Erum0810	*greA*	Transcription elongation factor GreA
Erum4700	*nusA*	N utilization substance protein A
Erum1670	*nusG*	Transcription antitermination protein NusG
Erum1400	*rho1*	Transcription termination factor 1
Erum7670	*rho2*	Transcription termination factor 2
Erum5850	*rpoA*	DNA-directed RNA polymerase alpha chain
Erum1720	*rpoB*	DNA-directed RNA polymerase beta chain
Erum1730	*rpoC*	DNA-directed RNA polymerase beta' chain
Erum3320	*rpoD*	RNA polymerase sigma-70 factor
Erum3960	*rpoH*	RNA polymerase sigma-32 factor
Erum2990	*rpoZ*	DNA-directed RNA polymerase omega chain
Erum8560		Probable nucleic acid independent RNA polymerase
Aminoacyl-tRNA synthetases (21)		
Erum1500	*alas*	Alanyl-tRNA synthetase
Erum4910	*argS*	Arginyl-tRNA synthetase
Erum6660	*asps*	Aspartyl-tRNA synthetase
Erum3250	*cysS*	Cysteinyl-tRNA synthetase
Erum7610	*gltX1*	Glutamyl-tRNA synthetase 1
Erum4310	*gltX2*	Glutamyl-tRNA synthetase 2
Erum0110	*glyQ*	Glycyl-tRNA synthetase alpha chain
Erum0120	*glyS*	Glycyl-tRNA synthetase beta chain
Erum7010	*hisS*	Histidyl-tRNA synthetase
Erum4870	*ileS*	Isoleucyl-tRNA synthetase
Erum3010	*leuS*	Leucyl-tRNA synthetase
Erum4220	*lysS*	Lysyl-tRNA synthetase
Erum7710	*metG*	Methionyl-tRNA synthetase
Erum1360	*pheS*	Phenylalanyl-tRNA synthetase alpha chain
Erum5830	*pheT*	Phenylalanyl-tRNA synthetase beta chain
Erum3440	*proS*	Prolyl-tRNA synthetase
Erum4540	*serS*	Seryl-tRNA synthetase
Erum8890	*thrS*	Threonyl-tRNA synthetase
Erum1120	*trpS*	Tryptophanyl-tRNA synthetase
Erum0620	*tyrS*	Tyrosyl-tRNA synthetase
Erum0780	*valS*	Valyl-tRNA synthetase
tRNA and aminoacyl-tRNA modification (17)		
Erum0540	*def1*	Probable deformylase 1
Erum1820	*def2*	Probable peptide deformy-lase 2
Erum2030	*fmt*	Methionyl-tRNA formyltrans-ferase
Erum3670	*gatA*	Glutamyl-tRNA(Gln) amidotransferase subunit A
Erum2850	*gatB*	Aspartyl/glutamyl-tRNA amidotransferase subunit B
Erum7910	*gatC*	Probable glutamyl-tRNA(Gln) amidotransferase subunit C
Erum4030	*ksgA*	Dimethyladenosine trans-ferase
Erum4370	*miaA*	Probable tRNA delta(2)-isopentenylpyrophoshate transferase
Erum0910	*pth*	Peptidyl-tRNA hydrolase
Erum4970	*rbn*	tRNA processing ribonuclease BN
Erum5750	*tgt*	Queuine tRNA-ribosyltrans-ferase
Erum8860	*trmD*	tRNA (Guanine-N(1)-) -methyltransferase
Erum0400	*trmE*	Probable tRNA modification GTPase
Erum2230	*trmU*	tRNA (5-methylaminomethyl-2-thiouridylate)-methyltrans-ferase

Table 4 (Continued)

Erum4240	*truA*	tRNA pseudouridine synthase A
Erum3520	*truB*	Probable tRNA pseudouridine synthase B
Erum6100		Probable tRNA/rRNA methyltransferase
Translation factors, modification of ribosomes and nascent peptides (16)		
Erum3190	*efp*	Probable elongation factor P
Erum7230	*frr*	Ribosome recycling factor
Erum1650	*fusA*	Elongation factor G
Erum5110	*infA*	Translation initiation factor IF-1
Erum4690	*infB*	Translation initiation factor IF-2
Erum8900	*infC*	Translation initiation factor IF-3
Erum4500	*prfA*	Peptide chain release factor 1
Erum3650	*prfB*	Peptide chain release factor 2
Erum4680	*rbfA*	Ribosome-binding factor A
Erum8850	*rimM*	Probable 16S rRNA processing protein
Erum3210	*rluC*	Ribosomal large subunit pseudouridine synthase C
Erum5330	*rluD*	Ribosomal large subunit pseudouridine synthase D
Erum0790	*smpB*	SsrA-binding protein
Erum5080	*tsf*	Elongation factor Ts
Erum1660	*tufA*	Elongation factor Tu-A
Erum6090	*tufB*	Elongation factor Tu-B
Ribosomal proteins (53)		
Erum1690	*rplA*	50S ribosomal protein L1
Erum6040	*rplB*	50S ribosomal protein L2
Erum6070	*rplC*	50S ribosomal protein L3
Erum6060	*rplD*	50S ribosomal protein L4
Erum5960	*rplE*	50S ribosomal protein L5
Erum5930	*rplF*	50S ribosomal protein L6
Erum6850	*rplI*	50S ribosomal protein L9
Erum1700	*rplJ*	50S ribosomal protein L10
Erum1680	*rplK*	50S ribosomal protein L11
Erum1710	*rplL*	50S ribosomal protein L7/L12
Erum7810	*rplM*	50S ribosomal protein L13
Erum5980	*rplN*	50S ribosomal protein L14
Erum5900	*rplO*	50S ribosomal protein L15
Erum6000	*rplP*	50S ribosomal protein L16
Erum5840	*rplQ*	50S ribosomal protein L17
Erum5920	*rplR*	50S ribosomal protein L18
Erum8870	*rplS*	50S ribosomal protein L19
Erum1370	*rplT*	50S ribosomal protein L20
Erum4830	*rplU*	50S ribosomal protein L21
Erum6020	*rplV*	50S ribosomal protein L22
Erum6050	*rplW*	50S ribosomal protein L23
Erum5970	*rplX*	50S ribosomal protein L24
Erum0920	*rplY*	Probable 50S ribosomal protein L25
Erum4820	*rpmA*	50S ribosomal protein L27
Erum5350	*rpmB*	50S ribosomal protein L28
Erum5991	*rpmC*	50S ribosomal protein L29
Erum7480	*rpmE*	50S ribosomal protein L31
Erum5740	*rpmF*	50S ribosomal protein L32
Erum2190	*rpmG*	50S ribosomal protein L33
Erum5791	*rpmH*	50S ribosomal protein L34
Erum1380	*rpmI*	50S ribosomal protein L35
Erum3950	*rpmJ*	50S ribosomal protein L36
Erum6120	*rpsA*	30S ribosomal protein S1
Erum5090	*rpsB*	30S ribosomal protein S2
Erum6010	*rpsC*	30S ribosomal protein S3
Erum1940	*rpsD*	30S ribosomal protein S4
Erum5910	*rpsE*	30S ribosomal protein S5
Erum6870	*rpsF*	30S ribosomal protein S6
Erum1640	*rpsG*	30S ribosomal protein S7
Erum5940	*rpsH*	30S ribosomal protein S8
Erum7820	*rpsI*	30S ribosomal protein S9
Erum6080	*rpsJ*	30S ribosomal protein S10
Erum5860	*rpsK*	30S ribosomal protein S11
Erum1630	*rpsL*	30S ribosomal protein S12
Erum5870	*rpsM*	30S ribosomal protein S13
Erum5950	*rpsN*	30S ribosomal protein S14
Erum3530	*rpsO*	30S ribosomal protein S15
Erum1320	*rpsP*	30S ribosomal protein S16
Erum5990	*rpsQ*	30S ribosomal protein S17
Erum6860	*rpsR*	30S ribosomal protein S18
Erum6030	*rpsS*	30S ribosomal protein S19
Erum0480	*rpsT*	30S ribosomal protein S20

(continued)

Table 4 (Continued)

Erum1530	*rpsU*	Possible 30S ribosomal protein S21
Degradation of proteins (18)		
Erum4660	*clpA*	ATP-dependent Clp protease, ATP-binding subunit
Erum2000	*clpP*	ATP-dependent Clp protease proteolytic subunit
Erum2010	*clpX*	ATP-dependent Clp protease ATP-binding subunit ClpX
Erum4060	*gcp*	o-Sialoglycoprotein endopeptidase
Erum7680	*hslV*	ATP-dependent protease HslV
Erum7690	*hslU*	ATP-dependent hsl protease ATP-binding subunit
Erum2020	*lon*	ATP-dependent protease La
Erum8160	*map*	Methionine aminopeptidase
Erum6380	*pepA*	Cytosol aminopeptidase
Erum3510		Possible glycoprotease
Erum5610		Possible carboxypeptidase
Erum6130		Probable peptidase
Erum7410		Probable zinc protease
Erum8050		Probable exported serine protease
Erum8090		Probable exported peptidase
Erum8100		Probable exported M16 family peptidase
Erum8220		Probable exported D-alanyl-D-alanine carboxypeptidase
Erum8250		Probable membrane-associated zinc metalloprotease
Cell processes (27)		
Cell division (8)		
Erum4490	*engB*	Probable GTP protein EngB
Erum8400	*ftsA*	Cell division protein FtsA
Erum8430	*ftsH*	Cell division protein FtsH
Erum2090	*ftsK*	Probable cell division protein FtsK
Erum6620	*ftsQ*	Probable cell division protein FtsQ
Erum8520	*ftsY*	Probable cell division protein FtsY
Erum8800	*ftsZ*	Cell division protein FtsZ
Erum6460	*gidA*	Glucose inhibited division protein A
Chromosome replication (2)		
Erum8830	*parA*	Chromosome partitioning protein ParA
Erum8840	*parB*	Chromosome partitioning protein ParB
Chaperones (12)		
Erum6400	*clpB*	Heat shock protein ClpB
Erum0130	*dnaJ*	Chaperone protein DnaJ
Erum5500	*dnaK*	Chaperone protein DnaK
Erum6420	*groEL*	60 kDa chaperonin GroEL
Erum6430	*groES*	10 kDa chaperonin GroES
Erum1130	*grpE*	GrpE protein
Erum4180	*hscB*	Possible co-chaperone protein HscB
Erum4190	*hscA*	Chaperone protein HscA
Erum2450	*htpG*	Chaperone protein HtpG
Erum4010	*pmbA*	Probable PmbA protein
Erum3500	*ppiD*	Probable peptidyl-prolyl cis-trans isomerase D
Erum7030		Probable disulphide oxidoreductase
Adaptation to atypical conditions (5)		
Erum3350	*cutA*	Probable periplasmic divalent cation tolerance protein CutA
Erum0440	*dksA*	Probable DnaK suppressor protein
Erum3050	*surE*	Acid phosphatase SurE
Erum5270	*sodB*	Superoxide dismutase [Fe]
Erum3480		Probable peroxiredoxin
pathogenicity-associated genes (14)		
Erum5260	*virB3*	Type IV secretion system protein VirB3
Erum5250	*virB4*	Type IV secretion system protein VirB4
Erum5240	*virB6*	Type IV secretion system protein VirB6
Erum0300	*virB8*	Type IV secretion system protein VirB8
Erum0290	*virB9*	Type IV secretion system protein VirB9

Table 4 (Continued)

Erum0280	*virB10*	Type IV secretion system protein VirB10	Erum2590		Probable ABC transporter, ATP-binding protein
Erum0270	*virB11*	Type IV secretion system protein VirB11	Erum5060		Probable ABC transporter, membrane-spanning protein
Erum0260	*virD4*	Type IV secretion system protein VirD4	Erum5280		Probable ABC transporter, membrane-spanning protein
Erum4410		Possible type IV secretion system protein	Erum6270		Probable ABC transporter, ATP-binding protein
Erum5210		Possible type IV secretion system protein	Erum6820		Probable ABC transporter, ATP-binding and membrane-spanning protein
Erum5220		Possible type IV secretion system protein	*Amino acids (2)*		
Erum5230		Possible type IV secretion system protein	Erum1130	*proP*	Proline/betaine transporter
Erum7530		Probable conjugal transfer protein	Erum4510		Probable sodium:dicarboxylate symporter(glutamate)
Erum7980		Possible type IV secretion system protein	*Proteins And Peptides (11)*		
Transporters (49)			Erum5430	*Ffh*	Signal recognition particle protein
ABC transporters (16)			Erum8780	*secA*	Preprotein translocase SecAsubunit
Erum7050	*ccmA*	Heme exporter protein A	Erum7430	*secB*	Probable protein-export protein SecB
Erum0450	*ccmB*	Possible heme exporter protein B	Erum8470	*secD*	Probable protein-export membrane protein SecD
Erum6750	*ccmC*	Heme exporter protein C	Erum0640	*secF*	Protein-export membrane protein SecF
Erum1190	*lolD*	Lipoprotein releasing system ATP-binding protein LolD	Erum1170	*secG*	Probable protein-export membrane protein SecG
Erum0860	*lolE*	Probable lipoprotein releasing system transmembrane protein LolE	Erum5890	*secY*	Preprotein translocase secY subunit
Erum5760	*pstB*	Probable phosphate ABC transporter, ATP-binding protein	Erum2560	*tatA*	Possible Sec-independent protein translocase membrane protein
Erum0580		Probable ABC transporter, ATP binding protein	Erum4720	*tatC*	Sec-independent protein translocase protein TatC
Erum1490		Possible ABC transporter, membrane-spanning protein	Erum1990	*tig*	Trigger factor
Erum1580		Probable ABC transporter, membrane-spanning protein	Erum7780		Probable preprotein translocase subunit YajC
Erum2550		Probable ABC transporter, ATP-binding protein	*Cations (9)*		
Erum2580		Probable ABC transporter, periplasmic solute binding protein	Erum0190	*corC*	Possible magnesium and cobalt efflux protein
			Erum1310	*fbpA*	Probable iron-binding periplasmic protein

(continued)

Table 4 (Continued)

Erum8410	*trkH*	Trk system potassium uptake-protein	**Regulatory functions (9)**		
Erum0460		Probable cation efflux system protein	Erum3200	*suhB*	Probable inositol-1-monophosphatase
Erum0950		Probable glutathione-regulated potassium-efflux system protein	Erum1000	*tldD*	TldD protein
Erum1780		Possible Na+/H+ antiporter subunit	Erum2120		Possible histidine kinase sensor component of a two-component regulatory system
Erum4600		Probable magnesium transporter	Erum3220		Possible response regulator component of a two-component regulatory system
Erum5530		Probable Na+/H+ antiporter subunit	Erum3360		Probable two component sensor kinase
Erum5550		Probable Na+/H+ antiporter subunit	Erum6610		Probable response regulator component of a two-component regulatory system
Other (11)			Erum6960		Probable histidine kinase sensor component of a two-component regulatory system
Erum6780	*bcr*	Probable bicyclomycin resistance protein	Erum7860		Probable response regulator component of a two-component regulatory system
Erum1590		Probable secretion protein	Erum8580		Possible transcriptional regulator
Erum2740		Probable integral membrane transport protein	**Phage related (3)**		
Erum2810		Probable integral membrane transport protein	Erum0200		Possible protease
Erum2820		Probable integral membrane transport protein	Erum0210		Possible genetic exchange protein
Erum3150		Probable integral membrane transport protein	Erum2660		Unknown
Erum4710		Probable integral membrane transport protein	**Membrane associated proteins (175)**		
Erum5810		Probable integral membrane transport protein	**Conserved hypothetical proteins (50)**		
Erum5820		Possible competence protein	**Some miscellaneous information, but no functional classification (63)**		
Erum7580		Probable integral membrane transport protein	**No similarity, no functional information (80)**		
Erum7800		Probable outer membrane efflux protein			

(1,115) of predicted open reading frames than the other two species, but a similar number of RNA species (42) and a similar G+C content (30.1%). *E. chaffeensis* has a higher percentage of coding sequence (79.7%), compared to both *E. ruminantium* (62%) and *E. canis* (73%) (Hotopp et al. 2006).

4.2.3 Genomic Insights into the Biology of *Ehrlichia*

Central Metabolic Pathways

Ehrlichia species are aerobic organisms that appear to lack a glycolytic pathway, since essential enzymes such as hexokinase, glucokinase, and phosphofructokinase are absent and a glucose transport system has not been found, indicating that the organism is unable to use glucose or fructose as a carbon source. In addition, there are no enzymes for the Entner-Douderoff pathway, which some microorganisms use as an alternative catabolic pathway for carbohydrates. *Ehrlichia* species have a lysine and arginine biosynthesis pathway that is not present in closely related *Anaplasma* species (Hotopp et al. 2006). In *E. ruminantium* there are transporters for proline (*proP*) and glutamate (Erum4510), as well as enzymes for the conversion of proline to glutamate, including pyrroline-5-carboxylate reductase (*proC*) and the bifunctional enzyme proline dehydrogenase/delta-1-pyrroline-5-carboxylate dehydrogenase (*putA*). It seems likely, therefore, that the primary carbon sources are proline and glutamate, a prediction supported by the observation that the proline consumption of *E. ruminantium*-infected mammalian cells is increased in comparison with uninfected cells (Josemans and Zweygarth 2002). *E. canis* has transporters and enzymes for the utilization of aspartate, proline, glutamate, and arginine, suggesting that these amino acids are the main energy and carbon source for this species.

Ehrlichia contain complete sets of genes for the enzymes of the tricarboxylic acid (TCA) pathway, as well as a putative glutamate dehydrogenase. This enzyme is used by many bacteria for the reversible oxidative deamination of glutamate to α-ketoglutarate and it probably provides the route by which glutamate is fed into the TCA cycle. There are complete sets of enzymes for the conversion of glutamate to fumarate and/or arginine, as well as enzymes for pathways from pyruvate to fructose-6-phosphate, which may be used in a partial gluconeogenesis pathway. There are also genes for all the enzymes in the nonoxidative branch of the pentose-phosphate pathway.

Ehrlichia contain biosynthetic pathways for both purine and pyrimidine nucleotides. These pathways are present in other organisms of the family Anaplasmataceae, such as *Wolbachia pipientis* (Wu et al. 2004) and *A. marginale* (Brayton et al. 2005), but absent from *Rickettsia* species such as *R. prowazekii* (Andersson et al. 1998) and *R. conorii* (Ogata et al. 2001). Since *Rickettsia* species grow free in the cytoplasm of their host cells it would be expected that host nucleotides would be easily available, and this may explain why these organisms have dispensed with the energy intensive de novo synthesis of nucleotides. In contrast, organisms in the family Anaplasmataceae replicate in an intracellular vacuole and they appear to have been obliged to retain their ability to synthesis nucleotides.

Ehrlichia have several genes encoding enzymes for fatty acid and phospholipid biosynthesis from intermediates of central metabolism, including those for phosphatidylglycerol and cardiolipin biosynthesis, but they appear to lack enzymes for the production or modification of unsaturated fatty acids. *E. ruminantium* does not have a full complement of genes for amino acid biosynthesis; there are complete pathways for the biosynthesis of arginine from glutamate and lysine from aspartate, as well as a pathway for the interconversion between proline, glutamate, and glutamine. The remaining amino acids are likely to be obtained from the host cell, although specific amino acid transporters have not been identified. There are, however, several ATP-binding cassette (ABC) transporters, some of which may have the ability to import a wide variety of substrates. Several pathways were identified for the biosynthesis of cofactors, such as biotin, coenzyme A, riboflavin, and dihydrofolate (DHF). Genes were not identified that would encode DHF reductase, thymidine kinase, or thymidylate synthase (thyA), but a gene was present for flavin-dependent thymidylate synthase (thyl) upon which the organism must rely for thymidylate synthesis. This situation is common among many pathogenic and free-living bacteria (Agrawal et al. 2004).

Genes for the biosynthesis of lipid A and peptidoglycan, which are essential components of cell wall lipopolysaccharide, are absent from *Ehrlichia* genomes, and there are incomplete sets of genes for the synthesis of peptidoglycan precursors. While the genes for diaminopimelate synthesis are present, those for murein sacculus synthesis, with the exception of dacF, are absent. The lack of essential cell wall components, which impart strength and structure to the cell membranes of gram negative bacteria, may explain the fragile nature of *Ehrlichia* cells, and they may well need to incorporate cholesterol from the host cell to compensate for the lack of cell wall components (Lin and Rikihisa 2003a). Among the Anaplasmataceae only *A. marginale* has genes for peptidoglycan biosynthesis (Brayton et al. 2005), although it is not known whether they are expressed. This suggests that *A. marginale* has horizontally re-acquired these genes after they had been lost from the common ancestor of the Anaplasmataceae.

Energy Metabolism

Ehrlichia have genes that code for putative enzyme complexes typical of aerobic respiration, including the ATP-synthase complex and the electron transfer complexes. In many microbial genomes, the genes encoding these components are clustered in a single highly conserved operon. In *E. ruminantium* the genes encoding the A, B, and C chains of the F_0 complex (*atpB*, *atpE*, and *atpF*) are arranged in the common order, but the catalytic core (F_1 subunit) genes are located in dispersed areas of the genome in three groups: *atpH* and *atpA*, *atpG*, and *atpD* and *atpC*. Genes coding for components of the NADH dehydrogenase complex are also dispersed in several clusters in *E. ruminantium*. Three clusters are located close to each other (*nuoG/H*, *nuoD/E*, and *nuoN/M/L/K/J/F*), and between these clusters, and another consisting of *nuoA/B/C*, there is the single gene *nuoI*. There are three additional genes closely related to *nuoM* (Erum1240, Erum5440, and Erum6700), which may be components of the NADH dehydrogenase complex, since all three have significant homologies to NADH-quinone oxidoreductase complex I chain M (EC 1.6.99.5) from *Rhodobacter capsulatus* or *Rickettsia prowazekii*. The genes *sdhA*, *sdhB*, *sdhC*, and *sdhD* are also present, enabling the synthesis of a complete succinate dehydrogenase enzyme, which links the TCA cycle to the aerobic electron transport chain.

Several proteins in the cytochrome bc_1 reductase complex, including ubiquinol–cytochrome *c* reductase iron–sulphur subunit (*petA*), cytochrome *b* (*petB*), and cytochrome c_1 (*petC*), are present in *Ehrlichia*, as are most subunits of the cytochrome oxidase complex (*coxA*, *coxB*, and *coxC*). There are also complete pathways for porphyrin biosynthesis, as well as several proteins responsible for cytochrome biosynthesis. These features support a central role in the metabolism of *Ehrlichia* for aerobic respiration and an electron transport system. ATP/ADP translocases have not been identified, however, indicating that these organisms do not make use of ATP from the host cell, unlike Chlamydiales and some other Rickettsiales (Greub and Raoult 2003). Since the rickettsial ATP/ADP translocases are thought to have originated in the common ancestor of mitochondria and Rickettsiales (Schmitz-Esser et al. 2004), their absence from *Ehrlicha*, *A. marginale* (Brayton et al. 2005), and *Wolbachia pipientis* (Wu et al. 2004) seems to be the result of subsequent loss.

Replication, Repair, and Recombination

All *Ehrlichia* species have a similar ~425 bp origin of replication, in an AT-rich region of the genome, that contains features including DnaA and IHF-binding sites, flanked by genes *hem*E and COG1253, and in *E. canis* and *E. chaffeensis* the region also includes one or more CtrA binding sites. In the case of *E. ruminantium*, the CtrA binding sites are either absent or are possibly mutated (Ioannidis et al. 2007). In common with several other intracellular organisms, *Ehrlichia* only possess a subset of the genes for DNA replication, which are present in free-living organisms. Five genes were identified in *E. ruminantium*, which could code for the core structure of a functional DNA polymerase III: *dnaE*, *dnaN*, *holB*, *dnaQ*, and *dnaZ*, putatively encoding the α, β, δ', ε, and γ chains, respectively. There is a gene for DNA polymerase I (*polA*), DNA repair genes such as *mutM*, *radA*, *radC*, and *nth*, and the transcription-repair coupling factor *mfd*. Mismatch-repair enzymes are limited to *mutS* and *mutL*, and of the normal ultraviolet-induced DNA damage repair system, *uvrABC*, only the subunit A gene is present. Several other replication and repair enzymes are absent from some of the *Ehrlichia* genomes, but not others. For example *dnaZ* (DNA polymerase III chain γ) and *mutM* (formamidopyrimidine DNA glycosylase) were absent from *E. canis* but present in *E. ruminantium*. The latter enzyme mediates recovery from mutagenesis or lethal injury, suggesting that *E. canis* may be under increased mutational pressure compared to other *Ehrlichia* species (Mavromatis et al. 2006).

E. ruminantium has several genes involved in homologous recombination, these include *rmuC*, *recA*, *recR*, *recF*, and a gene (Erum4920) similar to *recO* of *Mesorhizobium loti* (Massung et al. 2004). There is also a gene (Erum6250) coding for an enzyme similar to *recB*, although other genes of the *recBCD* complex have not been identified. The organism has no plasmids, phages, insertion sequences, or genes for pilus assembly, which mediate horizontal gene transfer in many bacteria (Thomas and Nielsen 2005). *E. ruminantium* does have, however, most of the genes which, in other bacteria, are required for the assembly of a channel spanning both inner and outer membranes, a channel that is required for bacterial transformation competence (Cascales and Christie 2003). The missing gene is *vir*B7, which codes for a protein used in *Bartonella* and *Brucella* species to stabilize other VirB proteins by disulphide cross-linking (Cascales and Christie 2003). All this suggests that *E. ruminantium* could be a naturally transformable bacterium, actively capable of acquiring foreign genes by DNA uptake followed by homologous recombination (Thomas and Nielsen 2005), which would provide an explanation for the extensive recombination that has been observed to occur between different genotypes of the organism in the field (Allsopp and Allsopp 2007). The most likely time and place for recombination to occur would seem to be in the vector tick during the 15 day period after ingestion of a blood meal from an infected animal. During this time, before they becomes established in gut epithelial cells, organisms remains extracellular within the tick (Kocan and Bezuidenhout 1987), and it is possible that naked DNA from degraded *E. ruminantium* chromosomes would be available to be taken up by viable organisms.

Transcription and Translation

The DNA-dependent RNA polymerase of *E. ruminantium* consists of four subunits (α, β, β′, and ω) encoded by *rpoA*, *rpoB*, *rpoC*, and rpoZ. Two initiation factors σ^{70} (*rpoD*) and σ^{32} (*rpoH*) were also identified. The *nusA*, *nusG*, *greA*, and *rho* genes involved in transcription elongation and termination are also present. There are two almost identical copies of the *rho* gene; *rho1* being 60 base pairs longer than *rho2* at the 5′ end. Several genes involved in RNA degradation were identified, including *rnpA* and *rnpB* (ribonuclease P), *rnhA*, *rnhB*, and *rnc*, (encoding ribonucleases HI, HII, and III, respectively).

The single copy ribosomal RNA (rRNA) genes have a much higher G+C content than the rest of the genome, 48.6%, 49.6%, and 45.8% for 16S, 5S, and 23S rRNA genes, respectively. Several genes involved in rRNA processing and modification are present, including *ksgA*, *rbfA*, *rimM*, and two pseudouridine synthetases, *rluC* and *rluD*. *E. ruminantium* possesses genes that encode a complete set of ribosomal proteins, except for the 50S ribosomal protein.

All three *Ehrlichia* genomes contain 36 tRNA genes with specificities for all 20 amino acids. Several genes for tRNA modification are present, including *truB*, *miaA*, *rnpA*, and *trmD*. Aminoacyl-transfer RNA (tRNA) synthetase genes for aminoacylation of nearly all amino acids are present, including two genes encoding glutamyl-tRNA synthetase (*gltX1* and *gltX2*). As in several other bacteria, the genes encoding glutaminyl-tRNA synthetase and asparaginyl-tRNA synthetase are absent. In *E. ruminantium*, the genes *gatA*, *gatB*, and *gatC*, which encode the three subunits of glutamyl-tRNA amidotransferase, are present, suggesting that the organism derives glutaminyl-$tRNA^{Gln}$ and asparaginyl-$tRNA^{Asn}$ by transamidation of misacylated glutamyl-$tRNA^{Gln}$ or aspartyl-$tRNA^{Asn}$. A putative tmRNA, responsible for tagging incomplete proteins on stalled ribosomes for proteolysis, is also present.

4.2.3.5 Transporters

Ehrlichia genomes contain numerous orthologs of genes involved in eubacterial membrane transport systems, including several ATP-binding cassette (ABC) transporters and several transporters involved in import and efflux of cations. In *E. ruminantium* Na^+/H^+ (Erum1780, Erum5530, and Erum5550) and K^+/H^+ (Erum0950) antiporters are present, and are probably involved in maintaining the pH of the *E. ruminantium* cell. There are also two transporters that could be involved in multidrug efflux: one is an integral membrane transport protein (Erum4710) and the other (Erum6780) is a member of the major facilitator superfamily (MFS), also called the uniporter–symporter–antiporter family, which are often involved in drug efflux systems. *Ehrlichia* organisms also have the basic secretion mechanisms that are found in free-living proteobacteria; these include common chaperones such as *dnaK*, *dnaJ*, *hslU*, *hslV*, *groEL*, *groES* and *htpG*, genes of the *secA*-dependent secretion system and the sec-independent secretion system *tat*.

Secretory System Genes

A type IV secretion system present in *E. ruminantium*, *E. canis*, and *E. chaffeensis* contains several orthologs of the *virB* gene operon, a system that has also been identified in *Anaplasma phagocytophilum* (Lohr et al. 2004). There are two clusters of virB genes: *virD4*, *virB8*, *virB9*, *virB10*, and *virB11* are grouped together, while the second locus consists of *virB3*, *virB4*, *virB6*, and three additional large genes, which probably encode type IV secretion proteins. The *virB1*, *virB2*, *virB5*, and *virB7* genes, as well as genes encoding the proteins VirA and VirG, responsible for regulating the expression of the virB locus in *A. tumefaciens*, appear to be missing (Thompson et al. 1988; Das and Pazour 1989). Genes encoding the known effector proteins VirD2, VirE2, and VirF are not present, but a putative *trbG* gene, involved in conjugal transfer of T-DNA in *A. tumefaciens*, is found to be located at a distance from the *virB* gene clusters. In view of the dispersed nature of many *Ehrlichia* genes, the maintenance of the *virB* operon structure may have considerable significance, and the system is known to be essential for the survival of other intracellular bacteria (Celli and Gorvel 2004).

Genomic Repeats

A striking and unusual feature of *Ehrlichia* genomes is that the percentage of repetitive DNA is much higher than that for most prokaryotes, amounting to 8.3% of the genome in the case of *E. ruminantium* and 3.6% for *E. chaffeensis*, which contributes to the relatively low percentage of coding DNA for these organisms. During the sequencing of all three *E. ruminantium* genomes, it was observed that the copy number of many of the repeats is actively variable, with the addition or deletion of repeats occurring primarily in noncoding regions, but also within CDSs (Collins et al. 2005; Frutos et al. 2006). This genome plasticity is an unusual characteristic of these organisms and it will be dealt with in a subsequent section.

Genomic Repeats in *E. ruminantium*

In *E. ruminantium* there are three types of repeats, simple sequence repeats (SSRs) of 1–5 bp, large tandem repeats (LTRs) of >5 bp, and dispersed repeats, including direct and inverted repeats. There are 126 SSRs of 1–5 bp, of which 13 were located within the promoter region upstream of the predicted start of genes, and three were located within ORFs close to the start codon. These 16 SSRs could play a role in promoter regulation or phase variation of surface-associated proteins, a function which SSRs have in *Campylobacter jejuni* (Parkhill et al. 2000). The remaining 110 *E. ruminantium* SSRs were neither within nor close to genes, and so their role is unlikely to be associated with gene regulation.

The *E. ruminantium* genome contains 158 LTRs, from six to several hundred basepairs in length, 50 of which occur within 31 genes, while 85 are located in noncoding regions. Three LTRs overlap genes at the 5' end, 18 at the 3' end, and two at the 3' end of one gene and the 5' end of the following gene. In four cases where LTRs overlap the beginning or end of a gene they produce eight pseudogenes, a quarter of the 32 pseudogenes identified in this genome. Of the 31 CDSs containing LTRs, 27 are predicted to encode membrane-associated proteins or are genes unique to *E. ruminantium*, suggesting that LTRs are particularly important to this organism. The most important mechanism thought to generate LTRs is slipped-strand mispairing (Levinson and Gutman 1987), and at four sites this appears to have occurred more frequently than at other tandem repeat sequences in the genome. Three of the LTRs have different 7 bp motifs with highly variable numbers of the repeated sequence motifs, and a higher G+C content than the rest of the genome. G+C rich hypervariable sequences have been shown readily to form secondary structures, which can cause the DNA polymerase to pause and may result in rapid generation of tandem repeats (Weitzmann et al. 1997). Erum1110 contains a G+C rich LTR repeated 56 times, and a homolog of this gene identified in another strain of the organism has the same LTR repeated 21 times (Barbet et al. 2001). Notably, the Erum1110 ortholog in *E. canis* (gp36) also has LTRs and is considered a secreted major immunoreactive protein that is present on the surface of dense-cored ehrlichial cells (Doyle et al. 2006). It has been suggested that the larger size of surface proteins containing a greater number of LTRs might provide a protective shield for other surface proteins that are less free to change (Citti et al. 1997).

Seventy-five dispersed repeats were identified in the *E. ruminantium* genome, ranging in size from 64 bp to almost 3 kbp, with the majority between 100 and 400 bp. The repeat units were from 75 to 100% identical and approximately equal numbers of direct and inverted repeats were present.

Genomic Repeats in *E. canis* and *E. chaffeensis*

Compared to *E. ruminantium*, *E. canis* and *E. chaffeensis* have a smaller number of short sequence repeats that are distributed evenly throughout the genome. A small group ($n = 12$) of proteins associated with host–pathogen interactions have LTRs within the open reading frames. Two of these proteins (gp140 and gp36) are considered to be major immunoreactive proteins of *E. canis* and have been well characterized (Yu et al. 2000a, b; Doyle et al. 2006). Their corresponding orthologs in *E. chaffeensis*, gp120 and gp47, have also been molecularly characterized (Yu et al. 1997; Doyle et al. 2006). The gp36/gp47 proteins were found to elicit strong humoral immune responses directed at the tandem repeat, and since they are secreted from the ehrlichial organisms they may be important virulence factors (Doyle et al. 2006). There is evidence that the *E. ruminantium* ortholog of gp36/gp47, Erum1110, functions as an ehrlichial adhesin and is associated with attachment to tick cells (De la Fuente et al. 2004). This gene will also be discussed later in relation to *E. ruminantium* genome evolution.

Genome Plasticity

The genomic libraries used for the *E. ruminantium* genome project (Collins et al. 2005) came from DNA that was prepared from pools of ehrlichial organisms purified from different culture passages of the same stock, representing several generations of the organism. At four different sites it was found that the copy numbers of LTRs were variable in different genomic clones, three of these were tandem repeats with different 7 bp motifs and the fourth was a 122 bp repeat. At these loci there were highly variable (5–20×) numbers of tandem repeats, a finding which was confirmed by performing PCR amplification across the repeat region of different clones within each locus. There were no sequence variations, other than occasional SNPs, in any other repetitive or nonrepetitive loci. The variable repeat regions were found to have a higher G+C content than normal, with marked strand asymmetry, making them ideal sites for tandem repeat generation by means of secondary structure formation (Weitzmann et al. 1997) and slipped-strand mispairing (Rocha 2003). When two further strains of *E. ruminantium* were sequenced (Frutos et al. 2006), it was concluded that the differences in genome size among the three strains was a result of differences in the numbers of repeats in the intergenic LTRs. This unusual situation has been aptly called "genome plasticity" and it appears to contribute to an ongoing evolutionary process in *Ehrlichia* species, which will be discussed in depth in Section "Genome Evolution".

The different repetitive sequences in *E. ruminantium* have homologs, but no paralogs, in all three genomes, and no orthologs in *E. canis* or *E. chaffeensis*. The *Ehrlichia* repeats are not similar to mobile genetic elements, which are common in *Rickettsia* and *Wolbachia*, although there are some similarities with LTRs in *A. marginale* (Brayton et al. 2005). It has been suggested that the mechanism that produces these repeats may be a specific feature of *E. ruminantium*, and possibly also of *A. marginale*, which allows an increased rate of mutation to occur in these organisms when they are placed under environmental stress (Frutos et al. 2007).

Although similar observable changes in genomic repeats were not reported during genome sequencing of *E. canis* and *E. chaffeensis*, plasticity has been documented in tandem repeat containing genes from both organisms and is described in detail in the genetic variability section. Genome sequencing of another strain of *E. chaffeensis* (Sapulpa) is currently underway, which may provide additional insight into genome plasticity in *E. chaffeensis*.

Genome Evolution

Of the 32 predicted pseudogenes in the *E. ruminantium* genome, 29 are associated with repeats and 25 appear to be truncated fragments of other genes. This observation suggests that most of the pseudogenes in *E. ruminantium* are the products of sequence duplication events, mostly in repeat regions. Duplications also appear to have resulted in the formation of new genes, and four examples are shown in Fig. 4. Figure 4a shows Erum1110 which contains, after a 421 bp 5′ region, a 151 1 bp region that consists of a 27 bp motif repeated 56 times. The 5′ region of Erum1110, plus an additional 65 bp upstream of the start codon, is >90% identical to the adjacent gene *Erum1100*, and this gene terminates where the 27 bp tandem repeat starts in its paralog Erum1110. An ortholog of Erum1110, containing 21.7 copies of the same 27 bp motif, was identified in another strain of *E. ruminantium* by screening of an expression library with *E. ruminantium* antisera (Barbet et al. 2001). It seems that either Erum1100 was duplicated and subsequently became fused with the tandem repeat, or that the nonrepetitive 5′ portion of Erum1110 was duplicated and became the independent gene Erum1100.

Figure 4b shows gene *Erum8190*, which appears to have been duplicated, and the duplicate has acquired a stop codon, creating two new genes, *Erum8170* and *Erum8180*. The fact that the two parts

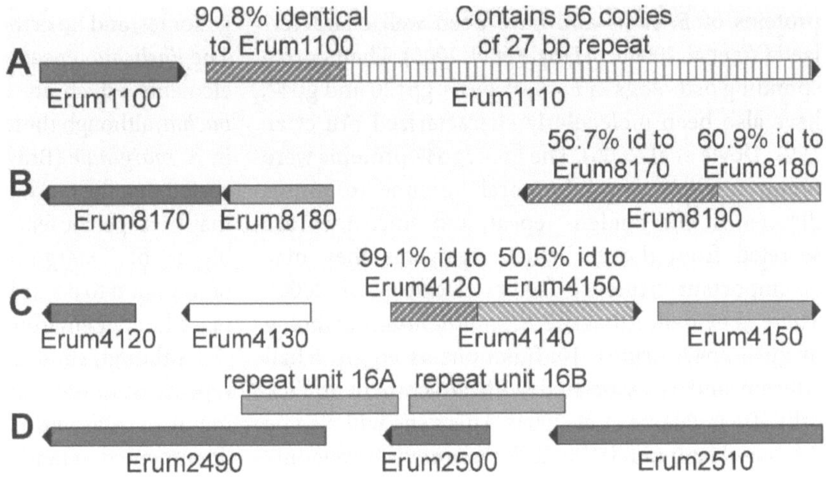

Fig. 4 *E. ruminantium* genes, which appear to have been generated as the result of sequence duplication events. Adapted from Collins et al. 2005

of *Erum1110* have similar identities to the two adjacent genes supports this hypothesis. The third example in Fig. 4c is rather different. Here two sections of *Erum4140* have very different levels of identity to the two adjacent genes *Erum4120* and *Erum4150*, each of which has homologs in other bacteria. It is unlikely, therefore, that these two genes are the result of the duplication of *Erum4140*, but rather that duplications occurred, first of *Erum4150* and much later of *Erum4120*, and that the paralogs subsequently later fused to generate the new gene *Erum4140*. *Erum4120* is believed to code for an iron–sulfur cofactor synthesis protein, and *Erum4150* for a cysteine desulfurase. If these functions are retained in *Erum4140*, the new gene product will be a bifunctional "Rosetta Stone" protein (Marcotte and Marcotte 2002).

The fourth example in Fig. 4d shows a direct repeat (repeat units 16A and 16B), which appears to have resulted in the duplication of the 3′ end of *Erum2490* creating the new gene *Erum2500*. This new gene has been found in seven strains of *E. ruminantium* originating in southern Africa, but not in five stocks originating in West Africa, suggesting that the duplication event occurred in southern Africa, where the species is believed to have originated, but some time after *E. ruminantium* had spread to other parts of the continent (Allsopp et al. 2003).

Gene Duplication Events, or Other Genome Evolution Events, in *E. canis* and *E. chaffeensis*

There appear to have been fewer gene duplication events in *E. canis* and *E. chaffeensis*; however, *E. canis* has three additional major outer membrane protein genes (p28/p30) found in a separate locus that appears to be the result of a recent duplication event (Ohashi et al. 1998a). Pseudogenes ($n = 17$) corresponding to only 11 full length functional genes are present in *E. canis*, but these pseudogenes are distinct from those of *E. ruminantium* and must therefore have occurred after the separation of the species (Mavromatis et al. 2006).

Membrane Protein Genes

The genomes of *Ehrlichia* species contain many CDSs, which are predicted to code for proteins with transmembrane helices and/or N-terminal signal peptides, which suggests that many of these proteins are secreted and/or are membrane bound. Some of these membrane proteins occur in families of paralogs, and a few of them were known prior to genome sequencing and have been relatively well characterized, but many were unknown before the completion of the genome sequences. Some of these protein families are thought to be involved in immune evasion (Collins et al. 2005; Mavromatis et al. 2006).

In the *E. ruminantium* genome, 28% (247) of CDSs other than pseudogenes are predicted to contain at least one transmembrane helix, 197 of which begin within the first 10 amino acids of the protein. Since it is difficult to differentiate between N-terminal signal peptides and N-terminal transmembrane helices (Yuan et al. 2003), it is likely that at least some of the 197 predicted N-terminal transmembrane domains are in fact signal sequences. Firm prediction of the existence of signal peptides have been made for 66 CDSs, 53 of which did not contain predicted transmembrane helices. Up to 263 CDSs could therefore have signal sequences, and so potentially up to 29.6% of all CDSs encode for membrane proteins. There are 37 CDSs containing 1–20 transmembrane helices, but without a signal peptide, three of which are pseudogenes, 24 are unknown, and 10 have orthologs in other bacteria. Similarly, a large number of genes in *E. canis* (310; 33%) are predicted to code for secreted proteins and/or contain at least one transmembrane helix. A majority (n = 179) have N-terminal signal peptides, but many (n = 39) lack transmembrane domains and are likely to be secreted. Disulphide bond formation between membrane proteins may be important in outer membrane structure; this has been reported in *A. marginale* and identification of a functional thio-disulphide oxidoreductase (DsbA) has been demonstrated in *E. canis* (McBride et al. 2002). In *E. canis*, 36 proteins have a high cysteine content and 15 have transmembrane domains or signal peptides, suggesting that they are components of the outer membrane. *E. chaffeensis* has 49 proteins that are associated with the cell envelope.

Several families of paralogous hypothetical membrane protein genes are present in *Ehrlichia*. In this discussion, CDSs are considered to be paralogous if they are of similar lengths, have similar features, occur in close proximity, and have a mean of pairwise amino acid identities, which do not fall below the 15–25% "twilight zone," below which a common origin is unlikely (Doolittle 1987). One such family encodes orthologous major outer membrane proteins (omps) of *E. ruminantium* (*map*1) (Van Heerden et al. 2004a), *E. chaffeensis* (omp/p28) (Ohashi et al. 1998b), and *E. canis* (p28/p30) (Ohashi et al. 1998a). MAP1 was the first individual protein of *E. ruminantium* to be sequenced (Van Vliet et al. 1994), although it came under investigation well before the genomics era because it is immunodominant and appears as a prominent band on Western blots of total *E. ruminantium* proteins (Rossouw et al. 1990). MAP1 varies in size from strain to strain (Barbet et al. 1994), is located on the surface of the organism (Jongejan et al. 1991b), and is recognised by immune antisera and by $CD4^+$ T lymphocytes from immune animals (Mwangi et al. 1998). The *map*1 multigene family consists of 16 paralogs having identities ranging from 13.3 to 66.5%, with a mean of 35.1%, and 14 of them are predicted to have signal peptides (Van Heerden et al. 2004a). Orthologs of the *map*1 family are the p28/p30 family in *E. canis* and the omp/p28 family in *E. chaffeensis* and several papers describe their identification (Reddy et al. 1998; Ohashi et al. 1998a; Yu et al. 1999a, b; McBride et al. 1999) and complete characterization of the loci (Yu et al. 2000a, b; Ohashi et al. 2001). In both *E. chaffeensis* and *E. canis*, the omp families have 22 paralogs arranged in a single locus, while *E. canis* also has three additional omp genes found in a separate locus (Ohashi et al. 2001). In each *Ehrlichia* genome, the multigene omp clusters lie upstream from the *sec*A gene and downstream from a hypothetical transcriptional regulator gene. At the 5′ end of the clusters the paralogs are linked by short intergenic spaces, while the intergenic spaces are longer at the 3′ ends. Although the functions of these families are not well understood, it is likely that they play a role in immune evasion (Van Heerden et al. 2004a). There is evidence that one of the *E. chaffeensis* P28 proteins and the *E. ruminantium* MAP1 protein is protective in mice (Nyika et al. 1998; Ohashi et al. 1998b), and this is discussed at length in Sect. 4.2.5.

The major omp genes appear to be differentially expressed in the tick in all three species of *Ehrlichia*, with one, and sometimes two, genes being the only ones expressed in the tick or in tick cell lines. The differentially expressed genes are the p28–14 gene of *E. chaffeensis* (Unver et al. 2002), the orthologous p30–10 gene of *E. canis* (Felek et al. 2003), and both the *map*1 and orthologous *map*1-1 genes of *E. ruminantium* (Postigo et al. 2007). In the case of *E. ruminantium*, the *map*1 and *map*1-1 genes are both upregulated when infected ticks begin feeding, but the upregulation of *map*1 itself occurs predominantly in the salivary glands of the tick immediately before the organisms are injected into the mammalian bloodstream (Postigo et al. 2007).

Two other potentially interesting membrane protein families of *E. ruminantium* should be mentioned, of 14 and 10 members, respectively. Based on codon usage analysis, seven adjacent members of the 14-member family were predicted to be present as the result of horizontal gene transfer (Merkl 2004). The 10 member family is split between two loci (Erum2750-Erum2800 and Erum3600-Erum3630). All members of these two families are predicted to have either signal peptides or N-terminal transmembrane domains, which could be signal peptides, although these are predicted to be noncleavable (Krogh et al. 2001). Both the *E. ruminantium* MAP1 protein (Jongejan et al. 1991b) and the *E. chaffeensis* P28 protein (Ohashi et al. 1998b) are known to be expressed on the surface of the organism, and the *map1* gene also has a signal peptide, but this is predicted to be cleavable. It seems likely that any of these membrane protein families could potentially be exposed on the parasite surface; however, there is currently very little information about their possible function.

Tandem Repeat and Ankyrin Repeat Containing Proteins of *Ehrlichia*

The genome of *E. canis* revealed a small group of tandem repeat containing proteins ($n = 12$) that are associated with host–pathogen interactions. Several of these proteins are well characterized major immunoreactive proteins and are surface exposed and secreted (Popov et al. 2000; Doyle et al. 2005a; McBride et al. 2007). Other common features include a bias of polar, hydrophobic, and acidic amino acids, such as serine, valine, and aspartate, respectively (Yu et al. 2000a, b; Doyle et al. 2005a; McBride et al. 2007; Nethery et al. 2007). Several of these proteins, including gp120/gp140, gp36/gp47, and gp19/VLPT, lack signal peptides but have been localized to the surface of the organism. They are also found outside the bacteria, associated with the morula matrix and/or the morula membrane, and are frequently present in soluble forms in the supernatants from infected cells (Popov et al. 2000; Doyle et al. 2005a; McBride et al. 2007). The *E. chaffeensis* gp120 was the first major immunoreactive protein of this organism to be molecularly characterized, and it was found to be differentially expressed on dense-cored cells, to be localized to the surface of the bacteria, and to be present extracellularly on the intra morula fibrils (Popov et al. 2000). The role of the gp120 is not completely understood, but it is a member of a small group of tandem repeat containing proteins that are associated with host–pathogen interactions. Transformation of a nonadherent *Escherichia coli* strain with the gp120 gene resulted in attachment to and invasion of mammalian host cells, indicating that the gp120 is an ehrlichial adhesin (Popov et al. 2000). Similarly, tandem repeat containing orthologs gp36/gp47 are differentially expressed on the surface of dense-cored ehrlichiae, are also secreted and associated with the intramorular fibrils, but are also associated with the morula membrane (Doyle et al. 2006). Furthermore, relatively large amounts of gp36 and gp47 are present in supernatants from *Ehrlichia* infected cells. The *E. ruminantium* ortholog of gp36/gp47, Erum1110, is associated with attachment to tick cells, suggesting that these proteins may also be important ehrlichial adhesins (De la Fuente et al. 2004). Most recently, a 19 kDa major immunoreactive protein of *E. canis* was identified and determined to be the ortholog of *E. chaffeensis* VLPT (McBride et al. 2007). This protein was present on both morphological forms of ehrlichiae and is also secreted to become associated with the morula fibrils and the morula membrane.

Another small group of proteins identified in *E. canis* ($n = 7$; Ecaj_0052, 0221, 0352, 0365, 0387, 0516, and 0627) have eukaryote-like ankyrin domains. One of these proteins (Ecaj_00365; gp200) has been molecularly characterized and is considered to be a major immunoreactive protein (McBride et al. 2003a; Nethery et al. 2007). Four dominant species-specific antibody epitopes have been mapped to acidic terminal domains of this protein, which is translocated to the host cell nucleus during infection (Nethery et al. 2007). The role of Ecaj_00365 in pathobiology is not understood, but it may bind DNA and alter host cell gene transcription. Other ankyrin domain containing genes have not been studied, and one (Ecaj_0221) contains both ankyrin domains and tandem repeats, suggesting that it may have an important role in ehrlichial pathology.

4.2.4 Molecular Diagnostics

For many years after the discovery of heartwater reliable diagnosis of the disease remained difficult. Clinical diagnosis was always imprecise, since many of the symptoms generally associated with the disease were nonspecific, hence a definitive diagnosis was usually made only after postmortem examination (Camus and Barré 1987). Even though diagnosis was difficult, the pathology of the disease was variable and it was not always easy to locate the organism microscopically. Reliance was generally placed on the demonstration of the pathogen in brain endothelial cells, where it is generally more numerous than in other tissues (Prozesky 1987b). None of the serological tests developed for *E. ruminantium* was entirely reliable either, and they all relied upon detection of the strongly immunodominant *E. ruminantium* outer membrane protein MAP1. The reason became evident when *Ehrlichia* genomic studies revealed the existence of orthologous families of immunodominant outer membrane proteins in other *Ehrlichia* species, as described in the previous section.

PCR-based diagnostic techniques have become the only reliable methods for *E. ruminantium* diagnosis, and the first genomic target to be identified especially for this purpose was the pCS20 genetic region (Waghela et al. 1991). A PCR test that targeted this region showed it to provide a specific test for *E. ruminantium*, giving no cross reactions with other *Ehrlichia* species, and the test has been extensively used to detect the organism in domestic animals, wild game, and ticks (Peter et al. 1995, 1999, 2000; Mahan et al. 1998, 2004; Allsopp et al. 1999a, b; Simbi et al. 2003). The pCS20 region contains two overlapping genes, ribonuclease III (*rnc*), and cytochrome c oxidase assembly protein (*cta*G), and the original clone has been shown to be chimeric (Van Heerden et al. 2004b). A more sensitive version of the original test has been developed, which utilizes a probe that targets only the *cta*G gene of the amplicon (Van Heerden et al. 2004b). There are sequence polymorphisms in the pCS20 region of different *E. ruminantium* strains, mostly single nucleotide differences, and they do not compromise the sensitivity or specificity of the test. *E. canis* and *E. chaffeensis* carry *rnc* and *cta*G genes overlapping in the same manner as in *E. ruminantium* but they have not yet been exploited for the diagnosis of those species.

The 16S RNA gene has been used for the diagnosis of *E. ruminantium*, and because of its widespread use as a taxonomic and phylogenetic tool it is a particularly useful target when previously unknown *Ehrlichia* species or strains are encountered in field surveys (Allsopp et al. 1998). Probes for the 16S RNA gene are difficult to use, because the sequence variations are small and the sensitivity of the test is inferior to that of the pCS20 test, and so it is not generally used for the routine diagnosis of *E. ruminantium*. The *map1* gene, which is extensively polymorphic, has been used as a diagnostic target for *E. ruminantium* when the objective is to characterize different antigenic variants (Allsopp et al. 2001, 2003; Martinez et al. 2004).

Human monocytotropic ehrlichiosis is a serious and potentially fatal disease that is usually diagnosed empirically based on clinical findings, but laboratory testing is necessary to confirm the diagnosis (Walker et al. 2004; Dumler et al. 2007). Similarly canine monocytic ehrlichiosis is routinely diagnosed on empirical clinical findings, which must later be confirmed with laboratory and serologic tests. Serological testing by immunofluorescence (IFA) is considered to be the best choice at the current time for both CME and HME (Waner et al. 2001; Walker 2005). Serological assays have limitations which, in the case of HME, can have a direct consequence on the outcome of the patient's illness due to delays in the administration of the appropriate therapy. Patients treated in the first week of illness, prior to the development of antibodies, have more favorable outcomes and less opportunity to develop severe disease manifestations (Eng et al. 1990). Unfortunately, since most patients seek medical attention 4 days after onset of illness, a large majority (67%) of those with HME may be missed, since most patients do not have IgG antibodies during the first week, and IgM does not substantially improve clinical diagnostic sensitivity (Childs et al. 1999). In addition, it is difficult to determine accurately the nature of the infecting agent based on serology, owing to the existence of closely related agents such as *E. ewingii* that induce cross-reactive antibodies. *E. ewingii* has not been cultivated in vitro, and serological diagnosis has been performed using *E. chaffeensis* antigens (Buller et al. 1999), but the reliability of this procedure is not known. Protein immunoblotting can be useful in classifying indeterminant cases (Buller et al. 1999), but lack of standardization remains a problem.

One of the major problems with IFA is cross-reactive antigens that are shared among ehrlichial pathogens as well as with conserved immunoreactive antigens shared by other bacteria. Molecular characterization and epitope mapping of major immunoreactive proteins of *E. chaffeensis* and *E. canis* have provided new opportunities to substantially improve the sensitivity and specificity of serologic tests for the HME and CME. Antigens that elicit an early antibody response in *E. canis*-infected dogs (McBride et al. 2003b), such as gp36 and gp19, have species-specific epitopes that can be used in diagnostic tests to substantially improve the specificity and sensitivity of serodiagnostic tests (Doyle et al. 2006; McBride et al. 2007). Similarly, the

E. chaffeensis gp120 protein and the orthologous *E. canis* gp140 protein provide sensitive and specific serological diagnoses (Yu et al. 1999a, b, 2000a, b).

Several variations in PCR-based detection of *Ehrlichia*, including nested PCR, reverse-transcriptase PCR, and real-time PCR, have been developed targeting numerous genes including 16S rRNA, VLPT, gp120, dsb, and p28 (McBride et al. 1996; Sumner et al. 1999; Gusa et al. 2001; Stich et al. 2002; Loftis et al. 2003; Olano et al. 2003a). Nested PCR, using the 16S rRNA gene as the target, is the most widely used method and has the analytical sensitivity to detect low levels of circulating ehrlichial organisms (Buller et al. 1999; Childs et al. 1999; Gusa et al. 2001; Paddock et al. 2001). Recently, a multicolour real-time PCR assay has been introduced, which is capable of detecting levels of *Ehrlichia* infection comparable to those detectable by nested PCR (Loftis and Levin 2004). A multiplex method of testing to discriminate among *Ehrlichia* of medical and veterinary importance (*E. chaffeensis*, *E. ewingii*, and *E. canis*) in a single reaction has been developed and this may be useful in clinical improvement of diagnostic capability and in surveillance to raise the level of diagnostic confirmation of HME and of ehrlichiosis caused by *E. ewingii* (Doyle et al. 2005b; Sirigireddy and Ganta 2005). The new methods improve analytical sensitivity and specificity, as well as reducing the time taken to attain a result, and in fact these are the only tests that can be used for the definitive detection of *E. ewingii* in clinical samples.

4.2.5 Vaccine Development

Most of the recent efforts to develop vaccines for *Ehrlichia* species have focussed on *E. ruminantium* because of the important economic losses that it causes in Africa and the Caribbean and the danger of it being introduced onto the American mainland (Sect. 4.1.8). The emergence of human ehrlichioses in the last decade, and the risk to public health which they pose, has recently elicited interest in the development of vaccines for humans. Such vaccines would be targeted for persons at increased risk of infection such as military personnel who train in endemic areas and others with increased potential for exposure due to recreational or occupational activities.

The only commercially available procedure for protective immunization against a species of *Ehrlichia* is an infection and treatment technique originally developed in South Africa to protect against heartwater (Neitz and Alexander 1945). The Ball 3 stock of *E. ruminantium* is currently used because it produces a temperature rise some days before the animal becomes seriously sick, thus allowing antibiotic treatment to be administered in a timely fashion (Van der Merwe 1987). Unfortunately, the Ball 3 vaccine confers only limited protection against virulent field challenge with classical heartwater-causing stocks like Welgevonden (Du Plessis et al. 1989), and a discussion of the differing immunogenicities of different *E. ruminantium* strains will be found in Section "*E. ruminantium* Immunogenicity". But apart from this failing there are other serious limitations to the Ball 3 vaccine, there is a need to maintain a continuous cold chain for storage of the live inoculum (Van der Merwe 1987), there is the risk of transmitting of other blood-borne pathogens, and both preparation and administration of the live vaccine are expensive (Mahan et al. 1999). Research conducted over the last two decades has been aimed at producing a more effective vaccine against *E. ruminantium*, and three experimental types of vaccine are being investigated, inactivated, attenuated, and recombinant.

E. ruminantium Immunogenicity

At one time it was believed that all strains of *E. ruminantium* were fully cross-protective and antigenically conserved (Du Plessis and Kumm 1971). However, when the infection and treatment method of immunization began to be widely used in South Africa, vaccine "breakthroughs" were observed and the existence of poor cross-protection between some strains became evident (Jongejan et al. 1988, 1991a; Du Plessis et al. 1989). Assessing the immunogenicity and cross-protection of stocks prior to the mid-1990s was very difficult, however, since there were no reliable methods available to genotype the stocks being used. After molecular genotyping became possible, at least one stock that had been used in cross-immunity experiments was shown to be genetically heterogeneous (Zweygarth et al. 2002). Nevertheless, the Welgevonden stock was initially found to provide a broad spectrum of protection against a range of other strains (Du Plessis et al. 1989) and this was subsequently confirmed in cross-immunity trials with quantified challenge material and molecularly characterized single genotype stocks (Collins et al. 2003). The stocks used in this study were Ball 3, Mara 87/7, Gardel, Welgevonden, Kwanyanga, and Blaauwkrans (Table 1). Blaauwkrans and Kwanyanga, isolated in the Eastern Cape province of South Africa, were pathogenic to small ruminants and had the same 16S genotype as Welgevonden. Batches of sheep were immunized with each stock and the animals were later challenged with 10 LD_{50} of either the homologous stock or one heterologous stock. The Welgevonden stock was the only one that provided complete cross-protection against challenge with any of the other stocks. The Kwanyanga, Gardel, and Blaauwkrans stocks provided little cross protection against heterologous challenge, while Mara 87/7 and Ball3 provided limited cross protection against heterologous challenge. It was also notable that animals immune to Welgevonden were protected against both Kwanyanga and Blaauwkrans, but the converse was not true.

It has already been noted in Sect. 4.2.1 that the pathogenicities and infectivities of *E. ruminantium* stocks do not correlate with their 16S genotype, and the results of the cross-protection experiments show that the same is true of immunogenicities. It is evident that there exists, within *E. ruminantium* and closely related organisms, a large reservoir of genetic diversity, only a fraction of which is known from the few well-characterized heartwater-causing stocks. The genes involved in antigenicity, infectivity, virulence, and pathogenicity are as likely as the core genes to recombine between different strains of the organism (Allsopp and Allsopp 2007). This has very important implications for vaccine development.

Inactivated *E. ruminantium* Vaccines

Inactivated heartwater vaccines consist of nonviable organisms, partially purified from chemically treated tissue culture material, which are inoculated together with an adjuvant. The first successful application of such material was in goats using the Gardel isolate, and 50–80% of the animals were protected against a homologous needle challenge, which killed 100% of the negative controls (Martinez et al. 1994). In a subsequent study, the Crystal Springs isolate was used in sheep, and in this case 50–100% of the animals were protected against a homologous needle challenge, which killed 60% of the negative controls (Mahan et al. 1995). When the needle challenge is with a heterologous strain, or when a tick challenge is involved, the inactivated vaccines are less protective (Mahan et al. 2001; Faburay et al. 2007). The protection levels are still lower in a field situation when the challenge is with naturally infected ticks carrying genotypes having differing immunogenicities. A summary of all the field trials conducted by one research group over a period of years showed that overall mortality levels of 71% in naive animals could be reduced to 36% by vaccination (Mahan et al. 2003). So although inactivated vaccines provide some levels of protection it is likely that considerable improvements in these levels will be required before inactivated heartwater vaccines become a commercially viable proposition.

Attenuated *E. ruminantium* Vaccines

The Senegal isolate of *E. ruminantium* was the first isolate to be attenuated in culture, and it conferred 100% protection on animals subjected to a homologous needle challenge (Jongejan 1991). The results were more variable when field trials were conducted, with reductions in mortality, as compared to unvaccinated controls, of 70–43% in one study (Gueye et al. 1994), and 100–25% in another (Faburay et al. 2007). The Welgevonden isolate has been shown, as discussed earlier, to provide cross-protection against a needle challenge with a range of other isolates (Collins et al. 2003), but initial attempts to attenuate this isolate by growing it in culture through hundreds of passages over several years were not successful (Gueye et al. 1994; Zweygarth et al. 1997). However, the Welgevonden isolate was recently attenuated by continuous passage in a canine macrophage-monocyte cell line, followed by readaption to bovine endothelial cells (Zweygarth et al. 2005). When the attenuated Welgevonden strain was used to infect sheep or goats, there were no adverse symptoms, except for a brief rise in body temperature, and the animals were subsequently found to be fully protected against a lethal needle challenge with the homologous isolate or any one of four other heterologous isolates (Zweygarth et al. 2005). This attenuated vaccine has not yet been tested in the field against natural tick challenge, but if it were to be successful in such trials it could provide a cheap and effective vaccine for use in endemic heartwater areas.

Recombinant *E. ruminantium* Vaccines

The fact that immunization with killed organisms can be successful indicates that the development of a subunit vaccine for *E. ruminantium* is possible. Such a vaccine could in principle be cheap and effective, and unlike the attenuated vaccine it could be used to stop an outbreak in a nonendemic area. The first attempts to develop a recombinant vaccine involved immunization with a plasmid clone expressing the *map*1 gene of *E. ruminantium*, and this protected mice against a lethal homologous challenge at levels ranging from 23 to 88% (Nyika et al. 1998). In further experiments, the naked DNA-induced immunity was boosted with MAP1 protein and as a result protection levels in mice were increased, from a range of 13–27% without boosting to a range of 53–67% with boosting (Nyika et al. 2002). Similar work has been done with *E. chaffeensis* using a recombinant version of the P28 protein and partial protection was obtained after homologous challenge (Ohashi et al. 1998b). Substantial divergence has been reported in *map*1 and *p28* genes among different isolates of *E. ruminantium* and *E. chaffeensis*, suggesting that to provide comprehensive protection using these genes would require the inclusion of multiple genes from various strains (Yu et al. 1999a, b, 2007). Conversely, the *p28/p30* genes of *E. canis* appear to be substantially conserved among geographically dispersed strains (McBride et al. 2000; Unver et al. 2001), which could facilitate the development of effective vaccines utilizing this antigen for CME. Several newly identified major protein orthologs (gp120/gp140, gp36/gp47, and gp200s) from *E. chaffeensis* and *E. canis* are immunodominant and are consistently recognized by antibodies in patients and animals that have recovered from HME and CME (Yu et al. 2000a, b; Doyle et al. 2006; McBridge et al. 2007; Nethery et al. 2007). The ability of these proteins to protect against homologous challenge has not been determined but it is possible that they could be of value for vaccine development.

Denatured MAP1 protein appears to confer no protection against *E. ruminantium* infection in ruminants (Van Kleef et al. 1993), and so the *map*1 gene might not be the best choice for recombinant vaccine experiments with this organism. The *E. ruminantium* genome contains 888 annotated genes from which to chose vaccine candidates (Collins et al. 2005), but there are no reliable strategies for identifying the genes that code for antigens which stimulate the protective T-cell response (Esteves et al. 2004). One attempt to overcome this difficulty involved selecting clones from *E. ruminantium* expression libraries on the following basis: firstly that their expression products were recognized by anti-*E. ruminantium* antibodies, and secondly that they stimulated proliferation of peripheral blood mononuclear cells (PBMC) from cattle immunized against *E. ruminantium* by infection and treatment (Barbet et al. 2001). Lysates of recombinant bacterial cultures expressing the selected genes were then used to immunize mice, and 58–89% survival was observed with some pools of recombinants. The levels of protection were therefore similar to those obtained with the *map*1 gene.

All reported *E. ruminantium* immunization trials performed in mice have given unpredictably variable results, and genes that have conferred immunity in mice have not been protective in ruminants (Louw et al. 2002; Collins et al. 2003). Vaccination trials conducted in sheep, however, have been shown to be reproducible. A cocktail of four *E. ruminantium* genes that may be components of an ABC transporter system were cloned in a DNA vaccine vector and used to immunize sheep, which were subsequently completely protected against a virulent needle challenge with both homologous and heterologous *E. ruminantium*-infected blood (Collins et al. 2003). Disappointingly, however, the cocktail vaccine protected poorly against a natural tick challenge in the field unless the immunity had been boosted by a challenge with *E. ruminantium*-infected blood. Further work on this experimental vaccine showed that each of the components of the cocktail was able to provide similar levels of protection to that of the whole cocktail (Pretorius et al. 2007), and the protection was shown to be specific to the *E. ruminantium* genes since other *E. ruminantium* ORFs tried in the same vaccine system failed to provide any protection. The protection against field challenge remained poor, however, and the apparently enhanced virulence of *E. ruminantium* organisms injected by the tick, as compared to those that are present in infected mammalian blood, has been discussed in Sect. 2.1.4.

4.2.6 Comparative Genomics of *Ehrlichia* Species

As noted in Sect. 4.2.2, three different strains of *E. ruminantium* have been sequenced: the type strain (Welgevonden) (Collins et al. 2005); a further sample of the Welgevonden strain (known as *Erwe*), which was transferred to Guadeloupe from South Africa in 1985 and then maintained for 18 years in tissue culture (Frutos et al. 2006); and the Gardel strain (known as *Erga*), which was isolated on Guadeloupe (Uilenberg et al. 1985). A detailed comparison made between these genomes (Frutos et al. 2007) shows that they are, as expected, very similar to each other, with a total of 888 orthologous CDSs present in all three strains. Different annotation criteria were used for *Erwe* and *Erga* as compared to the original Welgevonden, and so it was difficult to make entirely accurate comparisons of all the CDSs, but there appear to be no substantive differences in the ORFs of the two Welgevonden strains, although some intra-ORF repeats were of different lengths. In *Erga* 22 CDSs have no orthologs in the two Welgevonden strains, while these two strains contain 35 (common) CDSs that have no orthologs in *Erga*. Four CDSs were found to be mutated in *Erwe* as compared with the original Welgevonden, and these CDSs were intact in *Erga*, suggesting that the mutations had occurred during the 18 years (11–13 passages) under different cell culture conditions in isolation from its parent strain. An analysis of the synonymous vs. nonsynonymous (d_S/d_N) substitutions per site in *Erga* vs. the two Welgevonden strains showed only three CDSs with a d_S/d_N ratio biased towards nonsynonymous substitutions, two of which are annotated as pseudogenes. The d_S/d_N ratio for the rest of the homologous CDSs was biased towards synonymous substitutions, indicative of mutation constrained by the need to conserve protein function.

An estimate of 4.4–4.7 × 10^{-9} synonymous substitutions per site per year for free-living bacteria was obtained from a comparison of *E. coli* and *Salmonella typhimurium* (Ochman et al. 1999), and in the same study the authors noted that the endosymbiotic γ-proteobacterium *Buchnera* has synonymous substitution rates and 16S rRNA gene mutation rates, which are both approximately twice as fast as those of *E. coli* and *S. enterica*. They therefore proposed that this approximately doubled mutation rate was a general feature of intracellular bacteria, perhaps as a result of smaller populations promoting an easier fixation of mutations. In a separate comparative genomics study, this elevated estimate of synonymous substitution rates for intracellular bacteria was used to calculate the divergence times of the three *E. ruminantium* genomes (Hughes and French 2007), and the estimates obtained were 26,500–28,500 years ago for divergence of the two Welgevonden genomes, and 2.2 mya for the last common ancestor of all three. Since it is known that the two Welgevonden genomes were in fact separated 18 years (11–13 passages) ago (Frutos et al. 2006), this was taken as evidence that in *E. ruminantium* accelerated mutation rates follow exposure of the organism to changes of environment, a suggestion already made based on the unusual features of the repetitive sequences in this species as discussed in the section describing genome plasticity (Frutos et al. 2007). An alternative suggestion made by Hughes and French was that the original Welgevonden isolate contained two different genotypes, and that one of these was a minor constituent that had outgrown the other during the subsequent 18-year cultivation period of the sample after it had been taken to Guadeloupe (Hughes and French 2007). If this were the case it would be expected that at least some of the SNPs observed in the differing *Erwe* sequence would appear among the 25,648 gel readings in the sequencing database of the original Welgevonden genome (Collins et al. 2005). However, a careful search of this database by one of us (BAA) did not reveal any indication that this was a mixed genotype containing a subpopulation now represented by *Erwe*.

Experimental evidence that *E. ruminantium* mutates when maintained in in vitro culture was obtained during a transcription *analysis* of the *map*1 omp family in three in vitro cultured isolates of *E. ruminantium*, Gardel, *Senegal*, and Welgevonden (Bekker et al. 2005). Two subpopulations of the original Gardel isolate were examined: one had been transferred from Guadeloupe to Portugal, and thence to the Netherlands, and had undergone 45–66 passages in culture in bovine pulmonary artery endothelial cells; the other had been transferred from Guadeloupe to the UK and had undergone 12–15 passages in bovine aorta endothelial cells. Sequence analysis showed that, while in culture in the UK, the second subpopulation had undergone a rearrangement of a part of the *map*1 genomic region in which the *map*1-2 and *map*1-3 genes had recombined at a 14 bp site common to the two genes, resulting in a single hybrid gene and deletion of the original *map*1-2 gene. It is also noteworthy that the *map*1-1 gene sequences of the Welgevonden type strain and the Guadeloupe-cultured Welgevonden strain (*Erwe*) are also different (Frutos et al. 2006), and so an increase in mutation rate when in culture does seem to be the most likely explanation of the differences observed between the two Welgevonden genome sequences.

Mutation of *E. chaffeensis* and *E. canis* after laboratory passage has not been reported. For example, several genes of *E. chaffeensis* (the *VLPT* gene, the *gp120* gene, and the *p28/omp* multigene locus) have been independently sequenced from various different passages of the Arkansas strain grown in different laboratories and mutations have not been seen (Reddy et al. 1998; Yu et al. 2000a, b; Ohashi et al. 2001; Hotopp et al. 2006). However, differences have been noted among *E. chaffeensis* strains (Arkansas, Wakulla, Liberty, Jax, and St Vincent) with respect to *p28/omp* genes and others (Sumner et al. 1999; Yu et al. 1999a, b; Reddy and Streck 1999; Miura and Rikihisa 2007). Similar sequence heterogeneity has been reported of the *E. ruminantium map*1 genes in geographically dispersed isolates (Reddy et al. 1996; Allsopp et al. 2001).

Mutation rates in *Ehrlichia* species were considered during a comparative study of the three *E. ruminantium* genomes with that of *E. canis* and *A. marginale* (Mavromatis et al. 2006). These authors proposed 40 mya as the oldest divergence time of *E. canis* from *E. ruminantium*, based on the first appearance in the fossil record of the Metastriata ticks of which the two *Ehrlichia* species are parasites. This leads to an assessment that both 16S rRNA mutation rates and the fixation rate of gene duplication events have a lower limit, which is 5–6 times slower in *Ehrlichia* than in *E. coli* (Mavromatis et al. 2006), which appears to accord with the observation that generation times in *Buchnera* are about six times longer than those in *E. coli* (Clark et al. 1999). However, although the intuitive link between generation time and mutation rates has been documented in eukaryotes, there is evidence that no such link exists in prokaryotes (Maughan 2007), and indeed some workers have inferred that *Buchnera* has gene mutation rates that are approximately twice as fast as those of *E. coli* (Ochman et al. 1999). The lack of agreement between bacterial mutation rates determined experimentally over short terms (Lenski et al. 2003) and those inferred from species divergence times over long terms (Ochman et al. 1999) could be due to differences in mutational processes under different degrees of environmental stress (Bjedov et al. 2003; Ochman 2003). It is a matter of observation that in the case of the *E. ruminantium map*1 omp family mutations and rearrangements have taken place on the relatively short timescale of decades when the organisms are maintained in culture, which could indeed have put them under unusual stress. The lack of similar observations in other *Ehrlichia* species means, however, that the general question of mutation rates in *Ehrlichia* species is one which cannot be answered satisfactorily at this time.

Comparisons between the genomes of the three different *Ehrlichia* species show that *E. ruminantium* and *E. canis* exhibit almost complete synteny with each other (Mavromatis et al. 2006) and both also show a high degree of synteny with *E. chaffeensis*, except that *E. chaffeensis* has a symmetrical inversion close to the two *rho* termination factors, which are a feature of all the *Ehrlichia* genomes (Hotopp et al. 2006). It is interesting that a similar symmetrical *rho* inversion occurs between *A. marginale* and *A. phagocytophilum* (Hotopp et al. 2006) but not between the much more distantly related *E. ruminantium* (Collins et al. 2005) and *A. marginale* (Brayton et al. 2005). This suggests that the inversions occurred independently in the lines leading to *E. chaffeensis* and *A. phagocytophilum* after the separation of the two genera from their common ancestor.

A series of comparisons of clusters of orthologous genes from five species representing five different genera in the order Rickettsiales was recently performed (Hotopp et al. 2006). When comparing the genomes of *R. prowazekii*, *N. sennetsu*, *W. pipientis*, *A. phagocytophilum*, and *E. chaffeensis*, it was found that 423 ortholog clusters contained genes from all five organisms. As might have been expected, these were mostly genes for housekeeeping functions, but less intuitive was the finding that very few genes were shared between pairs of species from different genera. In fact four of the genomes had almost as many unique genes as five-way shared genes, and one (*A. phagocytophilum*) had more unique genes than shared genes. One or two genes in the unique clusters had a predicted function, several more were omp genes, but the overwhelming majority were genes of unknown function.

The five-way comparison found that the two Anaplasmataceae (*A. phagocytophilum* and *E. chaffeensis*) were the only organisms that had genes encoding thiamine biosynthetic proteins, a particular potassium transporter, a peptide deformylase, and several ankyrin repeat proteins, but the significance of these observations is currently not evident. One interesting observation was that *R. prowazekii*, *N. sennetsu*, and *W. pipientis* all contained a gene which included an N-terminal class II aldolase and adducin head domain, and this gene was not present in either of the two Anaplasmataceae. Since *Ehrlichia* and *Anaplasma* are the only members of the Rickettsiales, which are not transmitted transovarially in the arthropod vector, this observation suggests that the putative aldolase is essential for transovarial transmission. When *E. ruminantium* (Welgevonden and Gardel) was included in the orthologous cluster comparison, along with *E. chaffeensis*, it was found that only 53 orthologous clusters were specific to all the *Ehrlichia* species, of which five had a predicted function, eight were omp genes, and 40 were genes of unknown function. The three *Ehrlichia* genomes shared 709 genes, but each genome also contained unique genes, of which Welgevonden had four, Gardel 48, and *E. chaffeensis* 267.

It might have been expected that the comparison of different non-cross-protective strains of *E. ruminantium* might have provided some insight into the reasons for the immunogenic differences, but this has unfortunately not materialized. This highlights a common complaint about contemporary comparative genomics, which is that it generates lists of genes without leading to any in depth understanding of genomic function. These gene lists are the twenty-first century equivalents of the endless biodiversity lists, which dominated nineteenth and early twentieth century biology, and this is inevitably a phase through which comparative genomics must pass. Eventually deeper insights into genomic functions will emerge when functional genomics and proteomics have advanced to the stage where the functions of most genes are known. Currently these disciplines are in the relatively early stages of their development, and for *Ehrlichia* species very little has been done.

4.3 Future Scope

Given the experimental difficulties of working with the fragile and obligately intracellular *Ehrlichia* species, it is likely that future genomic research will be driven by the need to develop recombinant vaccines. In the case of *E. ruminantium*, an attenuated vaccine may become useful in the endemic area in Africa, but it is likely that a recombinant vaccine will be the best long-term solution, especially in the event that the disease spreads from the Caribbean to the American mainland. There are numerous fundamental aspects of *Ehrlichia* biology, which are poorly understood and which need clarification in order to support vaccine development, and some of the most important questions are the following: (a) what genetic and phenotypic changes occur, and where do they occur, during the developmental cycle in the tick; (b) what is the role of the tick in enhancing the potency of infecting elementary bodies; (c) what are the molecular mechanisms involved in persistent infections of natural mammalian hosts; (d) what are the ehrlichial virulence factors and what mechanisms control their expression in invertebrate and vertebrate hosts; (e) what is the nature of disease pathogenesis in natural and incidental hosts; (f) which ehrlichial proteins stimulate protective immunity and what are the protective immune mechanisms. Substantial progress has been made in some areas and less in others. The genome sequences of these most important *Ehrlichia* pathogens represents, however, a major comprehensive step towards being able to develop testable hypotheses about molecular mechanisms that will lead to answers for these important questions.

A rapidly growing range of mostly high throughput screening methods is being developed to determine the functions of individual genes and proteins in a global context, rather than one gene product at a time. In the case of *Ehrlichia* species the global context includes both the invertebrate and vertebrate hosts, which suggests that comprehensive answers to some of the aforementioned questions will not become available until genome sequences are available for these other two essential players in the parasitic cycle. The current frontier of *E. ruminantium* vaccine research is to look for vaccine candidate genes, but the algorithms used to predict these candidates are all narrowly focussed on single genes of the pathogen itself. We are unlikely to be able to make good predictions until we know how all the *E. ruminantium* genes and their products interact within the tick–host–pathogen triad. It is likely that answering these complex questions will be aided by future developments in functional genomics, proteomics, large scale transcript mapping, and genetic manipulation.

References

Agrawal N, Lesley SA, Kuhn P, Kohen A (2004) Mechanistic studies of a flavin-dependent thymidylate synthase. Biochemistry 43:10295–10301

Allsopp MT, Allsopp BA (2001) Novel *Ehrlichia* genotype detected in dogs in South Africa. J Clin Microbiol 39:4204–4207

Allsopp MT, Allsopp BA (2007) Extensive genetic recombination occurs in the field between different genotypes of *Ehrlichia ruminantium*. Vet Microbiol 124:58–65

Allsopp MT, Visser ES, du Plessis JL, Vogel SW, Allsopp BA (1997) Different organisms associated with heartwater as shown by analysis of 16S ribosomal RNA gene sequences. Vet Parasitol 71:283–300

Allsopp MT, Hattingh CM, Vogel SW, Allsopp BA (1998) Comparative evaluation of 16S, map1 and pCS20 probes for the detection of *Cowdria* and *Ehrlichia* species in ticks. Ann NY Acad Sci 849:78–84

Allsopp MT, Hattingh CM, Vogel SW, Allsopp BA (1999a) Evaluation of 16S, *map*1 and pCS20 probes for detection of *Cowdria* and *Ehrlichia* species. Epidemiol Infect 122:323–328

Allsopp MT, Theron J, Coetzee ML, Dunsterville MT, Allsopp BA (1999b) The occurrence of *Theileria* and *Cowdria* parasites in African buffalo (*Syncerus caffer*) and their associated *Amblyomma hebraeum* ticks. Onderstepoort J Vet Res 66:245–249

Allsopp MT, Dorfling CM, Maillard JC, Bensaid A, Haydon DT, van Heerden H, Allsopp BA (2001) *Ehrlichia ruminantium* major antigenic protein gene (*map*1) variants are not geographically constrained and show no evidence of having evolved under positive selection pressure. J Clin Microbiol 39:4200–4203

Allsopp MT, Van Heerden H, Steyn HC, Allsopp BA (2003) Phylogenetic relationships among *Ehrlichia ruminantium* isolates. Ann NY Acad Sci 990:685–691

Allsopp MT, Louw M, Meyer EC (2005) *Ehrlichia ruminantium*: An emerging human pathogen? Ann NY Acad Sci 1063:358–360

Allsopp MT, Van Strijp MF, Faber E, Josemans AI, Allsopp BA (2007) *Ehrlichia ruminantium* variants which do not cause heartwater found in South Africa. Vet Microbiol 120: 158–166

Anderson BE, Dawson JE, Jones DC, Wilson KH (1991) *Ehrlichia chaffeensis*, a new species associated with human ehrlichiosis. J Clin Microbiol 29:2838–2842

Andersson SG, Zomorodipour A, Andersson JO, Sicheritz-Ponten T, Alsmark UC, Podowski RM, Naslund AK, Eriksson AS, Winkler HH, Kurland CG (1998) The genome sequence of *Rickettsia prowazekii* and the origin of mitochondria. Nature 396:133–140

Andrew HR, Norval RA (1989) The carrier status of sheep, cattle and African buffalo recovered from heartwater. Vet Parasitol 34:261–266

Barbet AF, Semu SM, Chigagure N, Kelly PJ, Jongejan F, Mahan SM (1994) Size variation of the major immunodominant protein of *Cowdria ruminantium*. Clin Diagn Lab Immunol 1:744–746

Barbet AF, Whitmire WM, Kamper SM, Simbi BH, Ganta RR, Moreland AL, Mwangi DM, McGuire TC, Mahan SM (2001) A subset of *Cowdria ruminantium* genes important for immune recognition and protection. Gene 275:287–298

Barnewall RE, Rikihisa Y, Lee EH (1997) *Ehrlichia chaffeensis* inclusions are early endosomes which selectively accumulate transferrin receptor. Infect Immun 65:1455–1461

Bekker CP, Postigo M, Taoufik A, Bell-Sakyi L, Ferraz C, Martinez D, Jongejan F (2005) Transcription analysis of the major antigenic protein 1 multigene family of three in vitro-cultured *Ehrlichia ruminantium* isolates. J Bacteriol 187:4782–4791

Bezuidenhout JD (1987) Natural transmission of heartwater. Onderstepoort J Vet Res 54:349–351

Bezuidenhout JD, Paterson CL, Barnard BJ (1985) *In vitro* cultivation of *Cowdria ruminantium*. Onderstepoort J Vet Res 52:113–120

Bjedov I, Tenaillon O, Gerard B, Souza V, Denamur E, Radman M, Taddei F, Matic I (2003) Stress-induced mutagenesis in bacteria. Science 300:1404–1409

Brayton KA, Fehrsen J, de Villiers EP, van Kleef M, Allsopp BA (1997) Construction and initial analysis of a representative lambda ZAPII expression library of the intracellular rickettsia *Cowdria ruminantium*: cloning of *map*1 and three other *Cowdria* genes. Vet Parasitol 72:185–199

Brayton KA, De Villiers EP, Fehrsen J, Nxomani C, Collins NE, Allsopp BA (1999) *Cowdria ruminantium* DNA is unstable in a SuperCos1 library. Onderstepoort J Vet Res 66: 111–117

Brayton KA, Kappmeyer LS, Herndon DR, Dark MJ, Tibbals DL, Palmer GH, McGuire TC, Knowles DP Jr (2005) Complete genome sequencing of *Anaplasma marginale* reveals that the surface is skewed to two superfamilies of outer membrane proteins. Proc Natl Acad Sci USA 102:844–884

Breitschwerdt EB, Hegarty BC, Hancock SI (1998) Sequential evaluation of dogs naturally infected with *Ehrlichia canis*, *Ehrlichia chaffeensis*, *Ehrlichia equi*, *Ehrlichia ewingii*, or *Bartonella vinsonii*. J Clin Microbiol 36:2645–2651

Buhles WC Jr, Huxsoll DL, Ristic M (1974) Tropical canine pancytopenia: clinical, hematologic, and serologic response of dogs to *Ehrlichia canis* infection, tetracycline therapy, and challenge inoculation. J Infect Dis 130:357–367

Buller RS, Arens M, Hmiel SP, Paddock CD, Sumner JW, Rikhisa Y, Unver A, Gaudreault-Keener M, Manian FA, Liddell AM, Schmulewitz N, Storch GA (1999) *Ehrlichia ewingii*, a newly recognized agent of human ehrlichiosis. N Engl J Med 341:148–155

Cascales E, Christie PJ (2003) The versatile bacterial type IV secretion systems. Nat Rev Microbiol 1:137–149

Camus E (1992) Le portage asymptomatique de bovins et chèvres Créole guéris de la cowdriose en Guadeloupe. Rev Elev Med Vet Pays Trop 45:133–135

Camus E, Barré N (1987) Diagnosis of heartwater in the live animal: experiences with goats in Guadeloupe. Onderstepoort J Vet Res 54:291–294

Celli J, Gorvel JP (2004) Organelle robbery: *Brucella* interactions with the endoplasmic reticulum. Curr Opin Microbiol 7:93–97

Cheng C, Paddock CD, Reddy Ganta R (2003) Molecular heterogeneity of *Ehrlichia chaffeensis* isolates determined by sequence analysis of the 28-kilodalton outer membrane protein genes and other regions of the genome. Infect Immun 71:187–195

Childs JE, Paddock CD (2003) The ascendancy of *Amblyomma americanum* as a vector of pathogens affecting humans in the United States. Annu Rev Entomol 48:307–337

Childs JE, Sumner JW, Nicholson WL, Massung RF, Standaert SM, Paddock CD (1999) Outcome of diagnostic tests using samples from patients with culture-proven human monocytic ehrlichiosis: implications for surveillance. J Clin Microbiol 37:2997–3000

Citti C, Kim MF, Wise KS (1997) Elongated versions of Vlp surface lipoproteins protect *Mycoplasma hyorhinis* escape variants from growth-inhibiting host antibodies. Infect Immun 65:1773–1785

Clark MA, Moran NA, Baumann P (1999) Sequence evolution in bacterial endosymbionts having extreme base compositions. Mol Biol Evol 16:1586–1598

Codner EC, Caceci T, Saunders GK, Smith CA, Robertson JL, Martin RA, Troy GC (1992) Investigation of glomerular lesions in dogs with acute experimentally induced *Ehrlichia canis* infection. Am J Vet Res 53:2286–2291

Cole JR, Chai B, Farris RJ, Wang Q, Kulam-Syed-Mohideen AS, McGarrell DM, Bandela AM, Cardenas E, Garrity GM, Tiedje JM (2007) The ribosomal database project (RDP-II): introducing myRDP space and quality controlled public data. Nucl Acids Res 35:D169–D172

Collins NE, Pretorius A, van Kleef M, Brayton KA, Allsopp MT, Zweygarth E, Allsopp BA (2003) Development of improved attenuated and nucl acid vaccines for heartwater. Dev Biol (Basel) 114:121–136

Collins NE, Liebenberg J, de Villiers EP, Brayton KA, Louw E, Pretorius A, Faber FE, van Heerden H, Josemans A, van Kleef M, Steyn HC, van Strijp MF, Zweygarth E, Jongejan F, Maillard JC, Berthier D, Botha M, Joubert F, Corton CH, Thomson NR, Allsopp MT, Allsopp BA (2005) The genome of the heartwater agent *Ehrlichia ruminantium* contains multiple tandem repeats of actively variable copy number. Proc Natl Acad Sci USA 102:838–843

Cowdry EV (1925a) Studies on the etiology of heartwater 1. Observation of a rickettsia, *Rickettsia ruminantium* (n. sp.), in the tissues of infected animals. J Exp Med 42:231–252

Cowdry EV (1925b) Studies on the etiology of heartwater 2. *Rickettsia ruminantium* (n. sp.) in the tissues of ticks transmitting the disease. J Exp Med 42:253–274

Dame JB, Mahan SM, Yowell CA (1992) Phylogenetic relationship of *Cowdria ruminantium*, agent of heartwater, to *Anaplasma marginale* and other members of the order Rickettsiales determined on the basis of 16S rRNA sequence. Int J Syst Bacteriol 42:270–274

Das A, Pazour GJ (1989) Delineation of the regulatory region sequences of *Agrobacterium tumefaciens* virB operon. Nucleic Acids Res 17:4541–4550

Dawson JE, Anderson BE, Fishbein DB, Sanchez JL, Goldsmith CS, Wilson KH, Duntley CW (1991a) Isolation and characterization of an *Ehrlichia* sp. from a patient diagnosed with human ehrlichiosis. J Clin Microbiol 29:2741–2745

Dawson JE, Rikihisa Y, Ewing SA, Fishbein DB (1991b) Serologic diagnosis of human ehrlichiosis using two *Ehrlichia canis* isolates. J Infect Dis 163:564–567

Dawson JE, Ewing SA (1992) Susceptibility of dogs to infection with *Ehrlichia chaffeensis*, causative agent of human ehrlichiosis. Am J Vet Res 53:1322–1327

De la Fuente J, Garcia-Garcia JC, Barbet AF, Blouin EF, Kocan KM (2004) Adhesion of outer membrane proteins containing tandem repeats of *Anaplasma* and *Ehrlichia* species (Rickettsiales: Anaplasmataceae) to tick cells. Vet Microbiol 98:313–322

De Villiers EP, Brayton KA, Zweygarth E, Allsopp BA (1998) Purification of *Cowdria ruminantium* organisms for use in genome analysis by pulsed-field gel electrophoresis. Ann NY Acad Sci 849:313–320

De Villiers EP, Brayton KA, Zweygarth E, Allsopp BA (2000) Genome size and genetic map of *Cowdria ruminantium*. Microbiol 146:2627–2634

Deem SL (1998) A review of heartwater and the threat of introduction of *Cowdria ruminantium* and *Amblyomma spp.* ticks to the American mainland. J Zoo Wildl Med 29: 109–113

Dixon RW (1898) Heartwater experiments. Agric J Cape of Good Hope 12:754–760

Donatien A, Lestoquard F (1935) Existence en Algérie d'une rickettsia du chien. Bull Soc Pathol Exot 28:418–419

Doolittle RF (1987) Of URFs and ORFs, 1st edn. University Science Books, CA, USA

Doyle CK, Cardenas AM, Aguiar DM, Labruna MB, Ndip LM, Yu XJ, McBride JW (2005a) Molecular characterization of *E. canis* gp36 and *E. chaffeensis* gp47 tandem repeats among isolates from different geographic locations. Ann NY Acad Sci 1063:433–435

Doyle CK, Labruna MB, Breitschwerdt EB, Tang YW, Corstvet RE, Hegarty BC, Bloch KC, Li P, Walker DH, McBride JW (2005b) Detection of medically important *Ehrlichia* by quantitative multicolor TaqMan real-time polymerase chain reaction of the dsb gene. J Mol Diagn 7:504–510

Doyle CK, Nethery KA, Popov VL, McBride JW (2006) Differentially expressed and secreted major immunoreactive protein orthologs of *Ehrlichia canis* and *E. chaffeensis* elicit early antibody responses to epitopes on glycosylated tandem repeats. Infect Immun 74:711–720

Du Plessis JL (1970) Pathogenesis of heartwater. I. *Cowdria ruminantium* in the lymph nodes of domestic ruminants. Onderstepoort J Vet Res 37:89–95

Du Plessis JL (1982) Mice infected with a *Cowdria ruminantium*-like agent as a model in the study of heartwater. D Thesis, Univ of Pretoria, South Africa, pp 157

Du Plessis JL (1985) A method for determining the *Cowdria ruminantium* infection rate of *Amblyomma hebraeum*: effects in mice injected with tick homogenates. Onderstepoort J Vet Res 52:55–61

Du Plessis JL (1990) Increased pathogenicity of an *Ehrlichia*-like agent after passage through *Amblyomma hebraeum*: a preliminary report. Onderstepoort J Vet Res 57:233–237

Du Plessis JL, Kumm NA (1971) The passage of *Cowdria ruminantium* in mice. J S Afr Vet Med Assoc 42:217–221

Du Plessis JL, Malan L (1987) The application of the indirect fluorescent antibody test in research on heartwater. Onderstepoort J Vet Res 54:319–325

Du Plessis JL, Van Gas L, Olivier JA, Bezuidenhout JD (1989) The heterogenicity of *Cowdria ruminantium* stocks: cross-immunity and serology in sheep and pathogenicity to mice. Onderstepoort J Vet Res 56:195–201

Dumler JS, Barbet AF, Bekker CP, Dasch GA, Palmer GH, Ray SC, Rikihisa Y, Rurangirwa FR (2001) Reorganization of genera in the families Rickettsiaceae and Anaplasmataceae in the order Rickettsiales: unification of some species of *Ehrlichia* with *Anaplasma*, *Cowdria* with *Ehrlichia* and *Ehrlichia* with *Neorickettsia*, descriptions of six new species combinations and designation of *Ehrlichia equi* and 'HGE agent' as subjective synonyms of *Ehrlichia phagocytophila*. Int J Syst Evol Microbiol 51:2145–2165

Dumler JS, Madigan JE, Pusterla N, Bakken JS (2007) Ehrlichioses in humans: epidemiology, clinical presentation, diagnosis, and treatment. Clin Infect Dis 45:S45–S51

Edington A (1898) Heartwater. Agric J Cape of Good Hope 12:748–754

Eng TR, Harkess JR, Fishbein DB, Dawson JE, Greene CN, Redus MA, Satalowich FT (1990) Epidemiologic, clinical, and laboratory findings of human ehrlichiosis in the United States, 1988. J Am Med Assoc 264:2251–2258

Esteves I, Vachiery N, Martinez D, Totte P (2004) Analysis of *Ehrlichia ruminantium*-specific T1/T2 responses during vaccination with a protective killed vaccine and challenge of goats. Parasite Immunol 26:95–103

Ewing SA (1963) Observations on leukocytic inclusion bodies from dogs infeced with *Babesia canis*. J Am Vet Med Assoc 143:503–506

Faburay B, Geysen D, Ceesay A, Marcelino I, Alves PM, Taoufik A, Postigo M, Bell-Sakyi L, Jongejan F (2007) Immunisation of sheep against heartwater in The Gambia using inactivated and attenuated *Ehrlichia ruminantium* vaccines. Vaccine 25:7939–7947

Felek S, Greene R, Rikihisa Y (2003) Transcriptional analysis of p30 major outer membrane protein genes of *Ehrlichia canis* in naturally infected ticks and sequence analysis of p30-10 of *E. canis* from diverse geographic regions. J Clin Microbiol 41:886–888

Fleischmann RD, Adams MD, White O, Clayton RA, Kirkness EF, Kerlavage AR, Bult CJ, Tomb JF, Dougherty BA, Merrick JM, et al (1995) Whole-genome random sequencing and assembly of *Haemophilus influenzae* Rd. Science 269: 496–512

Frutos R, Viari A, Ferraz C, Morgat A, Eychenie S, Kandassamy Y, Chantal I, Bensaid A, Coissac E, Vachiery N, Demaille J, Martinez D (2006) Comparative genomic analysis of three strains of *Ehrlichia ruminantium* reveals an active process of genome size plasticity. J Bacteriol 188:2533–2542

Frutos R, Viari A, Vachiery N, Boyer F, Martinez D (2007) *Ehrlichia ruminantium*: genomic and evolutionary features. Trends Parasitol 23:414–419

Greub G, Raoult D (2003) History of the ADP/ATP-translocase-encoding gene, a parasitism gene transferred from a Chlamydiales ancestor to plants 1 billion years ago. Appl Environ Microbiol 69:5530–5535

Groves MG, Dennis GL, Amyx HL, Huxsoll DL (1975) Transmission of *Ehrlichia canis* to dogs by ticks (*Rhipicephalus sanguineus*). Am J Vet Res 36:937–940

Gueye A, Jongejan F, Mbengue M, Diouf A, Uilenberg G (1994) Essai sur le terrain d'un vaccin atténué contre la cowdriose. Rev Elev Med Vet Pays Trop 47:401–404

Guindon S, Gascuel O (2003) A simple, fast, and accurate algorithm to estimate large phylogenies by maximum likelihood. Syst Biol 52:696–704

Gusa AA, Buller RS, Storch GA, Huycke MM, Machado LJ, Slater LN, Stockham SL, Massung RF (2001) Identification of a p28 gene in *Ehrlichia ewingii*: evaluation of gene for use as a target for a species-specific PCR diagnostic assay. J Clin Microbiol 39:3871–3876

Haig DA (1952) Note on the use of the white mouse for the transport of strains of heartwater. J S Afr Vet Med Assoc 23:167–170

Hotopp JC, Lin M, Madupu R, Crabtree J, Angiuoli SV, Eisen J, Seshadri R, Ren Q, Wu M, Utterback TR, Smith S, Lewis M, Khouri H, Zhang C, Niu H, Lin Q, Ohashi N, Zhi N, Nelson W, Brinkac LM, Dodson RJ, Rosovitz MJ, Sundaram J, Daugherty SC, Davidsen T, Durkin AS, Gwinn M, Haft DH, Selengut JD, Sullivan SA, Zafar N, Zhou L, Benahmed F, Forberger H, Halpin R, Mulligan S, Robinson J, White O, Rikihisa Y, Tettelin H (2006) Comparative genomics of emerging human ehrlichiosis agents. PLoS Genet 2:e21

Hughes AL, French JO (2007) Homologous recombination and the pattern of nucleotide substitution in *Ehrlichia ruminantium*. Gene 387:31–37

Huxsoll DL, Hildebrandt PK, Nims RM, Ferguson JA, Walker JS (1969) *Ehrlichia canis* – the causative agent of a haemorrhagic disease of dogs? Vet Rec 85:587

Huxsoll DL, Hildebrandt PK, Nims RM, Amyx HL, Ferguson JA (1970) Epizootiology of tropical canine pancytopenia. J Wildl Dis 6:220–225

Ioannidis P, Hotopp JC, Sapountzis P, Siozios S, Tsiamis G, Bordenstein SR, Baldo L, Werren JH, Bourtzis K (2007) New criteria for selecting the origin of DNA replication in *Wolbachia* and closely related bacteria. BMC Genom 8:182

Jongejan F (1991) Protective immunity to heartwater (*Cowdria ruminantium* infection) is acquired after vaccination with *in vitro*-attenuated rickettsiae. Infect Immun 59:729–731

Jongejan F, Uilenberg G, Franssen FF, Gueye A, Nieuwenhuijs J (1988) Antigenic differences between stocks of *Cowdria ruminantium*. Res Vet Sci 44:186–189

Jongejan F, Thielemans MJ, Briere C, Uilenberg G (1991a) Antigenic diversity of *Cowdria ruminantium* isolates determined by cross- immunity. Res Vet Sci 51:24–28

Jongejan F, Thielemans MJ, De Groot M, van Kooten PJ, van der Zeijst BA (1991b) Competitive enzyme-linked immunosorbent assay for heartwater using monoclonal antibodies to a *Cowdria ruminantium*-specific 32-kilodalton protein. Vet Microbiol 28:199–211

Jongejan F, Zandbergen TA, van de Wiel PA, de Groot M, Uilenberg G (1991c) The tick-borne rickettsia *Cowdria ruminantium* has a *Chlamydia*-like developmental cycle. Onderstepoort J Vet Res 58:227–237

Josemans AI, Zweygarth E (2002) Amino acid content of cell cultures infected with *Cowdria ruminantium* propagated in a protein-free medium. Ann N Y Acad Sci 969: 141–146

Katavolos P, Armstrong PM, Dawson JE, Telford SR IIIrd (1998) Duration of tick attachment required for transmission of granulocytic ehrlichiosis. J Infect Dis 177:1422–1425

Kawahara M, Rikihisa Y, Lin Q, Isogai E, Tahara K, Itagaki A, Hiramitsu Y, Tajima T (2006) Novel Genetic Variants of *Anaplasma phagocytophilum*, *Anaplasma bovis*, *Anaplasma centrale*, and a Novel *Ehrlichia* sp. in Wild Deer and Ticks on Two Major Islands in Japan. Appl Environ Microbiol 72:1102–1109

Keefe TJ, Holland CJ, Salyer PE, Ristic M (1982) Distribution of *Ehrlichia canis* among military working dogs in the world and selected civilian dogs in the United States. J Am Vet Med Assoc 181:236–238

Kocan KM, Bezuidenhout JD (1987) Morphology and development of *Cowdria ruminantium* in *Amblyomma* ticks. Onderstepoort J Vet Res 54:177–182

Kocan KM, Bezuidenhout JD, Hart A (1987a) Ultrastructural features of *Cowdria ruminantium* in midgut epithelial cells and salivary glands of nymphal *Amblyomma hebraeum*. Onderstepoort J Vet Res 54:87–92

Kocan KM, Morzaria SP, Voigt WP, Kiarie J, Irvin AD (1987b) Demonstration of colonies of *Cowdria ruminantium* in midgut epithelial cells of *Amblyomma variegatum*. Am J Vet Res 48:356–360

Konstantinidis KT, Tiedje JM (2005) Genomic insights that advance the species definition for prokaryotes. Proc Natl Acad Sci USA 102:2567–2572

Krogh A, Larsson B, von Heijne G, Sonnhammer EL (2001) Predicting transmembrane protein topology with a hidden Markov model: application to complete genomes. J Mol Biol 305:567–580

Kumagai Y, Cheng Z, Lin M, Rikihisa Y (2006) Biochemical activities of three pairs of *Ehrlichia chaffeensis* two-component regulatory system proteins involved in inhibition of lysosomal fusion. Infect Immun 74:5014–5022

Lenski RE, Winkworth CL, Riley MA (2003) Rates of DNA sequence evolution in experimental populations of *Escherichia coli* during 20,000 generations. J Mol Evol 56:498–508

Levinson G, Gutman GA (1987) Slipped-strand mispairing: a major mechanism for DNA sequence evolution. Mol Biol Evol 4:203–221

Lin M, Rikihisa Y (2003a) *Ehrlichia chaffeensis* and *Anaplasma phagocytophilum* lack genes for lipid A biosynthesis and incorporate cholesterol for their survival. Infect Immun 71:5324–5331

Lin M, Rikihisa Y (2003b) Obligatory intracellular parasitism by *Ehrlichia chaffeensis* and *Anaplasma phagocytophilum* involves caveolae and glycosylphosphatidylinositol-anchored proteins. Cell Microbiol 5:809–820

Loftis AD, Levin ML (2004) Lack of susceptibility of guinea pigs and gerbils to experimental infection with *Ehrlichia chaffeensis*. Vector Borne Zoonotic Dis 4:319–322

Loftis AD, Massung RF, Levin ML (2003) Quantitative real-time PCR assay for detection of *Ehrlichia chaffeensis*. J Clin Microbiol 41:3870–3872

Loftis DA, Reeves WK, Spurlock JP, Mahan SM, Troughton DR, Dasch GA, Levin ML (2006) Infection of a goat with a tick-transmitted *Ehrlichia* from Georgia, U.S.A., that is closely related to *Ehrlichia ruminantium*. J Vector Ecol 31:213–223

Lohr CV, Brayton KA, Barbet AF, Palmer GH (2004) Characterization of the *Anaplasma marginale* msp2 locus and its synteny with the ompl/p30 loci of *Ehrlichia chaffeensis* and *E. canis*. Gene 325:115–111

Lounsbury CP (1900) Tick heartwater experiments. Agric J Cape of Good Hope 16:682–687

Louw E, Brayton KA, Collins NE, Pretorius A, Van Strijp F, Allsopp BA (2002) Sequencing of a 15-kb *Ehrlichia ruminantium* clone and evaluation of the *cpg*1 open reading frame for protection against heartwater. Ann NY Acad Sci 969:147–150

Louw M, Allsopp MT, Meyer EC (2005) *Ehrlichia ruminantium*, an emerging human pathogen – a further report. S Afr Med J 95:948–950

Mahan SM, Waghela SD, McGuire TC, Rurangirwa FR, Wassink LA, Barbet AF (1992) A cloned DNA probe for *Cowdria ruminantium* hybridizes with eight heartwater strains and detects infected sheep. J Clin Microbiol 30:981–986

Mahan SM, Andrew HR, Tebele N, Burridge MJ, Barbet AF (1995) Immunisation of sheep against heartwater with inactivated *Cowdria ruminantium*. Res Vet Sci 58:46–49

Mahan SM, Peter TF, Simbi BH, Burridge MJ (1998) PCR detection of *Cowdria ruminantium* infection in ticks and animals from heartwater-endemic regions of Zimbabwe. Ann NY Acad Sci 849:85–87

Mahan SM, Allsopp BA, Kocan KM, Palmer GH, Jongejan F (1999) Vaccine strategies for *Cowdria ruminantium* infections and their application to other ehrlichial infections. Parasitol Today 15:290–294

Mahan SM, Peter TF, Simbi BH, Kocan K, Camus E, Barbet AF, Burridge MJ (2000) Comparison of efficacy of American and African *Amblyomma* ticks as vectors of heartwater (*Cowdria ruminantium*) infection by molecular analyses and transmission trials. J Parasitol 86:44–49

Mahan SM, Smith GE, Kumbula D, Burridge MJ, Barbet AF (2001) Reduction in mortality from heartwater in cattle, sheep and goats exposed to field challenge using an inactivated vaccine. Vet Parasitol 97:295–308

Mahan SM, Barbet AF, Burridge MJ (2003) Development of improved vaccines for heartwater. Dev Biol (Basel) 114:137–145

Mahan SM, Simbi BH, Burridge MJ (2004) The pCS20 PCR assay for *Ehrlichia ruminantium* does not cross-react with the novel deer ehrlichial agent found in white-tailed deer in the United States of America. Onderstepoort J Vet Res 71:99–105

Maillard JC, Maillard N (1998) Historique du peuplement bovin et de l'introduction de la tique *Amblyomma variegatum* dans les îles françaises des Antilles: synthèse bibliographique. Ethnozootechnie 1:19–36

Marcotte CJ, Marcotte EM (2002) Predicting functional linkages from gene fusions with confidence. Appl Bioinform 1:93–100

Martinez D, Maillard JC, Coisne S, Sheikboudou C, Bensaid A (1994) Protection of goats against heartwater acquired by immunisation with inactivated elementary bodies of *Cowdria ruminantium*. Vet Immunol Immunopathol 41:153–163

Martinez D, Vachiery N, Stachurski F, Kandassamy Y, Raliniaina M, Aprelon R, Gueye A (2004) Nested PCR for detection and genotyping of *Ehrlichia ruminantium*: use in genetic diversity analysis. Ann NY Acad Sci 1026:106–113

Massung RF, Priestley RA, Levin ML (2004) Transmission route efficacy and kinetics of *Anaplasma phagocytophilum* infection in white-footed mouse, *Peromyscus leucopus*. Vector Borne Zoonotic Dis 4:310–318

Maughan H (2007) Rates of molecular evolution in bacteria are relatively constant despite spore dormancy. Evolution Int J Org Evol 61:280–288

Mavromatis K, Doyle CK, Lykidis A, Ivanova N, Francino MP, Chain P, Shin M, Malfatti S, Larimer F, Copeland A, Detter JC, Land M, Richardson PM, Yu XJ, Walker DH, McBride JW, Kyrpides NC (2006) The genome of the obligately intracellular bacterium *Ehrlichia canis* reveals themes of complex membrane structure and immune evasion strategies. J Bacteriol 188:4015–4023

McBride JW, Corstvet RE, Gaunt SD, Chinsangaram J, Akita GY, Osburn BI (1996) PCR detection of acute *Ehrlichia canis* infection in dogs. J Vet Diagn Invest 8:441–447

McBride JW, Yu X, Walker DH (1999) Molecular cloning of the gene for a conserved major immunoreactive 28-kilodalton protein of *Ehrlichia canis*: a potential serodiagnostic antigen. Clin Diagn Lab Immunol 6:392–399

McBride JW, Yu XJ, Walker DH (2000) A conserved, transcriptionally active p28 multigene locus of *Ehrlichia canis*. Gene 254:245–252

McBride JW, Ndip LM, Popov VL, Walker DH (2002) Identification and functional analysis of an immunoreactive DsbA-like thio-disulfide oxidoreductase of *Ehrlichia spp*. Infect Immun 70:2700–2703

McBride JW, Comer JE, Walker DH (2003a) Novel immunoreactive glycoprotein orthologs of *Ehrlichia spp*. Ann NY Acad Sci 990:678–684

McBride JW, Corstvet RE, Gaunt SD, Boudreaux C, Guedry T, Walker DH (2003b) Kinetics of antibody response to *Ehrlichia canis* immunoreactive proteins. Infect Immun 71:2516–2524

McBride JW, Doyle CK, Zhang X, Cardenas AM, Popov VL, Nethery KA, Woods ME (2007) Identification of a glycosylated *Ehrlichia canis* 19-kilodalton major immunoreactive protein with a species-specific serine-rich glycopeptide epitope. Infect Immun 75:74–82

McKusick VA, Ruddle FH (1987) A new discipline, a new name, a new journal. Genomics 1:1–2

McQuiston JH, Paddock CD, Holman RC, Childs JE (1999) The human ehrlichioses in the United States. Emerg Infect Dis 5:635–642

Merkl R (2004) SIGI: score-based identification of genomic islands. BMC Bioinform 5:22

Minjauw B, McLeod A (2003) Tick-borne diseases and poverty. The impact of ticks and tickborne diseases on the livelihood of small-scale and marginal livestock owners in India and eastern and southern Africa. Research Report, DFID Animal Health Programme, Centre for Tropical Veterinary Medicine, University of Edinburgh, Edinburgh, UK

Miura K, Rikihisa Y (2007) Virulence potential of *Ehrlichia chaffeensis* strains of distinct genome sequences. Infect Immun 75:3604–3613

Moshkovski SD (1937) Sur l'existence, chez le cobaye, d'une rickettsiose chronique déterminée par *Ehrlichia* (*Rickettsia*) *kurlovi* subg. nov. sp. nov. C R Soc Biol (Paris) 126:379–382

Moshkovski SD (1945) Cytotropic inducers of infection and the classification of the Rickettsiae with Clamydozoa. Adv Modern Biol 19:1–44

Moshkovski SD (1947) Comments by readers. Science 106:62

Mukhebi AW, Chamboko T, O'Callaghan CJ, Peter TF, Kruska RL, Medley GF, Mahan SM, Perry BD (1999) An assessment of the economic impact of heartwater (*Cowdria ruminantium* infection) and its control in Zimbabwe. Prev Vet Med 39:173–189

Mwangi DM, McKeever DJ, Nyanjui JK, Barbet AF, Mahan SM (1998) Major antigenic proteins 1 and 2 of *Cowdria ruminantium* are targets for T-lymphocyte responses of immune cattle. Ann NY Acad Sci 849:372–374

Neitz WO (1947) The transmission of heartwater by *Amblyomma pomposum*, Dönitz, 1909. S Afr Sci 1:83

Neitz WO (1967) The epidemiological pattern of viral, protophytal and protozoal zoonoses in relation to game preservation in South Africa. J S Afr Vet Med Assoc 38: 129–141

Neitz WO (1968) Heartwater. Bull Off Int Epizoot 70:329–336

Neitz WO, Alexander RA (1945) Immunization of cattle against heartwater and the control of the tick-borne diseases, redwater, gallsickness and heartwater. Onderstepoort J Vet Sci Anim Ind 20:137–158

Nethery KA, Doyle CK, Zhang X, McBride JW (2007) *Ehrlichia canis* gp200 contains dominant species-specific antibody epitopes in terminal acidic domains. Infect Immun 75:4900–4908

Nims RM, Ferguson JA, Walker JL, Hildebrandt PK, Huxsoll DL, Reardon MJ, Varley JE, Kolaja GJ, Watson WT, Shroyer EL, Elwell PA, Vacura GW (1971) Epizootiology of tropical canine pancytopenia in Southeast Asia. J Am Vet Med Assoc 158:53–63

Nyika A, Mahan SM, Burridge MJ, Mcguire TC, Rurangirwa F, Barbet AF (1998) A DNA vaccine protects mice against the rickettsial agent *Cowdria ruminantium*. Parasite Immunol 20:111–119

Nyika A, Barbet AF, Burridge MJ, Mahan SM (2002) DNA vaccination with *map1* gene followed by protein boost augments protection against challenge with *Cowdria ruminantium*, the agent of heartwater. Vaccine 20:1215–1225

Nyindo MB, Ristic M, Huxsoll DL, Smith AR (1971) Tropical canine pancytopenia: in vitro cultivation of the causative agent – *Ehrlichia canis*. Am J Vet Res 32:1651–1658

Oberem PT, Bezuidenhout JD (1987) Heartwater in hosts other than domestic ruminants. Onderstepoort J Vet Res 54:271–275

Ochman H (2003) Neutral mutations and neutral substitutions in bacterial genomes. Mol Biol Evol 20:2091–2096

Ochman H, Elwyn S, Moran NA (1999) Calibrating bacterial evolution. Proc Natl Acad Sci USA 96:12638–12643

Ogata H, Audic S, Renesto-Audiffren P, Fournier PE, Barbe V, Samson D, Roux V, Cossart P, Weissenbach J, Claverie JM, Raoult D (2001) Mechanisms of evolution in *Rickettsia conorii* and *R. prowazekii*. Science 293:2093–2098

Ohashi N, Unver A, Zhi N, Rikihisa Y (1998a) Cloning and characterization of multigenes encoding the immunodominant 30-kilodalton major outer membrane proteins of *Ehrlichia canis* and application of the recombinant protein for serodiagnosis. J Clin Microbiol 36:2671–2680

Ohashi N, Zhi N, Zhang Y, Rikihisa Y (1998b) Immunodominant major outer membrane proteins of *Ehrlichia chaffeensis* are encoded by a polymorphic multigene family. Infect Immun 66:132–139

Ohashi N, Rikihisa Y, Unver A (2001) Analysis of transcriptionally active gene clusters of major outer membrane protein multigene family in *Ehrlichia canis* and *E. chaffeensis*. Infect Immun 69:2083–2091

Olano JP, Hogrefe W, Seaton B, Walker DH (2003a) Clinical manifestations, epidemiology, and laboratory diagnosis of human monocytotropic ehrlichiosis in a commercial laboratory setting. Clin Diagn Lab Immunol 10:891–896

Olano JP, Masters E, Hogrefe W, Walker DH (2003b) Human monocytotropic ehrlichiosis, Missouri. Emerg Infect Dis 9:1579–1586

Olsen G, Woese C (1993) Ribosomal RNA: a key to phylogeny. FASEB J 7:113–123

Paddock CD, Childs JE (2003) *Ehrlichia chaffeensis*: a prototypical emerging pathogen. Clin Microbiol Rev 16:37–64

Paddock CD, Folk SM, Shore GM, Machado LJ, Huycke MM, Slater LN, Liddell AM, Buller RS, Storch GA, Monson TP, Rimland D, Sumner JW, Singleton J, Bloch KC, Tang YW, Standaert SM, Childs JE (2001) Infections with *Ehrlichia chaffeensis* and *Ehrlichia ewingii* in persons coinfected with human immunodeficiency virus. Clin Infect Dis 33:1586–1594

Parkhill J, Wren BW, Mungall K, Ketley JM, Churcher C, Basham D, Chillingworth T, Davies RM, Feltwell T, Holroyd S, Jagels K, Karlyshev AV, Moule S, Pallen MJ, Penn CW, Quail MA, Rajandream MA, Rutherford KM, van Vliet AH, Whitehead S, Barrell BG (2000) The genome sequence of the food-borne pathogen *Campylobacter jejuni* reveals hypervariable sequences. Nature 403:665–668

Parola P, Cornet JP, Sanogo YO, Miller RS, Thien HV, Gonzalez JP, Raoult D, Telford SR III, Wongsrichanalai C (2003) Detection of *Ehrlichia spp.*, *Anaplasma spp.*, *Rickettsia spp.*, and other eubacteria in ticks from the Thai–Myanmar border and Vietnam. J Clin Microbiol 41:1600–1608

Patel RG, Byrd MA (1999) Near fatal acute respiratory distress syndrome in a patient with human ehrlichiosis. South Med J 92:333–335

Perez M, Rikihisa Y, Wen B (1996) *Ehrlichia canis*-like agent isolated from a man in Venezuela: antigenic and genetic characterization. J Clin Microbiol 34:2133–2139

Perez M, Bodor M, Zhang C, Xiong Q, Rikihisa Y (2006) Human infection with *Ehrlichia canis* accompanied by clinical signs in Venezuela. Ann NY Acad Sci 1078:110–117

Peter TF, Deem SL, Barbet AF, Norval RA, Simbi BH, Kelly PJ, Mahan SM (1995) Development and evaluation of PCR assay for detection of low levels of *Cowdria ruminantium* infection in *Amblyomma* ticks not detected by DNA probe. J Clin Microbiol 33:166–172

Peter TF, Bryson NR, Perry BD, O'Callaghan CJ, Medley GF, Smith GE, Mlambo G, Horak IG, Burridge MJ, Mahan SM (1999) *Cowdria ruminantium* infection in ticks in the Kruger National Park. Vet Rec 145:304–307

Peter TF, Barbet AF, Alleman AR, Simbi BH, Burridge MJ, Mahan SM (2000) Detection of the agent of heartwater, *Cowdria ruminantium*, in *Amblyomma* ticks by PCR: validation and application of the assay to field ticks. J Clin Microbiol 38:1539–1544

Peter TF, Burridge MJ, Mahan SM (2002) *Ehrlichia ruminantium* infection (heartwater) in wild animals. Trends Parasitol 18:214–218

Pienaar JG (1970) Electron microscopy of *Cowdria* (*Rickettsia*) *ruminantium* (Cowdry, 1926) in the endothelial cells of the vertebrate host. Onderstepoort J Vet Res 37:67–78

Popov VL, Chen SM, Feng HM, Walker DH (1995) Ultrastructural variation of cultured *Ehrlichia chaffeensis*. J Med Microbiol 43:411–421

Popov VL, Han VC, Chen SM, Dumler JS, Feng HM, Andreadis TG, Tesh RB, Walker DH (1998) Ultrastructural differentiation of the genogroups in the genus *Ehrlichia*. J Med Microbiol 47:235–251

Popov VL, Yu XJ, Walker DH (2000) The 120 kDa outer membrane protein of *Ehrlichia chaffeensis*: preferential expression on dense-core cells and gene expression in *Escherichia coli* associated with attachment and entry. Microb Pathog 28:71–80

Postigo M, Taoufik A, Bell-Sakyi L, de Vries E, Morrison WI, Jongejan F (2007) Differential transcription of the major antigenic protein 1 multigene family of *Ehrlichia ruminantium* in *Amblyomma variegatum* ticks. Vet Microbiol 122:298–305

Pretorius A, Collins NE, Steyn HC, van Strijp F, van Kleef M, Allsopp BA (2007) Protection against heartwater by DNA immunisation with four *Ehrlichia ruminantium* open reading frames. Vaccine 25:2316–2324

Provost A, Bezuidenhout JD (1987) The historical background and global importance of heartwater. Onderstepoort J Vet Res 54:165–169

Prozesky L (1987a) Heartwater. The morphology of *Cowdria ruminantium* and its staining characteristics in the vertebrate host and in vitro. Onderstepoort J Vet Res 54:173–176

Prozesky L (1987b) The pathology of heartwater. III. A review. Onderstepoort J Vet Res 54:281–286

Prozesky L, Du Plessis JL (1987) Heartwater. The development and life cycle of *Cowdria ruminantium* in the vertebrate host, ticks and cultured endothelial cells. Onderstepoort J Vet Res 54:193–196

Prozesky L, Bezuidenhout JD, Paterson CL (1986) Heartwater: an in vitro study of the ultrastructure of *Cowdria ruminatium*. Onderstepoort J Vet Res 53:153–159

Reardon MJ, Pierce KR (1981) Acute experimental canine ehrlichiosis. I. Sequential reaction of the hemic and lymphoreticular systems. Vet Pathol 18:48–61

Reddy GR, Streck CP (1999) Variability in the 28-kDa surface antigen protein multigene locus of isolates of the emerging disease agent *Ehrlichia chaffeensis* suggests that it plays a role in immune evasion. Mol Cell Biol Res Commun 1:167–175

Reddy GR, Sulsona CR, Harrison RH, Mahan SM, Burridge MJ, Barbet AF (1996) Sequence heterogeneity of the major antigenic protein 1 genes from *Cowdria ruminantium* isolates from different geographical areas. Clin Diagn Lab Immunol 3:417–422

Reddy GR, Sulsona CR, Barbet AF, Mahan SM, Burridge MJ, Alleman AR (1998) Molecular characterization of a 28 kDa surface antigen gene family of the tribe Ehrlichiae. Biochem Biophys Res Commun 247:636–643

Rikihisa Y (1991) The tribe Ehrlichieae and ehrlichial diseases. Clin Microbiol Rev 4:286–308

Rocha EP (2003) An appraisal of the potential for illegitimate recombination in bacterial genomes and its consequences: from duplications to genome reduction. Genom Res 13:1123–1132

Rossouw M, Neitz AW, de Waal DT, du Plessis JL, van Gas L, Brett S (1990) Identification of the antigenic proteins of *Cowdria ruminantium*. Onderstepoort J Vet Res 57: 215–221

Schmitz-Esser S, Linka N, Collingro A, Beier CL, Neuhaus HE, Wagner M, Horn M (2004) ATP/ADP translocases: a common feature of obligate intracellular amoebal symbionts related to Chlamydiae and Rickettsiae. J Bacteriol 186:683–691

Shpynov S, Fournier PE, Rudakov N, Tarasevich I, Raoult D (2006) Detection of members of the genera *Rickettsia*, *Anaplasma*, and *Ehrlichia* in ticks collected in the asiatic part of Russia. Ann NY Acad Sci 1078:378–383

Simbi BH, Peter TF, Burridge MJ, Mahan SM (2003) Comparing the detection of exposure to *Ehrlichia ruminantium infection* on a heartwater-endemic farm by the pCS20 polymerase chain reaction assay and an indirect MAP1-B enzyme linked immunosorbent assay. Onderstepoort J Vet Res 70:231–235

Simpson BC, Lindsay MS, Morris JR, Muirhead FS, Pollock A, Prichard SG, Stanley HG, Thirlwell GR, Hunter AG, Bradley J (1987) Protection of cattle against heartwater in Botswana: comparative efficacy of different methods against natural and blood-derived challenges. Vet Rec 120:135–138

Sirigireddy KR, Ganta RR (2005) Multiplex detection of *Ehrlichia* and *Anaplasma* species pathogens in peripheral blood by real-time reverse transcriptase-polymerase chain reaction. J Mol Diagn 7:308–316

Skotarczak B (2003) Canine ehrlichiosis. Ann Agric Environ Med 10:137–141

Spreull J (1904) Heartwater inoculation experiments. Agric J Cape of Good Hope 24:433–442

Stackebrandt E, Goebel B (1994) Taxonomic note: a place for DNA-DNA reassociation and 16S rRNA sequence analysis in the present species definition in bacteriology. Int J Syst Bacteriol 44:846–849

Standaert SM, Yu T, Scott MA, Childs JE, Paddock CD, Nicholson WL, Singleton J Jr, Blaser MJ (2000) Primary isolation of *Ehrlichia chaffeensis* from patients with febrile illnesses: clinical and molecular characteristics. J Infect Dis 181:1082–1088

Stich RW, Rikihisa Y, Ewing SA, Needham GR, Grover DL, Jittapalapong S (2002) Detection of *Ehrlichia canis* in canine carrier blood and in individual experimentally infected ticks with a p30-based PCR assay. J Clin Microbiol 40:540–546

Sumner JW, Childs JE, Paddock CD (1999) Molecular cloning and characterization of the *Ehrlichia chaffeensis* variable-length PCR target: an antigen-expressing gene that exhibits interstrain variation. J Clin Microbiol 37:1447–1453

Teglas M, Matern E, Lein S, Foley P, Mahan SM, Foley J (2005) Ticks and tick-borne disease in Guatemalan cattle and horses. Vet Parasitol 131:119–127

Thomas CM, Nielsen KM (2005) Mechanisms of, and barriers to, horizontal gene transfer between bacteria. Nat Rev Microbiol 3:711–721

Thompson DV, Melchers LS, Idler KB, Schilperoort RA, Hooykaas PJ (1988) Analysis of the complete nucleotide sequence of the *Agrobacterium tumefaciens* virB operon. Nucl Acids Res 16:4621–4636

Uilenberg G (1982) Experimental transmission of *Cowdria ruminantium* by the Gulf coast tick *Amblyomma maculatum*: danger of introducing heartwater and benign African theileriasis onto the American mainland. Am J Vet Res 43:1279–1282

Uilenberg G (1983) Heartwater (*Cowdria ruminantium* infection): current status. Adv Vet Sci Comp Med 27: 427–480

Uilenberg G, Camus E, Barré N (1985) Quelques observations sur une souche de *Cowdria ruminantium* isolée en Guadeloupe (Antilles Françaises). Rev Elev Med Vet Pays Trop 38:34–42

Unver A, Perez M, Orellana N, Huang H, Rikihisa Y (2001) Molecular and antigenic comparison of *Ehrlichia canis* isolates from dogs, ticks, and a human in Venezuela. J Clin Microbiol 39:2788–2793

Unver A, Rikihisa Y, Stich RW, Ohashi N, Felek S (2002) The omp-1 major outer membrane multigene family of *Ehrlichia chaffeensis* is differentially expressed in canine and tick hosts. Infect Immun 70:4701–4704

Van Amstel SR, Oberem PT (1987) The treatment of heartwater. Onderstepoort J Vet Res 54:475–479

Van de Pypekamp HE, Prozesky L (1987) Heartwater. An overview of the clinical signs, susceptibility and differential diagnoses of the disease in domestic ruminants. Onderstepoort J Vet Res 54:263–266

Van der Merwe L (1987) The infection and treatment method of vaccination against heartwater. Onderstepoort J Vet Res 54:489–491

Van Heerden H, Collins NE, Brayton KA, Rademeyer C, Allsopp BA (2004a) Characterization of a major outer membrane protein multigene family in *Ehrlichia ruminantium*. Gene 330:159–168

Van Heerden H, Steyn HC, Allsopp MT, Zweygarth E, Josemans AI, Allsopp BA (2004b) Characterization of the pCS20 region of different *Ehrlichia ruminantium* isolates. Vet Microbiol 101:279–291

Van Heerden J (1982) A retrospective study on 120 natural cases of canine ehrlichiosis. J S Afr Vet Assoc 53:17–22

Van Kleef M, Neitz AW, De Waal DT (1993) Isolation and characterization of antigenic proteins of *Cowdria ruminantium*. Rev Elev Med Vet Pays Trop 46:157–164

Van Vliet AH, Jongejan F, van der Zeijst BA (1992) Phylogenetic position of *Cowdria ruminantium* (Rickettsiales) determined by analysis of amplified 16S ribosomal DNA sequences. Int J Syst Bacteriol 42:494–498

Van Vliet AH, Jongejan F, van Kleef M, van der Zeijst BA (1994) Molecular cloning, sequence analysis, and expression of the gene encoding the immunodominant 32-kilodalton protein of *Cowdria ruminantium*. Infect Immun 62:1451–1456

Waghela SD, Rurangirwa FR, Mahan SM, Yunker CE, Crawford TB, Barbet AF, Burridge MJ, McGuire TC (1991) A cloned DNA probe identifies *Cowdria ruminantium* in *Amblyomma variegatum* ticks. J Clin Microbiol 29:2571–2577

Walker DH (2005) *Ehrlichia* under our noses and no one notices. Arch Virol Suppl (19):147–156

Walker DH, Ismail N, Olano JP, McBride JW, Yu XJ, Feng HM (2004) *Ehrlichia chaffeensis*: a prevalent, life-threatening, emerging pathogen. Trans Am Clin Climatol Assoc 115: 375–384

Waner T, Harrus S, Jongejan F, Bark H, Keysary A, Cornelissen AW (2001) Significance of serological testing for ehrlichial diseases in dogs with special emphasis on the diagnosis of canine monocytic ehrlichiosis caused by *Ehrlichia canis*. Vet Parasitol 95:1–15

Wayne LG, Brenner DJ, Colwell RR, Grimont PAD, Kandler O, Krichevsky MI, Moore LH, Moore WEC, Murray RGE, Stackebrandt ESMP, Truper HG (1987) Report on the *ad hoc* committee on reconciliation of approaches to bacterial systematics. Int J Syst Bacteriol 37:463–464

Weitzmann MN, Woodford KJ, Usdin K (1997) DNA secondary structures and the evolution of hypervariable tandem arrays. J Biol Chem 272:9517–9523

Wu M, Sun LV, Vamathevan J, Riegler M, Deboy R, Brownlie JC, McGraw EA, Martin W, Esser C, Ahmadinejad N, Wiegand C, Madupu R, Beanan MJ, Brinkac LM, Daugherty SC, Durkin AS, Kolonay JF, Nelson WC, Mohamoud Y, Lee P, Berry K, Young MB, Utterback T, Weidman J, Nierman WC, Paulsen IT, Nelson KE, Tettelin H, O'Neill SL, Eisen JA (2004) Phylogenomics of the reproductive parasite *Wolbachia pipientis* wMel: a streamlined genome overrun by mobile genetic elements. PLoS Biol 2:E69

Yabsley MJ, Little SE, Sims EJ, Dugan VG, Stallknecht DE, Davidson WR (2003) Molecular variation in the variable-length PCR target and 120-kilodalton antigen genes of *Ehrlichia chaffeensis* from white-tailed deer (*Odocoileus virginianus*). J Clin Microbiol 41:5202–5206

Yu XJ, Crocquet Valdes P, Walker DH (1997) Cloning and sequencing of the gene for a 120-kDa immunodominant protein of Ehrlichia chaffeensis. Gene 184:149–154

Yu XJ, Crocquet-Valdes PA, Cullman LC, Popov VL, Walker DH (1999a) Comparison of *Ehrlichia chaffeensis* recombinant proteins for serologic diagnosis of human monocytotropic ehrlichiosis. J Clin Microbiol 37:2568–2575

Yu XJ, McBride JW, Walker DH (1999b) Genetic diversity of the 28-kilodalton outer membrane protein gene in human isolates of *Ehrlichia chaffeensis*. J Clin Microbiol 37:1137–1143

Yu X, McBride JW, Zhang X, Walker DH (2000a) Characterization of the complete transcriptionally active *Ehrlichia chaffeensis* 28 kDa outer membrane protein multigene family. Gene 248:59–68

Yu XJ, McBride JW, Diaz CM, Walker DH (2000b) Molecular cloning and characterization of the 120-kilodalton protein gene of *Ehrlichia canis* and application of the recombinant 120-kilodalton protein for serodiagnosis of canine ehrlichiosis. J Clin Microbiol 38:369–374

Yu XJ, McBride JW, Walker DH (2007) Restriction and expansion of *Ehrlichia* strain diversity. Vet Parasitol 143:337–346

Yuan Z, Davis MJ, Zhang F, Teasdale RD (2003) Computational differentiation of N-terminal signal peptides and transmembrane helices. Biochem Biophys Res Commun 312:1278–1283

Yunker CE (1996) Heartwater in sheep and goats: a review. Onderstepoort J Vet Res 63:159–170

Zhang JZ, McBride JW, Yu XJ (2003) L-selectin and E-selectin expressed on monocytes mediating *Ehrlichia chaffeensis* attachment onto host cells. FEMS Microbiol Lett 227: 303–309

Zhang JZ, Popov VL, Gao S, Walker DH, Yu XJ (2007) The developmental cycle of *Ehrlichia chaffeensis* in vertebrate cells. Cell Microbiol 9:610–618

Zweygarth E, Vogel SW, Josemans AI, Horn E (1997) In vitro isolation and cultivation of *Cowdria ruminantium* under serum-free culture conditions. Res Vet Sci 63:161–164

Zweygarth E, Josemans AI, Van Strijp MF, Van Heerden H, Allsopp MT, Allsopp BA (2002) The Kümm isolate of *Ehrlichia ruminantium*: in vitro isolation, propagation and characterization. Onderstepoort J Vet Res 69: 147–153

Zweygarth E, Josemans AI, Van Strijp MF, Lopez-Rebollar L, Van Kleef M, Allsopp BA (2005) An attenuated *Ehrlichia ruminantium* (Welgevonden stock) vaccine protects small ruminants against virulent heartwater challenge. Vaccine 23:1695–1702

5 *Cryptosporidium*

Guan Zhu[1](✉), Shinichiro Enomoto[2], Jason M. Fritzler[1], Mitchell S. Abrahamsen[2], and Thomas J. Templeton[3]

[1] Department of Veterinary Pathobiology, College of Veterinary Medicine & Biomedical Sciences, and Faculty of Genetics Program, Texas A&M University, College Station, TX 77843-4467, USA, gzhu@cvm.tamu.edu

[2] Department of Veterinary and Biomedical Science, College of Veterinary Medicine, University of Minnesota, St. Paul, MN 55108, USA

[3] Department of Microbiology and Immunology, Weill Cornell Medical College, Weill Graduate School of Medical Sciences of Cornell University, New York, NY 10021, USA

5.1 *Cryptosporidium* and Cryptosporidiosis

5.1.1 Introduction

The publication of this book coincidentally marks the 100-year anniversary (1907–2007) of the first description of *Cryptosporidium* by E. E. Tyzzer (Tyzzer 1907). However, the importance of *Cryptosporidium* was not fully recognized until roughly 70 years later; indeed, less than 30 articles were published before 1980, of which most were simplified descriptions of *Cryptosporidium* found in farm and laboratory animals. Since then the number of articles published has steadily increased, and today nearly 6,000 articles have been indexed in the PubMed database that either directly or indirectly involve *Cryptosporidium* or cryptosporidiosis (Fig. 1). In the past two decades, the research scope has rapidly expanded from simple descriptions of morphology and life cycles to extensive and comprehensive investigations at the molecular and genomic levels on its biology and host–pathogen interactions.

Cryptosporidium infects humans in addition to animals of veterinary interest. The first case of human cryptosporidiosis was reported in 1976, and was described as an acute, self-limiting enterocolitis in a 3-year-old child (Nime et al. 1976). Two months later, the first cases of cryptosporidiosis in immunocompromised patients were also reported (Meisel et al. 1976), and physicians began to recognize *Cryptosporidium* as a significant opportunistic pathogen causing chronic and life-threatening diarrhea in AIDS patients. The massive outbreak of *Cryptosporidium* in Milwaukee, Wisconsin in 1993 escalated this parasite to headline news in the United States and around the world. In that outbreak, more than 403,000 people were affected due to the contamination of city water supply system with this parasite, and approximately 100 deaths occurred among the elderly and immunocompromised population. *Cryptosporidium* may also be transmitted via direct contact between and among humans and animals, or by food contamination, although the majority of outbreaks of cryptosporidiosis in humans are associated with either drinking or recreational water such as swimming pools or water parks. Because *Cryptosporidium* contamination in water supplies is difficult to control and causes a devastating effect to communities, this parasite is listed as one of the water-borne, category B priority agents in the NIH and CDC biodefense research programs in the United States.

Like humans, newborn or young animals are more likely to suffer from cryptosporidiosis. Although *Cryptosporidium* may infect all animals, more attention has been paid to the infections in cattle, which is largely due to the high prevalence in this livestock species, its economic importance, and the potential of the parasite as a reservoir for human infections. Acute diarrhea is a major symptom among many animals, including calves, foals, piglets, deer, and goat kids, but not in others such as rodents, in which only a mild or no diarrhea is observed. Reduced weight gain and wasting are the major problems in animals suffering from cryptosporidiosis, but death can also occur. For example, the TAMU strain of *C. parvum* (one of the commonly used laboratory strains) originated from a

Genome Mapping and Genomics in Animal-Associated Microbes
V. Nene, C. Kole (Eds.)

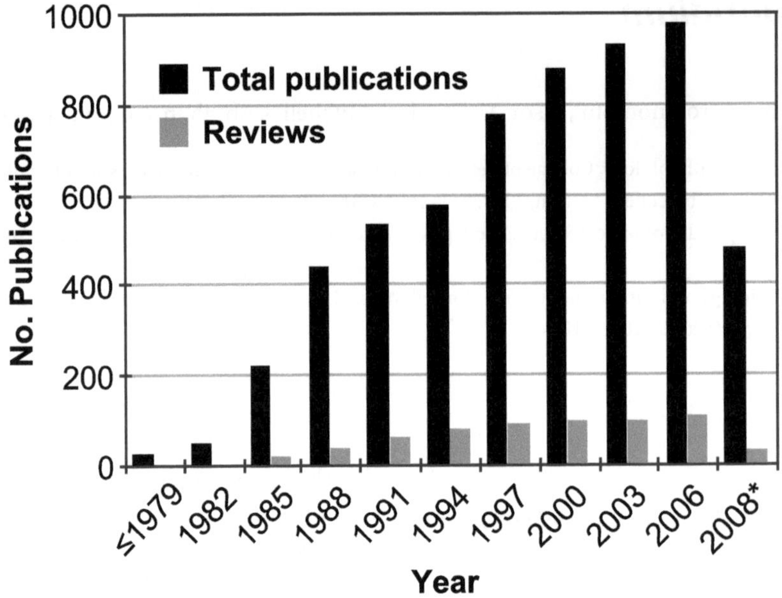

Fig. 1 Number of articles indexed by PubMed containing words *Cryptosporidium* or cryptosporidiosis. Years are in 3-year increments except for 1980 and earlier. *As of 5 April 2008

foal that died from severe diarrhea. Companion animals such as dogs and cats may also suffer from cryptosporidiosis, but it is largely unknown if their close association with humans contributes to the transmission of *Cryptosporidium*. In addition to mammals, *Cryptosporidium* infection in birds and reptiles can also be severe, such as causing morbidity and mortality in snakes. The list of victims is not limited to land animals, as fish are also prone to cryptosporidiosis. A recent study demonstrated the association between reduced growth and *C. molnari* infection in Spanish gilthead sea bream and European sea bass (Sitja-Bobadilla et al. 2005). Thus *Cryptosporidium* not only impacts human health and welfare, but can also cause significant economic losses to farmers and ranchers, pet owners, and even fish farmers.

Within the *Cryptosporidium* genus, *C. parvum* is the most widely studied species due to its importance in both public and animal health, as well as the early success in propagation of this species in calves and mice. A second species, *C. hominis*, is morphologically indistinguishable from *C. parvum* and was previously considered as the human genotype (or Type 1 vs. zoonotic Type 2) of *C. parvum*, but recently renamed as a separate species based on host specificity (i.e., predominantly a parasite of humans) and molecular divergence from Type 2 *C. parvum* (Morgan-Ryan et al. 2002). The validity of *C. hominis* as a distinct species is debated, as no evidence has shown that this species is reproductively separated from *C. parvum*. In fact, some investigators have observed recombination between *C. parvum* and *C. hominis* under both experimental and natural conditions (Widmer 2004). Moreover, genome and metabolic comparisons, described in the following sections, have not identified a "smoking gun" that supports species distinctions of the two pathogens. Nevertheless, the name change from *C. parvum* Type I to *C. hominis* has promoted the awareness of cryptosporidiosis both within and outside the parasitology research communities.

5.1.2 Taxonomic Position

Cryptosporidium belongs to the Phylum Apicomplexa, based upon the morphological criteria of an apical complex, described further below, which is unique to this group of protozoans. The parasite is further described within Class Conoidasida (also known as Coccidia), Order Eucoccidiorida, Suborder Eimeriorina, and family Cryptosporiidiae (Levine 1985; Brends 1989–2005; Tzipori 1998). The taxonomic status of *Cryptosporidium* has been the subject of debate for several years, and recent molecular phylogeny has placed this genus as an early branch at the base of the Apicomplexa, perhaps as a sister clade to the gregarines (Bull et al. 1998; Carreno et al. 1999; Zhu et al. 2000a; Hijjawi et al. 2002; Barta and Thompson 2006). More recently, the genome sequencing projects for *C. parvum* and *C. hominis* have revealed that this genus differs from other apicomplexans in many metabolic pathways, including the complete absence of apicoplast and mitochondrial genomes; thus further supporting the notion that *Cryptosporidium* is highly divergent from the traditional coccidian species (Zhu et al. 2000b; Abrahamsen et al. 2004; Xu et al. 2004).

Tyzzer first established the genus *Cryptosporidium* when he described the *Cryptosporidium* type species *C. muris* (Tyzzer 1907). A few years later, *C. parvum* was introduced based on a morphological size and location of infection that differed from *C. muris* (Tyzzer 1907, 1912). For many years other apicomplexan organisms, such as *Sarcocystis*, were wrongly assigned to the *Cryptosporidium* genus based upon observed similarities in their life strategies (Triffitt 1925; Bearup 1954; Dubey and Pande 1963; Anderson et al. 1968; Duszynski 1969; Pande et al. 1972; Xiao et al. 2004). It was not until the mid- to late-1960s that *Cryptosporidium* was confirmed to have a unique attachment organelle (Hampton and Rosario 1966; Jervis et al. 1966; Vetterling et al. 1971); and this cellular feature has since become the defining taxonomic unit of the genus and family (Levine 1961, 1984; Upton 2000; Xiao et al. 2004). Within this taxonomic structure, focus turned to naming species based on host specificity. However, cross-transmission studies soon showed that most isolates readily transmit across host species, thus effectively ending efforts to distinguish species further than the general group, *C. parvum* (Xiao et al. 2004). At that time, only a few species (e.g., *C. meleagridis*, *C. wrairi*, and *C. felis*) were agreed upon as new species due to biological differences from the previously established *C. parvum* and *C. muris* species.

The ongoing controversy in *Cryptosporidium* taxonomy basically boils down to a difficulty in defining the criteria that define a biological species (Egyed et al. 2003; Xiao et al. 2004). Since the 1990s, a mixture of methods has been used to define species and genotypes; for example, some are based solely on molecular data, whereas others have relied upon morphologic and biologic methods without sequence data. Molecular phylogeny reconstruction based on the 18S small subunit ribosomal RNA (SSU rRNA), 70-kDa heat shock protein (HSP70), and *Cryptosporidium* oocyst wall protein (COWP) genes have clearly demonstrated genetic variability within the genus (Egyed et al. 2002, 2003). The correct identification of a parasite, and understanding the range of genetic variation within the parasite group, is not only of interest in deciphering taxonomy, but also crucial to the development of vaccines, drugs, and new diagnostic methods (McManus and Bowles 1996). Recent work suggests that the *Cryptosporidium* species can be fully characterized by a polyphasic approach that is supported by morphologic, biologic, and genetic data (Egyed et al. 2003; Xiao et al. 2004). Modern molecular techniques have helped to clear the confusion of *Cryptosporidium* taxonomy and, although there is still some debate, there are currently 15 named valid *Cryptosporidium* species (Table 1).

5.1.3 Life Cycle

Like intestinal coccidia such as *Eimeria* and *Isospora*, *Cryptosporidium* undergoes a monoxenous life cycle; that is, it completes its entire life cycle within a single host. The complex life cycle has been detailed for *C. parvum* (Current and Reese 1986; Fayer and Ungar 1986; Thompson et al. 2005), which entails both sexual and asexual cycles and six main developmental stages (Fig. 2). The endogenous life cycle begins following ingestion of only the exogenous stage, the sporulated oocyst. This typically occurs via the fecal–oral route and can be through direct or indirect person-to-person contact including sexual activity, animal-to-animal contact, animal-to-human contact, water-borne ingestion of drinking or recreational water, food-borne, and possibly air-borne transmission (Fayer et al. 2000). Depending on the *Cryptosporidium* species, oocysts vary in size and shape ranging from 4.3 to 8.3 mm in length by 4.4–7.4 mm in width (Table 1), and are ovoid to elliptical in shape with shape indices (length/width) ranging from 1.0 to 1.4 (O'Donoghue 1995). Similar to other coccidia, the *Cryptosporidium* oocyst wall is composed of two distinct layers; however, it is unique in that it contains a suture at one end that dissolves during excystation, allowing an opening to form at one end of the oocyst.

Table 1 The currently recognized 15 *Cryptosporidium* species along with their respective oocyst dimensions, primary host(s), and site of infection[a]

Species	Oocyst sizes (mm)	Primary host	Primary site of infection	References
C. parvum	4.5 × 5.5	Mammals	Small intestine	(Tyzzer 1912)
C. hominis	4.5 × 5.5	Humans	Small intestine	(Morgan-Ryan et al. 2002)
C. andersoni	5.5 × 7.4	Cattle	Abomasum	(Lindsay et al. 2000)
C. muris	5.6 × 7.4	Rodents[b]	Stomach	(Tyzzer 1907)
C. wrairi	4.4 × 5.3	Guinea pigs	Small intestine	(Vetterling et al. 1971)
C. felis	4.5 × 5.0	Cats	Small intestine	(Iseki 1979)
C. canis	5.0 × 4.7	Dogs	Small intestine	(Fayer et al. 2001)
C. suis	5.1 × 4.4	Pigs	Small, large intestine	(Ryan et al. 2004)
C. meleagridis	4.3 × 4.9	Turkeys[b]	Small intestine	(Slavin 1955)
C. baileyi	4.6 × 6.2	Chickens[b]	Large intestine, bursa, respiratory system	(Current et al. 1986)
C. galli	8.3 × 6.3	Birds	Proventriculus	(Ryan et al. 2003)
C. serpentis	5.3 × 6.1	Snakes	Stomach	(Levine 1980)
C. saurophilum	4.7 × 5.0	Lizards	Stomach, small intestine	(Koudela and Modry 1998)
C. molnari	4.7 × 4.5	Fish	Stomach, small intestine	(Alvarez-Pellitero and Sitja-Bobadilla 2002)
C. bovis	4.8 × 4.6	Cattle	Abomasum[c]	(Fayer et al. 2005)

[a] Data obtained from Smith et al. 2005, Xiao et al. 2004, and other listed references
[b] Species that have also been found in humans, mainly immunocompromised individuals
[c] Speculative data

Each ingested oocyst excysts to release four haploid sporozoites, which then invade and parasitize epithelial cells within the small intestine or colon (Reduker et al. 1985). In contrast to other coccidia, sporozoites are free within the oocyst and are not surrounded by a sporocyst (Spano and Crisanti 2000). They are typically fusiform (banana shaped) and are 3.5–4.2 μm long by 0.53–0.6 μm wide (Reduker et al. 1985). *Cryptosporidium* sporozoites possess typical apicomplexan cellular architectures, such as the organelles rhoptries, micronemes, and electron-dense granules. They possess apical rings, but lack polar rings, micropores, and the conoid that are present in sporozoites of other coccidia (Fayer 1997). Invasion of the host cell by the sporozoite involves the release of several materials that are housed in the apical complex at the anterior pole of the sporozoite (Tetley et al. 1998), and are believed to participate in a wide array of events during the invasion process.

Sporozoites invade host cells and reside within a host cell membrane-derived parasitophorous vacuolar membrane (PVM). The parasite remains adjacent to the surface of the columnar host intestinal epithelial cell, a highly polar cellular location which is often referred to being intracellular but extracytoplasmic (O'Donoghue 1995; Marshall et al. 1997; Umemiya et al. 2005). As the parasite develops to a trophozoite within this niche, the parasitophorous vacuole becomes highly invaginated at the host cell interface to form a "feeder organelle" that is suspected to mediate the uptake of nutrients from the host cell (Fayer 1997; Spano and Crisanti 2000; Umemiya et al. 2005).

Asexual multiplication of the parasite within host epithelial cells is referred to as type I merogony, and culminates in nuclear division to form eight haploid merozoites assembled into a single type I meront. Mature type I merozoites are similar to merozoites of other coccidia. Once released from the PV they can either undergo another cycle of type

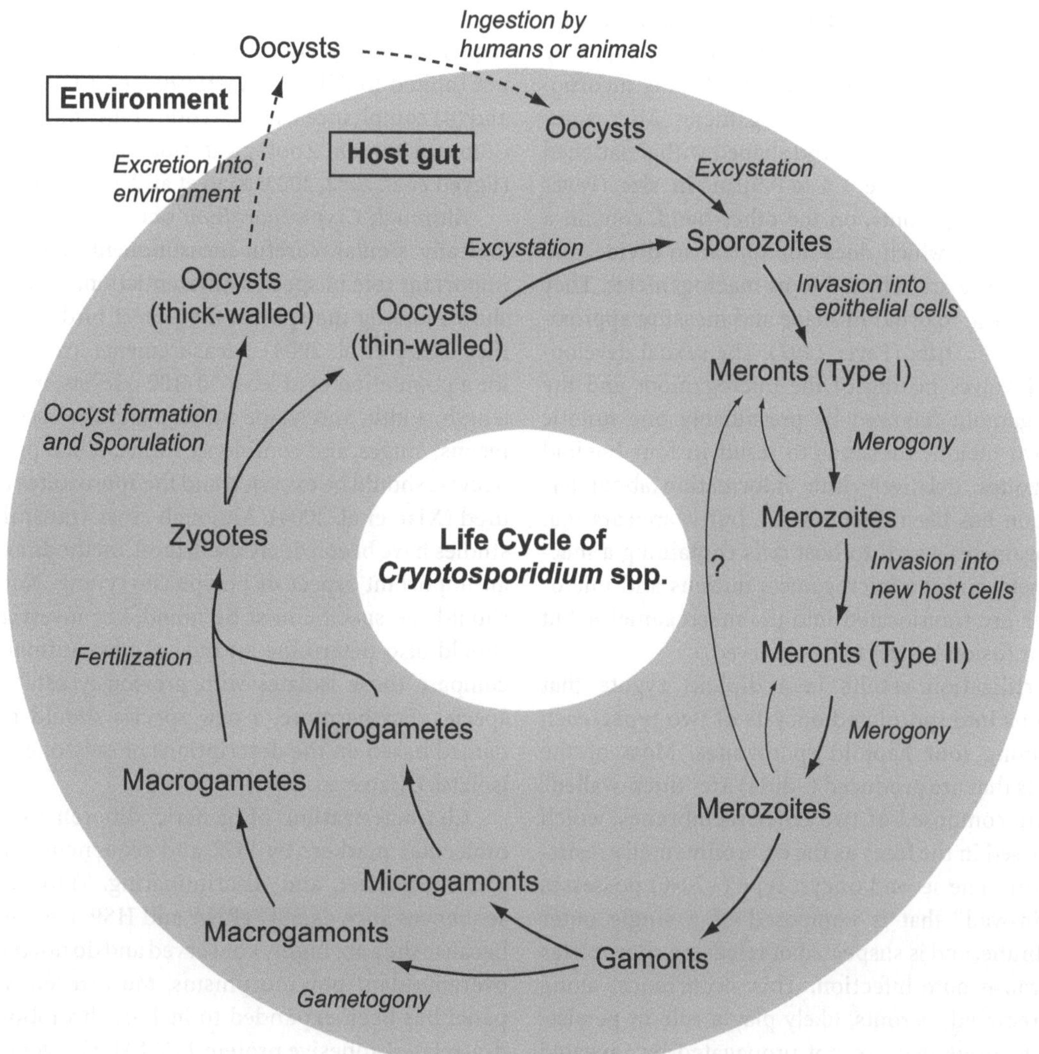

Fig. 2 Diagram of the *Cryptosporidium* life cycle. Environmentally stable oocysts are ingested by a host to start the life cycle. Typically, after two rounds of merogony, gametogony occurs in which the micro- and macrogametes are formed. The only sexual stage in the life cycle occurs during the fertilization of the gametes, which produces a zygote. Oocysts are formed from zygotes, sporulation occurs within the gut, and sporulated oocysts are either excreted in the feces or undergo excystation to start another round of the life cycle

I merogony or further develop through a type II merogony cycle. Those that are cycled back through type I merogony likely contribute to the persistent infection that is observed in cryptosporidiosis (Current and Reese 1986). Type II merogony results in the formation of four haploid merozoites within a type II meront. Like type I merozoites, the type II merozoites are able to undergo another round of type II merogony, but it is not known if type II merozoites are able to initiate another cycle of type I merogony.

Type II merozoites rarely initiate multiple rounds of merogony and instead develop into the sexual reproductive stages of the parasite, known as gamonts. Preliminary reports have suggested that the extracellular forms of *Cryptosporidium* closely resemble extracellular gamont stages of the gregarines (Hijjawi et al. 2002, 2004; Rosales et al. 2005). Gamonts further differentiate into either microgamonts (male) or macrogamonts (female). As microgamonts develop, they become multinucleate and each nucleus is incorporated into as many as 16 microgametes. Microgametes, on the one hand, are rod shaped with a flattened anterior end, and are 1.4 × 0.5 μm^2 in size (Fayer 1997). Macrogamonts, on the other hand, contain a large nucleus, which does not appear to divide, and when mature are referred to as macrogametes. They are spherical to ovoid in shape and measure approximately 4.0–6.0 μm (Fayer 1997). The sexual development involves mating of the microgamont and the macrogamont followed by presumably one mitotic and two meiotic divisions to result in four haploid sporozoites. Relatively little information about fertilization has been documented, but it appears that microgametes attach to host cells containing a macrogamete, and the microgamete nucleus and microtubules are translocated into the macrogametes, but nuclear fusion has yet to be observed.

Fertilization results in a diploid zygote that develops into sporulated oocysts of two types, each containing four haploid sporozoites. Most of the oocysts that are produced (~80%) are "thick-walled" and are composed of two outer membranes, which are passed in the feces as the environmentally resistant form. The second oocyst type (~20%) possesses a "thin-wall" that is composed of a single outer membrane, and is suspected of releasing sporozoites that cause auto-infection. This occurrence, along with recycled meronts, likely play a role in persistent infections that are not propagated by repeated ingestion of oocysts, such as the life-threatening nonresolving cryptosporidiosis that is observed in AIDS patients (Current and Reese 1986; Fayer 1997; Marshall et al. 1997).

5.1.4 *Cryptosporidium* Genotypes

For the polyphasic typing of *Cryptosporidium* species, it has been proposed that at least four main requirements should be fulfilled: (1) morphological measurements of oocysts should be made, particularly an index of size and shape; (2) demonstration of natural and experimental host specificity, organ location, pathogenicity, prepatent and patent periods, and intensity of oocyst shedding; (3) genetic characterization such as differences in nucleotide sequences of well characterized genes, including but not limited to SSU rRNA, HSP70, and COWP genes; and (4) compliance with the code of the International Commission on Zoological Nomenclature (ICZN) (Egyed et al. 2002, 2003; Xiao et al. 2004, 2000b).

Although *Cryptosporidium* oocysts are morphologically similar, careful measurement can play an important role in species differentiation, as has been shown among the species that infect birds and reptiles (Xiao et al. 2004). Measurements are required for a population of at least 20–100 oocysts, including length, width, and shape indices, and the associated means, ranges, and confidence limits. When possible, oocysts should be excysted and the sporozoites measured (Xiao et al. 2004). Although cross-transmission studies have been widely used, these methods are still an important aspect of polyphasic typing. Not only should the specific host be found, but investigators should also determine a range of host animals and compare those isolates with previously established species. Furthermore, a new species should not be named based on the descriptions of only one or two isolates (Xiao et al. 2004).

Characterization of genetic differences using molecular markers by PCR and sequencing is simple, cheap, fast, and discriminating. Traditionally, sequences such as SSU rRNA and HSP70 were used because they are highly conserved and do not possess overabundant polymorphisms. More recently, this panel has been expanded to include thrombospondin-related adhesive protein 1 (TRAP-C1) genes and tubulin, as well as the noncoding sequences such as the internal transcribed spacer 1 (ITS-1) and microsatellites (Widmer et al. 1998; Aiello et al. 1999; Rochelle et al. 1999; Sulaiman et al. 1999; Caccio et al. 1999, 2000; Feng et al. 2000; Morgan-Ryan et al. 2001; Xiao et al. 2004, 2000a). Additionally, the genome sequence databases for *C. parvum* and *C. hominis* provide a wide spectrum of primer design for polymorphism detection.

5.1.5 Treatment

Immunocompetent individuals rarely require treatment to recover from cryptosporidiosis. Antidiarrheal agents in combination with rehydration therapy are typically used to treat the diarrhea and associated fluid and electrolyte depletion. Individuals with an underlying immunosuppressed condition are at an increased risk for severe complications due to *Cryptosporidium* infection. However, an effective drug to treat cryptosporidiosis is lacking, and there is no approved drug in the United States to treat cryptosporidiosis in animals, although halofuginone lactate (trade name Halocur) is approved in some European countries for use in cattle and sheep. Currently, the only approved drug in the United States for the treatment of cryptosporidiosis in humans is nitazoxanide (NTZ) (White 2003; Fox and Saravolatz 2005), which is approved by the Federal Drug Administration (FDA) under the trade name Alinia (Romark Laboratories, www.romarklabs.com). NTZ is currently approved for use in both children and adults, but not for patients of any age who have AIDS.

NTZ is a nitrothiazolyl-salicylamide that has been reported to show broad-spectrum parasiticidal activity against protozoa, nematodes, trematodes, cestodes, some anaerobic bacteria, and viruses (Rossignol and Maisonneuve 1984; Romero Cabello et al. 1997; Tzipori and Ward 2002; Hemphill et al. 2006; Pankuch and Appelbaum 2006). The success of NTZ as an anti-protozoan agent led to clinical trials of this drug for cryptosporidiosis. Although the mechanism of action against *Cryptosporidium* is not clearly understood, it may target the bifunctional pyruvate: $NADP^+$ oxidoreductase (PNO) that is encoded in the *C. parvum* genome, as is the case with some other organisms (Ctrnacta et al. 2006; Rotte et al. 2001) (for further discussion see Sect. 5.4, “Parasite Targets for New Therapeutics”).

Despite lacking complete efficacy, NTZ has been shown to greatly reduce oocyst excretion associated with the resolution of diarrhea in immunocompetent patients. In one uncontrolled study, NTZ was associated with a 95% reduction in or eradication of oocyst excretion in 58% of the patients, and was concurrent with complete resolution of diarrhea in 57% of these patients (Doumbo et al. 1997). Similarly, a prospective randomized, placebo-controlled, double-blind study that was composed of adults and children reported that 80% of the patients treated with NTZ for 7 days showed symptomatic improvement and 67% had eradication of oocyst excretion compared with 41% (symptomatic improvement) and 22% (oocyst eradication), respectively, of the placebo group (Rossignol et al. 2001). One report suggested that NTZ improved the resolution of diarrhea, oocyst eradication, and mortality in HIV-seronegative, but not HIV-seropositive children (Amadi et al. 2002). However, another study showed that, when treated with NTZ, 59% of AIDS patients displayed eradication of oocyst excretion and an improvement of clinical symptoms associated with cryptosporidiosis (Rossignol 2006).

Paromomycin is an aminoglycoside antibiotic that displays activity against a wide range of organisms, and is one of the few older antibiotics remaining in clinical use (Tzipori and Ward 2002). This drug is approved by the FDA in tablet form under the trade name of Humatin, but may also be found as Aminosidine, Neomycin E, Hydroxymycin, and Catenulin among several others. Paromomycin is a protein synthesis inhibitor that was thought to act on the aminoacyl tRNA site of ribosomes. However, a recent report indicates that it may inhibit maturation of tRNA (Tekos et al. 2003), but its mechanism of action against *Cryptosporidium* is presently unknown. Paromomycin alone has shown relatively little efficacy in treating cryptosporidiosis in AIDS patients. One study showed that only 47% of patients treated with paromomycin displayed improvement in clinical symptoms compared with 36% of those receiving a placebo control (Hewitt et al. 2000). Another prospective, double-blind, placebo controlled study showed no benefit in using paromomycin vs. placebo (Hewitt et al. 2000).

Combination therapy of cryptosporidiosis with paromomycin plus azithromycin (Smith et al. 1998), or with antiretroviral drugs (Hommer et al. 2003), shows some promise in having greater efficacy than treatment with either drug alone. In a second regimen, the protein synthesis inhibitor azithromycin in combination with paromomycin was given to AIDS patients for 4 weeks, followed by paromomycin alone for 8 weeks, and resulted in greatly reduced clinical symptoms and oocyst excretion. Short-term azithromycin treatment for cryptosporidiosis in AIDS patients was associated with clinical improvement, but oocyst excretion remained positive for most patients (Kadappu et al. 2002). Antiretroviral therapy for AIDS patients, in attempts to increase CD4 counts and restore immune system function, can sometimes lead to partial recovery from cryptosporidiosis, but further treatment with paromomycin after antiretroviral treatment is usually of greater benefit.

Perhaps the most efficient means to prevent cryptosporidiosis, other than a healthy immune system, are public health measures to avoid exposure to the environmentally resistant oocyst. Because of the lack of greatly effective therapy against *Cryptosporidium* infection, preventative measures for immunosuppressed individuals include exhaustive hand washing, avoidance of human or animal feces, avoidance of recreational water, and insurance of a safe drinking water supply.

5.2 Genome Sequencing, Mapping, and Resources

5.2.1 History of Genome Sequencing Projects

Preliminary analyses of the *Cryptosporidium* genome began in the early 1990s with the isolation of partial 18S rRNA gene sequences, as well as the random amplification of DNA fragments with unknown functions. Progress remained at a slow pace until several small scale genome survey projects were initiated, such as a random genomic sequence survey (GSS) that was undertaken at the University of Minnesota (Liu et al. 1999). This modest attempt of 432 clones yielded 324 kb (257 kb unique) of sequence covering an estimated 2.5% of the genome. This sample was largely composed of previously unknown genes, and thus boosted the total number of known *C. parvum* genes by over 100-fold. The analysis also revealed that the *C. parvum* genome was highly compact, that is, gene rich (~2 kb per gene), and had less than 1% repetitive sequences. An Expressed Sequence Tag (EST) project was initiated at the University of California, San Francisco (UCSF), for the sporozoite stage, the only life stage that can be isolated with purity and abundance (Strong and Nelson 2000b). This analysis netted 567 EST; which was again a quantum leap compared to what was known but was nearing the saturation limit of the EST project and confirmed the expectation that the number of genes expressed in the sporozoites were limited. In a complementary effort, the UCSF group also isolated genomic sequences, and obtained an additional 1,507 sequences, including 80% that were novel (Strong and Nelson 2000a).

The aforementioned modest efforts in random sequencing of genomic DNA were efficient and supported the feasibility of a complete genome sequencing project for *Cryptosporidium*. Three sequencing projects were initiated: a group centered in Minnesota, which completed the nucleotide sequence for *C. parvum* (Abrahamsen et al. 2004); a group at Virginia Commonwealth University (VCU), which determined the genome sequence for *C. hominis* (Xu et al. 2004); and a third project at Medical Research Council (MRC), UK, completed the nucleotide sequence for chromosome 6 of *C. parvum* (Bankier et al. 2003). The *C. parvum* genome sequencing projects culminated at 13× coverage, 9.1 Mb estimated genome sequence size, 30% G+C content, 21 contigs, six gaps, 3,807 protein coding genes, five ribosomal RNA (rRNA) genes, six 5S RNA genes, and 45 tRNA genes. Similarly, the *C. hominis* genome reached 12× coverage, 9.16 Mb, 30% GC, 1,413 contigs, 246 gaps, 3,994 protein coding genes, five rRNA genes, six 5S RNA genes, and 45 tRNA genes.

5.2.2 General Features of *Cryptosporidium* Genomes

The *Cryptosporidium* genome is highly compact (~9 Mb within eight chromosomes) in comparison to the *Plasmodium* (~23 Mb) and *Toxoplasma* (~80 Mb) genomes. *Cryptosporidium* lacks both mitochondrial and apicoplast genomes, in notable contrast to other apicomplexans. The proteome of *Cryptosporidium* is predicted to contain approximately 3,800 protein coding genes, far fewer than the ~5,300 genes that are predicted for the proteome of *P. falciparum*. This difference reflects the extreme parasitic lifestyle of *Cryptosporidium*; for instance, it lacks a TCA cycle and the oxidative phosphorylation complexes I–IV, as well as metabolic pathways for the de novo synthesis of amino acids and nucleotides. Additionally, the parasite lacks large amplified families of antigenically variant proteins, such as those encoded by the *var*, *rifin*, and *stevor* genes, which totalled approximately 250 genes in *P. falciparum*.

The acquisition of complete genome nucleotide sequence clarified speculations on the status of the *Cryptosporidium* mitochondria and apicoplast, two organelles that are present in all other apicomplexans examined so far. First, it confirms that

Cryptosporidium truly lacks a plastid genome, as previously reported (Zhu et al. 2000b), along with apicoplast-associated pathways. This indicates that the chloroplast-derived organelle is not available to serve as a drug target. Second, it reveals the absence of mitochondrial genome, but the presence of a number of nuclear-encoded proteins that possess typical mitochondrial targeting sequences, indicating the presence of a relict mitochondrion in *Cryptosporidium*. These proteins include elements of a protein import apparatus (e.g., TOM40 and TIM17), heat shock proteins (e.g., HSP70 and HSP65), solute transporters, Fe–S cluster assembling proteins (e.g., nifS, nifU, frataxin, and ferredoxin), and an alternative oxidase (AOX) (Riordan et al. 2003; LaGier et al. 2003; Abrahamsen et al. 2004; Roberts et al. 2004; Slapeta and Keithly 2004; Suzuki et al. 2004; Xu et al. 2004). Although the relict mitochondrion clearly has a greatly reduced function, the scope of the remaining role of this organelle remains to be elucidated. Because both organellar genomes were undoubtedly present in the ancestral apicomplexan, the lineage-specific loss in *Cryptosporidium*, together with the parasite's highly streamlined metabolism, highlight the molecular and evolutionary divergence of this pathogen from other apicomplexans.

Introns are rare in *Cryptosporidium*, estimated to be in only ~5% of the genes. However, there is a surprising exception. The COWP family was expanded by identification of paralogs in the completed genome (Spano et al. 1997; Templeton et al. 2004b). When the transcripts for COWP were sequenced, COWP7 had eight exons with seven introns while all other COWPs had no introns. At the time of annotation of *C. parvum* genome, a total number of 98 introns were mapped to the genome contigs, which include seven introns at the COWP locus, 18 introns identified by aligning ESTs to the contigs, and additional 72 introns identified from gapped BLAST alignments. In the latter case, cDNA products were amplified and sequenced to define exact intron/exon boundaries, as exampled for the RAD50 locus (cgd1_2410) (Fig. 3). Figure 4 shows the consensus as a logo diagram. The donor site GTAAGTT is more rigid compared with intron-rich organisms and is consonant with the smaller splicing machinery that depends on exact annealing of splicing RNA.

Another surprising observation is that the single intron within the β-tubulin transcripts does not splice efficiently in the late stages of *C. parvum* cultured in vitro. However, the unspliced transcripts were not found in vivo (Cai et al. 2004). More recently, an additional 17 examples of inefficient splicing were found, and these aberrant splicing events were also detected in vitro in bovine cells, but not in vivo (unpublished observations).

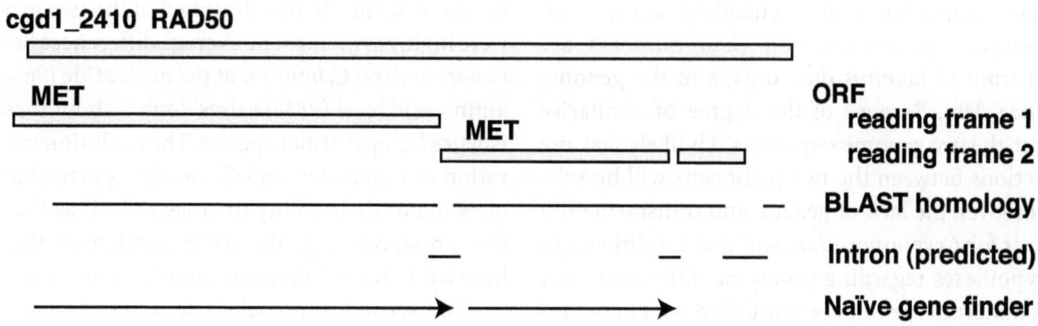

Fig. 3 Exon mapping at RAD50 locus. The top bar indicates the coding sequence for the *C. parvum* RAD50 gene. The other lines and bars indicate relative location of these entities mapped to the coding sequence. ORF, open reading frame larger than 50 amino acids; MET, the left most in frame methionine codon; BLAST homology, alignment to BLAST hits to RAD50; Intron, approximate location of intron inferred from BLAST and ORF; Naïve gene finder, in silico prediction of two genes from gene finder programs trained with splicing sequences of other organisms; Transcript mapped, coding sequence deduced after alignment of cDNA sequence

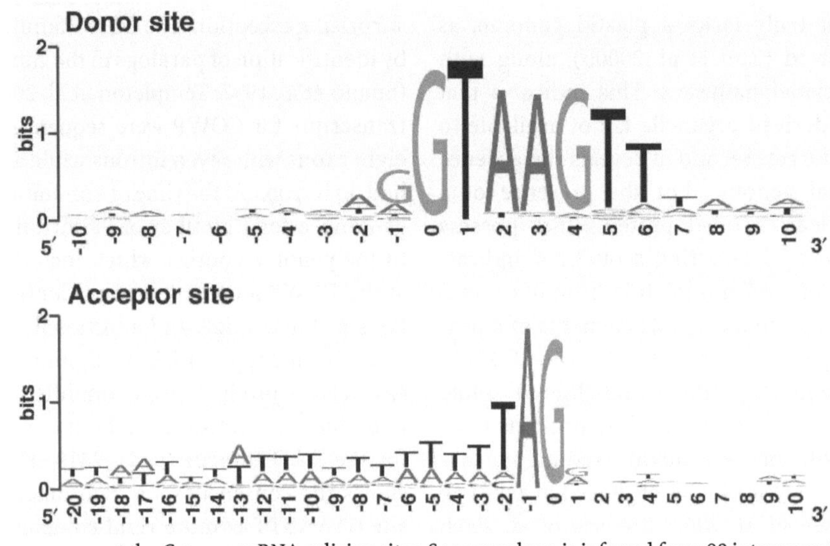

Fig. 4 Consensus sequences at the *C. parvum* RNA splicing sites. Sequence logo is inferred from 98 introns and indicated in DNA. Position 0 indicates the start and the end of intron. Y-Axis, 2 bits indicates 100% consensus. Generated by "Weblogo 3" (http://weblogo.berkeley.edu)

5.2.3 Comparison Between *C. parvum* and *C. hominis* Genomes

Comparison of the predicted proteomes of *C. parvum* and *C. hominis* has not revealed genotypic differences that might account for the disparate host specificities and pathogenicities observed between these closely related *Cryptosporidium* isolates. However, this analysis is compromised by the incompleteness of the *C. hominis* genome nucleotide sequence database. For example, numerous protein-encoding genes with assigned enzyme commission (EC) numbers are absent from *C. hominis* due to gaps in the genome sequence data. Because of the degree of similarity between the two genome sequences, it is likely that any distinctions between the two pathogens will be subtle. Moreover, the lack of genetic and transformation methods for *Cryptosporidium* will make it difficult to test hypotheses regarding genotypic differences that might underpin species-specific differences in pathogenicity. Finally, the null hypothesis should always be borne in mind, namely, that the difference between *C. parvum* and *C. hominis* do not constitute speciation, rather they represent different pathogenic poles of a spectrum within a single species. With these caveats in mind three initial methods might be proposed regarding approaches to delineate the phenotypic differences of *C. parvum* and *C. hominis.*

The first method is to determine genome-level differences by synteny comparisons. In a preliminary analysis using a hitchhike-assembly approach, *C. hominis* contigs were randomly broken into 2-kb fragments and mapped onto *C. parvum* contigs to find noncolinear regions between the two species (unpublished studies). Fragmentation to 2 kb enabled investigators to distinguish 20 kb with 85% similarity (divergence) and 1 kb with 99% similarity (no synteny) by looking at the top few BLAST hits. However, this analysis did not uncover a single region unique to C. *hominis*, thus demonstrating that the two genomes were highly syntenic. The average differences between *C. parvum* and *C. hominis* at the nucleotide (96%) and amino acid level (98%) raises doubts about these two isolates being distinct species. The evolutionary separation of *C. parvum* and *C. hominis*, particularly the question of their ability to cross, as well as the selective pressures (e.g., the AIDS pandemic) that may have led to their divergence remains unknown.

The second approach is to assess genome-wise gene expression differences between these two species. However, early attempts to use microarray technique to study gene expression profiles of intracellular stages of *C. parvum* were hindered by technical difficulties in obtaining satisfactory signal-to-noise ratios. Therefore, an ambitious project is now underway at the University of Minnesota to use real-time quantitative RT-PCR (qRT-PCR) to study the expression of all genes at various developmental stages of both *C. parvum* and *C. hominis*. Further details of the protocol, primer sequences, and the definition of the estimators are (or will be) available at http://cryptogenome.umn.edu/transcriptome/.

The third choice is to compare amino acid sequence diversities of orthologous extracellular and cytoplasmic proteins to summarize distinctions that might confer functional changes. Differences among amino acid sequences may have arisen from selection following either functional pressure or host immune pressure. Alternatively, diversity may arise via random variation in the absence of selective pressures. Preliminary analyses of amino acid sequence diversity in different classes of extracellular proteins indicates that amino acid polymorphisms are common, and examples of abundant polymorphisms within a protein might indicate positive selection (Table 2). The mucin-like proteins exhibit a range of polymorphisms, with some members differing in up to 16% of the residues between *C. parvum* and *C. hominis*. However, other mucins appear to be highly conserved, outside of the stretches of threonines, and are no more polymorphic than intracellular enzymes such as pyrophosphate-dependent 6-phosphofructokinase or glucose isomerase (both having 98% identity across orthologs). Members of the cluster of the CpLSP family are highly divergent in sequence between paralogs with less than 22% identity between adjacent paralogs, whereas the diversity between orthologs is approximately 96% over 1,622 amino acids in length (in the example of cgd7_3870 and its ortholog in *C. hominis*, Chro.70432). Thus, it is possible for adjacent genes to significantly diverge within a cluster of amplified genes.

Currently, there is no sequence that is unique to *C. hominis* that is not found in *C. parvum*. However, BLAST searches against the NR protein database will occasionally return a hit for *C. hominis* only. This does not necessarily mean that these proteins are unique to *C. hominis*; more likely that the GenBank (NR) protein database contains only coding sequences having defined gene predictions, and the annotation

Table 2 Examples of *Cryptosporidium* extracellular (EC) proteins and domains[a]

Protein or protein family	Description	*Cryptosporidium* gene identifiers
Insl	*Plasmodium* falcilysin-like (insulin degrading enzyme-like proteases)	cgd1_1680, cgd1_3840, cgd2_920, cgd2_930, cgd2_2760, cgd2_4270, cgd3_4170, cgd3_4180, cgd3_4190, cgd3_4200, cgd3_4210, cgd3_4220, cgd3_4240, cgd3_4250, cgd3_4260, cgd3_4270, cgd3_4280, cgd4_4240, cgd6_5510
Mucins	Contain long tracts of contiguous Thr and ser residues, predicted O-linked glycoproteins	cgd2_400, cgd2_430, cgd2_3140, cgd2_3290, cgd3_720, cgd3_1540, cgd4_770, cgd5_340, cgd5_460, cgd5_2060, cgd6_40, cgd6_710, cgd6_5410, cgd6_5400, cgd7_4020, cgd7_4660, cgd8_1160, cgd8_2800, cgd8_3520
MEDLE	Telomeric family, high conservation between paralogs suggests recent duplication events	cgd5_3580, cgd5_3590, cgd7_1380, cgd7_1390
WYLE	Lineage-specific family, telomeric	cgd5_2740, cgd8_3530, cgd8_3540, cgd8_3560, cgd8_3570, cgd8_3590
FLGN	Lineage-specific family, telomeric	cgd3_4360, cgd4_4470, cgd4_4480, cgd5_3600, cgd6_5460, cgd8_50
GGC	Lineage-specific family, telomeric	cgd2_2610, cgd5_3570, cgd7_5500, cgd8_1740
HCD	Lineage-specific family	cgd4_3570, cgd4_3580, cgd4_3590, cgd4_3600, cgd5_1420, cgd6_5340
SKSR	Lineage-specific family	cgd1_140, cgd1_150, cgd1_160, cgd1_3580, cgd3_10, cgd4_4490, cgd8_20, cgd8_30, cgd8_40
*Cp*LSP	Cluster of adjacent genes that encode large secreted proteins, lineage-specific	cgd5_1440, cgd7_3800, cgd7_3810, cgd7_3820, cgd7_3830, cgd7_3840, cgd7_3860, cgd7_3870

(Continued)

Table 2 (Continued)

Protein or protein family	Description	*Cryptosporidium* gene identifiers
COWP	Highly cysteine-rich oocyst wall proteins found in *Cryptosporidium*, *Toxoplasma*, and gregarines	cgd4_670, cgd4_3090, cgd6_200, cgd6_210, cgd6_2090, cgd7_1800, cgd7_5150, cgd8_3350
Hedgehog-type HINT domain	Lineage-specific, predicted lateral transfer of a domain similar to the carboxy-terminal autoprocessing domain of the animal Hedgehog proteins	cgd7_5290
LCCL domain family	Apicomplexan-specific multidomain family that is united by the common presence of an LCCL domain	cgd7_1730 (*CpCCp1/Cpa*135), cgd7_300 (CpCCp2), cgd2_790 (CpCCp3)
TRAP superfamily	Predicted functional ortholog of *Plasmodium* TRAP or *Toxoplasma* MIC2	cgd1_3500 (TRAP-C1), cgd5_4470
MAM+ Cu amine oxidase	Fusion of a MAM adhesive domain with a copper amine oxide domain	cgd3_3430
SCP domain proteins	Domain similar to found in animal sperm coat protein (SCP-1), the Glioma pathogenesis related protein, and the wasp Venom allergen 5	cgd7_4310, cgd5_2020
ToxI/ShkT domain protein	Contains a domain that is similar to the sea anemone toxin metridin and fused to animal metal proteases, plant prolyl hydroxylases. Vastly expanded in the genome of *C. elegans*	cgd5_3420 (termed TRAP-C2, although it is not a member of the TRAP superfamily)
Notch domain proteins	Contain a domain with approximately six cysteines that is thus far found only in the C-terminal part of the extracellular regions of the animal Notch proteins	cgd8_2800, cgd5_3420 (TRAP-C2), cgd6_670

[a] Additional examples of proteins and domains can be found in the supplementary table within reference (Templeton et al. 2004a)

team with the Minnesota project opted for a rigorous definition, which excluded submission of a handful of genes for which start codons or intron structures were unknown.

5.2.4 Genome Databases and Resources

Complete *C. parvum* and *C. hominis* genome nucleotide sequences and their annotations are available at the NCBI databases (http://www.ncbi.nlm.nih.gov/genomes/leuks.cgi). Whole chromosome assemblies and their maps are additionally available for *C. parvum* (Fig. 5). The original two databases for the *Cryptosporidium* sequencing projects (http://cryptogenome.umn.edu and http://www.hominis.mic.vcu.edu/) are also accessible, with the caveat that they were built to facilitate communication during the annotation phase of the projects, and the information there is not actively updated.

The most up-to-date and comprehesive database dedicated to *Cryptosporidium* bioinformatics is the CryptoDB website (http://CryptoDB.org) (Heiges et al. 2006). This database is a member of the ApiDB (http://ApiDB.org) family that serves as a common gateway to the apicomplexan databases, which also include PlasmoDB (*Plasmodium*) and ToxoDB (*Toxoplasma*). Searches, queries, and comparisons involving Apicomplexa may be more convenient from the ApiDB entry point (Aurrecoechea et al. 2006). Within CryptoDB, it is possible to browse or download sequences and annotations for either short segments or the whole genome. Bioinformatic analyses of all protein coding genes are periodically performed, such as BLAST against NR and InterPro (a growing suite of bioinfomatic analyses performed at different sites), and are made available through a user friendly interface. Publications related

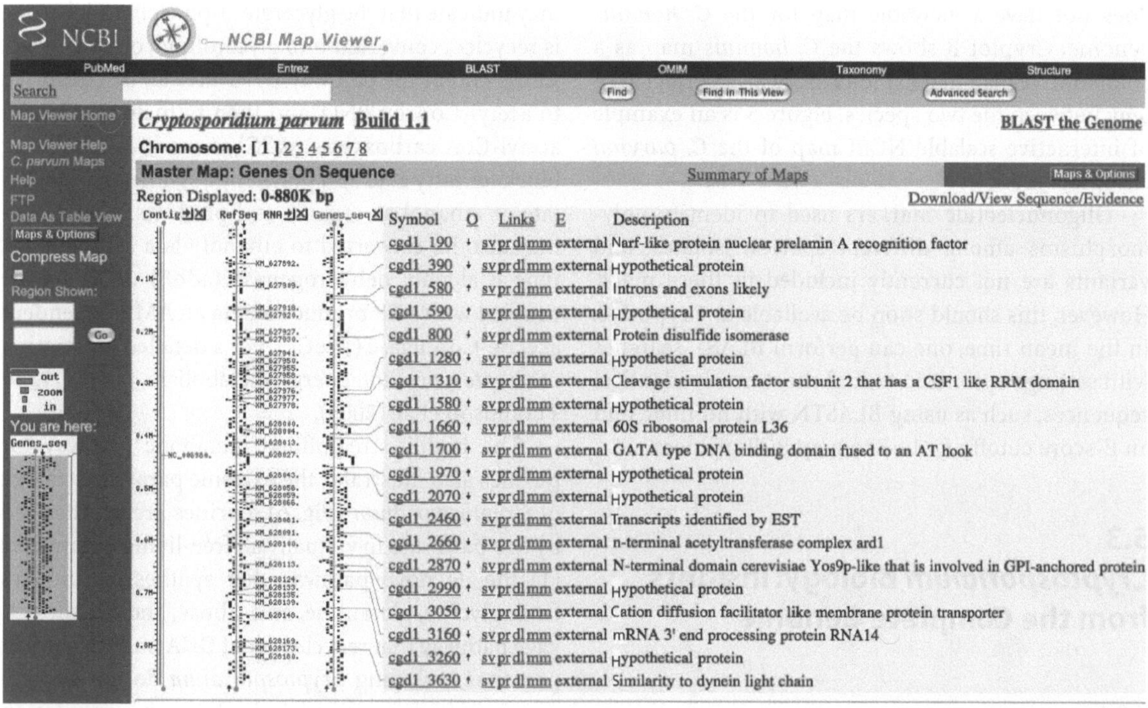

Fig. 5 Detailed maps of all *Cryptosporidium parvum* chromosomes are available at NCBI (http://www.ncbi.nlm.nih.gov/mapview/maps.cgi?taxid=5807). Only map for chromosome 1 is shown here as an example

to *Cryptosporidium* are also curated and organized with appropriate links to the genome. Various bioinformatic analysis tools are available and configured to run on the most current *Cryptosporidium* genome nucleotide sequence, and analysis results are returned with live links to other relevant information. Data mining can be performed manually, one gene at a time from a webbrowser, or performed automatically in batch format through their web services.

Many enzymes that are mapped to metabolic pathways can be accessed and compared with those of other apicomplexans via the "Metabolic Pathways" link within CryptoDB. Additionally, *C. parvum* and *C. hominis* metabolic pathways are also accessible from the Kyoto Encyclopedia of Genes and Genomes (KEGG)'s pathway databases (http://www.genome.ad.jp/kegg/pathway.html).

Resources that are not directly associated with *Cryptospordium* but may be useful in finding specific groups of proteins are the following: (1) MEROPS (http://merops.sanger.ac.uk/) for proteases; (2) TransportDB (http://www.membranetransport.org/) for transporters; and (3) Protein Kinase Resource (http://www.kinasenet.org/pkr/). MEROPS, for example, searches new sequences in the NR database that are similar to a seed set of well characterized proteases. Bidirectional best hits are scanned for the presence of known active site residues and classified. The membership tends to be exclusive and is not intended to be comprehensive, and degenerate paralogs are often omitted. Searches are not run on the latest sequence, and they may be limited to protein sequence databases. These compilations and the search/classification scheme make a great starting point for compiling ones own list.

5.2.5
Genome Maps

The best genome maps are available at the National Center for Biotechnology Information (NCBI) (e.g., http://www.ncbi.nlm.nih.gov/mapview/maps.cgi?taxid=5807) and CryptoDB (http://CryptoDB.org/cgi-bin/gbrowse/cryptodb/). Both maps are scalable, dynamically drawn by the server, and gene annotations are linked to additional useful information. The two maps differ in that the NCBI map is based on reference sequences (RefSeq), and the CryptoDB map is based on the most current assembly. Although NCBI does not have a viewable map for the *C. hominis* genome, CryptoDB shows the *C. hominis* map as a comparative map that exploits the nearly perfect synteny between the two species. Figure 5 is an example of interactive scalable NCBI map of the *C. parvum* chromosome 1.

Oligonucleotide markers used to identify polymorphisms among different species, isolates, and variants are not currently included in these maps. However, this should soon be available at CryptoDB. In the mean time, one can perform BLAST searches with settings optimized to find short nearly identical sequences, such as using BLASTN with no filter, and an E-score cutoff of <1e-8 to map PCR primers.

5.3 *Cryptosporidium* Biology: Insights from the Complete Genome

5.3.1 Streamlined Metabolism

Annotation of the *C. parvum*-predicted proteome revealed that the parasite adapted to an extremely parasitic life style by employing a highly simplified metabolism. The parasite has discarded the entire apicoplast and associated pathways, and retained only a fraction of the mitochondrial functions. It possesses no de novo synthetic pathways, and appears to rely on the host to supply all nutrients, including amino acids, nucleotides, and fatty acids (Abrahamsen et al. 2004).

For example, the gluconeogenesis and pentose phosphate pathways are absent in *Cryptosporidium*. Glucose is broken down to pyruvate, with minimal branching and no movement in the reverse direction as a synthetic step. The second step in glycolysis is handled by two pyrophosphate-dependent phosphofructokinase (PPi-PFK) genes. The replacement of PPi-PFK over the ATP-PFK likely reflects an adaptation to conserve ATP. Further down the glycolysis pathway, there is possible involvement of at least two and possibly three phosphoglycerate mutases that are encoded in the *Cryptosporidium* genome sequence. Two of these are tandem duplications of nearly identical enzymes, while the third enzyme appears to be similar to a fructose bisphosphatase. The tandem duplication may indicate that the glycerate-3-phosphate substrate is recycled, converted into glycerolipid, or consumed. At the end of the pathway, pyruvate can be converted to acetyl-CoA by PNO, and then to malonyl-CoA by acetyl-CoA carboxylase (ACC) to serve as a building block for fatty acid and polyketide synthesis, or to lactate or ethanol as organic end products. Acetyl-CoA may also be converted to ethanol via a bifunctional, type E alcohol dehydrogenase (adhE) or to acetate coupled with ATP production via an AMP-dependent acetate-CoA ligase (AceCL). For a detailed illustration of *Cryptosporidium* energy metabolism, see Fig. 12 in Thompson et al. (2005).

The highly streamlined metabolic pathway for purines also illustrates the extreme parasitic strategy of *Cryptosporidium* (Fig. 6). Purines are synthesized by two pathways in virtually all free-living organisms: (1) the de novo pathway that synthesizes purines from glycine, glutamine, and ribose, and (2) the salvage pathway that recycles spent DNA and RNA. Most parasites including *Cryptosporidium* do not synthesize purines de novo and, indeed, genome annotation indicates that the parasite lacks all the enzymes for this portion of the pathway. It was anticipated that *Cryptosporidium* relies on scavenging from the host via purine phosphoribosyltransferase (PRT). However, *Cryptosporidium* lacks any purine PRT, and its purine pathway consists of the minimal four enzymes to convert adenosine to GMP, which differs from *T. gondii* that has both adenosine-PRT and HXGPRT. The IMPDH is the preferred pathway compared to HXGPRT in both human and parasite, and humans use IMPDH in cells that need large amounts of GMP, such as dividing lymphocytes. *T. gondii* appears to prefer IMPDH, as inhibition of this enzyme by 6-thioxanthine (Pfefferkorn et al. 2001) results in growth arrest despite the active salvage enzyme HXGPRT.

Three enzymes, adenosine kinase (AK), adenylate deaminase (ADA), and GMP synthetase (GMPS), are of eukaryotic origin, whereas the IMPDH was likely acquired from a bacteria. Both the IMPDH and GMPS reactions use glutamine as a substrate, which results in release of glutamate. Interestingly, *Cryptosporidium* has no glutamate synthase (GLTS) that competes for the glutamine. Glutamine synthase (GS) and GLTS provide two important functions in most organisms: (1) GS and GLTS in tandem assimilate ammonia, and (2) GS and GLTS maintain homeostasis of ammonia, glutamine, and glutamate

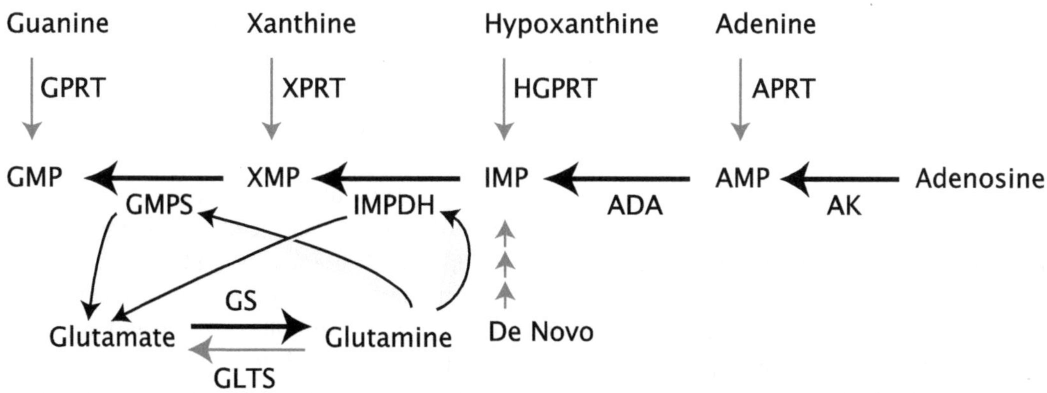

Fig. 6 Streamelined purine metabolism in *Cryptosporidium. Bold black arrows* indicate enzymes retained in *Cryptosporidium. Light weight gray arrows* indicate enzymes absent from *Cryptosporidium*, normally found in many organisms. The *multiple arrows* indicate multistep reaction for de novo synthesis. The *curved arrows* illustrate the competing demand for glutamine in three reactions, IMPDH, GMPS, and GLTS. The four GPRT, XPRT, HGPRT, and APRT reactions may or may not be performed by dedicated enzymes. Abbreviations: *GPRT* guanine phosphoribosyltranferase, *XPRT* xanthine phosphoribosyltranferase, *HGPRT* hypoxanthine phosphoribosyltranferase, *APRT* adenine phosphoribosyltranferase, *GMP* guanosine monophosphate, *XMP* xanthosine monophosphate, *IMP* inosine monophosphate, *AMP* adenosine monophosphate, *GMPS* guanosine monophosphate synthetase, *IMPDH* inosine monophosphate dehydrogenase, *ADA* adenylate deaminase, *AK* adenosine kinase, *GS* glutamine synthase, *GLTS* glutamate synthase

in the cell, thereby regulating all nitrogen metabolisms. *Cryptosporidium* lacks all genes except GS in concordance with loss of all de novo synthesis of amino acids and nucleotides.

Cryptosporidium also lacks plastid-associated Type II fatty acid synthesis responsible for synthesizing fatty acid de novo, but possesses a 25-kb Type I fatty acid synthase (*CpFAS1*) gene encoding a 900 kDa protein for elongating fatty acids (Zhu et al. 2004, 2000c). Related to CpFAS1 is a 40-kb polyketide synthase (*CpPKS1*) gene that encodes a 1,500 kDa protein consisting of 29 enzymatic functions that serve an unknown function (Fritzler and Zhu 2007; Zhu et al. 2002). Type I FAS and PKS are not present in *Plasmodium*, but recently found in *Toxoplasma* and *Eimeria* genome contigs (Zhu 2004). Polyketides are found in distantly related dinoflagellates (Snyder et al. 2003, 2005; Kubota et al. 2006; Rein and Snyder 2006), and may have a conserved function in coccidia.

5.3.2 Expanded Families of Transporters

Although *Cryptosporidium* has extensively downsized its genome and streamlined its metabolism, it has expanded smaller families of proteins, including transporters (discussed here), proteases, and predicted surface proteins involved in resistance to environmental factors, host–pathogen interaction, and immune evasion (discussed in following sections). The most notablly expanded transporters are those responsible for the uptake of nutrients. There are up to 20 genes encoding sugar transporters, along with 11 genes encoding amino acid transporters, and represent a much greater expansion than is found in *P. falciparum*, which possesses two transporters for sugar and one for amino acids (Gardner et al. 2002).

The parasite genome also encodes about 19 ABC transporters that may be involved in transporting various molecules across membranes (e.g., metabolites, lipids and sterols, and drugs). One ABC transporter has been previously localized at the host cell–parasite boundary (Perkins et al. 1999), and a few more have also been cloned localized in sporozoites (Zapata et al. 2002). At least four ABC genes belong to the sbmA family that are typically involved in the transport of peptide antibiotics in bacteria or multidrug resistance in pathogens or cancer cells, but whether these pumps are involved in the natural drug resistance of this parasite to some drugs is yet undetermined. Additionally, the *C. parvum* genome also encodes at least 7 P-type ATPases (CpATPases) that are typically involved in transporting cations (six genes) or phospholipids (one gene). However, there is only one V-type ATPase, for which all subunits including the largest subunit with seven transmembrane regions can be found in the parasite genomes.

5.3.3 Lineage-Specific Expansion of Proteases

At least 50 genes have been annotated as proteases in the *Cryptosporidium* genome (v.s. 31 protease genes in *P. falciparum*) (Abrahamsen et al. 2004). Several protease families apparently have undergone lineage-specific expansions, which include an aspartyl protease (six genes), subtilisin-like protease, cryptopain-like cysteine protease (five genes), and a *Plasmodium* falcilysin-like protease (19 genes). Falcilysin-like proteases are homologuous to insulin degrading enzymes that typically contain a Zn-chelating "HXXEH" active site motif. However, nine of the *Cryptosporidium* falcilysin genes lack this motif, suggesting that their encoded proteins may be enzymatically inactive, but reused for specific protein–protein interactions on the cell-surface. This notion is partly supported by the presence of secretory signal peptide sequences in the *Cryptosporidium* falcilysins that are absent among *Plasmodium* orthologs. The dramatic expansion of this family suggests that either the proteins have distinct cleavage specificities or their diversity may be related to evasion of a host immune response.

5.3.4 *Cryptosporidium*-Specific Amplified Gene Families

Complete genome nucleotide sequence information is now available for *Plasmodium* (Gardner et al. 2002), *Theileria* (Gardner et al. 2005; Pain et al. 2005), *Cryptosporidium* (Abrahamsen et al. 2004; Xu et al. 2004), as well as the ciliate *Tetrahymena* (Eisen et al. 2006). Annotation of the *Cryptosporidium* genome sequence revealed over 200 proteins that have no similarity to other proteins in GenBank and are predicted to enter the secretion pathway, based upon the presence of a signal peptide sequence. Many of these proteins have paralogs, indicating duplication or greater gene amplification in the genome. The abundance of lineage-specific "hypothetical" proteins is of course not unique to *Cryptosporidium* and has been observed for all prokaryotic and eukaryotic organisms that have available genome sequence information. As an example of lineage-specific secreted proteins, *Cryptosporidium* possesses numerous large anonymous proteins, some of which are duplicated or amplified, including cgd2_2550, cgd3_3370, cgd2_340, cgd7_2340, cgd7_4280, and cgd1_590.

Annotation of *Plasmodium falciparum* showed that several antigenically variant gene families, including the *var*, *rifin*, and *stevor* genes, are predominantly localized to telomeric or sub-telomeric regions of the chromosome. Partitioning of the genome in this manner may allow recombination or gene conversion events to occur, thus creating diversity, without endangering the integrity of "housekeeping" protein-encoding genes that are localized to the more central portions of the chromosomes. *Cryptosporidium* also has families of genes that are lineage-specific and predominantly localized to telomeres, including the SKSR, FGLN, and MEDLE families (Abrahamsen et al. 2004), and a tandem array of 11 genes within the 20-member family of insulinase-like proteases that have homologs in *P. falciparum*. The degree of amplification of *Cryptosporidium* extracellular protein-encoding genes is far less than the profound degree that is found in *P. falciparum*, perhaps because *Cryptosporidium* does not encounter substantial antibody-mediated immune pressure. *Cryptosporidium* also possesses additional lineage-specific gene families that are not localized to telomeres, including the GCC, HCP, and *Cp*LSP families, and several mucin paralogs. None of the amplified gene families have been studied in *Cryptosporidium* and the functional or immune pressures that have driven their amplification is not understood.

5.3.5 The Surface Protein Repertoire of *Cryptosporidium*

Over 300 proteins are predicted to be either secreted from the parasite or localized to the parasite cell surface (Abrahamsen et al. 2004; Templeton et al. 2004a). These predictions were based upon the presence of a signal peptide sequence, the absence of known intracellular targeting or retention motifs, and the absence of similarity to known cytoplasmic or organellar proteins. The total number of *Cryptosporidium* extracellular proteins is less than half of the repertoire that is predicted for *Plasmodium*; a difference that can be largely accounted for by the absence of large families of antigenically variant proteins (e.g., the *Plasmodium falciparum* families PfEMP1, RIFIN, and STEVOR). Only a few percent of the 300-plus proteins have been studied at the bench, and little is known other than structural predictions of these *Cryptosporidium* extracellular proteins. The catalog of surface proteins can be simplified into protein classes that are based upon structural and phylogenetic criteria, such as (1) membrane association via GPI-anchors or transmembrane domain(s); (2) architectures of single or multiple globular domains; (3) metazoan or bacterial domains that might have arisen via lateral transfer; (4) threonine- or serine-rich regions that identify glycosylated mucin-like proteins; (5) presence within the *Cryptosporidium* genome sequence as either single copy or gene families; and (6) the presence or absence of orthologs in related taxa. The following sections attempt to introduce the range of extracellular proteins, exclusive of transporters, channels, and proteins that have predicted enzymatic activities.

5.3.6 Mucin-Like Proteins

Toxoplasma and *Cryptosporidium* additionally share a class of mucin-like proteins that are predicted to be highly O-linked glycosylated. As a class these proteins are not found in *Plasmodium* and *Theileria*, and might confer adaptations of coccidians to the harsh environment of the intestine, or compose parasite surface coats that aid in the evasion of mucosal immunity. The component of a predicted metabolic pathway that mediates O-linked glycosylation is present in *Cryptosporidium* and *Toxoplasma*, but is absent in *Plasmodium* (Stwora-Wojczyk et al. 2004; Templeton et al. 2004a). This predicted pathway contains several glycosyltransferases, including WcaK-like glycosyltransferases, which are not found in *Plasmodium*. The pathway is also not present in *Tetrahymena*, and may have entered the apicomplexan lineage following the split with the ciliates. *Cryptosporidium* possesses over 30 proteins that contain stretches of 50 or more consecutive Thr or Ser residues. The proteins may have functional roles, such as the predicted involvement of the *Cryptosporidium* mucins, GP900, gp40, and other Gal/GalNAc-modified proteins in the adhesion of sporozoites to intestinal epithelium (Barnes et al. 1998; Cevallos et al. 2000; Chen and LaRusso 2000; Bonnin et al. 2001). GP900 is localized to micronemes, in addition to the cell surface (Bonnin et al. 2001), and possesses an acidic cytoplasmic tail and carboxy-terminal proximal tryptophan residue which together are reminiscent of the TRAP superfamily proteins (described later). Some of the mucins are amplified within paralogous families, whereas many do not have any distinguishable ancestry. The latter class of unique mucins includes a protein, cgd3_720, that is 11,700 AA long and is composed of 17 repeats of a 600 AA long cysteine-rich domain. This mucin contains a stretch of 308 consecutive thr residues. Stretches of thr residues are not confined to lineage-specific proteins; for example, a protein that is composed of six consecutive Notch domains contains a section of 48 thr residues.

5.3.7 The TRAP Family of Motility and Invasion Proteins

Apicomplexans derive their name from a unique apical secretory apparatus that confers the capacity of gliding motility and the invasion of host cells (Sibley 2004). Batteries of proteins within two apical secretory organelles, the micronemes and rhoptries, mediate cell surface adhesion to host substrates, as well as the invasion of specific target cells and the establishment of the parasitophorous vacuole (PV) (Carruthers and Sibley 1997). Gliding motility in the apicomplexans is mediated by a phylum-specific superfamily of extracellular proteins, termed TRAP, members of which couple extracellular adhesion with intracellular interaction to an actinomyosin motility apparatus. The canonical members of the family include TRAP and CTRP in *Plasmodium* and MIC2 in *Toxoplasma*. The TRAP architecture consists of a signal peptide, followed by one or more vWA and TSP1 domains, a transmembrane domain, and a short acidic cytoplasmic domain that contains a conserved tryptophan residue near the carboxy terminus. *Cryptosporidium* possesses at least one TRAP protein, termed TRAP-C1, and perhaps two additional members, *Cp*TSP7 (Deng et al. 2002) and GP900. In contrast to the *Plasmodium* and *Toxoplasma* versions, *Cryptosporidium* TRAP family members lack vWA domains and instead possess two Apple domains in the extracellular adhesive region.

5.3.8 The LCCL Domain Family

The LCCL domain superfamily contains the most conserved multidomain extracellular proteins in the Apicomplexa (Fig. 7). Four members of the family (in *C. parvum* termed *Cp*CCp1 through *Cp*CCp3, and CpFNPA) have orthologs in all apicomplexans that have been examined (Delrieu et al. 2002; Pradel et al. 2004; Tosini et al. 2004; Templeton et al. 2004a). Their domain architectures are composed of multiple metazoan- and bacterial-like domains and are predicted to confer manifold adhesion to lipids, polysaccharides, and proteins. The conservation of this family across the Apicomplexa suggests that they serve conserved functions in all pathogens in the clade. The LCCL domain superfamily members are expressed in gametocytes and emerging gametes in *Plasmodium*, whereas studies to date indicate that they are expressed in the sporozoite stage in *C. parvum*. Targeted gene disruption of *Plasmodium* LCCL domain genes results in a phenotype that manifests at the oocyst stage, and perhaps expression during this stage will be the unifying feature of the family members in the Apicomplexa.

5.3.9 The Oocyst Wall Proteins

As described earlier, the phylogentic relationship among *Cryptosporidium*, *Plasmodium*, and *Toxoplasma* is under much debate (Carreno et al. 1999; Zhu et al. 2000a; Barta and Thompson 2006). In this taxonomic discussion it should be considered that *Cryptosporidium* shares with *Toxoplasma*, to the exclusion of *Plasmodium* and *Theileria*, numerous extracellular proteins that might indicate a common coccidian ancestry. Foremost of these proteins is the oocyst wall protein (OWP; in *Cryptosporidium*, COWP) family that is replete in both pathogens, including nine members in *Cryptosporidium* (Templeton et al. 2004b; Spano et al. 1997). OWP proteins are also found in gregarines (Omoto et al. 2004), thus indicating a molecular feature that is likely common to all apicomplexans that utilize external cyst stages. The OWP proteins are composed of a dramatic periodicity in which a cysteine is found every 10–12 residues. The high cysteine content may form intra- and inter-molecular disulfide bonds that contribute to the profound stability of the oocyst wall. Indeed, the coccidian oocyst stage is exceptionally durable in the external environment and is resistant to chemical disinfection to the extent that bleach treatment does not affect the viability of the enclosed sporozoites (Korich et al. 1990).

5.3.10 The Role of Lateral Gene Transfer in the Origin of Apicomplexan Extracellular Proteins

The lateral gene transfer of extracellular domains has likely played a significant role in shaping the ability of the apicomplexans to physically interact with their hosts (Templeton et al. 2004a). *Cryptosporidium* possesses over 30 proteins that are composed of complex "mix and match" architectures of multiple domains that are predicted to have metazoan or bacterial origins [see Table 2 for examples; see also the supplementary table in Templeton et al. (2004a)]. These domains include LH, LCCL, SR, Pentraxin, Ricin, Discoidin, NEC, SCP, Sushi, Apple, MAM, Tox1, and Notch. Complex domain architectures in extracellular proteins appear to be a general feature of the apicomplexans, and are not found in other protists such as the kinetoplastids or diplomonads. Although each apicomplexan lineage contains similar or greater numbers of this mix and match class of proteins, few are conserved as orthologs in multiple species. This may indicate that complex multidomain architectures have a measure of plasticity in their domain organization, such that two different domain architectures can confer the same function despite a nonorthologous relationship, or that the different apicomplexan species have diverged such that they share few functional orthologs. In this regard it is remarkable that some highly complex domain architectures are conserved across genus boundaries, such as the TRAP-C2 (a misnomer, since the protein is not within the TRAP superfamily) protein that is found in *Cryptosporidium* (cgd5_3420) and *Toxoplasma* (42.m03281), or the CCp family members that are conserved in all apicomplexans (Fig. 7).

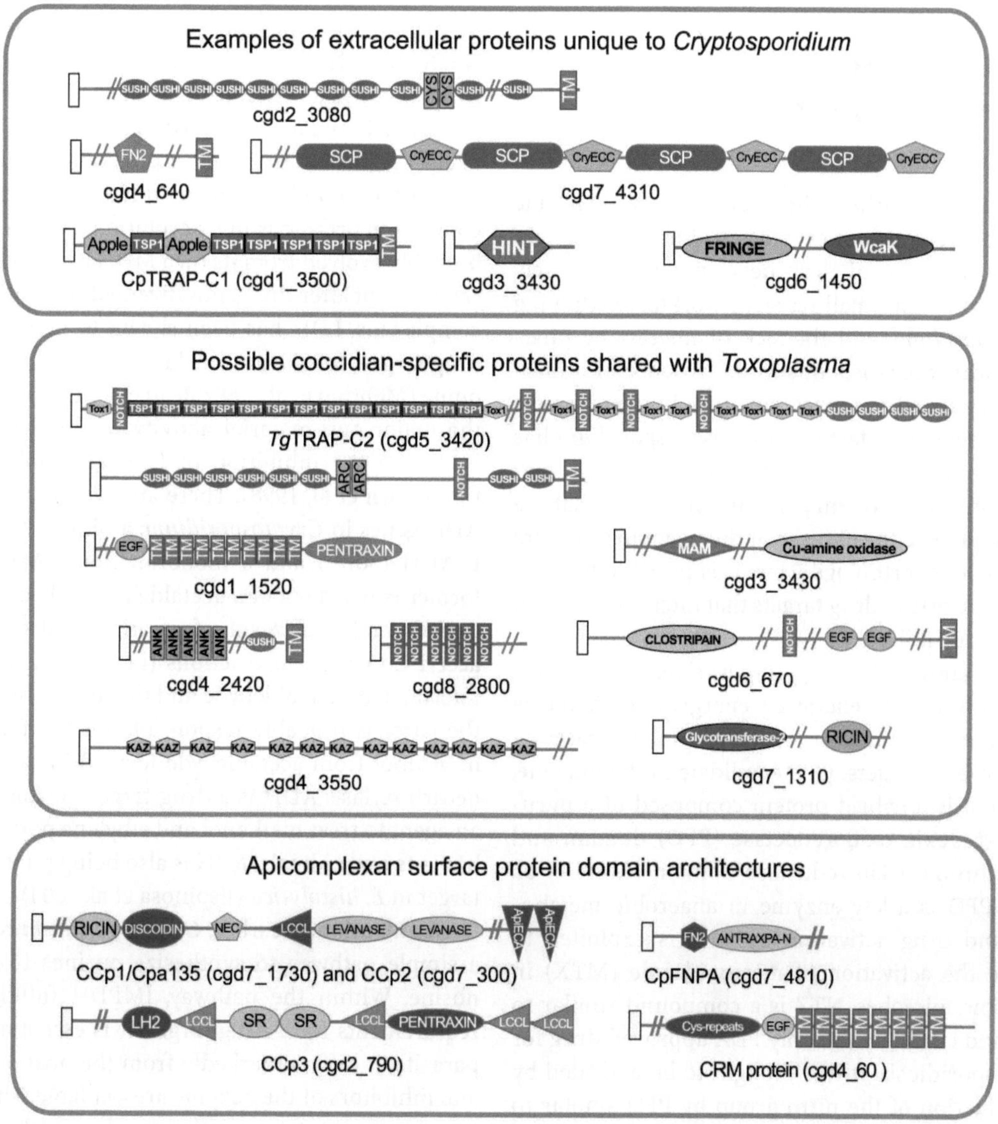

Fig. 7 Examples of multidomain extracellular proteins and their distribution within the Apicomplexa. The bottom panel illustrates phylum-specific proteins that are conserved throughout the Apicomplexa. The white boxes indicate signal peptide sequences; TM, transmembrane domain. Domains not described in Table 2 or the text include Sushi, also termed CCP or complement control protein repeat; *CYS* a cysteine rich domain specific to *Cryptosporidium*; *FN2* fibronectin type 2 domain found within a variety of plasma proteins; Apple, found in coagulation and other plasma proteins; *TSP1* thrombospondin type 1 domain; *ARC* a small cysteine rich domains similar to versions in *Archaeoglobus fulgidus*; *EGF* epidermal growth factor-like domain; Pentraxin, similar to domain found in the serum acute phase response proteins and sperm acrosomal protein; *ANK* ankyrin repeat; Clostripain, a secreted protease similar to caspase-hemoglobinase; *KAZ* Kazal repeat found in protease inhibitors; *RICIN* predicted carbohydrate-binding domain; *NEC* a globular domain present in the globular head of collagen; Levanase, a domain that is found in bacterial secreted levanases and glucosidases; *ApiECA* a cysteine-rich repeat found only in CCp1 and CCp2; *SR* found in macrophage scavenger receptor and sea urchin speract; *LCCL* a domain first identified in Limulus factor C, also Coch-5b2 and Lgl1; *ANTRAXPA-N* a domain found in the N-terminal region of the Anthrax PA toxin; *LH2* a domain found in lipoxygenase

5.4 Parasite Targets for New Therapeutics

As described earlier, the landscape of available drugs for treatment of cryptosporidiosis is relatively bleak. The future of drug development is severely restrained by the small research workforce studying *Cryptosporidium*, and the lack of interest by pharmaceutical companies due to the absence of a market in developed countries. Fortunately, the complete genome sequence database for *Cryptosporidium* has allowed annotation of the metabolic pathways, and the identification of enzymes and pathways that are either unique to the parasite or divergent in structure from the host orthologs. Thus it is possible to identify and prioritize drug targets that might confer high therapeutic indices.

It is known that *Cryptosporidium* relies solely on glycolysis for generating energy, and therefore a number of unique enzymes within the pathway may serve as targets. One candidate is the enzyme, PNO, that is a hybrid protein composed of a pyruvate ferredoxin oxidoreductase (PFO) domain and a cytochrome P450 reductase domain (Rotte et al. 2001). PFO is a key enzyme in anaerobic metabolism, and drug activation by PFO is exploited to confine the activation of metronidazole (MTX) in anaerobic microbes. NTZ is a compound similar to MTX and currently the only FDA approved drug for cryptosporidiosis. It was thought to be activated by the reduction of the nitro group by PFO similar to MTX. Recent studies indicate no evidence for the nitro group reduction, but rather demonstrated that NTZ is deacytylated into tizoxanide in the liver. NTZ blocks the formation of a normal parasitophorous vacuole of the coccidian parasite, *Neospora caninum* (Esposito et al. 2005). It is also hypothesized that a parasite cell signaling system is affected by NTZ.

Other glycolytic enzymes include PPi-PFK, lactate dehydrogenase (LDH), and alcohol dehydrogenase (ADH), all of which differ structurally from host orthologs. PPi-PFK is a rate-limiting enzyme that differs significantly from ATP-PFK. It can be inhibited by biphosphonates and has been proposed as a drug target in *Entamoeba histolytica* (Byington et al. 1997a, b). Both LDH and ADH catalyze the formation of organic end products, which is critical in maintaining the carbon flow and recycling of NAD(P)H to $NAD(P)^+$ in cells that rely solely or heavily on glycolysis (Thompson et al. 2005). All apicomplexan LDH and malate dehydrogenase (MDH) genes likely originated from an α-proteobacterial MDH. *Cryptosporidium* LDH further differs from other apicomplexan orthologs in its evolutionary history, as it probably evolved from its own MDH by a gene duplication event after this genus diverged from other apicomplexans. LDH has been shown to be one of the drug targets for the anti-malarial compound chloroquine (Menting et al. 1997; Read et al. 1999), although the major anti-malarial activity of this compound is due to the inhibition of haematin polymerization (Dorn et al. 1998). There are two bacterial-type ADH genes in *Cryptosporidium*: a bifunctional, type E ADH (adhE) and a monofunctional ADH2. The former is fused with an acetaldehyde dehydrogenase domain and catalyzes the formation of alcohol from acetyl-CoA via two reactions (i.e., the formation of intermediate acetaldehyde and that of alcohol), while the latter is probably responsible for the formation of alcohol from acetaldehyde formed by a pyruvate decarboxylase. ADH is a drug target for fomepizole, an agent to treat methanol and ethylene poisoning in humans and animals. ADH is also being pursued as a target in *E. histolytica* (Espinosa et al. 2004).

As mentioned earlier, *Cryptosporidium* relies on a simple pathway to synthesize purines from adenosine. Within the pathway, IMPDH fulfills three requirements for a drug target: it is essential to the parasite, it differs markedly from the host ortholog, and inhibitors of the enzyme are available. The parasite does not possess redundancies in the purine pathway that will allow by-pass of this enzyme. In vitro inhibition by mycophenolic acid (MPA) and ribavirin inhibits parasite growth (Striepen et al. 2004). Importantly, the *Cryptosporidium* IMPDH is likely of bacterial origin and differs in structure from the host IMPDH (Striepen et al. 2002). Three widely used drugs on the market target inhibition of IMPDH; namely, MPA, ribavirin, and mizoribin. Additional compounds are in the pipeline from industrial interest in immunosupressants and antivirals (Pankiewicz et al. 2004); and consequently large libraries of compounds are available that may have specificity towards *Cryptosporidium* IMPDH. Moreover, adenosine kinase (AK) is responsible for the first reaction in scavenging purines. Its activity could be inhibited by 4-nitro-6-benzylthioinosine, a compound that demonstrates therapeutic promise against the related parasite *T. gondii* (Galazka et al. 2006). Within the pyrimidine scavenging pathway, *Cryptosporidium* thymidine kinase (TK) is of bacterial origin, thus likely to have structural distinctions that will allow targeting with specific drugs. However, because this parasite may also utilize cystidine, uridine, or uracil via parasite-encoded uridine kinase (UK) and uracil phosphoribosyltransferase (UPRT) (Striepen and Kissinger 2004), the essentiality of *Crptosporidium* TK needs to be carefully confirmed before being pursued as a drug target.

Another enzyme that might be considered as a drug target is glutamine synthetase (GS) that converts glutamate produced by GMP synthetase (GMPS) back to glutamine. *Cryptosporidium* GS is a bacterial type I enzyme and is structurally distinct from the type II enzymes found in other eukaryotes. The metabolic role of the GS appears to be limited to conversion of glutamate to glutamine and is not the universal gateway for all nitrogenous compounds. An extracellular form of GS has been shown to be a critical virulence factor in *Mycobacterium* (Harth and Horwitz 1999). Many amino acid analogs that inhibit GS have shown potency towards inhibiting the extracellular form of GS in *Mycobacterium*, and protection against *Mycobacterium* infection in vitro and in vivo (Harth and Horwitz 2003). The extracellular GS in *Mycobacteria* has emerged as a drug target because it may overcome the multidrug resistance that is conferred by highly active ABC transporters. Interestingly, the *Cryptosporidium* GS has an amino-terminal extension that is predicted to be a secretory signal.

Fatty acid synthesis is the target for a number of antibacterial agents, including isoniazid, triclosan, and thiolactomycin. *Cryptosporidium* lacks a plastid-specific Type II FAS, but possesses both type I FAS and PKS that prefer medium to long chain fatty acids as substrates (Zhu et al. 2002, 2004, 2000c; Fritzler and Zhu 2007). The FAS inhibitor cerulenin was able to block the growth of *C. parvum* in vitro and inhibit the activity of recombinant FAS (Zhu et al. 2004, 2000c). Other enzymes include fatty acid-CoA ligase (ACL, also termed fatty acyl-CoA synthetase), the cytosolic acetyl-CoA carboxylase (ACC), and probably the long-type acyl-CoA binding protein (ACBP). Please refer to a recent review for a more detailed description of these enzymes (Zhu 2004).

References

Abrahamsen MS, Templeton TJ, Enomoto S, Abrahante JE, Zhu G, Lancto CA, Deng M, Liu C, Widmer G, Tzipori S, Buck GA, Xu P, Bankier AT, Dear PH, Konfortov BA, Spriggs HF, Iyer L, Anantharaman V, Aravind L, Kapur V (2004) Complete genome sequence of the apicomplexan, *Cryptosporidium parvum*. Science 304:441–445

Aiello AE, Xiao L, Limor JR, Liu C, Abrahamsen MS, Lal AA (1999) Microsatellite analysis of the human and bovine genotypes of *Cryptosporidium parvum*. J Eukaryot Microbiol 46:46S–47S

Alvarez-Pellitero P, Sitja-Bobadilla A (2002) *Cryptosporidium molnari* n. sp. (Apicomplexa: Cryptosporidiidae) infecting two marine fish species, *Sparus aurata* L. and *Dicentrarchus labrax* L. Int J Parasitol 32:1007–1021

Amadi B, Mwiya M, Musuku J, Watuka A, Sianongo S, Ayoub A, Kelly P (2002) Effect of nitazoxanide on morbidity and mortality in Zambian children with cryptosporidiosis: a randomised controlled trial. Lancet 360:1375–1380

Anderson DR, Duszynski DW, Marquardt WC (1968) Three new coccidia (Protozoa: Telosporea) from kingsnakes, *Lampropeltis* spp., in Illinois, with a redescription of *Eimeria zamenis* Phisalix, 1921. J Parasitol 54:577–581

Aurrecoechea C, Heiges M, Wang H, Wang Z, Fischer S, Rhodes P, Miller J, Kraemer E, Stoeckert CJ Jr, Roos DS, Kissinger JC (2006) ApiDB: integrated resources for the apicomplexan bioinformatics resource center. Nucl Acids Res 35 (Database issue):D427–D430

Bankier AT, Spriggs HF, Fartmann B, Konfortov BA, Madera M, Vogel C, Teichmann SA, Ivens A, Dear PH (2003) Integrated mapping, chromosomal sequencing and sequence analysis of *Cryptosporidium parvum*. Genom Res 13:1787–1799

Barnes DA, Bonnin A, Huang JX, Gousset L, Wu J, Gut J, Doyle P, Dubremetz JF, Ward H, Petersen C (1998) A novel multidomain mucin-like glycoprotein of *Cryptosporidium parvum* mediates invasion. Mol Biochem Parasitol 96:93–110

Barta JR, Thompson RC (2006) What is *Cryptosporidium*? Reappraising its biology and phylogenetic affinities. Trends Parasitol 22:463–468

Bearup AJ (1954) The coccidia of carnivores of Sydney. Aust Vet J 30:185–186

Bonnin A, Ojcius DM, Souque P, Barnes DA, Doyle PS, Gut J, Nelson RG, Petersen C, Dubremetz JF (2001) Characterization of a monoclonal antibody reacting with antigen-4 domain of gp900 in *Cryptosporidium parvum* invasive stages. Parasitol Res 87:589–592

Brends SJ (1989–2005) Systema Naturae 2000 The Taxonomicon. Universal Taxonomic Services, Amsterdam, The Netherlands. [http://sn2000.taxonomy.nl/Taxonomicon/]. Access date: Nov 23, 2005

Bull S, Chalmers R, Sturdee AP, Curry A, Kennaugh J (1998) Cross-reaction of an anti-*Cryptosporidium* monoclonal antibody with sporocysts of Monocystis species. Vet Parasitol 77:195–197

Byington CL, Dunbrack RL, Jr., Cohen FE, Agabian N (1997a) Molecular modeling of phosphofructokinase from *Entamoeba histolytica* for the prediction of new antiparasitic agents. Arch Med Res 28(Spec No):86–88

Byington CL, Dunbrack RL, Jr., Whitby FG, Cohen FE, Agabian N (1997b) *Entamoeba histolytica*: computer-assisted modeling of phosphofructokinase for the prediction of broad-spectrum antiparasitic agents. Exp Parasitol 87:194–202

Caccio S, Homan W, Camilli R, Traldi G, Kortbeek T, Pozio E (2000) A microsatellite marker reveals population heterogeneity within human and animal genotypes of *Cryptosporidium parvum*. Parasitology 120(Pt 3):237–244

Caccio S, Homan W, van Dijk K, Pozio E (1999) Genetic polymorphism at the beta-tubulin locus among human and animal isolates of *Cryptosporidium parvum*. FEMS Microbiol Lett 170:173–179

Cai X, Lancto CA, Abrahamsen MS, Zhu G (2004) Intron-containing beta-tubulin transcripts in *Cryptosporidium parvum* cultured in vitro. Microbiology 150:1191–1195

Carreno RA, Martin DS, Barta JR (1999) *Cryptosporidium* is more closely related to the gregarines than to coccidia as shown by phylogenetic analysis of apicomplexan parasites inferred using small-subunit ribosomal RNA gene sequences. Parasitol Res 85:899–904

Carruthers VB, Sibley LD (1997) Sequential protein secretion from three distinct organelles of *Toxoplasma gondii* accompanies invasion of human fibroblasts. Eur J Cell Biol 73:114–123

Cevallos AM, Bhat N, Verdon R, Hamer DH, Stein B, Tzipori S, Pereira ME, Keusch GT, Ward HD (2000) Mediation of *Cryptosporidium parvum* infection in vitro by mucin-like glycoproteins defined by a neutralizing monoclonal antibody. Infect Immun 68:5167–5175

Chen XM, LaRusso NF (2000) Mechanisms of attachment and internalization of *Cryptosporidium parvum* to biliary and intestinal epithelial cells. Gastroenterology 118:368–379

Ctrnacta V, Ault JG, Stejskal F, Keithly JS (2006) Localization of pyruvate:NADP+ oxidoreductase in sporozoites of *Cryptosporidium parvum*. J Eukaryot Microbiol 53:225–231

Current WL, Reese NC (1986) A comparison of endogenous development of three isolates of *Cryptosporidium* in suckling mice. J Protozool 33:98–108

Current WL, Upton SJ, Haynes TB (1986) The life cycle of *Cryptosporidium baileyi* n. sp. (Apicomplexa, Cryptosporidiidae) infecting chickens. J Protozool 33:289–296

Delrieu I, Waller CC, Mota MM, Grainger M, Langhorne J, Holder AA (2002) PSLAP, a protein with multiple adhesive motifs, is expressed in *Plasmodium falciparum* gametocytes. Mol Biochem Parasitol 121:11–20

Deng M, Templeton TJ, London NR, Bauer C, Schroeder AA, Abrahamsen MS (2002) *Cryptosporidium parvum* genes containing thrombospondin type 1 domains. Infect Immun 70:6987–6995

Dorn A, Vippagunta SR, Matile H, Jaquet C, Vennerstrom JL, Ridley RG (1998) An Assessment of drug-haematin binding as a mechanism for inhibition of haematin polymerization by quinoline antimalarials. Biochem Pharmacol 55:727–736

Doumbo O, Rossignol JF, Pichard E, Traore HA, Dembele TM, Diakite M, Traore F, Diallo DA (1997) Nitazoxanide in the treatment of cryptosporidial diarrhea and other intestinal parasitic infections associated with acquired immunodeficiency syndrome in tropical Africa. Am J Trop Med Hyg 56:637–639

Dubey JP, Pande BP (1963) Observations on the coccidian oocysts from Indian Jungle cat (*Felis chaus*). Ind J Microbiol 3:103–108

Duszynski DW (1969) Two new coccidia (Protozoa: Dimeriidae) from Costa Rican lizards with a review of the *Eimeria* from lizards. J Protozool 16:581–585

Egyed Z, Sreter T, Szell Z, Beszteri B, Dobos-Kovacs M, Marialigeti K, Cornelissen AW, Varga I (2002) Polyphasic typing of *Cryptosporidium baileyi*: a suggested model for characterization of cryptosporidia. J Parasitol 88:237–243

Egyed Z, Sreter T, Szell Z, Varga I (2003) Characterization of *Cryptosporidium* spp. – recent developments and future needs. Vet Parasitol 111:103–114

Eisen JA, Coyne RS, Wu M, Wu D, Thiagarajan M, Wortman JR, Badger JH, Ren Q, Amedeo P, Jones KM, Tallon LJ, Delcher AL, Salzberg SL, Silva JC, Haas BJ, Majoros WH, Farzad M, Carlton JM, Smith RK Jr., Garg J, Pearlman RE, Karrer KM, Sun L, Manning G, Elde NC, Turkewitz AP, Asai DJ, Wilkes DE, Wang Y, Cai H, Collins K, Stewart BA, Lee SR, Wilamowska K, Weinberg Z, Ruzzo WL, Wloga D, Gaertig J, Frankel J, Tsao CC, Gorovsky MA, Keeling PJ, Waller RF, Patron NJ, Cherry JM, Stover NA, Krieger CJ, del Toro C, Ryder HF, Williamson SC, Barbeau RA, Hamilton EP, Orias E (2006) Macronuclear genome sequence of the ciliate *Tetrahymena thermophila*, a model eukaryote. PLoS Biol 4:e286

Espinosa A, Clark D, Stanley SL Jr (2004) *Entamoeba histolytica* alcohol dehydrogenase 2 (EhADH2) as a target for anti-amoebic agents. J Antimicrob Chemother 54:56–59

Esposito M, Stettler R, Moores SL, Pidathala C, Muller N, Stachulski A, Berry NG, Rossignol JF, Hemphill A (2005) In vitro efficacies of nitazoxanide and other thiazolides against *Neospora caninum* tachyzoites reveal antiparasitic activity independent of the nitro group. Antimicrob Agents Chemother 49:3715–3723

Fayer R (1997) *Cryptosporidium* and cryptosporidiosis. CRC Press, Boca Raton, Florida, USA

Fayer R, Morgan U, Upton SJ (2000) Epidemiology of *Cryptosporidium*: transmission, detection and identification. Int J Parasitol 30:1305–1322

Fayer R, Santin M, Xiao L (2005) *Cryptosporidium bovis* n. sp. (Apicomplexa: Cryptosporidiidae) in cattle (Bos taurus). J Parasitol 91:624–629

Fayer R, Trout JM, Xiao L, Morgan UM, Lai AA, Dubey JP (2001) *Cryptosporidium canis* n. sp. from domestic dogs. J Parasitol 87:1415–1422

Fayer R, Ungar BL (1986) *Cryptosporidium* spp. and cryptosporidiosis. Microbiol Rev 50:458–483

Feng X, Rich SM, Akiyoshi D, Tumwine JK, Kekitiinwa A, Nabukeera N, Tzipori S, Widmer G (2000) Extensive polymorphism in *Cryptosporidium parvum* identified by multilocus microsatellite analysis. Appl Environ Microbiol 66:3344–3349

Fox LM, Saravolatz LD (2005) Nitazoxanide: a new thiazolide antiparasitic agent. Clin Infect Dis 40:1173–1180

Fritzler JM, Zhu G (2007) Functional characterization of the acyl-[acyl carrier protein] ligase in the *Cryptosporidium parvum* giant polyketide synthase. Int J Parasitol 37(3-4):307–316

Galazka J, Striepen B, Ullman B (2006) Adenosine kinase from *Cryptosporidium parvum*. Mol Biochem Parasitol 149:223–230

Gardner MJ, Bishop R, Shah T, de Villiers EP, Carlton JM, Hall N, Ren Q, Paulsen IT, Pain A, Berriman M, Wilson RJ, Sato S, Ralph SA, Mann DJ, Xiong Z, Shallom SJ, Weidman J, Jiang L, Lynn J, Weaver B, Shoaibi A, Domingo AR, Wasawo D, Crabtree J, Wortman JR, Haas B, Angiuoli SV, Creasy TH, Lu C, Suh B, Silva JC, Utterback TR, Feldblyum TV, Pertea M, Allen J, Nierman WC, Taracha EL, Salzberg SL, White OR, Fitzhugh HA, Morzaria S, Venter JC, Fraser CM, Nene V (2005) Genome sequence of *Theileria parva*, a bovine pathogen that transforms lymphocytes. Science 309:134–137

Gardner MJ, Hall N, Fung E, White O, Berriman M, Hyman RW, Carlton JM, Pain A, Nelson KE, Bowman S, Paulsen IT, James K, Eisen JA, Rutherford K, Salzberg SL, Craig A, Kyes S, Chan MS, Nene V, Shallom SJ, Suh B, Peterson J, Angiuoli S, Pertea M, Allen J, Selengut J, Haft D, Mather MW, Vaidya AB, Martin DM, Fairlamb AH, Fraunholz MJ, Roos DS, Ralph SA, McFadden GI, Cummings LM, Subramanian GM, Mungall C, Venter JC, Carucci DJ, Hoffman SL, Newbold C, Davis RW, Fraser CM, Barrell B (2002) Genome sequence of the human malaria parasite *Plasmodium falciparum*. Nature 419:498–511

Hampton JC, Rosario B (1966) The attachment of protozoan parasites to intestinalepithelial cells of the mouse. J Parasitol 52:939–949

Harth G, Horwitz MA (1999) An inhibitor of exported *Mycobacterium tuberculosis* glutamine synthetase selectively blocks the growth of pathogenic mycobacteria in axenic culture and in human monocytes: extracellular proteins as potential novel drug targets. J Exp Med 189:1425–1436

Harth G, Horwitz MA (2003) Inhibition of *Mycobacterium tuberculosis* glutamine synthetase as a novel antibiotic strategy against tuberculosis: demonstration of efficacy in vivo. Infect Immun 71:456–464

Heiges M, Wang H, Robinson E, Aurrecoechea C, Gao X, Kaluskar N, Rhodes P, Wang S, He CZ, Su Y, Miller J, Kraemer E, Kissinger JC (2006) CryptoDB: a *Cryptosporidium* bioinformatics resource update. Nucl Acids Res 34:D419–D422

Hemphill A, Mueller J, Esposito M (2006) Nitazoxanide, a broad-spectrum thiazolide anti-infective agent for the treatment of gastrointestinal infections. Expert Opin Pharmacother 7:953–964

Hewitt RG, Yiannoutsos CT, Higgs ES, Carey JT, Geiseler PJ, Soave R, Rosenberg R, Vazquez GJ, Wheat LJ, Fass RJ, Antoninievic Z, Walawander AL, Flanigan TP, Bender JF (2000) Paromomycin: no more effective than placebo for treatment of cryptosporidiosis in patients with advanced human immunodeficiency virus infection. AIDS Clinical Trial Group. Clin Infect Dis 31:1084–1092

Hijjawi NS, Meloni BP, Ng'anzo M, Ryan UM, Olson ME, Cox PT, Monis PT, Thompson RC (2004) Complete development of *Cryptosporidium parvum* in host cell-free culture. Int J Parasitol 34:769–777

Hijjawi NS, Meloni BP, Ryan UM, Olson ME, Thompson RC (2002) Successful in vitro cultivation of *Cryptosporidium andersoni*: evidence for the existence of novel extracellular stages in the life cycle and implications for the classification of Cryptosporidium. Int J Parasitol 32:1719–1726

Hommer V, Eichholz J, Petry F (2003) Effect of antiretroviral protease inhibitors alone, and in combination with paromomycin, on the excystation, invasion and in vitro development of *Cryptosporidium parvum*. J Antimicrob Chemother 52:359–364

Iseki M (1979) *Cryptosporidium felis* sp. n. (Protozoa: Eimeriorina) from the domestic cat. Jpn J Parasitol 28:285–307

Jervis HR, Merrill TG, Sprinz H (1966) Coccidiosis in the guinea pig small intestine due to a *Cryptosporidium*. Am J Vet Res 27:408–414

Kadappu KK, Nagaraja MV, Rao PV, Shastry BA (2002) Azithromycin as treatment for cryptosporidiosis in human immunodeficiency virus disease. J Postgrad Med 48:179–181

Korich DG, Mead JR, Madore MS, Sinclair NA, Sterling CR (1990) Effects of ozone, chlorine dioxide, chlorine, and monochloramine on *Cryptosporidium parvum* oocyst viability. Appl Environ Microbiol 56:1423–1428

Koudela B, Modry D (1998) New species of *Cryptosporidium* (Apicomplexa: Cryptosporidiidae) from lizards. Folia Parasitol 45:93–100

Kubota T, Iinuma Y, Kobayashi J (2006) Cloning of polyketide synthase genes from amphidinolide-producing dinoflagellate *Amphidinium* sp. Biol Pharm Bull 29:1314–1318

LaGier MJ, Tachezy J, Stejskal F, Kutisova K, Keithly JS (2003) Mitochondrial-type iron-sulfur cluster biosynthesis genes (IscS and IscU) in the apicomplexan *Cryptosporidium parvum*. Microbiology 149:3519–3530

Levine ND (1961) Protozoan parasites of domestic animals and of man. Burgess, Minneapolis

Levine ND (1980) Some corrections of coccidian (Apicomplexa: Protozoa) nomenclature. J Parasitol 66:830–834

Levine ND (1984) Taxonomy and review of the coccidian genus *Cryptosporidium* (protozoa, apicomplexa). J Protozool 31:94–98

Levine ND (1985) Phylum II. Apicomplexa Levine, 1970. In: Lee JJ, SH Hunter, EC Bovee (ed.) An illustrated guide to the Protozoa. Allen Press, Lawrence, KS, pp 322–374

Lindsay DS, Upton SJ, Owens DS, Morgan UM, Mead JR, Blagburn BL (2000) Cryptosporidium andersoni n. sp. (Apicomplexa: Cryptosporiidae) from cattle, Bos taurus. J Eukaryot Microbiol 47:91–95

Liu C, Vigdorovich V, Kapur V, Abrahamsen MS (1999) A random survey of the *Cryptosporidium parvum* genome. Infect Immun 67:3960–3969

Marshall MM, Naumovitz D, Ortega Y, Sterling CR (1997) Waterborne protozoan pathogens. Clin Microbiol Rev 10:67–85

McManus DP, Bowles J (1996) Molecular genetic approaches to parasite identification: their value in diagnostic parasitology and systematics. Int J Parasitol 26:687–704

Meisel JL, Perera DR, Meligro C, Rubin CE (1976) Overwhelming watery diarrhea associated with a cryptosporidium in an immunosuppressed patient. Gastroenterology 70:1156–1160

Menting JG, Tilley L, Deady LW, Ng K, Simpson RJ, Cowman AF, Foley M (1997) The antimalarial drug, chloroquine, interacts with lactate dehydrogenase from *Plasmodium falciparum*. Mol Biochem Parasitol 88:215–224

Morgan-Ryan UM, Fall A, Ward LA, Hijjawi N, Sulaiman I, Fayer R, Thompson RC, Olson M, Lal A, Xiao L (2002) *Cryptosporidium hominis* n. sp. (Apicomplexa: Cryptosporidiidae) from *Homo sapiens*. J Eukaryot Microbiol 49:433–440

Morgan-Ryan UM, Monis P, Possenti A, Crisanti A, Spano F (2001) Cloning and phylogenetic analysis of the ribosomal internal transcribed spacer-1 (ITS1) of Cryptosporidium wrairi and its relationship to *C. parvum* genotypes. Parassitologia 43:159–163

Nime FA, Burek JD, Page DL, Holscher MA, Yardley JH (1976) Acute enterocolitis in a human being infected with the protozoan *Cryptosporidium*. Gastroenterology 70:592–598

O'Donoghue PJ (1995) *Cryptosporidium* and cryptosporidiosis in man and animals. Int J Parasitol 25:139–195

Omoto CK, Toso M, Tang K, Sibley LD (2004) Expressed sequence tag (EST) analysis of *Gregarine* gametocyst development. Int J Parasitol 34:1265–1271

Pain A, Renauld H, Berriman M, Murphy L, Yeats CA, Weir W, Kerhornou A, Aslett M, Bishop R, Bouchier C, Cochet M, Coulson RM, Cronin A, de Villiers EP, Fraser A, Fosker N, Gardner M, Goble A, Griffiths-Jones S, Harris DE, Katzer F, Larke N, Lord A, Maser P, McKellar S, Mooney P, Morton F, Nene V, O'Neil S, Price C, Quail MA, Rabbinowitsch E, Rawlings ND, Rutter S, Saunders D, Seeger K, Shah T, Squares R, Squares S, Tivey A, Walker AR, Woodward J, Dobbelaere DA, Langsley G, Rajandream MA, McKeever D, Shiels B, Tait A, Barrell B, Hall N (2005) Genome of the host-cell transforming parasite *Theileria annulata* compared with *T. parva*. Science 309:131–133

Pande BP, Bhatia BB, Chauhan PP (1972) A new genus and species of cryptosporidiid Coccidia from India. Acta Vet Acad Sci Hung 22:231–234

Pankiewicz KW, Patterson SE, Black PL, Jayaram HN, Risal D, Goldstein BM, Stuyver LJ, Schinazi RF (2004) Cofactor mimics as selective inhibitors of NAD-dependent inosine monophosphate dehydrogenase (IMPDH) – the major therapeutic target. Curr Med Chem 11:887–900

Pankuch GA, Appelbaum PC (2006) Activities of tizoxanide and nitazoxanide compared to those of five other thiazolides and three other agents against anaerobic species. Antimicrob Agents Chemother 50:1112–1117

Perkins ME, Riojas YA, Wu TW, Le Blancq SM (1999) CpABC, a *Cryptosporidium parvum* ATP-binding cassette protein at the host-parasite boundary in intracellular stages. Proc Natl Acad Sci USA 96:5734–5739

Pfefferkorn ER, Bzik DJ, Honsinger CP (2001) *Toxoplasma gondii*: mechanism of the parasitostatic action of 6-thioxanthine. Exp Parasitol 99:235–243

Pradel G, Hayton K, Aravind L, Iyer LM, Abrahamsen MS, Bonawitz A, Mejia C, Templeton TJ (2004) A multidomain adhesion protein family expressed in *Plasmodium falciparum* is essential for transmission to the mosquito. J Exp Med 199:1533–1544

Read JA, Wilkinson KW, Tranter R, Sessions RB, Brady RL (1999) Chloroquine binds in the cofactor binding site of *Plasmodium falciparum* lactate dehydrogenase. J Biol Chem 274:10213–10218

Reduker DW, Speer CA, Blixt JA (1985) Ultrastructure of *Cryptosporidium parvum* oocysts and excysting sporozoites as revealed by high resolution scanning electron microscopy. J Protozool 32:708–711

Rein KS, Snyder RV (2006) The biosynthesis of polyketide metabolites by dinoflagellates. Adv Appl Microbiol 59:93–125

Riordan CE, Ault JG, Langreth SG, Keithly JS (2003) *Cryptosporidium parvum* Cpn60 targets a relict organelle. Curr Genet 44:138–147

Roberts CW, Roberts F, Henriquez FL, Akiyoshi D, Samuel BU, Richards TA, Milhous W, Kyle D, McIntosh L, Hill GC, Chaudhuri M, Tzipori S, McLeod R (2004) Evidence for mitochondrial-derived alternative oxidase in the apicomplexan parasite *Cryptosporidium parvum*: a potential antimicrobial agent target. Int J Parasitol 34:297–308

Rochelle PA, Jutras EM, Atwill ER, De Leon R, Stewart MH (1999) Polymorphisms in the beta-tubulin gene of

Cryptosporidium parvum differentiate between isolates based on animal host but not geographic origin. J Parasitol 85:986–989

Romero Cabello R, Guerrero LR, Munoz Garcia MR, Geyne Cruz A (1997) Nitazoxanide for the treatment of intestinal protozoan and helminthic infections in Mexico. Trans R Soc Trop Med Hyg 91:701–703

Rosales MJ, Cordon GP, Moreno MS, Sanchez CM (2005) Extracellular like-gregarine stages of *Cryptosporidium parvum*. Acta Trop 95:74–78

Rossignol JF (2006) Nitazoxanide in the treatment of acquired immune deficiency syndrome-related cryptosporidiosis: results of the United States compassionate use program in 365 patients. Aliment Pharmacol Ther 24:887–894

Rossignol JF, Ayoub A, Ayers MS (2001) Treatment of diarrhea caused by *Cryptosporidium parvum*: a prospective randomized, double-blind, placebo-controlled study of Nitazoxanide. J Infect Dis 184:103–106

Rossignol JF, Maisonneuve H (1984) Nitazoxanide in the treatment of *Taenia saginata* and *Hymenolepis nana* infections. Am J Trop Med Hyg 33:511–512

Rotte C, Stejskal F, Zhu G, Keithly JS, Martin W (2001) Pyruvate: NADP+ oxidoreductase from the mitochondrion of Euglena gracilis and from the apicomplexan *Cryptosporidium parvum*: a biochemical relic linking pyruvate metabolism in mitochondriate and amitochondriate protists. Mol Biol Evol 18:710–720

Ryan UM, Monis P, Enemark HL, Sulaiman I, Samarasinghe B, Read C, Buddle R, Robertson I, Zhou L, Thompson RC, Xiao L (2004) *Cryptosporidium suis* n. sp. (Apicomplexa: Cryptosporidiidae) in pigs (*Sus scrofa*). J Parasitol 90:769–773

Ryan UM, Xiao L, Read C, Sulaiman IM, Monis P, Lal AA, Fayer R, Pavlasek I (2003) A redescription of *Cryptosporidium galli* Pavlasek, 1999 (Apicomplexa: Cryptosporidiidae) from birds. J Parasitol 89:809–813

Sibley LD (2004) Intracellular parasite invasion strategies. Science 304:248–253

Sitja-Bobadilla A, Padros F, Aguilera C, Alvarez-Pellitero P (2005) Epidemiology of *Cryptosporidium molnari* in Spanish gilthead sea bream (Sparus aurata L.) and European sea bass (Dicentrarchus labrax L.) cultures: from hatchery to market size. Appl Environ Microbiol 71:131–139

Slapeta J, Keithly JS (2004) *Cryptosporidium parvum* mitochondrial-type HSP70 targets homologous and heterologous mitochondria. Eukaryot Cell 3:483–494

Slavin D (1955) *Cryptosporidium meleagridis* (sp. nov.). J Comp Pathol 65:262–266

Smith HV, Nichols RAB, Grimason AM (2005) *Cryptosporidium* excystation and invasion: getting to the guts of the matter. Trends Parasitol 21:133–142

Smith NH, Cron S, Valdez LM, Chappell CL, White AC Jr (1998) Combination drug therapy for cryptosporidiosis in AIDS. J Infect Dis 178:900–903

Snyder RV, Gibbs PD, Palacios A, Abiy L, Dickey R, Lopez JV, Rein KS (2003) Polyketide synthase genes from marine dinoflagellates. Mar Biotechnol (NY) 5:1–12

Snyder RV, Guerrero MA, Sinigalliano CD, Winshell J, Perez R, Lopez JV, Rein KS (2005) Localization of polyketide synthase encoding genes to the toxic dinoflagellate Karenia brevis. Phytochemistry 66:1767–1780

Spano F, Crisanti A (2000) *Cryptosporidium parvum*: the many secrets of a small genome. Int J Parasitol 30:553–565

Spano F, Puri C, Ranucci L, Putignani L, Crisanti A (1997) Cloning of the entire COWP gene of *Cryptosporidium parvum* and ultrastructural localization of the protein during sexual parasite development. Parasitology 114(Pt 5):427–437

Striepen B, Kissinger JC (2004) Genomics meets transgenics in search of the elusive *Cryptosporidium* drug target. Trends Parasitol 20:355–358

Striepen B, Pruijssers AJ, Huang J, Li C, Gubbels MJ, Umejiego NN, Hedstrom L, Kissinger JC (2004) Gene transfer in the evolution of parasite nucleotide biosynthesis. Proc Natl Acad Sci USA 101:3154–3159

Striepen B, White MW, Li C, Guerini MN, Malik SB, Logsdon JM Jr., Liu C, Abrahamsen MS (2002) Genetic complementation in apicomplexan parasites. Proc Natl Acad Sci USA 99:6304–6309

Strong WB, Nelson RG (2000a) Gene discovery in *Cryptosporidium parvum*: expressed sequence tags and genome survey sequences. Contrib Microbiol 6:92–115

Strong WB, Nelson RG (2000b) Preliminary profile of the *Cryptosporidium parvum* genome: an expressed sequence tag and genome survey sequence analysis. Mol Biochem Parasitol 107:1–32

Stwora-Wojczyk MM, Kissinger JC, Spitalnik SL, Wojczyk BS (2004) O-glycosylation in *Toxoplasma gondii*: identification and analysis of a family of UDP-GalNAc:polypeptide N-acetylgalactosaminyltransferases. Int J Parasitol 34:309–322

Sulaiman IM, Lal AA, Arrowood MJ, Xiao L (1999) Biallelic polymorphism in the intron region of beta-tubulin gene of *Cryptosporidium* parasites. J Parasitol 85:154–157

Suzuki T, Hashimoto T, Yabu Y, Kido Y, Sakamoto K, Nihei C, Hato M, Suzuki S, Amano Y, Nagai K, Hosokawa T, Minagawa N, Ohta N, Kita K (2004) Direct evidence for cyanide-insensitive quinol oxidase (alternative oxidase) in apicomplexan parasite *Cryptosporidium parvum*: phylogenetic and therapeutic implications. Biochem Biophys Res Commun 313:1044–1052

Tekos A, Prodromaki E, Papadimou E, Pavlidou D, Tsambaos D, Drainas D (2003) Aminoglycosides suppress tRNA processing in human epidermal keratinocytes in vitro. Skin Pharmacol Appl Skin Physiol 16:252–258

Templeton TJ, Iyer LM, Anantharaman V, Enomoto S, Abrahante JE, Subramanian GM, Hoffman SL, Abrahamsen MS, Aravind L (2004a) Comparative analysis of apicomplexa and genomic diversity in eukaryotes. Genom Res 14:1686–1695

Templeton TJ, Lancto CA, Vigdorovich V, Liu C, London NR, Hadsall KZ, Abrahamsen MS (2004b) The *Cryptosporidium* oocyst wall protein is a member of a multigene family and has a homolog in *Toxoplasma*. Infect Immun 72:980–987

Tetley L, Brown SM, McDonald V, Coombs GH (1998) Ultrastructural analysis of the sporozoite of *Cryptosporidium parvum*. Microbiology 144 (Pt 12):3249–3255

Thompson RC, Olson ME, Zhu G, Enomoto S, Abrahamsen MS, Hijjawi NS (2005) *Cryptosporidium* and cryptosporidiosis. Adv Parasitol 59:77–158

Tosini F, Agnoli A, Mele R, Gomez Morales MA, Pozio E (2004) A new modular protein of *Cryptosporidium parvum*, with ricin B and LCCL domains, expressed in the sporozoite invasive stage. Mol Biochem Parasitol 134:137–147

Triffitt MJ (1925) Observations on two new species of coccidia parasitic in snakes. Protozoology 1:19–26

Tyzzer EE (1907) A sporozoan found in the peptic glands of the common mouse. Proc Soc Exp Biol Med 5:12–13

Tyzzer EE (1912) *Cryptosporidium parvum* (sp. nov.), a coccidium found in the small intestine of the common mouse. Arch Protisenkd 26:394–412

Tzipori S (1998) Cryptosporidiosis: laboratory investigations and chemotherapy. Adv Parasitol 40:187–221

Tzipori S, Ward H (2002) Cryptosporidiosis: biology, pathogenesis and disease. Microbes Infect 4:1047–1058

Umemiya R, Fukuda M, Fujisaki K, Matsui T (2005) Electron microscopic observation of the invasion process of *Cryptosporidium parvum* in severe combined immunodeficiency mice. J Parasitol 91:1034–1039

Upton SJ (2000) Suborder Eimeriorina Leger, 1911. In: Lee JJ, Leedale GF, Bradbury P (ed) The Illustrated Guide to the Protozoa, 2nd edn. Society of Protozoologists, Allen Press, Lawrence, KS, USA

Vetterling JM, Jervis HR, Merrill TG, Sprinz H (1971) *Cryptosporidium wrairi* sp. n. from the guinea pig *Cavia porcellus*, with an emendation of the genus. J Protozool 18:243–247

White AC Jr (2003) Nitazoxanide: an important advance in anti-parasitic therapy. Am J Trop Med Hyg 68:382–383

Widmer G (2004) Population genetics of *Cryptosporidium parvum*. Trends Parasitol 20:3–6 (discussion 6)

Widmer G, Tchack L, Chappell CL, Tzipori S (1998) Sequence polymorphism in the beta-tubulin gene reveals heterogeneous and variable population structures in *Cryptosporidium parvum*. Appl Environ Microbiol 64:4477–4481

Xiao L, Fayer R, Ryan U, Upton SJ (2004) *Cryptosporidium* taxonomy: recent advances and implications for public health. Clin Microbiol Rev 17:72–97

Xiao L, Limor J, Morgan UM, Sulaiman IM, Thompson RC, Lal AA (2000a) Sequence differences in the diagnostic target region of the oocyst wall protein gene of *Cryptosporidium* parasites. Appl Environ Microbiol 66:5499–5502

Xiao L, Morgan UM, Fayer R, Thompson RC, Lal AA (2000b) *Cryptosporidium* systematics and implications for public health. Parasitol Today 16:287–292

Xu P, Widmer G, Wang Y, Ozaki LS, Alves JM, Serrano MG, Puiu D, Manque P, Akiyoshi D, Mackey AJ, Pearson WR, Dear PH, Bankier AT, Peterson DL, Abrahamsen MS, Kapur V, Tzipori S, Buck GA (2004) The genome of *Cryptosporidium hominis*. Nature 431:1107–1112

Zapata F, Perkins ME, Riojas YA, Wu TW, Le Blancq SM (2002) The *Cryptosporidium parvum* ABC protein family. Mol Biochem Parasitol 120:157–161

Zhu G (2004) Current progress in the fatty acid metabolism in *Cryptosporidium parvum*. J Eukaryot Microbiol 51:381–388

Zhu G, Keithly JS, Philippe H (2000a) What is the phylogenetic position of *Cryptosporidium*? Int J Syst Evol Microbiol 50 (Pt 4):1673–1681

Zhu G, LaGier MJ, Stejskal F, Millership JJ, Cai X, Keithly JS (2002) *Cryptosporidium parvum*: the first protist known to encode a putative polyketide synthase. Gene 298:79–89

Zhu G, Li Y, Cai X, Millership JJ, Marchewka MJ, Keithly JS (2004) Expression and functional characterization of a giant Type I fatty acid synthase (CpFAS1) gene from *Cryptosporidium parvum*. Mol Biochem Parasitol 134: 127–135

Zhu G, Marchewka MJ, Keithly JS (2000b) *Cryptosporidium parvum* appears to lack a plastid genome. Microbiology 146(Pt 2):315–321

Zhu G, Marchewka MJ, Woods KM, Upton SJ, Keithly JS (2000c) Molecular analysis of a Type I fatty acid synthase in *Cryptosporidium parvum*. Mol Biochem Parasitol 105: 253–260

6 *Theileria*

Richard P. Bishop[1](✉), David O. Odongo[1], David J. Mann[2], Terry W. Pearson[3], Chihiro Sugimoto[4], Lee R. Haines[3,5], Elizabeth Glass[6], Kirsty Jensen[6], Ulrike Seitzer[7], Jabbar S. Ahmed[7], Simon P. Graham[1,8], and Etienne P. de Villiers[1]

[1] The International Livestock Research Institute (ILRI), P.O. Box 30709, Nairobi 00100, Kenya, r.bishop@cgiar.org
[2] Department of Biochemistry, Imperial College, South Kensington, London SW7 2AZ, UK
[3] Department of Biochemistry and Microbiology Petch Building, University of Victoria, Victoria, British Columbia, Canada V8W 3P6
[4] Department of Veterinary Science, University of Hokkaido, Sapporo, Japan
[5] Present address: Liverpool School of Tropical Medicine, Pembroke Place, Liverpool, L3 5QA UK
[6] The Roslin Institute and Royal (Dick) School of Veterinary Studies, University of Edinburgh, Roslin Biocentre, Roslin, Midlothian, EH25 9PS, UK
[7] Division of Veterinary Infection Biology and Immunology, Parkallee 22, Research Center Borstel, 23845 Borstel, Germany
[8] Present address: Virology Department, Veterinary Laboratories Agency, Woodham Lane, New Haw, Addlestone, Surrey, KT15 3NB, United Kingdom

6.1 Introduction

6.1.1 The Genus *Theileria*

Theileria are tick-transmitted intracellular protozoa that induce a spectrum of disease syndromes in domestic livestock including large and small ruminants. *Theileria* are also present in many wild ruminant species, which are presumed to represent the ancestral mammalian hosts. *Theileria* can be responsible for fatalities in the wildlife "reservoirs," although infections in wild ruminants typically do not result in obvious clinical symptoms. *Theileria* are unique among protozoa, in that certain species are capable of immortalizing either lymphocytes or cells of the monocyte/macrophage lineage in the mammalian species that they infect, in a reversible manner. Another novel aspect of *Theileria* biology is that the cell cycle is synchronized with that of the mammalian host. *Theileria* has been included within a "subphylum" designated the Apicomplexa, which includes malaria parasites in the genus *Plasmodium*. This classification is based on the shared possession of an apical complex containing secretory organelles involved in invasion, or establishment, in the cells of the mammalian and invertebrate hosts. However, the evolutionary and functional equivalence of the apical complex between different subphylum members is unclear. Unlike other apicomplexa, *Theileria* exists free in the cytoplasm of the host cell and is not enclosed within a parasitophorous vacuole of host cell origin. In *Theileria* the apical complex is believed to be involved in the establishment of the host cell, rather than the entry process as in other apicomplexa. In common with several other Apicomplexa including *Plasmodium* and *Toxoplasma*, *Theileria* contains an apicoplast, a chloroplast-like organelle believed to be derived from an algal symbiont, with its own circular genome. Sequence data indicate that *Theileria* is only distantly related to *Plasmodium*. However, certain aspects of both the life cycle and the immune response of the mammalian host to infection are similar between *Theileria* and *Plasmodium*.

Theileria species that infect cattle and small ruminants are transmitted by ixodid ticks of the genera *Rhipicephalus*, *Amblyomma*, *Hyalomma*, and *Haemaphysalis*. Globally, the most economically important species are *T. parva* and *T. annulata* that infect cattle in sub-Saharan Africa, and Europe, North Africa, and South Asia, respectively, frequently resulting in lethal outcomes, particularly in exotic *Bos taurus* animals. The disease history, biology, life cycle, and control strategies for these two highly pathogenic *Theileria* species have been the subject of regular reviews over the past two decades (Irvin and Morrison 1987; Norval

Genome Mapping and Genomics in Animal-Associated Microbes
V. Nene, C. Kole (Eds.)

et al. 1991; Preston et al. 1999; Bishop et al. 2004). The *Theileria orientalis* complex, which occurs primarily in eastern Asia, is less pathogenic but causes anemia in cattle. A considerable amount of data has been generated in recent years on the phylogeny, distribution, and biology of pathogenic small ruminant *Theileria*, whose economic importance has probably been underestimated in the past.

The genome sequences of *T. parva* and *T. annulata* have recently been determined (Gardner et al. 2005; Pain et al. 2005) and transcriptional analyses performed using massively parallel signature sequencing (MPSS) and expressed sequence tag (EST) approaches for the schizont stages of *T. parva* and *T. annulata*, respectively (Bishop et al. 2005; Pain et al. 2005). Preliminary data are also available from a pilot *T. parva* proteomics project. Functional genomic analyses of the interaction of *Theileria* with the tick vector and the mammalian host have also been initiated. The emphasis of this chapter will be on the nature and biological implications of the *T. parva* and *T. annulata* genome sequences, the application of the genome data for candidate antigen identification and parasite population genetics, together with a summary of recent functional genomics research.

6.1.2 Economic Importance of *Theileria*

The most widely quoted estimate for the economic impact of ticks and tick-borne diseases (TBD) in the livestock sector, of which losses induced by *Theileria* represent a significant component, is US$ 17 million (De Castro et al. 1997). Formal economic analyses have been performed for *T. parva* that is transmitted predominantly by *Rhipicephalus appendiculatus* ticks in East Africa and by *R. zambeziensis* in southern Africa. The disease was estimated to have been responsible for $170 million of economic loss in 1989 alone (Mukhebi and Perry 1992) and limits introduction of more productive exotic (*Bos taurus*) and crossbred cattle in much of eastern, central, and southern Africa. The likely primary ancestral reservoir host of *T. parva* is the African Cape buffalo (*Syncerus caffer*), in which the parasite typically does not cause apparent disease, despite frequent infection with multiple genotypes (reviewed by Bishop et al. 2002). *T. annulata* is presumed to originate from Asian water buffalo (*Bubalus bubalis*), in which infection frequently results in mild disease; however, fatalities have been observed in water buffalo (Osman and Al Gaabary 2007). *T. annulata* is transmitted by several *Hyalomma* tick species and is responsible for tropical theileriosis from southern Europe to East Asia, a vast region in which 250 million cattle, and also large numbers of water buffalo, with varying degrees of introgression of *Bos indicus* genetic material, are at risk. It has been estimated that in India alone *T. annulata* may be responsible for US$800 million of economic loss (Brown 1997). The *T. orientalis*/*T. sergenti* complex, although relatively mildly pathogenic, is thought to be responsible for significant economic loss in Japan (Sugimoto 1997).

Small ruminants have considerable global economic significance. Two out of three of the estimated three billion ruminants on the planet are sheep and goats (Steinfeld et al. 2006). A report of FAO Expert Consultants in 1997 endorsed "the global significance of the small ruminant livestock industry both from the viewpoint of national economies and for individual resource-poor farmers in semi-arid zones." Diseases of small ruminants are among the least recognized problems in veterinary science; particularly in the case of ticks and tick-borne diseases. As highlighted in the report of the FAO Expert Consultants "even where it seems that the potential impact from introduction of indigenous breeds in enzootic areas is slight, there may be a great loss of potential production because of the difficulties in promoting up-grading schemes using more productive but susceptible exotic breeds."

6.1.3 *Theileria* Taxonomy

Phylogenetic analysis based on ribosomal RNA gene sequences demonstrates that *Theileria* is most closely related to *Babesia*, a genus of tick-borne protozoa infective to erythrocytes but not to leukocytes of mammals, including domestic livestock (Allsopp et al. 1994). Analyses based on ribosomal RNA are unambiguous in the separation of *T. parva* and *T. annulata*, although considerable intraspecific molecular polymorphism is observed at other loci (Bishop et al. 2000a; Gubbels et al. 2002; Taylor et al. 2003). However, molecular data based on 18S rDNA sequencing have recently provided new insights into the phylogeny of Asian *Theileria* species infective to both cattle and small ruminants (Chansiri et al. 1999).

T. sergenti and T. orientalis in Cattle

Taxonomy of the "benign" *Theileria* group parasites of cattle, found in East Asia and Siberia that result in limited pathology following infection, remains confusing. Morphologically indistinguishable, genetically related parasites that induce pathology in cattle and water buffalo assigned to the taxon *T. orientalis* are found in Asia and other parts of the world. The distribution of *T. buffeli*, which is closely related to *T. orientalis*, appears to be worldwide, but this species is less frequently associated with clinical disease. In addition, there are two genetic types within *T. orientalis*, currently named Chitose and Ikeda, which are distinguishable by differences in the 18S small subunit ribosomal RNA genes. These two genotypes of *T. orientalis* may represent different subspecies or species, but there is currently no consensus in the taxonomy of *T. orientalis*. Mixed infections of the different types of *T. orientalis* and *T. buffeli* in cattle are common and in vitro methods for culturing the intra-erythrocytic piroplasm or intra-lymphocytic schizont stages have not been established, creating difficulty in the selection of clonal parasites for use in taxonomic studies.

Theileria Infections in Domestic Small Ruminants

Small ruminant theilerioses in sheep and goats are caused by a variety of *Theileria* species. The parasites are transmitted trans-stadially by a range of tick species. A minimum of six *Theileria* species infects small ruminants. *T. separata*, *T. ovis*, and *T. recondita* are nonpathogenic, whereas *T. lestoquardi*, *Theileria* sp. 1 (China), and *Theileria* sp. 2 (China) are pathogenic. Inferred phylogenetic trees indicate that the 18S rRNA sequences of the Chinese parasites fall inside the clade consisting of *Theileria* species, but are clearly divergent to *T. lestoquardi*. The nonpathogenic species are *T. separata* that infects sheep in eastern and southern Africa and is transmitted by *Rhipicephalus evertsi* (Schnittger et al. 2003). *T. ovis* from sheep and goats from Africa, Europe, and Asia are transmitted primarily by *R. evertsi* (Neitz 1972; Uilenberg 1981) and also by *Rhipicephalus bursa* (Aktas et al. 2006). *T. ovis* isolates originating from entirely different geographic regions, including sub-Saharan Africa and West Asia, cluster in a phylogenetic tree based on 18S rRNA genes, suggesting that *T. ovis* represent a single taxon (Schnittger et al. 2003; Sparagano et al. 2006). The identity and validity of the third nonpathogenic species *T. recondita* is unclear. Recently, 18S sequence data have revealed additional genotypes among small ruminant *Theileria* isolated from asymptomatic animals in Spain and eastern Turkey (Nagore et al. 2004; Altay et al. 2007). The pathogenic small ruminant species include *T. lestoquardi* (previously known as *T. hirci*), a highly pathogenic *Theileria* species for sheep and to lesser degree for goats. The distribution includes northern Africa, southern Europe, and the Middle East (Hooshmand-Rad and Hawa 1973b; Salih et al. 2003). *T. lestoquardi* is trans-stadially transmitted by *Hyalomma anatolicum anatolicum* (Hooshmand-Rad and Hawa 1973a). Recently, two *Theileria* species have been identified in China. Biological and phylogenetic studies based on the analysis of the small subunit ribosomal RNA gene (SSUrRNA) have revealed that one of these parasites is closely related to *Theileria buffeli* (Schnittger et al. 2000) and divergent from *T. lestoquardi* and *T. annulata*. This parasite was designated as *Theileria* spp. (China) (Yin et al. 2002). Further analyses showed that besides the previously described *Theileria* species (China) another distinct pathogenic and phylogenetically unrelated *Theileria* parasite occurs in the same region. Both are transmitted by the same tick vector *Haemaphysalis qinghaiensis*. Accordingly, they were first designated as Theileria sp. 1 (China) and Theileria sp. 2 (China), and have recently been named *T. luwenshuni* and *T. uilenbergi*, respectively (Yin et al. 2007). Very recently, the transmission of *Theileria luwenshuni* by *Haemaphysalis longicornis* has also been demonstrated (Li et al. 2007). Both parasites are pathogenic particularly for lambs and imported exotic animals, with a mortality rate of up to 75.4%.

A similar parasite to *Theileria luwenshuni* has been reported to occur in Spain and has been designated as *Theileria* sp. OT1 showing a 99.6% similarity with *Theileria luwenshuni*. In contrast to the Chinese parasite, *Theileria* spp. OT1 from Spain seems to be of relatively low pathogenicity (Nagore et al. 2004). Recently, 18S sequence data have revealed additional genotypes among small ruminant infective *Theileria* isolated from asymptomatic animals in eastern Turkey (Altay et al. 2007).

6.1.4 Epidemiology and Pathology of *Theileria* Infections

Both *T. parva* and *T. annulata* schizonts, the life cycle stage that is the major cause of pathology in *Theileria* infections (Irvin and Morrison 1987), induce a transformation-like phenotype in nucleated mammalian host cells and are therefore described as lymphoproliferative diseases. In *T. annulata* infections, anemia in the intra-erythrocytic stage is also important in pathology, although the mechanisms responsible remain unclear (Campbell and Spooner 1999). Several other nontransforming *Theileria* species are also responsible for disease in cattle, primarily as a result of anemia induced by the intra-erythrocytic stage. *T. orientalis* and *T. buffeli* cause disease in cattle in East Asia. *Theileria mutans* may be responsible for disease in cattle in sub-Saharan Africa, particularly in mixed infections with other pathogens, but the economic impact of this species has yet to be fully assessed. Recent studies in Uganda (Oura et al. 2004a) and southern Sudan (Salih et al. 2007) using reverse-line blotting indicated a high prevalence of both *T. mutans* and *T. velifera* in cattle. It was suggested that due to the apparent absence of concurrent *T. parva* infections in the former study, *T. mutans* might induce clinical theileriosis in some cattle in Uganda. Immunosuppresion due to concurrent trypanosome infections may also promote theileriosis in these situations. Most wild bovids in Africa are infected by one or more species of *Theileria* and the epidemiological situation can be complex, with East African cattle potentially being infected with up to five *Theileria* species at one time (Norval et al. 1992) in addition to tick-transmitted rickettsial pathogens (Oura et al. 2004a). Cattle can be infested with several species of tick and more than one species of *Theileria* can also be transmitted by the same tick. For example, the morphologically very similar sporozoites of *T. parva* and the normally nonpathogenic *T. taurotragi* can occur together in *R. appendiculatus* salivary glands, although they are distinguishable by molecular methods, such as using oligonucleotides that hybridize to ribosomal RNA (Bishop et al. 1994). In most cases, the disease caused by *T. lestoquardi* is severe in sheep, leading to high mortality rates. The symptoms include high fever, anorexia and emaciation, diarrhea, and enlarged superficial lymph nodes. The pathological findings recorded for small ruminant theilerioses are generally similar to those described for tropical theileriosis (Neitz 1953; Hooshmand-Rad and Hawa 1973b; Tageldin et al. 1992).

Additional *Theileria* species that are infective to livestock and possibly pathogenic in specific circumstances continue to be discovered. The range of *T. buffeli* has been extended to Africa (Ngumi et al. 1994). A *Theileria* species originally described from schizont-infected lymphocyte culture isolates from Cape buffalo (Conrad et al. 1987; Allsopp et al. 1993) has recently been found in cattle subjected to challenge by ticks that are presumed to have fed on buffalo (Bishop and Musoke, unpublished data). *Theileria* infections have also been associated with bovine fatalities in the United States, and the presence of *Theileria* was confirmed in both *Amblyomma americanum* and *Dermacentor variabilis* by PCR amplification using primers derived from small subunit rRNA (SSU rRNA) genes sequences (Chae et al. 1999).

6.1.5 *Theileria* Life Cycle and Cell Biology

Theileria life-cycle stages in the tick vector and bovine host have been reviewed previously (Norval et al. 1992; Shaw 2002; Bishop et al. 2004) and the reader is referred to these articles for additional details. The life cycles of all species are generally similar (Shaw 2002). The three-host brown ear tick *Rhipicephalus appendiculatus* is the major vector of *T. parva*, which has a restricted tick and mammalian host range. There is no laboratory animal model. *R. appendiculatus* is classified as a three-host tick, because the larvae, nymphs, and adults feed on different hosts, which are not necessarily cattle. Transmission is trans-stadial; larval or nymphal instars of the tick acquire infections from a blood meal; and the pathogen is then transmitted to a new host, after moulting, by nymphs or adults, respectively. Transmission of *Theileria* has not been shown to be transovarian, unlike the related protozoan *Babesia* that is transmitted by the one host tick *Boophilus. Hyalomma anatolicum*, the major vector of *T. annulata*, is a two-host tick, with larval and nymphal instars feeding on the same host. Sporogony takes place in the tick salivary glands in response to a combination of increased temperature and components present within the blood meal. Infective sporozoites are released from about day 3 to day 7 of tick feeding. One infected cell may contain 40–50,000 sporozoites, but the mechanics of tick feeding are thought to result in a phased release of sporozoites from the salivary glands (Shaw and Tilney 1995). A tick with a single infected acinar cell can potentially cause severe disease and death. However, the severity of

East Coast fever (ECF) is sporozoite dose dependent, and individual animals exhibit different thresholds of susceptibility. Some animals undergo a mild disease reaction and recover due to development of protective cellular immune responses.

Once introduced into the host, *T. parva* sporozoites invade a restricted range of cattle and buffalo cells. In vitro these include purified B cells or T cells ($CD4^+$, $CD8^+$ and $CD4^-/CD8^-$) (Baldwin et al. 1988). In addition, *T. parva* can also enter afferent lymph veiled cells and monocytes (Shaw et al. 1993), although these cell types cannot subsequently be immortalized. The entry process is probably energy-dependent and is likely to involve specific receptors located on the host cells (Shaw 1997, 2002). *T. annulata* infects cells of the monocyte/macrophage lineage, although it can also enter and immortalize B cells. It has been suggested that *T. annulata* modulates antigen presentation by macrophages inducing aberrant responses by $CD4^+$ and $CD8^+$ T cells that may promote parasite survival (Campbell and Spooner 1999).

Other species, particularly *T. taurotragi*, originally isolated from eland (*Taurotragus oryx*), appear to have a much wider host cell range, in vitro (Stagg et al. 1983). Like viruses transmitted by *R. appendiculatus*, components of tick saliva appear to promote *T. parva* pathogen transmission by facilitating sporozoite entry into host leukocytes (Shaw et al. 1993). Unlike other apicomplexans, *Theileria* sporozoites are not motile. They do not possess all components of the typical apical complex described for other apicomplexa, entry into the host cells is not orientation specific, and the rhoptries and microspheres are involved in establishment in the host cell subsequent to entry, rather than in the entry process (Shaw et al. 1991). Unlike other apicomplexa, *Theileria* sporozoites also appear to lack morphologically distinguishable micronemes (Shaw et al. 1991). In *Theileria* infections, the newly internalized parasite is surrounded by a host cell membrane, which is rapidly (<15 min) removed by the parasite, thereby evading certain intracellular host cell defense mechanisms. In this respect, *Theileria* is similar to a phylogenetically diverse range of intracellular pathogens, including *Listeria, Shigella*, and *Trypanosoma cruzi* (Andrews and Webster 1991), presumably as a result of independent evolution of this solution to the problem of intracellular survival. Rapid dissolution of the host cell membrane by *T. parva* is closely correlated with the discharge of the contents of the rhoptries and microspheres. The major surface antigens of *T. parva* sporozoites – a 67 kDa protein and bovine MHC class I molecules, are implicated in the entry process, as monoclonal antibodies (mAbs) to these proteins inhibit sporozoite entry (Dobbelaere et al. 1984; Musoke et al. 1984; Shaw et al. 1991). Because of the restricted host cell specificity, it seems unlikely that MHC class I molecules constitute the sole recognition site of sporozoites. Sporozoite organelles that discharge their contents after entry are likely to contain a microtubule nucleating factor, as the free intracellular sporozoites are rapidly surrounded by a network of microtubules.

The *Theileria* schizont is a multinucleated, syncytial body. Association of the schizont with the host cell nuclear spindle ensures that daughter host cells remain infected during cytokinesis (Carrington et al. 1995; Shiels et al. 2006). Although the parasite and host cells divide in synchrony, schizont DNA synthesis occurs as the host cell enters mitosis and is immediately followed by division when the host cell is in metaphase (Irvin et al. 1982). The schizont stage of both *T. parva* and *T. annulata* is the major cause of pathology and disease. Merogony occurs within infected lymphocytes (Shaw and Tilney 1992) and merozoites are released by host cell rupture. The merozoites invade red blood cells (RBCs) in a manner similar to that described for sporozoite entry into lymphocytes (Shaw and Tilney 1995), and differentiate into piroplasms, which, like the intra-lymphocytic schizont, lie free in the erythrocyte cytoplasm. There is very little multiplication of the piroplasm stage of *T. parva*, although intra-erythrocytic multiplication of the piroplasm occurs in *T. annulata* infections. Other species of *Theileria*, especially the nontransforming *T. mutans, T. buffeli*, and *T. sergenti*, exhibit high levels of intra-erythrocytic merogony, resulting in anemia and pathology due to RBC lysis. Ticks become infected after feeding on blood containing infected RBCs. Gametogenesis to generate gametes, which are apparently not morphologically distinguishable as mega- and microgametes, followed by syngamy occur in the tick gut. Gametes fuse to form a zygote that invades the cells of the gut epithelium. Kinetes are released into haemolymph and invade cells of the salivary gland. Sporogony is initiated by stimuli received following tick attachment to the mammalian host and appears closely linked with the moulting cycle of the tick. All stages of the life cycle are haploid except for the zygote in the tick gut.

The life cycle of *Theileria lestoquardi* is broadly similar to that of *T. parva* and *T. annulata*. After being inoculated by the ticks, the sporozoites invade the host mononuclear cells rapidly within a few minutes. The sporozoites differentiate into a multinucleate schizont and the infected cells are immortalized and expand clonally. The transformed host cell divides rapidly and at a regular rate and can be cultured in vitro (Hooshmand-Rad and Hawa 1975). Subsequently, merogony occurs in a proportion of schizont-infected cells. Maturation of the nuclei takes place with the ultimate formation of the merozoite stage. This development destroys the host cell and the merozoites are free to invade the erythrocytes. An interesting biological difference from *T. parva* and *T. annulata* in one species of small ruminant-infective *Theileria* is that the schizonts of the *Theileria* China sp.1 are rarely found in the host cell cytoplasm and that most appear to be located outside the cells (Yin et al. 2003).

6.2 Physical Mapping of the *T. parva* Genome

A physical map of the *T. parva* genome, the first to be produced for any protozoan parasite, was developed using a linking clone strategy to order the 33 *Sfi*I fragments within the genome. Initially, agarose-embedded genomic DNA was digested with the eight base-pair recognition site restriction enzyme *Sfi*I and size fractionated on pulsed field gradient gel electrophoresis (CHEF) gels (Morzaria and Young 1992). The innovative strategy employed to determine which fragments were linked to one another involved production of a genomic library from *T. parva* DNA that had been partially digested using *Sau*IIIA, followed by digestion of this library with the restriction enzyme *Sfi*I, physical separation, and subsequent purification of the *Sfi*I linearized DNA fragments from uncut plasmid DNA on low melting point agarose gels. The cloned sequences containing *Sfi*I sites were then used to probe Southern blots of CHEF fractionated *Sfi*I-digested *T. parva* genomic fragments to determine linkage. The quality and accuracy of the empirically generated *Sfi*I restriction map are clearly demonstrated by the fact that, although there are scaling errors, in that not all the PFGE fragment sizes are 100% equivalent in size to an *Sfi*I map derived by in silico analysis from the *T. parva* genome sequence, these differences are minor. The differences are due to the fact that fragment sizes deduced from mobility on pulsed field gradient gels are not entirely accurate. In addition, there is an 8.0 kb region that was almost certainly overlooked because the size of the *Sfi*I fragment was too small for CHEF gel analysis. However, the overall concordance between the physical map and that derived from identifying the location of *Sfi*I sites in the genome sequence is remarkably good.

6.3 The *Theileria parva* and *T. annulata* Genome Sequences

The complete *T. parva* and *T. annulata* genomes were determined using shotgun sequencing approaches. The annotated genomes can be accessed at *T. annulata* (http://www.genedb.org/genedbannulata/index.jsp) and the *T. parva* (http://www.tigr.org/tbd/e2k1/tpa1) websites. Both genomes are approximately 8.3 Mbp in size and contain only four chromosomes (Table 1). The *T. parva* sequence was derived from the Muguga stock that was originally isolated in the 1960s and has been maintained by experimental passage through ticks and cattle since that time. Chromosomes 1 and 2 were completely sequenced. Chromosome 4 was sequenced to completion with the exception of a small region encoding a highly repetitive sequence that could not be assembled. Chromosome 3 had a gap of >60 kbp located 570 kbp from the right telomere, containing a large paralogous gene family named Tpr, an acronym for *T. parva* repeat (Baylis et al. 1991).

Initial annotation predicted proteomes of 4,035 and 3,792 for *T. parva* and *T. annulata*, respectively (Gardner et al. 2005; Pain et al. 2005). The *T. parva* nuclear genome therefore encodes 7% more putative proteins than *T. annulata* and 20% fewer than *P. falciparum*, which has approximately 5,000 predicted genes. Both *T. parva* (Table 1) and *T. annulata* exhibit higher gene density, a greater proportion of genes with introns, and shorter intergenic regions (Table 1) than other apicomplexa. Putative functions were assigned to 38% of the predicted proteins on the basis of BLAST identities over the complete predicted coding sequence (Table 1). It should, however, be noted that this figure

Table 1 Comparison of *Theileria parva* and *Theileria annulata* nuclear genome with the genomes of other apicomplexan protozoa

Features	Apicomplexan genomes			
	T. parva	*T. annulata*	*P. falciparum*	*C. parvum*
size (bp)	8,308,027	8,351,610	22,853,764	9,100,000
Number of chromosomes	4	4	14	8
Total G+C content (%)	34.1	32.5	19.4	30
Number of protein encoding genes	4,035	3,792	5,268	3,807
Number of hypothetical proteins	2,498	2,122	3,208	925
Mean gene length (bp)	1,407	1,606	2,283	1,795
Percent coding	68.4	72.8	52.6	75.3
Genes with introns (%)	73.6	70.6	53.9	5
Mean intergenic length (bp)	405	396	1,694	566
G+C content intergenic regions (%)	26.1	22.5	13.6	23.9
tRNA genes	47	47	43	45
5S rRNA genes	3	3	3	6
rRNA units	2	2	7	5

Gene length excludes introns. Source of data for *Plasmodium falciparum* (Gardner et al. 2002), for *Cryptosporidium parvum* (Abrahamsen et al. 2004), and *Theileria annulata* (Pain et al. 2005)

excludes significant domain matches that did not cover the complete open reading frame (ORF).

Comparative analysis reveals that a number of metabolic pathways present in *Plasmodium* are absent in *Theileria*, presumably explaining the reduced coding potential. Enzymes that appear to be absent on the basis of database searches include those involved in the biosynthesis of porphyrin, polyamines, and shikimic acid. Enzymes involved in purine salvage and interconversion of many amino acids also appear to be lacking. Although virtually all glycolytic and TCA enzymes are present, the mechanistic connection between these pathways is unclear and there is as yet no proof that they are actually functional in ATP synthesis (Gardner et al. 2005).

Like *Plasmodium* and *Toxoplasma*, but unlike *Cryptosporidium*, *Theileria* possess an apicoplast believed to be derived from an algal symbiont. As in *Arabidopsis thaliana* and other Apicomplexa, proteins encoded in the nuclear genome are translocated into the apicoplast through a bipartite targeting system, although the predicted number is considerably lower in *Theileria* than the 551 predicted for *Plasmodium*. Phylogenomics analysis of four apicomplexan genomes suggests that the evolutionary history of the apicoplast and nuclear genomes is complex, with genes predicted to be of both algal nuclear, chloroplast (cyanobacterial) and proteobacterial origin identified (Huang et al. 2004). In addition, some genes currently targeted to the apicoplast do not appear to be ancestral plant genes. The coding potential of the *Theileria* apicoplast suggests that it is involved in isoprenoid biosynthesis, but not type II fatty acid or heme biosynthesis, unlike *Plasmodium* and *Toxoplasma* (Gardner et al. 2005). Although approximately 60% of the *Theileria* apicoplast proteins had a significant degree of identity with those of *Plasmodium*, this ranged from a minimum of 27% to a maximum of 61%. The limited extent of apicoplast protein identity was consistent with the relatively distant phylogenetic relationship of *Theileria* and *Plasmodium* indicated by comparison of the nuclear genomes. Approximately 40 of the 100 apicoplast targeted proteins shared between *Theileria* and *Plasmodium* were of unknown predicted function according to BLAST analyses, suggesting that the apicoplast may have as yet unknown additional biological roles.

The Tpr gene family of *T. parva* comprises a tandemly arrayed locus on chromosome 3, which is estimated to be approximately 100 kb in size, together with 11 additional dispersed copies. The tandemly arrayed locus could not be fully assembled but contained two unordered contigs of 41 and 13 kbp.

Additionally, 7 kb at one end of the tandem array is 100% identical in sequence to an earlier Tpr sequence derived from a bacteriophage lambda genomic library (Baylis et al. 1991) and is contiguous with a 570 kb contig comprising one end of chromosome 3. The Tar gene family of *T. annulata* contains Tpr-homologous C-terminal domains, but the Tar genes are dispersed within the genome and located on all four chromosomes. In a remarkable contrast with *T. parva* there is apparently no tandemly arrayed Tar locus in the *T. annulata* genome. Unlike the majority of hypervariable gene families in parasitic protozoa (Barry et al. 2003), neither the Tpr or Tar multicopy gene families are telomere-associated. Within the longest (41 kb) *T. parva* contig, seven of the Tpr ORFs appeared to be "complete" since they contained in-frame ATG codons <50 amino acids from the N-terminus of the protein. The other eight were "incomplete" since they lacked in-frame ATG codons within 50 amino acids of the presumptive N-terminus, suggesting that if they are expressed, the mechanism of translation must be unusual, or that there is genomic rearrangement. This confirmed the unusual organization of the Tpr locus based on limited sequence data described previously (Baylis et al. 1991). The Tpr proteins are the most rapidly evolving *T. parva* protein family, with radical differences frequently apparent in the N-terminal domain (Bishop et al. 1997). The exceptional N-terminal variability is reminiscent of the *var* genes of *P. falciparum*, which encode membrane proteins (PfEMP1) that are expressed on the erythrocyte surface and mediate capillary adherence, preventing clearance of infected cells by the spleen of the host (reviewed by Borst et al. 1995). The conserved C-terminal ends of the Tpr ORFs contain predicted membrane-associated helices. In addition, two of the ORFs are predicted to contain signal anchors. This is consistent with insertion of the C-terminal Tpr domain into either the erythrocyte or parasite membrane. However, the Tpr ORFs lack signal sequences that can be predicted using standard algorithms; therefore, if they are secreted the mechanism must be unusual.

Neither the *T. annulata* or the *T. parva* genomes contained obvious candidate "oncogenes," as indicated by homology searches of the public databases, suggesting that the mechanism of "reversible transformation" of the eukaryotic host cell induced by *Theileria* may differ from those previously described for virus-induced tumors. By comparison with *Plasmodium* spp., the *Theileria* genomes are relatively rich in introns, but noncoding repetitive sequences are infrequent and intergenic sequences are generally short. The genome sequence reveals that *T. parva* has two ribosomal gene coding units, which are identical in sequence; therefore, unlike *P. falciparum* (McCutchan et al. 1988), *Theileria* does not express distinct ribosomal RNAs that are differentially expressed between arthropod vector and mammalian host. The *T. annulata* genome also contains two single copy ribosomal loci that are syntenic with those of *T. parva* (Pain et al. 2005).

6.3.1 Telomere-Associated Sequences

The subtelomeric regions of *T. parva* between the $CCCTA_{3-4}$ telomeric repeat and the first protein-encoding gene are short, 2.9 kbp on average with no other repeats. This is in contrast to the highly complex arrays of repeats in *P. falciparum* of up to 30 kbp (Gardner et al. 2005). The genome sequences confirm that chromosomal ends of both *Theileria* spp. contain a conserved ~140-bp sequence immediately adjacent to the telomeric repeat that was originally identified in five telomeres of *T. parva* (Sohanpal et al. 2000). The noncoding subtelomeric regions are characterized by a complex arrangement especially in the case of *T. annulata*, where from the terminus inward, a succession of paired guanine–cytosine (GC)-rich subtelomeric repeats (TaSrpt1 and TaSrpt2) are followed by a single-copy sequence TaSR3 at all chromosome ends (Fig. 1a). No such repeats are found in *T. parva* subtelomeres. The conserved *T. parva* terminal sequence (TpSrpt1, ~140 bp) is shared by all chromosomal ends, followed by a thymine-rich region (TpSR2), then by a region shared by many but not all *T. parva* subtelomeres (TpSR3). These subtelomeric regions exhibit 70–100% sequence similarity between different chromosomes. As in other eukaryotic pathogens, these features may facilitate interchromosomal recombination.

Most parasitic protozoa contain arrays of telomere associated open reading frames (ORFs) that may be involved in interaction with the mammalian host (Barry et al. 2003). The genomes of both *Theileria* species encode tandem arrays of genus-specific,

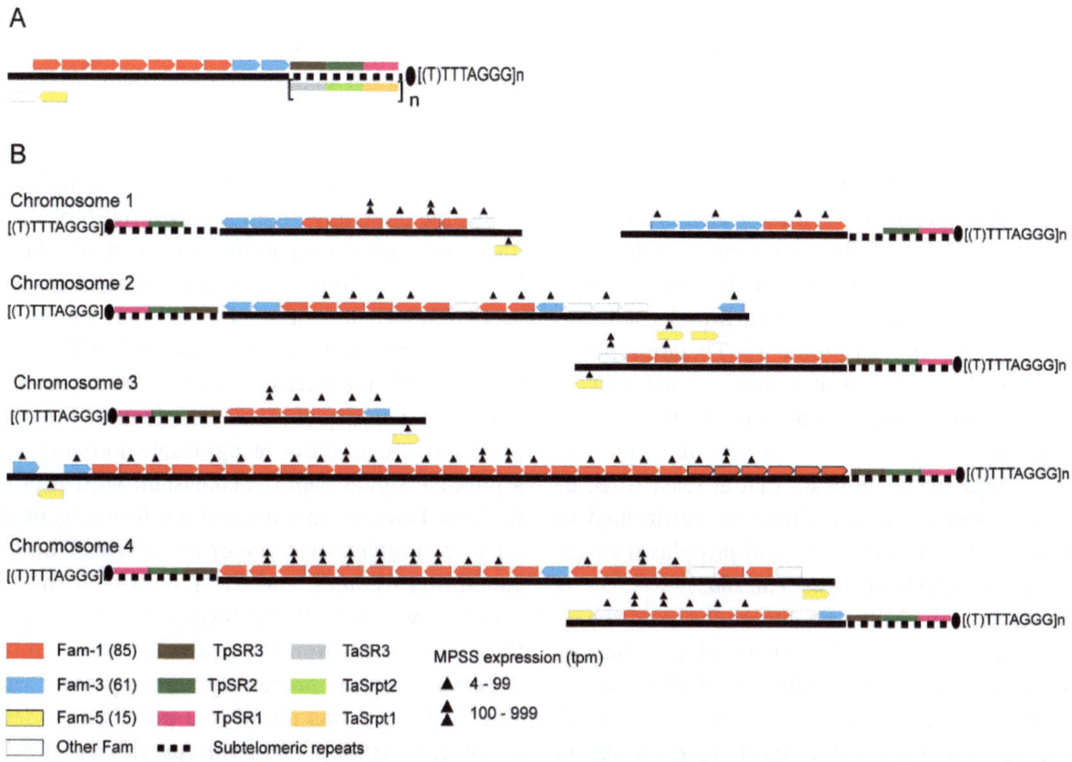

Fig. 1 Organization of *Theileria* parva and *Theileria annulata* telomeres and telomere-associated genes. **(a)** Organization of a representative telomere (not to scale) of Theileria annulata. **(b)** Organization of the eight telomeres of *Theileria parva* Muguga (not to scale) and expression of telomere-associated genes according to MPSS. Each of the four *T. parva* chromosomes are illustrated, showing the relative position of the ORFs on the forward (*top*) and reverse (*bottom*) strands with the multigene families color-coded. The left telomere is shown above the right. Genes within three multicopy families identified using TRIBE-MCL are color-coded, with the copy number in the genome indicated. *Arrowheads* indicate the transcriptional orientation; *triangles* above each ORF indicate the MPSS expression level in two quantitative categories; 4–99, 100–999. The *dotted black line* represents species-specific noncoding regions. Abbreviations are as follows: *Fam-1-3* multicopy coding families 1–3, *Srpts* subtelomeric repeats, *SR* subtelomeric regions

hypervariable gene families that map adjacent to the telomeres (Fig. 1a, b). There is limited sequence identity between the ORFs comprising these telomere proximal gene families between the two *Theileria* species, although the general properties of the genes are broadly similar. There are two major families of telomere-associated proteins, many of which contain predicted signal peptides and are therefore presumably secreted. Following analysis using the TRIBE-MCL program, these have been designated gene families 1 and 3 (Pain et al. 2005) and their organization are shown in Fig. 1. In the specific *T. annulata* example telomere shown in Fig. 1a, family 3 is telomere proximal, while family 1, which is typified by the presence of Pro/Gln-rich repeats, is located closer to the centromere. However, the precise arrangement and location of these two families with respect to the telomeres is variable among different telomeres. Family 5 is also telomere-associated (Fig. 1b in Pain et al. 2005). The organization of the telomere-associated ORFs is shown in detail for all eight *T. parva* telomeres (Fig. 1b). The family 3 ORFs are centromere-proximal at three of the eight *T. parva* telomeres, while family 1 is centromere proximal at the other five telomeres. Some of the constituent ORFs in these families are polymorphic in sequence, or in certain cases the entire ORF is present/absent when individual *T. parva* isolates are compared (Bishop et al. 2000b). As mentioned, the largest subtelomeric family (family 1) is rich in the amino acids Pro–Gln and these have been designated the sequence variable secreted proteins (SVSPs). The boundary between subtelomeric gene families and housekeeping genes is defined by adenosine 5-triphosphate-binding cassette (ABC) transporter genes (family 5) in the opposite coding orientation in *T. parva* and *T. annulata*. The substrate specificity of these transporters is as yet unknown, although the conservation of location and orientation is striking in both *Theileria* species. Application of MPSS to create an initial catalogue of the schizont transcriptome of *Theileria parva* (Bishop et al. 2005) indicates that at least four of the subtelomeric ABC transporters are transcribed in *T. parva*. In *T. annulata*, stage-specific expressed sequence tags (ESTs) indicate that at least three of these transporters are constitutively transcribed in the macroschizont, merozoite, and piroplasm stages in the mammalian host (Pain et al. 2005).

Analysis of the MPSS signatures for 56 of the 85 members of family 1 and 13 of the 61 members of family 3 indicated low to medium level MPSS signatures (4–606 tpm, representing 47% of ORFs in both families) indicates active, but selective, transcription (Fig. 1b). There was variation in the transcriptional activity between the different telomeres. For example, a signature was identified from only one of the seven potentially detectable telomere-proximal ORFs (one of the eight ORFs adjacent to a telomere did not contain a *Dpn*II site and would therefore not have been detectable using this technique). The transcriptionally active ORF was located on chromosome 1. These differences may be stochastic rather than stable and the expression profile could differ at another time point, or in different cell lines, as observed previously with *P. falciparum* (Ganesan et al. 2002). In support of this hypothesis, cDNA clones from *T. parva* have previously been isolated corresponding to the fifth ORF centromeric to the 3′ telomere of chromosome 1 (Bishop et al. 2000b) and the seventh ORF centromeric to the 5′ telomere of chromosome 2. The latter was also detected using northern blotting (Bishop et al. 2002). However, these two genes had relatively low MPSS signatures, 18 and 39 tpm, respectively. The telomeres of chromosome 4 were the most transcriptionally active, with ORFs 2–10 transcribed at the 5′ end and 4–9 at the 3′ end, nine of these at >100 tpm. As in the case of transcription of the *var* genes in *P. falciparum* erythrocytes, assessed using microarray analysis (Le Roch et al. 2003), several signatures derived from internal copies of the two subtelomeric protein families were detected. Among the 56 expressed telomeric ORFs, 33 (59%) have signal sequences, consistent with a role in pathogen–host interaction.

Almost all members of the major *Theileria* spp., specific subtelomeric protein family members incorporate varying numbers (1–54) of a highly polymorphic domain with an average length of 70 residues, designated frequently associated in *Theileria* (FAINT). This domain had already been described as DUF529 (Bateman et al. 2004), but the genome sequences revealed over 900 copies in 166 *T. annulata* proteins and in equivalent numbers of *T. parva* proteins. The majority of the FAINT domain-containing proteins have no other recognizable domains except a putative signal peptide, consistent with export to the host. However, members of the *TashAT* gene cluster of *T. annulata* has one or more FAINT domains containing AT-hook and PEST motifs on the same protein (Swan et al. 1999; Shiels et al. 2004), although these are absent from the equivalent gene family in *T. parva*. The only previously known protein with a similar domain was found in a protein encoded by a nontransforming *Theileria* (previously classified as *Babesia equi*), which also invades leukocytes and develops to a macroschizont stage (Mehlhorn and Schein 1998). Members of gene families 3 and 5 also occur internally in the genome in both *T. parva* and *T. annulata*, an arrangement that possibly allows expansion and diversification of these protein families within the subtelomeres, with the internal copies representing reservoirs.

An interesting feature of the genome that is potentially relevant to survival of *Theileria* in the ixodid vector is the presence of the enzymes trehalose-6 phosphate synthetase and trehalose phosphatase in the genome. It is tempting to speculate that this disaccharide, accumulation of which has been associated with survival from complete dehydration in insect larvae (Watanabe et al. 2002), may play a role in desiccation resistance of the parasite during the lengthy periods when the tick is waiting to feed on the vertebrate host.

6.3.2 Comparative Genomics of *T. parva* and *T. annulata*

Comparative genomic analysis between *T. parva* and *T. annulata* shows strong conservation of synteny across nonsubtelomeric regions of all four chromosomes with only a few inversions of small sequence blocks and no interchromosomal rearrangements (Pain et al. 2005). The *T. parva* and *T. annulata* protein encoding and noncoding sequence domains located at chromosome ends are described in more detail in Sect. 6.3.1. *T. parva* and *T. annulata* have 60 and 34 species-specific single copy genes, respectively (Pain et al. 2005). Syntenic clusters were distributed uniformly along each chromosome except for the subtelomeric regions, which contain species-specific gene families, although the overall organization is generally conserved. Large interruptions to synteny corresponded to the insertion of or deletions of genes and are usually part of multicopy gene families such as the *Tpr* genes and their related counterparts *Tar* in *T. annulata*. The *Tpr* and *Tar* gene family form the second largest family in both of these genomes (Pain et al. 2005). As previously mentioned, the Tpr and Tar loci exhibit a major difference between the two species. The *Tpr* genes form an array on *T. parva* chromosome 3 located at a large inversion point in the genome, located ~570 kbp from a telomere. The array appears to be lacking in *T. annulata*. Eleven additional dispersed copies of *Tpr*, also of varying length, contain a 268-amino acid membrane-associated helical domain typical of the *Tpr* family. *Tar* genes are dispersed throughout the four chromosomes in *T. annulata* and cause small interruptions in synteny. In *T. parva*, the majority of the *Tpr* genes form a single array on chromosome 3 located at a large inversion point. By contrast in *T. annulata* the *Tar* genes are dispersed throughout the four chromosomes, resulting in only small interruptions in synteny. It is likely that the tandem organization of the *Tpr* genes in *T. parva* facilitates gene conversion, resulting in decreased sequence divergence between different Tpr copies, since the dispersed copies of *Tpr* and *Tar* are more divergent from one another than those within the array. However, sequence homogenization of the tandem copies apparently also results in exceptionally rapid divergence of the array sequences between geographically separated populations of *T. parva* through concerted evolution (Bishop et al. 1997). Comparison of the *T. parva* genome with whole-genomes of apicomplexans such as *P. falciparum* clone 3D7 (Gunderson et al. 1987), *P. yoelii* (Carlton et al. 2002), and *C. parvum* (Abrahamsen et al. 2004) reveals limited synteny (Gardner et al. 2005). Only 435 microsyntenic regions containing 1,279 orthologs were observed between *P. falciparum* and *T. parva*, consisting of groups of 2–11 orthologs conserved in position between the two genomes (Gardner et al. 2005).

6.3.3
Current Status of the *Theileria orientalis* Genome Sequencing Project

A genome project for *T. orientalis* has been initiated at the Graduate School of Veterinary Medicine, Hokkaido University, Japan, in collaboration with other Japanese institutions including Institute of Physical and Chemical Research, National Institute of Genetics, and Institute of Medical Science, Tokyo University, together with The Institute for Genomic Research, USA. The project is supported by a grant from the Ministry of Education, Culture, Sports, Science, and Technology, Japan.

The Ikeda-type parasites were chosen for genome sequencing, because their distribution is limited to the eastern Asian countries including Japan, Korea, and the north-eastern part of China, whereas the Chitose type is found all over the world. In addition, clinical cases that are often severe occur in the region where the Ikeda-type is present.

Whole-genome shotgun sequencing generated 5× coverage of the genome and gap closure has been completed. The sequence data have been assembled into four contigs that represent four chromosomes without gaps, in addition to apicoplast and mitochondrial genomes. In addition, sequences of 27,478 reads from full-length cDNA libraries constructed by the oligo-capping and vector-trapper methods using mRNA obtained from the intra-erythrocytic stage have been mapped on the genome sequence, which contribute to improving the quality of annotation. An overview of the genome indicates that *T. orientalis* contains chromosomal DNA of 8.9 Mbp, which is about 0.6 Mbp larger than genomes of the two *Theileria* species that were previously sequenced. The total G+C content is 41.6%, which is significantly lower than those of *T. parva* and *T. annulata* genomes (32.5 and 34.1, respectively). Annotation of the genome is now underway and is expected to be completed in 2008.

Since *T. orientalis* is a nonlymphocyte-transforming *Theileria* parasite, a key focus of comparative genome analysis with *T. parva* and *T. annulata* will be to elucidate molecular mechanisms underlying lymphocyte transformation. Schizonts of the benign *Theileria* group parasites appear transiently in the lymph nodes, spleen, and liver after sporozoite invasion of the mammalian host, but no intra-lymphocytic multiplication has been reported. The host cell type for schizont multiplication has not yet been identified and may not be within lymphocyte lineages. The full-length cDNA data will be valuable for obtaining accurate information about transcription starting sites and their structure. The cDNAs will also allow candidates for vaccine development targeting the intraerythrocytic stage to be identified.

6.4 Population Genetics and Molecular Epidemiology of *T. parva* and *T. annulata*

A useful application of the *T. parva* and *T. annulata* genome sequences has been the ability to identify short tandemly repeated sequences, polymorphic variable number tandem repeats (VNTRs), whose copy number varies between parasite isolates and clones (Oura et al. 2003; Weir et al. 2007). Such sequences can be used as genome-wide markers for analysis of population genetics and monitoring of some aspects of live vaccine deployment in the field. Most of the informative polymorphic VNTR sequences identified in *T. parva* (60) and *T. annulata* (10) are minisatellites, with relatively few microsatellites. The majority are located in intergenic regions and may therefore be selectively neutral, while others are located in ORFs that may encode proteins (Oura et al. 2003; Weir et al. 2007). In addition to the strictly defined VNTRs that occur as "perfect" repeats directly in tandem, there are also an unusual group of AT-rich "imperfect" quasi-tandem repeats that comprise two distinct sets of conserved motifs that appear to be linked in the genome (Bishop et al. 1998).

In common with earlier analyses based on the use of multicopy probes and restriction polymorphisms in genes encoding antigens (reviewed by Bishop et al. 2002), the application of the *Theileria* VNTR sequences to both blood samples and schizont-infected lymphocyte culture isolates reveals very extensive genetic diversity in *Theileria*. For example, application of 12 VNTR markers revealed 84 multilocus genotypes in 104 blood samples from cattles from three geographical regions in Uganda (Oura et al. 2005). Application of a panel of 30 markers to schizont-infected lymphocyte culture isolates from Kenya revealed 183 distinct alleles, with 3–11 alleles per locus (Odongo et al. 2006). A single animal in Uganda was infected with at least eight different multilocus genotypes of *T. parva* (Oura et al. 2004b). The extent of diversity appears to be even greater in *T. annulata* with 8–27 alleles per locus observed in an initial study, with an average of 18 identified from 75 schizont-infected cell culture isolates (Weir et al. 2007).

The population genetic structure revealed by initial application of the VNTR markers to *T. annulata* isolates from Turkey and Tunisia (Weir et al. 2007) and *T. parva* isolates from Uganda and Kenya (Oura et al. 2005; Odongo et al. 2006) appears complex. This probably reflects differences in epidemiological parameters such as transmission intensity, combined with anthropogenic factors such as cattle mobility and disease control strategies. Differences in breed resistance and sexual recombination and other processes of molecular diversification in the parasite represent additional factors. It is also important to note that the methodologies for data collection and analysis were not standardized between these studies. Evidence was found for geographical sub-structuring of populations between Turkish and Tunisian *T. annulata*, perhaps due to trade barriers, as well as physical separation of populations (Weir et al. 2007) and also between parasite populations occurring in different cattle breeds within Uganda (Oura et al. 2005). By contrast, analysis of molecular variance indicated minimal substructuring relative to geographical origin within three regions of Kenya (Odongo et al. 2006). Linkage disequilibrium (LD) between alleles was observed in *T. parva* in both Kenya and Uganda. These observations were consistent with recent experimental work using the Marikebuni stock of *T. parva* (Katzer et al. 2006), suggesting that sexual recombination occurs as the stock is passaged through ticks and cattle, but is not entirely random with respect to the set of molecular markers used in this study. When multiple isolates with identical genotypes were treated as a single sample in the Ugandan study, an epidemic population structure was predicted. It was suggested that this situation represented an intermediate between the extremes of clonality and panmixia (random mating), resulting from the superimposition of clonal expansion of certain genotypes within sexually reproducing populations. In the case of *T. annulata* in Tunisia, the population was in linkage equilibrium, an observation consistent with panmixia. However, in *Plasmodium* populations, even quite high levels of inbreeding may still allow linkage equilibrium between different alleles in *Plasmodium* populations (reviewed by Paul et al. 1995).

6.5 Genetic Recombination in *T. parva*

Genetic mapping of *Theileria* is technically difficult, because only schizont-infected leukocytes can be cultured, whereas recombination occurs in the gut of the tick vector and the tick gut stages are difficult to analyze directly. As in the case of *Plasmodium* species in the mosquito, *T. parva* undergoes an obligatory sexual cycle in the tick vector. While the mammalian life cycle stages are haploid, there is a transient diploid phase in the tick gut after fusion of gametes (Gauer et al. 1995). In *T. parva*, according to direct measurements of the DNA content of single cells by fluorescence microscopy, a post zygotic meiosis occurs in the gut prior to differentiation of the kinete, rapidly restoring the haploid condition. In *T. annulata*, by contrast, the kinete appears to remain diploid and meiotic division may occur at later stage during sporogony in the salivary glands (Gauer et al. 1995). The stage at which meiosis occurs has not yet been confirmed for *T. annulata*.

An experimental cross was performed between two *T. parva* stocks, Muguga and Uganda, by coinfection of an animal (Morzaria et al. 1993), application of ticks at peak piroplasm parasitaemia, and selection of ticks with single infected salivary gland acinar cells. Putatively recombinant acini were identified using PCR amplification followed by hybridization with stock-specific oligonucleotide probes designed from the Tpr locus (Allsopp et al. 1989; Bishop et al. 1993). This indicated that 38% of these acini represented mixtures containing genetic material from both parents (Morzaria et al. 1993; Bishop et al. 2002). Sporozoites were then used to infect bovine lymphocytes at limiting dilutions to generate clonal recombinant progeny. Detailed analysis of a Muguga–Uganda clone generated in this experiment revealed a mixed genotype following hybridization with several multicopy probes. In particular, according to data from polymorphic *Eco*RI fragments recognized by a telomeric probe on Southern blots, three of the eight telomeres in the putative recombinant appeared to be of Muguga origin and three to be of Uganda origin. The other two telomeric fragments were not polymorphic between the parents and therefore uninformative. There was no evidence from these blots for major rearrangements near the telomeres as has been demonstrated to occur in *P. falciparum* (Hinterberg et al. 1994). The cloned recombinant parasite and parental stocks have recently been analyzed at higher resolution using a subset of 53 mini- and microsatellite markers from the panel described by Oura et al. (2003). In this study, the PCR products were fractionated on metaphor agarose gels. These data were combined with sequence information derived from four *Sfi*I linking clones that were polymorphic between the parental stocks and also identification of the origin of specific telomeres by reference to pulsed field gradient (CHEF) gels. This revealed a minimum of eight physical crossover events between polymorphic VNTR loci, two on chromosome 1, two on chromosome 2, three on chromosome 3, and one on chromosome 4 (Fig. 2). This represents a minimum estimate of the extent of physical recombination, since the genome-wide coverage of the markers is still at relatively low resolution. In addition, an apparently novel minisatellite allele absent from both parental genotypes (the parental stocks exhibited no detectable heterogeneity at this particular VNTR locus) was observed in cloned Muguga/Uganda recombinant on chromosome 4.

As mentioned in the previous section, further evidence for recombination in *T. parva* has been provided by genotyping of clones derived from the *T. parva* Marikebuni stock by in vitro transformation of lymphocytes with sporozoites following tick passage (Katzer et al. 2006). The *T. parva* stock has previously been shown to be genetically and phenotypically heterogeneous (Goddeeris et al. 1990). Genotyping of 231 clones using a panel of VNTR markers, including those described by Oura et al. (2003), plus additional ones generated during the study, revealed 48 multilocus genotypes (MLGs), among the 231 clones. The

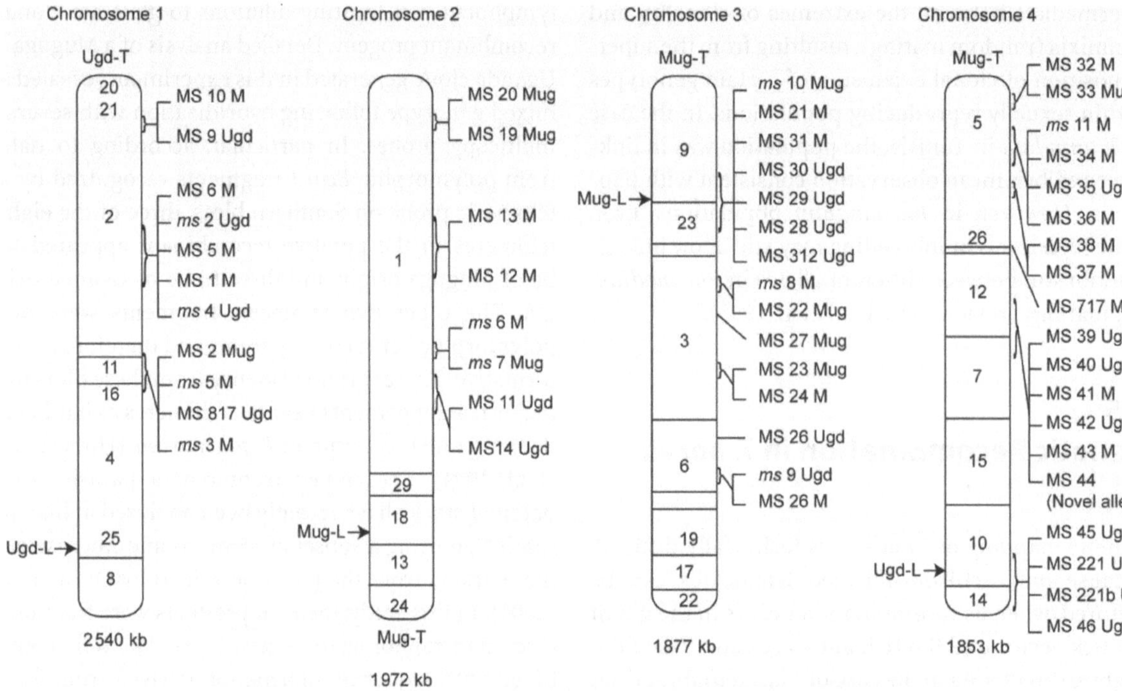

Fig. 2 Analysis of a cloned *Theileria parva* parasite generated from an experimental cross between *T. parva* Muguga and *T. parva* Uganda. The recombinant was isolated following coinfection of cattle, tick isolation at peak parasitaemia, in vitro infection of sporozoites from dissected salivary glands into bovine lymphocytes, and dilution cloning of resulting cell lines. A diagrammatic representation of the four *T. parva* chromosomes is shown. The parental origin of the telomeres is marked above or below the chromosome (in those instances where this was unambiguously determined by reference to blots of pulsed field gradient gels of *Sfi*I-digested DNA hybridized with a telomeric probe). The parental origin and approximate genomic location of informative mini- and microsatellite markers is indicated on the right hand side, where M is Muguga and U is Uganda. The origins of regions containing *Sfi*I sites determined from the sequence of the *Sfi*I linking clones are marked on the left hand side of the figure

assortment of the VNTR alleles was consistent with sexual recombination, although 75% of the clones represented a single MLG and many markers were inbred with respect to this genotype. A subsequent passage through a different *R. appendiculatus* tick stock reduced the number of MLGs observed to 18 from 142 clones analyzed, with the dominant MLG observed initially still comprising 75%. This data is consistent with recombination in *T. parva*, although as mentioned previously assortment of the markers apparently does not occur entirely at random. The study also suggests that there may be selection for specific parasite genotypes within the tick vector.

The extent of recombination in *T. parva* field populations was unknown until recent field studies using VNTR markers, although, as in *Plasmodium* (Paul and Day 1998; Awadalla et al. 2001), there is genetic heterogeneity at a variety of loci between individuals and populations in the field, and this was considered probably attributable at least partially to sexual recombination. The extent of recombination would ideally be determined by direct examination of kinetes using PCR analysis of single copy allelic markers, as has been done for *Plasmodium* oocytes from mosquito populations in the field. In one study, a high percentage of apparently recombinant *P. falciparum* zygotes was found at a Tanzanian study site (Babiker et al. 1994). However, such studies will be technically more difficult for *Theileria* infections in ticks. A factor that has yet to be fully addressed for *Theileria* or other protozoan parasites is the extent to which additional processes of diversification, including somatic mutation, genome rearrangements, and gene conversion during the asexual multiplication phases, in addition to classical recombination during the sexual cycle, generate diversity in parasite populations.

6.6 The Transcriptome of *Theileria parva* Schizonts as Revealed Using MPSS

Analysis of the *T. parva* transcriptome using MPSS identified a total of 1,095,000 MPSS signatures (Bishop et al. 2005). Expressed genes were grouped into three arbitrary categories, 4–99 tpm, 100–999 tpm, and >1,000 tpm, and displayed a relatively uniform expression pattern within all chromosomes, with no striking clusters of expression, and a unimodal distribution of logged abundances, similar to the lognormal distribution typical of microarray datasets (Bishop et al. 2005). These signatures provided evidence for consistent transcription of 2,533 of the predicted 4,036 protein coding genes. This represented 73% of those potentially detectable by MPSS. Four hundred and five (16%) of expressed genes contain sequences encoding a predicted signal peptide, among which the majority (75%) have no significant identity to genes in the public databases and were annotated as encoding-hypothetical or conserved-hypothetical proteins. Genes encoding four heat shock proteins, including Hsp90 and Hsp70, were among the 20 most highly expressed schizont genes according to MPSS and this observation was consistent with proteomics analysis of abundant *T. parva* schizont proteins (see Table 2 and the subsequent section of this chapter on proteomics). In *T. parva* infected bovine lymphocytes, Hsp90 has been indirectly shown to accesses the bovine cytosol as it represents a target of cytotoxic T lymphocyte (CTL) responses against the schizont (Graham et al. 2006; see also the section on antigen identification in this chapter). The p67 sporozoite surface antigen that has been shown to be protective against *T. parva* challenge in cattle (Musoke et al. 1992) was transcribed in the schizont at 22 tpm. This is consistent with a report that the p67 protein could be detected in a very small percentage of schizont-infected lymphocytes by immunohistochemical staining using an anti-p67 monoclonal antibody (Honda et al. 1998).

Antisense transcripts have been detected in many prokaryotic organisms and increasingly in eukaryotes and are speculated to perform a variety of functional roles (Vanhee-Brossollet and Vaquero 1998), including down-regulation of gene expression. In *T. parva*, 231 antisense signatures were observed in the MPSS dataset. Interestingly, 88 antisense signatures originating from 85 genes were observed that lacked a corresponding sense transcript (Bishop et al. 2005). The absence of a corresponding sense signature in these genes may be the result of regulatory silencing by the antisense transcript. There was no clear indication of a common functional role among this set of genes and alternative explanations for the lack of a sense MPSS signature are also possible.

Table 2 Abundant *T. parva* Muguga schizont proteins fractionated using two-dimensional gel electrophoresis and identified using Liquid Chromatography-Tandem Mass Spectrometry compared with transcript levels according to MPSS

Protein ref #	Locus	Putative function	MPSS (tpm)
603.1	TP01_0934	Endoplasmin	694
603.3	TP01_0937	Cell division cycle protein 48	17
603.4	TP02_0244	Heat shock protein 90	5,974
603.8	TP04_0683	Heat shock protein 70	2,617
603.15	TP02_0148	Heat shock protein 70	52,256
603.33	TP04_0066	Heat shock protein 60	778
603.46	TP01_0389	ATP beta chain	5
603.61	TP04_0700	Enolase, putative	2,488
603.64	TP04_0383	Glyceraldehyde-3-phosphate	4,330
603.70	TP03_0447	Nucleosome assembly protein 1	648
603.74	TP04_0748	40S ribosomal protein SA	1,360
603.87	TP04_0036	Thioredoxin peroxidase 1	1,199

Post-transcriptional control has been demonstrated in *T. parva* (Nene et al. 2000) and in certain life cycle stages of *P. falciparum* (Wirth 2002). The relative importance of transcriptional and post-transcriptional processes in contributing to the proteome is unclear. Ten of the 12 abundant *T. parva* proteins had corresponding transcript levels of >500 tpm, and were among the most highly transcribed proteins identified by MPSS analysis. This was particularly evident for heat shock proteins, with an Hsp70 homolog with a transcription level of 52,686 tpm, and the polymorphic immunodominant molecule, a highly expressed protein in schizonts (Toye et al. 1991), was also among the 20 most abundant signatures in the MPSS dataset. Thus there was strong evidence for a degree of concordance between high levels of gene transcription and protein expression, but also evidence for low levels of transcripts from many genes that may not be translated.

Because of the depth of coverage of MPPS (Brenner et al. 2000; Reinartz et al. 2002), numerous transcripts expressed at low levels were detected in *T. parva*. This suggests that transcription may not be stringently controlled (Vega-Palas et al. 2000; Gunasekera et al. 2004), at least in certain genomic domains. This could also explain the lack of identifiable specific-transcription factors in both *T. parva* and *P. falciparum* (Wirth 2002; Gardner et al. 2005). The average intergenic distance in *T. parva* is only 405 bp (Table 1), and it is unknown whether this length of sequence is sufficient for binding of a complete protein transcriptional regulatory complex. Expression data for the gene encoding the p67 antigen and a down-stream hypothetical coding sequence separated by only 93 bp have MPSS tpm values of 22 and 34, respectively, suggesting that their initial transcription may not be independent.

Only 81 and 77 transcription-associated proteins in *T. parva* and *T. annulata*, respectively, were identified, indicating that like *Plasmodium*, *Theileria* encodes low numbers of transcriptional regulators detectable by sequence homology with such proteins in other species. The identification of two transcriptional proteins, a TATA-box binding protein (PF00352) and a transcription factor TFIIB (PF00382), suggests that the basal transcriptional apparatus in *Theileria* are similar to those observed in free-living organisms. Since few transcription-associated proteins (TAP) were identified in *P. falciparum*, it has been suggested that post-translational mechanisms predominantly govern the regulation of protein expression (Coulson et al. 2004) and this may also be true for *Theileria*.

6.6.1 Transcription of the Multicopy *Tpr* and *Tar* Loci

The Tpr ORFs within the tandem array on *T. parva* chromosome 3, previously mentioned, appear to be absent from *T. annulata* and have been shown to be transcribed in the intra-erythrocytic piroplasm stage of *T. parva* (Baylis et al. 1991; Bishop et al. 1997). Transcripts derived from these ORFs were absent from the MPSS schizont dataset. This represents the only currently known example of genes that appear to be specifically transcribed only in the erythrocytic stage of *T. parva*. *T. parva* piroplasms do not replicate in the bovine host, and their primary function is assumed to be in transmission of *T. parva* to the tick vector. Attempts to identify a protein corresponding to the *Tpr* ORFs in *T. parva* piroplasm-infected erythrocytes have been unsuccessful (Bishop et al. in press). However, it is tempting to speculate that the products of this locus, expressed in the erythrocyte or the tick gut stages, play a role in transmission to the vector, since *T. parva* piroplasms have no other known function.

EST data indicates transcription of the *Tar* genes that exhibit a degree of sequence homology with *Tpr*, particularly within the C-terminus, and are widely dispersed in the *T. annulata* genome in the schizont stage (Pain et al. 2005). Interestingly, the MPSS data also indicate that some of the dispersed copies of the *Tpr* family in *T. parva*, but not those located with the tandem array, are also transcribed in the schizont. The transcriptional differences may indicate different functional roles for differentially organized subtypes within the *Tpr* gene family.

6.7 A Snapshot of the *T. parva* Proteome

With the completion of the genome sequence of *T. parva* in 2005 (Gardner et al. 2005), it will now be possible to determine the complete proteome of each of the life cycle stages of this parasite. Thus, mass spectrometry techniques, such as "shotgun proteomics" (Nesvizhskii and Aebersold 2005), can be used to identify most, if not all, of the expressed proteins. Of further value, iTRAQ (http://docs.appliedbiosystems.com/pebiodocs/00113379.pdf) and other related techniques can be used to identify proteins that are differentially expressed in different life cycle stages. However, to first obtain a snapshot of the major proteins of *T. parva* and to help with the annotation of the *T. parva* genome, we elected to identify parasite proteins using a combination of high-resolution two-dimensional gel electrophoresis and mass spectrometry. Proteins from enriched preparation of *T. parva* macroschizonts (a life cycle stage that is relatively easily obtained in large quantities) were separated by gel electrophoresis followed by peptide mass mapping using Matrix Assisted Laser Desorption Ionization Time of Flight (MALDI-TOF) mass spectrometry and peptide sequence determination by liquid chromatography - tandem mass spectrometry LC-MS/MS. Both public databases (i.e., NCBInr) and the ILRI-TIGR *T. parva* private database were used to identify parasite and bovine proteins (Table 3). The proteome data were useful for qualification/validation of some of the vaccine candidates selected by the ILRI *Theileria* Vaccine Project, by complementing genome and transcriptome information. In addition, confirmation of gene predictions of the *T. parva* genome sequence database was made possible by positive identification of specific expressed protein gene products.

Proteome analysis is most easily performed on pure materials since contaminating proteins can confound and confuse analysis, especially when annotation of sequences is incomplete and a high percentage of a genome contains genes thought to encode "hypothetical proteins." The latter is especially true for many parasites. We thus used highly purified macroschizonts of *T. parva*. This life cycle stage infects and transforms bovine lymphocytes. Macroschizonts were purified from an infected bovine T lymphocyte cell line (Baumgartner et al. 1999). Macroschizonts (10^9) were immediately lysed in a highly denaturing solubilization buffer that was specifically designed for rapid and complete denaturation of proteins (Anderson and Anderson 1978). The "high pH urea mix" eliminates proteolysis by rapid denaturation in a reducing environment, high levels of SDS and high pH above that favorable for enzyme activity, and allows solubilization of both cytoplasmic and membranep roteins.

Table 3 Summary of proteins identified from peptides derived from purified *T. parva* macroschizonts

Protein ref #	Database ref #	Protein spot identification
1	TP01_0934	Endoplasmin precursor
3	TP01_0937	Cell cycle protein 48
4	TP02_0244	Heat shock protein 90
6	TP02_0244	Heat shock protein 90
7	Bovine	Calregulin/calreticulin
8	TP04_0683	Unknown
9[a]	**TP04_0683**	**Unknown**
		Heat shock protein 70
10[a]	**TP04_0683**	**Unknown**
		Heat shock protein 70
15	TP02_0148	Heat shock protein 70
22	Bovine	68 kDa neurofilament
33	TP04_0066	Heat shock protein 60
39	Bovine	Disulphide isomerase
42	Bovine	Serine hydroxymethyl transferase
46	TP01_0389	ATP beta chain
50	Bovine	Alpha enolase
55	Bovine	Alpha actin
56	Bovine	Alpha actin
57	Bovine	Alpha actin
61	TP04_0700	Enolase 2-phosphoglycerate dehydratase
64	TP04_0383	Glyceraldehyde-3-phosphate
65	TP04_0383	Glyceraldehyde-3-phosphate
66	TP04_0383	Glyceraldehyde-3-phosphate
67	Bovine	Glyceraldehyde-phosphate-dehydrogenase
70	530.m02700	Nucleosome protein 1
74	TP04_0748	40S protein
87	TP04_0036	Unknown
90	Bovine	Cofilin

[a] Analysis of protein spots 9 and 10 by mass spectrometry identified peptides from two proteins, indicative of an overlap of protein charge trains on the 2D gel. There were more peptides derived from the protein identified as TP04_0683 (in bold) than from the heat shock protein 70

Macroschizont lysates yielded remarkably good 2D separation profiles as can be seen in the representative gel shown in Fig. 3. In more than five separate gel runs, there was no evidence of protein degradation and the spot constellation profiles remained similar when 2D gels, prepared from independent samples over several months, were compared. In-gel tryptic digests of proteins in individual cored spots were performed and the subsequent extracted peptides were analyzed by mass spectrometry.

Initially we set out to identify *T. parva* proteins by peptide mass mapping using MALDI-TOF mass spectrometry. The proteome analysis was performed by separation of proteins by 2D gel electrophoresis followed by in-gel destaining of proteins in cored spots, reduction, alkylation, tryptic digestion, and subsequent identification by screening the tryptic peptide masses against private *T. parva* databases supplied by TIGR. The generation of tryptic peptides identified by peptide mass fingerprinting was successful as we obtained good peptide masses from most of the protein spots that we selected. Forty of the most abundant protein spots were screened by this method. However, at that time, with limited resources and under-developed software, only two proteins were unequivocally identified: a *T. parva* 70 kDa heat shock protein and bovine serum albumin (BSA), a probable contaminant from the medium used to culture the infected bovine T cell line. Because of the incompatibility between specific search algorithms (i.e., MASCOT/MS-FIT) and the incomplete, early versions of the genome sequence databases, the identification of the other protein spots was impossible. The heat shock protein and bovine serum albumin (BSA) were identified only because their correct sequence information was previously submitted to the public nonredundant protein database (NCBInr). Searching of the early versions of the *T. parva* databases with MS generated peptide masses proved to be fraught with difficulties; therefore, we decided that sequence information, rather than peptide masses, would be of greater use for database screening.

In total, 100 new spots were cored from an optimized 2D gel (Fig. 3) and tryptic peptides from them were screened by MALDI-TOF MS to determine

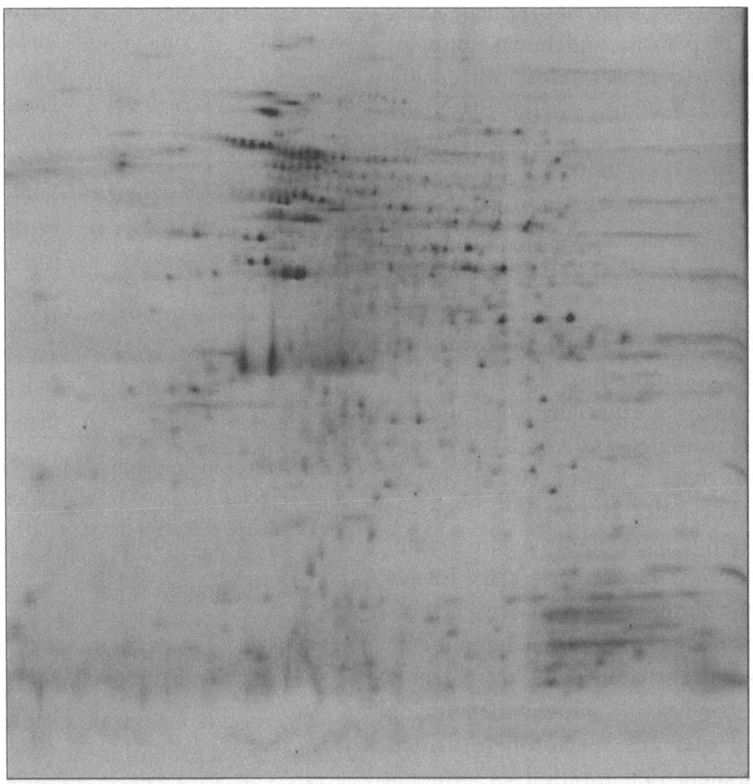

Fig. 3 Two-dimensional protein profile of *Theileria parva* macroschizonts. Proteins from 10^9 macroschizonts were solubilized in urea-mix (30 µl), separated in the first dimension using pH 3–10 isoelectric focusing tube gels and electrophoresed in 10–16.5% polyacrylamide gradient gels in the second dimension. Proteins were stained with colloidal Coomassie brilliant blue G-250. The acid end of the gel is shown to the left

digest completion and to eliminate redundant proteins from being submitted for LC-MS/MS analysis. In some cases, peptide mass mapping resulted in an unambiguous protein identification and the sample was not processed further. The peptides from the remaining unidentified proteins were separated by nanobore HPLC and gated for MS fragmentation. The data collected were used to screen the public NCBInr database as well as the in-house TIGR *T. parva* databases, including more recent versions loaded with advanced annotations. A summary of the protein identification results is presented in Table 2.

Of 31 major spots analyzed, proteins in 17 spots were identified unequivocally as *T. parva* molecules, while 10 spots were of bovine origin. Unfortunately, we were not able to identify four proteins that yielded excellent peptide sequence data. Despite the effort taken to purify the macroschizonts, 10 of the 17 proteins identified (59%) were of bovine origin, suggesting that these proteins were impurities from the bovine lymphocytes or from the foetal bovine serum used in cell culture. It is possible, however, that these bovine host proteins could have been sequestered by the parasites themselves and are not simply unwanted extraneous passengers from the purification procedure.

Given that there is now an improved annotation of the *T. parva* genome and better software available, it would be desirable to perform a more complete proteomic analysis of the main *T. parva* life cycle stages in order to gain additional insights into parasite biology. Ideally this would involve peptide sequence information obtained by tandem mass spectrometry and BLAST searching of new and more complete databases, including those of many parasites that were previously unavailable.

6.8 Transcriptional Analyses of Host–Pathogen Interactions of *T. annulata* Infections in Bovine Macrophages

There has been a recent expansion of genomic resources for both, cattle, with the recent generation of a 7.1-fold coverage of the bovine genome sequence in 2006 (see Green et al. 2007), as well as genome sequences for many livestock pathogens. These resources provide the means to investigate host–pathogen interaction in novel and more comprehensive ways. In addition, livestock transcriptomics will allow new approaches for controlling pathogens that currently continue to limit livestock production across the globe (see special issues edited by Glass and Coussens 2005 and Burton and Rosa 2006). However, although tick-borne diseases of livestock remain high on the list of major constraints to livestock productivity, specific genomic resources for investigating livestock-tick–pathogen interactions remain relatively modest. Progress is reviewed in Jensen et al. (2007). For example, there are now many genomic resources at least for cattle (see http://www.hgsc.bcm.tmc.edu/projects/bovine) and to a lesser extent for related species such as sheep, goats, and buffalo (Fadiel et al. 2005; Womack 2005). As already described, the *T. annulata* genome sequence has been determined and is publicly available on geneDB website (see http://www.genedb.org/genedb/annulata).

The use of transcriptional analyses to investigate livestock-tick–pathogen interactions is still in the very early phases. A number of immune-focused microarrays for investigating the host response have been developed for use in cattle with the initial arrays mainly consisting of cDNA clones derived from various publicly available libraries (reviewed in McGuire and Glass 2005). A major factor to take into consideration is to have relevant gene transcripts represented on the array, and many of the early bovine arrays were limited in this respect. The host transcriptome changes profoundly upon infection, yet most of the various bovine EST collections were mainly derived from nonimmune tissues, and included little from tissue or cells that were modulated following infection. To investigate transcriptional changes in bovine macrophages infected with *T. annulata*, a macrophage-focused cDNA microarray was created (Jensen et al. 2006b) that was derived from clones originating from a pool of mRNA from macrophages under a wide range of conditions and time points (Jensen et al. 2006a). Of particular relevance to the current project, these conditions included infection with *T. annulata* a parasite that predominately infects cells of the monocyte lineage (Glass et al. 1989). This microarray has been used to investigate the initial interactions between *T. annulata* and its host cell, the macrophage.

As with host responses to pathogen invasion in other systems, bovine host cell transcription is extensively remodeled following infection with *T. annulata. T. annulata* and *T. parva*, which predominately infects T lymphocytes (reviewed by Dobbelaere and Heussler 1999), are unique amongst intracellular parasites in their ability to immortalize the host cell. Both parasites differentiate into a multinucleated schizont form, which appears to exist free in the cytoplasm without being surrounded by a host membrane (Dobbelaere and Heussler 1999). The process is entirely reversible and depends on the presence of viable schizonts, since after elimination of the parasite by drug treatment the cells return to a resting state or undergo apoptosis. Additionally, both parasites alter the phenotype and function of their respective bovine hosts constitutively activating several signaling pathways and transcription factors (reviewed by (Dobbelaere and Heussler 1999; Dobbelaere and Küenzi 2004; Shiels et al. 2006)). Previous studies investigating single gene expression changes indicated that *T. annulata* infected macrophages adopt an activated cell profile with some dendritic cell-like properties – up-regulated bovine major histocompatibility complex (BoLA) class II and CD2 transcription, as well as increased levels of proinflammatory cytokines (McGuire et al. 2004; Glass and Jensen 2007) accompanied by enhancement of dendritic-cell function such as antigen presentation (Glass and Spooner 1990). At the same time, macrophage specific markers, such as CD14, are down-regulated and these changes are accompanied by impairment of macrophage-associated functions such as phagocytosis, nitric oxide generation, and oxidative burst activity (Sager et al. 1997). Although it is clear that the orchestration of the host gene expression changes are related to the ability of both parasites to highjack the host cell machinery, including the sequestration of the IκB kinase signalosome to the parasite surface (Dobbelaere and Küenzi 2004), the detailed interactions of specific host and parasite proteins is as yet not clearly defined. The best-characterized candidate transforming proteins identified to date are proteins encoded by the *TashAT* multicopy gene family of *T. annulata* that contain AT-hook binding motifs. These localize to the bovine cell nucleus, and the expression levels of several family members are correlated with leukocyte proliferation (Shiels et al. 2006).

A global analysis of transcription opens the possibility of a relatively unbiased approach that does not rely on previous discoveries and hypotheses. Because of the wealth of data generated, such studies have the potential to identify patterns of expression and ultimately regulatory processes that would not be apparent from single-gene studies. Thus, to explore the host response to *T. annulata* in more detail, a global approach was undertaken using the macrophage-focused microarray. At early time points following *T. annulata* sporozoite entry over 1,400 genes are differentially expressed (out of a possible 5,000 on the microarray), of which over 500 are potentially parasite-specific as their expression did not change the following activation of macrophages with lipopolysaccharide and interferon-γ (Jensen et al. 2008; Jensen et al., in preparation). Genes previously shown to change their expression patterns following *T. annulata* infection including interleukin (IL)-1β and IL-6 were confirmed by the microarray analysis but many other pathways were also affected, including toll-like receptor (TLR), mitogen-activated protein kinase, and Jak-STAT signaling pathways, apoptosis, and regulation of actin cytoskeleton. In addition, about 25% of the differentially expressed genes were of unknown function, with either no available annotation or unique to the microarray.

An approach that can add value to the dissection of host–pathogen interaction is to compare the transcriptional profiles of genetically resistant and susceptible animals. Cattle and other livestock exhibit breed differences in their resistance to both ticks and tick-borne pathogens (reviewed by Glass and Jensen 2007). A *B. indicus* breed was shown to be relatively resistant to experimental infection with *T. annulata* when compared to a *B. taurus* breed and there was evidence that these differences were likely to be caused by alterations in macrophages (Glass et al. 2005). Profiling the response of macrophages from each breed has provided novel insights into how *T. annulata* interacts with the host (Jensen et al. 2008) at the molecular level. Over 150 genes were differentially expressed and the majority of these were inherent differences between the two sets of macrophages. Many genes identified do not as yet have an assigned function, highlighting the need for enhanced annotation of the bovine genome. Of the known genes, many code for cell surface proteins, including BoLA class II and signal inhibitory–regulatory protein alpha

(SIRPA) that are expressed at significantly higher levels in macrophages of the susceptible breed. Both of these molecules interact with corresponding ligands on T lymphocytes, which are abnormally activated during infection in the susceptible host (reviewed in Glass and Jensen 2007), suggesting that the breed differences may be due to mechanisms by which infected macrophages interact with other cells. Intriguingly, the most highly differentially expressed gene was *TLR10*, for which no ligand has of yet been identified.

With the completion of the *T. annulata* and *T. parva* genome sequencing projects (Gardner et al. 2005; Pain et al. 2005), it is now also possible to investigate the changes in the parasite transcriptome as it adapts to the mammalian host environment. An oligonucleotide based *T. annulata* whole-genome array is currently being generated and will be used in conjunction with the macrophage array (Jensen et al. 2006b) to provide further insights into how the *T. annulata* parasite is able to manipulate its host genome.

6.9 Application of Genomics to Understand the Biology of *Theileria*–Mammalian Host Cell Interaction

6.9.1 Insights into the Interaction of *Theileria* with Bovine Leukocytes Derived from Analysis of the Virtual Proteomes of *T. parva* and *T. annulata*

Most research on *Theileria* has concentrated on the pathogenic, intracellular schizont stages of *T. parva* and *T. annulata*, which infect cells of the lymphoid and myeloid lineages and activate host signal transduction and cell cycle pathways. There are several unusual aspects of *T. parva* infections in mammalian cells. First, as mentioned previously in this chapter, infection of lymphocytes effectively transforms cells so that they continually cycle in an uncontrolled manner. This is highly reminiscent of cancer where cells accumulate oncogenic mutations that enforce proliferation under conditions in which normal cells would remain quiescent (reviewed by Dobbelaere and Heussler 1999). Most transformation events brought about by oncogene acquisition are accompanied by a hypermutation phenotype resulting in genomic instability and the Darwinian selection of more malignant cells. The transformation resulting from *Theileria* infection is distinct from this in that it is reversible: as mentioned previously, curing the host cell of the parasite results in the lymphocytes to the nontransformed state (or the host cells undergo apoptosis). This suggests that the parasite is maintaining the transformed state through an unstable intermediate either secreted or surface bound, which presumably interacts with host cell proteins and deregulates host cell proliferation. In addition, it has recently become evident that oncogene-induced unscheduled proliferation triggers a DNA damage-like response resulting in cell cycle arrest (Mooi and Peeper 2006); *Theileria* clearly subverts this host defence response during the instigation of the hyper-proliferative phenotype.

The second fascinating aspect of *T. parva*–host cell interactions from the cell cycle perspective is that host and parasite cell division is coordinated. The standard cell cycle of eukaryotic cells is traditionally divided into four phases. In the first growth (G1) phase, the cell increases in size and assesses the prevailing conditions as to whether division should occur; if conditions are not favorable for proliferation (e.g., if growth factors or space are limiting), then the cell enters a reversible cell cycle arrest known as quiescence, whereas under favorable conditions, the cell commits to enter the DNA synthetic (S) phase in which the genome is replicated. Following genome replication, the cell passes through a second growth phase in which it prepares for the subsequent mitotic (M) phase in which the chromosomes are segregated and cytokinesis occurs resulting in two daughter cells. The division of the intracellular schizonts appears linked to that of the host cell: parasite S phase occurs during host cell G2 phase. The schizont G2 phase is comparatively short so that nuclear division occurs before the host cell completes the mitotic metaphase (Jura et al. 1985). Thus, upon host cell division, the parasite has already divided, enhancing the chances of infection of both daughter lymphocytes. The close association between schizonts and the host mitotic spindle may also enhance segregation of the parasite between daughter cells. The surface of schizonts appears decorated with microtubules that are captured by the spindle during mitosis, favoring infection of both daughter cells (reviewed in Morrissette and Sibley 2002). It seems likely that parasite/host interactions are key to the coordination of the cell cycles of the two organisms and the segregation of the parasite to both daughter cells during M phase.

Transformation to a cancerous phenotype can occur by the acquisition of many distinct oncogenes. It has been clear for many years that expression of more than a single oncogene is required in many cases – the concept of cooperating oncogenes was established in the early 1980s. It may be noteworthy, however, in the context of a *Theileria* infection of lymphocytes that a number of leukemias arise with the activation of a single oncogene (Chin and De Pinho 2000). The best characterized of these cancers is chronic myelogenous leukemia (CML), in which the *abl* tyrosine kinase gene is fused with the *bcr* coding sequence, resulting in the expression of the *bcr-abl* fusion protein. Inhibition of the tyrosine kinase activity of *bcr-abl* leads to the suppression of the transformed phenotype and often the apoptosis of the afflicted cell. This situation is highly similar to that observed in *Theileria* infected lymphocytes and illustrates the idea that cellular transformation of lymphocytes can result from the introduction of a single oncogenic agent in contrast to most other cell types that require cooperation from multiple agents. Host pathways modulated by *Theileria* infection include those involving Src tyrosine kinases, JNK amino terminal kinase, phosphatidylinositol 3-kinase and casein kinase II, resulting in induction of a variety of host transcription factors. It is thought that although the host cell tropisms of *T. parva* and *T. annulata* differ, the signal transduction pathways that the parasite modulates may be similar in both species (Dobbelaere and Kuenzi 2004). One well-defined interaction identified by molecular biological experimental approaches is the activation of the host cell transcription factor NF-κB. It has been demonstrated that *T. parva* schizonts assemble a large aggregate of IκB kinase (IKK) on their surface, leading to activation of the kinase and phosphorylation and subsequent proteolytic destruction of the NF-κB inhibitor IκB (Heussler et al. 2002). *Theileria* therefore appears to avoid apoptosis of the host cell through direct subversion of the normal signaling host pathway for regulation of apoptosis. However, the *Theileria* surface proteins that nucleate the IKK complex have not yet been identified. *T. parva* infection has also been shown to deregulate host leukocyte signaling pathways (Galley et al. 1997; Dobbelaere and Heussler 1999). Other pro-proliferative and/or anti-apoptotic signaling pathways have been reported to be activated in *T. parva* infected lymphocytes and these pathways are similarly upregulated in vertebrate cancer. For example, the activities of phosphatidylinositol 3-kinase (PI3K) activity and its down-stream effector protein kinase B (PKB, also known as Akt) are elevated in infected cells and these enhanced activities are lost when host cells are cured of the parasite infection using an anti-theileriocidal drug. In mammalian cancers, elevated PI3K and PKB activity is frequent. There are several comprehensive reviews describing the phenotypic alterations induced in bovine leukocytes following *Theileria* infection (Dobbelaere and Heussler 1999; Heussler 2002; Dobbelaere and Küenzi 2004) and the reader is referred to these for additional information.

One mechanism of host cell transformation could involve reprogramming host transcriptional responses. A previously identified group of candidates identified in *T. annulata* are families of DNA binding proteins for which there is evidence for targeting to the host nucleus. These include *TashHN*, *SuAT1*, and *TashAT* gene families. The *TashAT* multicopy gene family, which contain AT hook-binding motifs associated with proteins that enhance transcription in eukaryotes, may be one contributor to the transformed phenotype associated with *T. annulata* (Swan et al. 1999, 2001). The *T. parva* genome also encodes homologous families of genes, although the *T. parva* TpHNs lack AT hook motifs (Pain et al. 2005), suggesting that the mechanism of action of these genes may perhaps be genus-specific. AT hook proteins (reviewed in Reeves 2001) are often expressed at elevated levels in human tumors (Fedele et al. 2001) and when overexpressed can enhance transformation and metastasis (Wood et al. 2000). Clearly, the ability to modulate host gene transcription would be invaluable in reprogramming host physiology. The genome sequence will enable investigation of the copy number and potential functional diversity of the *TashAT* gene family. Other known function proteins that may be involved in host cell transformation include kinases and phosphatases, heat shock proteins, cyclophilins, and glutaredoxin homologues; however, no direct evidence yet exists regarding their roles (Pain et al. 2005).

6.9.2
Analysis of the *T. parva* Predicted Proteome in Relation to Cell Cycle Regulation and Modulation of Mammalian Host Cell Function

Eukaryotic cell division is regulated largely by serine and threonine phosphorylation catalyzed by a family of cyclin-dependent kinases (cdks). These enzymes consist of a catalytic subunit (the cdk) and an activating cyclin subunit; active cyclin/cdk holoenzymes form in restricted binary combinations, so that at least in vertebrates, not every cyclin can activate every cdk. Cyclins are synthesized at specific points in the cell cycle and are rapidly degraded in a stage-specific manner so that a particular cdk is only active during the window of stable cyclin expression. Inhibitory proteins such as $p21^{Cip1}$, which bind to the cyclin and the cdk subunits to physically block phospho-transfer, can also regulate cdks. $p21^{Cip1}$ is under transcriptional control of p53, such that activation of p53 by, for example, unscheduled cell proliferation can lead to up-regulation of $p21^{Cip1}$ levels, resulting in cyclin/cdk inhibition and hence cell cycle arrest. Additionally, cdks are also subject to regulatory phosphorylation: phosphorylation of a threonine in the T-loop is required for full activation, whilst inhibitory phosphorylation can occur through modification of threonine and tyrosine residues within the GXGTYG motif that is involved in coordinating the phosphates derived from ATP. In higher eukaryotes, cell cycle control is exerted chiefly during the G1 phase through cdk-mediated phosphorylation of the Retinoblastoma protein, (pRb). Hypophosphorylated pRb binds to and inhibits the E2F transcription factor whose activity is required for S phase entry. Cdk-mediated hyperphosphorylation disrupts the pRb-E2F interaction, promoting S phase entry.

6.9.3
Predicted Cyclin-Dependent Kinases in the *T. parva* Genome

Within the *T. parva* genome, there are five predicted cyclin-dependent kinase homologues. Serine/threonine kinases typically possess a series of highly conserved amino acid motifs as well as 15 key residues that are conserved in (virtually) all protein kinases (Hanks and Hunter 1995). Figure 4 shows the conserved regions of the predicted *T. parva* cdks aligned with the human cdks. The key conserved residues are highlighted. Nonconserved residues in the predicted *T. parva* cdks are mostly also present in protein serine/threonine kinases from other species, indicating that these substitutions are unlikely to result in loss of catalytic ability. The most divergent of the putative *T. parva* cdks has 12 of the 15 core conserved residues present and contains an atypically long carboxy terminus. Comparison of predicted *T. parva* cdks with those of other species and with other major protein kinase subfamilies are shown in Table 4. A summary of the key features of each putative *T. parva* cdk is shown in Table 5. The PSTAIRE region is involved in both catalysis and cyclin binding. None of the cdks has a predicted nuclear localization signal as assessed using PredictNLS (http://cubic.bioc.columbia.edu/predictNLS/). The GxGTYG motif is a motif conserved in almost all nucleotide-binding proteins and is involved in coordinating the phosphates of ATP. Two of the *T. parva* cdks diverge within this sequence. There is an example from another organism of an active kinase with an SXGXXG motif akin to that of one of these two, but the other appears unique. The TY residues within the GXGTYG motif are both capable of phosphorylation in the context of most eukaryotic cdks. Phosphorylation of the T and particularly the Y leads to inhibition of ATP binding; all but one of the putative cdks possesses this Y. This motif is dephosphorylated by the dual specificity phosphatase cdc25 in higher eukaryotes. However, the *T. parva* genome does not contain any predicted cdc25 homologues.

6.9.4
Predicted Cyclins in the *T. parva* Genome

While the cdk family of proteins shows substantial similarity with their counterparts in other organisms, the same cannot be said of the putative cyclins in the *T. parva* genome. This is not entirely unexpected given the fact that cyclin-like folds can be achieved with highly variant amino acid sequences – for example, in mammals the TATA binding protein (TBP) has no primary sequence conservation with cyclins but nevertheless assumes a highly similar tertiary structure (Bagby et al. 1995). BLAST analyses indicate

A

```
                 a a  a                   b   b
cdk10           VKEFEKLN-------RIGEGTYGIVYRARDTQTD---EIVALKKVRMD------KEKDG- 49
TP04_0551       VECFKCLN-------KISEGTYGTVYRALELKTG---KIVALKHIKYHD----VQWKEG- 136
cdk6            DQQYECVA-------EIGEGAYGKVFKARDLKNGG--RFVALKRVRVQ------TGEEG- 53
cdk4            TSRYEPVA-------EIGVGAYGTVYKARDPHSG---HFVALKSVRVP------NGGGGG 46
cdk5            MQKYEKLE-------KIGEGTYGTVFKAKNRETH---EIVALKRVRLD------DDDEG- 43
TP01_0728       MRRYHKME-------KIGEGTYGVVYKAQN-NHG---EICALKKIRVE------EEDEG- 42
cdk3            MDMFQKVE-------KIGEGTYGVVYKAKNRETG---QLVALKKIRLD------LEMEG- 43
cdk2            MENFQKVE-------KIGEGTYGVVYKARNKLTG---EVVALKKIRLD------TETEG- 43
cdk1            MEDYTKIE-------KIGEGTYGVVYKGRHKTTG---QVVAMKKIRLE------SEEEG- 43
cdk9            VSKYEKLA-------KIGQGTFGEVFKARHRKTG---QKVALKKVLME------NEKEG- 58
TP03_0140       LKNFVKIH-------QVGQGAYGDVWLAEDIVNK---KPVALKKLKLN------EEREG- 223
cdk11           AERERVEDLFEYEGCKVGRGTYGHVYKARRKDG-----KDEKEYALKQ------IEGTG- 58
cdk8            SERERVEDLFEYEGCKVGRGTYGHVYKAKRKDG-----KDDKDYALKQ------IEGTG- 58
cdk7            AKRYEKLD-------FLGEGQFATVYKARDKNTNQIVAIKKIKLGHRS------EAKDG- 54
TP01_0781       DKRFTPVGK------HLGEGTYGQVIKAMDTLTGKMVAIKKVKNIEYKKGVTKDRQLVGM 138
TP04_0446       INPFKILSERRSK----GKVYMGKYLCDSNKNKNNYEDVVVRVVNLKLTN---AGKNDG- 98
                                                                          *

                   ccccccd
cdk10           --IPISSLREITLLLRLR---HPNIVELKEVVVG-------------------------- 78
TP04_0551       --FPLTNLREISILLQLN---HPNILSVKEIVTN-------------------------- 165
cdk6            --MPLSTIREVAVLRHLETFEHPNVVRLFDVCTVSRTDR--------------------- 90
cdk4            GGLPISTVREVALLRRLEAFEHPNVVRLMDVCATSRTDR--------------------- 85
cdk5            --VPSSALREICLLKELK---HKNIVRLHDVLHS-------------------------- 72
TP01_0728       --IPSTAIREISLLKELH---HPNIVWLRDVIHS-------------------------- 71
cdk3            --VPSTAIREISLLKELK---HPNIVRLLDVVHN-------------------------- 72
cdk2            --VPSTAIREISLLKELN---HPNIVKLLDVIHT-------------------------- 72
cdk1            --VPSTAIREISLLKELR---HPNIVSLQDVLMQ-------------------------- 72
cdk9            --FPITALREIKILQLLK---HENVVNLIEICRTKASPYNRCKG---------------- 97
TP03_0140       --FPKNAIREILLLNSLK---HKNIVNLLGICYSKSYSTSLLSDGLSSSKEELNEDDHKS 278
cdk11           --ISMSACREIALLRELK---HPNVIALQKVFLS-------------------------- 87
cdk8            --ISMSACREIALLRELK---HPNVISLQKVFLS-------------------------- 87
cdk7            --INRTALREIKLLQELS---HPNIIGLLDAFGH-------------------------- 83
TP01_0781       VGIHFTTLRELKVMTELS---HENLMGLVAVYVK-------------------------- 169
TP04_0446       --IPTSSLREMSFMKMIN---HPNVVKYYGAQII-------------------------- 127
                  .  .  **: .:  :    * *::

cdk10           ------------------------------------------------------------
TP04_0551       ------------------------------------------------------------
cdk6            ------------------------------------------------------------
cdk4            ------------------------------------------------------------
cdk5            ------------------------------------------------------------
TP01_0728       ------------------------------------------------------------
cdk3            ------------------------------------------------------------
cdk2            ------------------------------------------------------------
cdk1            ------------------------------------------------------------
cdk9            ------------------------------------------------------------
TP03_0140       KFNPKNPDKSKSAYGHTPSYSRRGSHDKGRDFHRSHREARNKPSNADQNKNDEDKNKNKN 338
cdk11           ------------------------------------------------------------
cdk8            ------------------------------------------------------------
cdk7            ------------------------------------------------------------
TP01_0781       ------------------------------------------------------------
TP04_0446       ------------------------------------------------------------
```

Fig. 4 Comparison of *Theileria parva* and *Homo sapiens cdk kinases*. **A.** CLUSTAL W (1.82) multiple sequence alignment of *H. sapiens* cdks 1–11 and the five predicted *T. parva* cdks that are referred to by the nomenclature using in the *T. parva* annotation (Gardner et al. 2005). Marked *in dark grey* and labeled are the conserved motifs (**a**), are the 15 residues conserved in almost all protein kinases (Hanks and Hunter 1995) with the subset of amino acids proposed to form the catalytic triad labeled (**b**). In *H. sapiens* cdk1 sequence, the PSTAIRE region is labeled (**c**), with the WP dipeptide highlighted with *light grey* and labeled (**d**), the T-loop threonine marked (**e**), and the CAK binding motif marked (**f**) **B.** Cladogram tree of the data from A generated using a Neighbor Joining algorithm

```
cdk10        ------NHLESIFLVMGYCEQDLASLLE-------------NMPTPFSEAQVKCIVLQVL 119
TP04_0551    ------KKHDQFYMVMEYVEHELKTLLE-------------ENRPNFTLSERKCLLKQLL 206
cdk6         --------ETKLTLVFEHVDQDLTTYLD-----------KVP-EPGVPTETIKDMMFQLL 130
cdk4         --------EIKVTLVFEHVDQDLRTYLD-----------KAP-PPGLPAETIKDLMRQFL 125
cdk5         --------DKKLTLVFEFCDQDLKKYFD-----------SCN--GDLDPEIVKSFLFQLL 111
TP01_0728    --------EKCLTLVFEYLDQDLKKLLD-----------ACD--GGLEPTTAKSFLYQIL 110
cdk3         --------ERKLYLVFEFLSQDLKKYMD-----------STP-GSELPLHLIKSYLFQLL 112
cdk2         --------ENKLYLVFEFLHQDLKKFMD-----------ASA-LTGIPLPLIKSYLFQLL 112
cdk1         --------DSRLYLIFEFLSMDLKKYLD-----------SIPPGQYMDSSLVKSYLYQIL 113
cdk9         ----------SIYLVFDFCEHDLAG-------------LLSNVLVKFTLSEIKRVMQMLL 134
TP03_0140    RDKDKKNDRENVWMVFEYLPFDLSGYIEALRDPHEKYDKLARPSVWLSIGEIKTIMRQLF 398
cdk11        ------HSDRKVWLLFDYAEHDLWHIIKFHRA-----SKANKKPMQLPRSMVKSLLYQIL 136
cdk8         ------HADRKVWLLFDYAEHDLWHIIKFHRA-----SKANKKPVQLPRGMVKSLLYQIL 136
cdk7         ------KSN--ISLVFDFMETDLEVIIKDN-------------SLVLTPSHIKAYMLMTL 122
TP01_0781    --------ESFINIVMDVMASDLKKVVD--------------AKIRLTEPNVKCIMSQIL 207
TP04_0446    --------DNNLFIVTEYLEYNLIEYMERKHN----EFECITPCKSLLRNEVMKIMFDLL 175
                     . ::       :*                    .         :   :

                             a a  a                       d a
cdk10        RGLQYLHRNFIIHRDLKVSNLLMT----DKGC--------VKTADFGLAR---------- 157
TP04_0551    DGINYLHQNWVMHRDLKTTNILYN----NSGL--------VKICDFGMAR---------- 244
cdk6         RGLDFLHSHRVVHRDLKPQNILVT----SSGQ--------IKLADFGLAR---------- 168
cdk4         RGLDFLHANCIVHRDLKPENILVT----SGGT--------VKLADFGLAR---------- 163
cdk5         KGLGFCHSRNVLHRDLKPQNLLIN----RNGE--------LKLADFGLAR---------- 149
TP01_0728    RGISYCHDHRILHRDLKPQNLLIN----REGV--------LKLADFGLAR---------- 148
cdk3         QGVSFCHSHRVIHRDLKPQNLLIN----ELGA--------IKLADFGLAR---------- 150
cdk2         QGLAFCHSHRVLHRDLKPQNLLIN----TEGA--------IKLADFGLAR---------- 150
cdk1         QGIVFCHSRRVLHRDLKPQNLLID----DKGT--------IKLADFGLAR---------- 151
cdk9         NGLYYIHRNKILHRDMKAANVLIT----RDGV--------LKLADFGLAR---------- 172
TP03_0140    RALNYCHKNNVLHRDLKTANLLMD----QDGV--------IKLADFGLAR---------- 436
cdk11        DGIHYLHANWVLHRDLKPANILVMGEGPERGR--------VKIADMGFAR---------- 178
cdk8         DGIHYLHANWVLHRDLKPANILVMGEGPERGR--------VKIADMGFAR---------- 178
cdk7         QGLEYLHQHWILHRDLKPNNLLLD----ENGV--------LKLADFGLAK---------- 160
TP01_0781    TGLSVLHSSSFAHRDLSPANIFID----NFGV--------CKIADFGLARRTVNPPIFRE 255
TP04_0446    KAVSSIHTMKVFHRNLKPENIFVDCDVIVDGTRLIYNFKSLKIGDOFAMGRLTG------- 228
              .:   *    . **::.  *::            *        *  *:.:.:

                            e              aa               a    a
cdk10        ---AYGVPVK----PMTPKVVTLWYRAPELLLGTTTQTTSIDMWAVGCILAELLA-HRPL 209
TP04_0551    ---KFGVPIR----KYTHNVVTHWYRAPELFLGEPYYTEKTDVWSIGCIFAELIL-SRPL 296
cdk6         ---IYSFQM-----ALTSVVVTLWYRAPEVLLQSS-YATPVDLWSVGCIFAEMFR-RKPL 218
cdk4         ---IYSYQM-----ALTPVVVTLWYRAPEVLLQST-YATPVDMWSVGCIFAEMFR-RKPL 213
cdk5         ---AFGIPVR----CYSAEVVTLWYRPPDVLFGAKLYSTSIDMWSAGCIFAELANAGRPL 202
TP01_0728    ---AFAIPVR----SYTHEVVTLWYRAPDVLMGSKKYSTAVDIWSVGCIFAEMIN-GVPL 200
cdk3         ---AFGVPLR----TYTHEVVTLWYRAPEILLGSKFYTTAVDIWSIGCIFAEMVT-RKAL 202
cdk2         ---AFGVPVR----TYTHEVVTLWYRAPEILLGCKYYSTAVDIWSLGCIFAEMVT-RRAL 202
cdk1         ---AFGIPIR----VYTHEVVTLWYRSPEVLLGSARYSTPVDIWSIGTIFAELAT-KKPL 203
cdk9         ---AFSLAKNSQPNRYTNRVVTLWYRPPELLLGERDYGPPIDLWGAGCIMAEMWT-RSPI 228
TP03_0140    ---FLPHGKG----LLTNRVVTLWYRSPELLLGSESYDFSVDLWSAGCIMSELVS-GSHI 488
cdk11        ---LFNSPLKPLA-DLDPVVVTFWYRAPELLLGARHYTKAIDIWAIGCIFAELLT-SEPI 233
cdk8         ---LFNSPLKPLA-DLDPVVVTFWYRAPELLLGARHYTKAIDIWAIGCIFAELLT-SEPI 233
cdk7         ---SFGSPNR----AYTHQVVTRWYRAPELLFGARMYGVGVDMWAVGCILAELLL-RVPF 212
TP01_0781    CTDLETMELNASRERMTSKVVTLWYRAPELLMGAECYHFACDLWSVGCIFAELLS-GKPL 314
TP04_0446    -LMEYSPEETKERYQSFRECRRLFYRAPELILRCDSYDKSIDLWSIGVIFYEIAC-NDIL 286
                                   :**.*::::          *:*. * *: *:       :
```

Fig. 4 (Continued)

```
              ffff         ff                                  gg
cdk10       LPGTSE---------IHQIDLIVQLLGTP-----------------SENIWP----GFSKL 240
TP04_0551   FMGTND---------ADTLDKIFRLCGSP-----------------TEENWP----GFSKL 327
cdk6        FRGSSD---------VDQLGKILDVIGLP-----------------GEEDWP----RDVAL 249
cdk4        FCGNSE---------ADQLGKIFDLIGLP-----------------PEDDWP----RDVSL 244
cdk5        FPGNDV---------DDQLKRIFRLLGTP-----------------TEEQWP----SMTKL 233
TP01_0728   FPGISE---------QDQLKRIFKILGTP-----------------SVDSWP----QVVNL 231
cdk3        FPGDSE---------IDQLFRIFRMLGTP-----------------SEDTWP----GVTQL 233
cdk2        FPGDSE---------IDQLFRIFRTLGTP-----------------DEVVWP----GVTSM 233
cdk1        FHGDSE---------IDQLFRIFRALGTP-----------------NNEVWP----EVESL 234
cdk9        MQGNTE---------QHQLALISQLCGS------------------ITPEVWP----NVDNY 259
TP03_0140   FAADKE---------SLILKLICEYLGLPDEADL--------KYLRTLPLWN----DKNSN 528
cdk11       FHCRQEDIKTSNPFHHDQLDRIFSVMGFP-----------------ADKDWE----DIRKM 273
cdk8        FHCRQEDIKTSNPYHHDQLDRIFNVMGFP-----------------ADKDWE----DIKKM 273
cdk7        LPGDSD---------LDQLTRIFETLGTP-----------------TEEQWP----DMCSL 243
TP01_0781   FPGTNE---------IDQLGKIYNILGTP------------------EPTWP----EVTKL 344
TP04_0446   FKGINE---------ISLVWDIFNVTGFPDHTSLNSLSMDLYLKWNSVILPNKPIDIEHV 337
            :                 :  *    *

                                                          a
cdk10       PLVGQYSLRKQPYN-------------NLKHKFPWLSEAGLRLLHFLFMYDPKKRATAGD 287
TP04_0551   PGVVSNKFQIHKYSPSFESVFKVGIMGGMVHGSTCMTELGLDLLKKMLNIDPNQRISAKD 387
cdk6        P--RQAFHSKSAQP--------------IEKFVTDIDELGKDLLLKCLTFNPAKRISAYS 293
cdk4        P--RGAFPPRGPRP--------------VQSVVPEMEESGAQLLLEMLTFNPHKRISAFR 288
cdk5        PDYKPYPMYPATTS--------------LVNVVPKLNATGRDLLQNLLKCNPVQRISAEE 279
TP01_0728   PAYNPDFSYYEKQS--------------WSSIVPKLNESGIDLISRMLQLDPVQRISAKE 277
cdk3        PDYKGSFPKWTRKG--------------LEEIVPNLEPEGRDLLMQLLQYDPSQRITAKT 279
cdk2        PDYKPSFPKWARQD--------------FSKVVPPLDEDGRSLLSQMLHYDPNKRISAKA 279
cdk1        QDYKNTFPKWKPGS--------------LASHVKNLDENGLDLLSKMLIYDPAKRISGKM 280
cdk9        ELYEK--LELVKGQKR---------KVKDRLKAYVRDPYALDLIDKLLVLDPAQRIDSDD 308
TP03_0140   PLHPERIGSIITRQRE---------FEKIFTKVNQLGKDGWDLLKTLFSWSPSTRITAKQ 579

cdk11       PEYPTLQKDFRRTTYANSS-----LIKYMEKHKVKPDSKVFLLLQKLLTMDPTKRITSEQ 328
cdk8        PEHSTLMKDFRRNTYTNCS-----LIKYMEKHKVKPDSKAFHLLQKLLTMDPIKRITSEQ 328
cdk7        PDYVTFKSFPG---------------IPLHHIFSAAGDDLLDLIQGLFLFNPCARITATQ 288
TP01_0781   PLYTQYSFSKPKD---------------LSTLFPHANSVTLDLLSQLLKLNPNERISAKK 389
TP04_0446   ISSSKCSEEYNSLCNEYSKSGQLEAFDKLVRFSTIMGAACTKLLFEFISVLPSNRPNITQ 397
                                                      *:   :   *  *
```

B

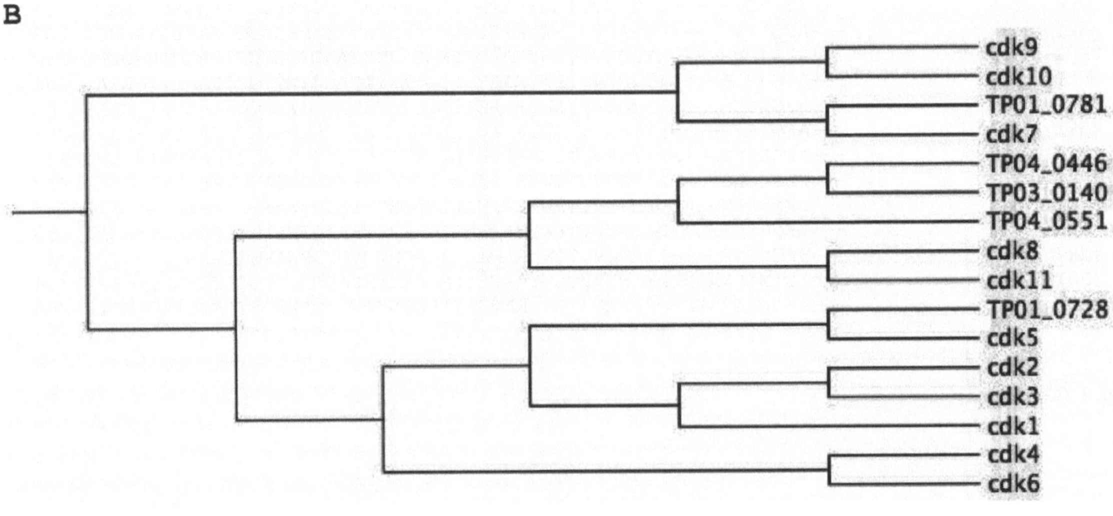

Fig. 4 (Continued)

Table 4 Summary of various cyclin-dependent kinase % identities (% similarities) with the predicted *T. parva* cdks generated using the bl2seq programme (http://www.ncbi.nlm.nih.gov/blast/bl2seq/wblast2.cgi)

Comparison	TP01_0728	TP04_0551	TP03_0140	TP01_0781	TP04_0446
HsCdk1	60 (75)	41 (58)	28 (41)	38 (54)	27 (51)
HsCdk2	63 (77)	43 (60)	31 (45)	41 (58)	30 (51)
HsCdk3	61 (75)	43 (63)	32 (44)	42 (58)	31 (51)
HsCdk4	45 (60)	37 (56)	36 (51)	35 (52)	26 (45)
HsCdk5	61 (76)	40 (57)	28 (43)	40 (55)	29 (46)
HsCdk6	45 (62)	38 (56)	29 (43)	37 (52)	26 (48)
HsCdk7	44 (64)	37 (56)	28 (40)	40 (54)	24 (45)
HsCdk8	36 (57)	33 (51)	30 (49)	31 (48)	27 (45)
HsCdk9	39 (58)	37 (53)	29 (44)	33 (48)	24 (42)
HsCdk10	41 (61)	41 (59)	32 (48)	39 (52)	28 (44)
HsCdk11	43 (66)	45 (65)	27 (40)	37 (54)	27 (46)
TaCRK3	41 (57)	35 (53)	34 (49)	84 (90)	28 (44)
PfPK5	74 (85)	44 (61)	27 (40)	40 (57)	28 (49)
PfPK6	37 (56)	34 (50)	34 (50)	32 (51)	27 (47)
Pfmrk	39 (60)	38 (53)	33 (50)	49 (64)	27 (47)
Pfcrk1	39 (59)	54 (68)	34 (53)	37 (52)	26 (42)
Pfcrk3	36 (55)	35 (49)	50 (68)	39 (52)	28 (44)
Pfcrk4	33 (52)	27 (44)	30 (48)	28 (43)	21 (44)
Pf23612235	24 (35)	35 (48)	30 (53)	35 (49)	24 (39)
TgPK1	42 (61)	54 (71)	26 (40)	36 (54)	29 (47)
TgPK2	74 (84)	42 (60)	28 (42)	37 (57)	28 (46)
TcCRK1	56 (71)	38 (59)	30 (40)	36 (52)	27 (46)
LmCRK3	53 (71)	42 (62)	28 (40)	39 (57)	30 (52)

Selected cdks from Human (Hs), *Theileria annulata* (Ta), *Plasmodium falciparum* (Pf), *Toxoplasma gondii* (Tg), *Trypanasoma cruzi* (Tc), and *Leishmania mexicana* (Lm) are compared

Table 5 Summary of conserved motifs within predicted *T. parva* cdk homologues

Gene	Kinase conserved residues	Cyclin-binding motif	GxGTYG	T160	CAK-binding motif	WP	Predicted NLS
Hs cdk2	15	PSTAIRE	GEGTYG	T	GDSEID	Yes	No
TP01_0728	14	PSTAIRE	GEGTYG	T	GISEQD	Yes	No
TP04_0551	14	PLTNLRE	SEGTYG	T	GTNDAD	Yes	No
TP03_0140	14	PKNAIRE	GQGAYG	T	ADKESL	No	No
TP01_0781	14	HFTTLRE	GEGTYG	T	GTNEID	Yes	No
TP04_0446	11	PTSSLRE	GKVYMG	R	GINEIS	No	No

three or possibly four cyclins in the genome of *T. parva*, although the overall level of sequence identity is moderate (approximately 25%). The sequence identities of the putative *T. parva* cyclins are shown in Table 6. Figure 5 shows all *T. parva* cyclins identified using the Pfam search tool that have threshold scores ("trusted" or "predicted") against the Cyclin N domain family (Pfam00134). MRAIL motifs are not evident in any of these cyclins; in mammalian cells this region is associated with substrate selection through interaction with an RXL motif common to many cdk substrates (Schulman et al. 1998). There are also no evident destruction boxes (Hunt 1991) in any of the *T. parva* cyclins, although three have putative PEST sequences that may destabilize these proteins (Rechsteiner and Rogers 1996). The putative

Table 6 Summary of cyclin % identities (% similarities) with the predicted *T. parva* cyclins generated using the bl2seq programme (http://www.ncbi.nlm.nih.gov/blast/bl2seq/wblast2.cgi)

Comparison	TP01_1216	TP02_0190	TP04_0577	TP04_0316
Hscyclin A1	0 (0)	0 (0)	0 (0)	0 (0)
Hscyclin B1	0 (0)	0 (0)	0 (0)	0 (0)
Hscyclin C	26 (41)	0 (0)	30 (45)	0 (0)
Hscyclin D1	0 (0)	0 (0)	0 (0)	0 (0)
Hscyclin E1	0 (0)	0 (0)	0 (0)	0 (0)
Hscyclin F	0 (0)	0 (0)	0 (0)	0 (0)
Hscyclin G	0 (0)	0 (0)	0 (0)	0 (0)
Hscyclin H	0 (0)	0 (0)	27 (47)	0 (0)
Hscyclin I	0 (0)	0 (0)	0 (0)	0 (0)
Dmcyclin J	0 (0)	0 (0)	0 (0)	0 (0)
Hscyclin K	29 (47)	0 (0)	0 (0)	0 (0)
Hscyclin L1	48 (77)	0 (0)	23 (42)	0 (0)
Hscyclin M3	0 (0)	0 (0)	0 (0)	0 (0)
Hscyclin T	23 (43)	0 (0)	0 (0)	23 (43)
HsCables/ik3-1	0 (0)	23 (43)	0 (0)	0 (0)
Hsp35	0 (0)	0 (0)	0 (0)	0 (0)
Cyclin box	0 (0)	0 (0)	0 (0)	0 (0)
Cyclin box	0 (0)	0 (0)	30 (50)[a]	0 (0)
ScPHO80	0 (0)	0 (0)	0 (0)	26 (49)
Tb cyc1	0 (0)	0 (0)	0 (0)	0 (0)
Tb cyc2	0 (0)	0 (0)	0 (0)	41 (66)
Pfcyclin1	0 (0)	0 (0)	30 (51)	0 (0)
Pfcyclin2	0 (0)	0 (0)	0 (0)	22 (34)
Pfcyclin3	0 (0)	0 (0)	0 (0)	39 (57)
Pfcyclin4	37 (53)	0 (0)	29 (49)	0 (0)
Pfcables/ik3-1	0 (0)	29 (50)	24 (44)	21 (39)
Pyy23489708	37 (56)	0 (0)	39 (60)[b]	0 (0)
Pyy23490373	0 (0)	0 (0)	0 (0)	38 (59)
Pyy23479686	0 (0)	0 (0)	27 (47)	0 (0)
Pyy23481355	0 (0)	33 (53)	0 (0)	0 (0)
Pyy23482209	0 (0)	0 (0)	33 (51)	0 (0)
TcCYC4	0 (0)	0 (0)	0 (0)	36 (57)
TcCYC5	0 (0)	0 (0)	0 (0)	0 (0)

Selected cyclins from Human (Hs), *Drosophila melanogaster* (Dm), *Saccharomyces cerevisiae* (Sc), *Trypanasoma brucei* (Tb), *Trypanasoma cruzi* (Tc), *Plasmodium falciparum* (Pf) and *Plasmodium yoelii yoelii* (Pyy) are compared

[a] With "Expect" set at 10,000

[b] Indicates less than 75 amino acids contributed

cdk regulators in *T. parva* are very divergent at the primary sequence level. However, there are parallels with the cyclins of other organisms. In particular, one putative cyclin has identity to the Cables proteins of higher eukaryotes. In mammalian cells, Cables is believed to regulate cdk activity through the promotion of inhibitory phosphorylation of the tyrosine in the GXGXYG motif (Zuckerberg et al. 2000; Wu et al. 2001). Four out of five of the putative *T. parva* cdks contain this tyrosine residue. Given the lack of similarity between the *T. parva* cyclins and their mammalian counterparts, it is difficult to suggest the pairings between the cyclins and cdks found in the *T. parva* genome with any degree of certainty.

6.9.5 Sequence Homologues of Additional Parasite-Encoded Cell Cycle Regulators

The retinoblastoma protein has no obvious homolog in the *T. parva* genome or in those of other Apicomplexa that have been sequenced (*P. falciparum, P. yoelii*, and *Cryptosporidium parvum*). Although there are three reported E2F homologs in *C. parvum*, these are not conserved in either *T. parva* or *P. falciparum*, suggesting divergence in basic cellular processes even within the subphylum Apicomplexa. Homologues of p53, $p14^{ARF}$, MDM2 detectable by database searching are all apparently absent from the *T. parva* and *P. falciparum* genomes. In addition, as in *Plasmodium* there are few homologs of specific transcriptional regulators, although nonspecific DNA binding proteins are present. This is consistent with the hypothesis of limited transcriptional control at many loci implied by the high percentage of schizont loci with detectable transcripts according to the MPSS data (Bishop et al. 2005). Overall, the analysis of the complete suite of proteins involved in the cell cycle suggests the existence of novel regulatory processes in *Theileria*.

6.9.6 Potential Modulators of the Host Cell Identified in the *Theileria* Genomes

Deregulation of signal transduction pathways is likely to be a key component of *T. parva*-induced host cell proliferation. Studies of mammalian tumor syndromes have shown that atypical interaction between different signal transduction pathways is extensive, and importantly, it is often extremely difficult to separate cause from effect in the context of host cell activation. Initial inspection of the *T. parva* and *T. annulata* genome sequences does not provide obvious previously unrecognized insights as to which host proteins might be the primary targets in the induction of host cell "transformation." Perhaps surprisingly, there is a lack of obvious homologs of known oncogenes. However, the *T. annulata* genome does contain phospoholipase A2 genes, proteins that have been implicated in transformation in other eukaryotic systems, and unusually encode predicted signal peptides, suggesting secretion into the host cell (Pain et al. 2005). When searching the genome of *T. parva* for predicted proteins potentially involved in host cell activation/transformation and/or parasite–host cell cycle coordination, the sequences of the *T. annulata* and *P. falciparum* genomes are an important resource. Because of their close evolutionary relationship, a credible hypothesis would be that proteins important in host cell functional manipulation would be conserved between *T. parva* and *T. annulata*, but not in *P. falciparum*, which does not induce a transformation-like phenotype in the nucleated host cell. Candidate host cell function-modulatory proteins should also be theoretically expressed within the schizont life cycle stage. In addition, these proteins would be predicted to be exposed to the host cell environment, and are thus likely to be either secreted or transmembrane proteins. The majority of signal peptides that target proteins to the classical secretory pathway are readily detectable in silico. It is, therefore, possible to generate a mini-proteome that is highly enriched for the secreted component of the total predicted *T. parva* proteome in silico. Recently, a bioinformatics approach using a similar set of assumptions to those specified earlier, including absence from the *Plasmodium* genome, specific expression in schizonts, and possession of a predicted signal peptides, has been used to define 244 *Theileria*-unique candidates that may be involved in host–cell interaction as a basis for further functional analysis (Shiels et al. 2006).

Of potential relevance to parasite–host cell cycle coordination, the *T. parva* proteome contains several putative candidates for proteins that may be involved in interaction with host microtubules, which are sequestered on the surface of the schizonts in the infected lymphocyte. Of particular note are predicted secreted versions of Tau-like proteins and EMAP115, which are absent from *Plasmodium* (Gardner et al. 2005). In addition, *T. parva* encodes a cdc-48-like AAA-adenosine triphosphatase (ATPase) with a predicted signal peptide. It is possible that this molecule could be a modulator of host cell mitosis through induction of disassembly of the host cell spindle, which is one of its multiple known functions.

```
TP01_1216
Domain Start End    Bits   Evalue Alignment mode
Cyclin_N     1      129    -26.8  0.028      l       s

#HMM       *->diyaylrelEeeylplpppdyldqqpqpdsidinpkMRaiLvDWLveVhekfkLlpeTLyLaVnyl
#MATCH          ++ +e  +++     +  l                          + +    ++L+  T     ++l
#SEQ     1    --MKNKNEIDIKS----HQMLLAFGS----------------ELIQKAGILLQLHAVTIASGQSIL

#HMM          DRFLskkpvpklkvprkkLQLvGvtaLlIAsKyEEiksdvypPevedfvyita......sDnqayt
#MATCH        + F++ + +      +  ++     +++ ++A+K+EE       + ++e+++ i +   +   +  n++y+
#SEQ          HKFYAYH-S----LKDFNIRDTSASCCFLACKLEEN-----HRKLEQVAKIFEflkyyeDENKCYK

#HMM          .........kkeilrMEkliLktLkfdls<-*
#MATCH         ++++++  kk il+ Ek+iL     f+l
#SEQ          yspenenmlKKKILEIEKEILIGFAFRLD  129


TP02_0190
Domain Start End    Bits          Evalue Alignment mode
Cyclin_N     191    319    -37.5         0.26              ls

#HMM       *->diyaylrelEeeylplpppdyldqqpqpdsidinpkMRaiLvDWLveVhekfkLlpeTLyLaVnyl
#MATCH          +  l+e     +   ++++ +l+  +       +    k   +i  + +      ++++p T   +a  ++
#SEQ    191   KLHPILFESNTLF--RQKHPWLH-PT-----LSFTKLCKIKYNVMSLPNLIHHVDPSTSAIAWTLI

#HMM          DRFLskkpvpklkvprkkLQLvGvtaLlIAsKyEEiksdvypPeved..fvyitasDnqaytkkei
#MATCH         R        v      ++ + +L     ++   +A K+  +       +  ev ++ +     t    ++  ++k i
#SEQ          ERLVLNG-V----LTKFNRKLFSSVCYILAYKFNQD----FEYEVINeiLTIFT--REKNVDAKSI

#HMM          lrMEkliLktLkfdls<-*
#MATCH           E++i    L+f l
#SEQ          FYNEMKIFALLEFSLK    319


TP04_0577
Domain Start End    Bits          Evalue Alignment mode
Cyclin_N     53     160    -21.0         0.0089        ls

#HMM       *->diyaylrelEeeylplpppdyldqqpqpdsidinpkMRaiLvDWLveVhekfkLlpeTLyLaVnyl
#MATCH        +i ++      Eee    l       y++ q ++     +n +                   +L p +   ++   ++
#SEQ     53   RIPSF----EEE---LWLIKYYSLQLSQ---FVNAN----------------NLKPSVKETSLVLF

#HMM          DRFLskkpvpklkvprkkLQLvGvtaLlIAsKyEEiksdvypPevedfvyitasDnqaytkkeilr
#MATCH         RF+ ++ +          ++     +  t+   +A K+E++       ++   ++ + y       n        + +
#SEQ          NRFYLRR-S----LLQYDPRIIMFTCITLATKLEDMW---RSVYIDKLLYKI--NN--LNITKVFE

#HMM          MEkliLktLkfdls<-*
#MATCH        ME   +  ++L+f+l+
#SEQ          MEAIVCDVLNFNLN  160

TP04_0316
Domain Start End    Bits          Evalue Alignment mode
Cyclin_N     10     137    86.4          9.3e-23           ls

HMM        *->dellesisrvlerllalndstaeksssasiplepSQKQSSAHFSVNTHSfdivtesstklttalsak
#MATCH        d+ + ++ +vl    ++   +d    e +s
#SEQ     10   DDFIRALGVVLTKIV--SDVVPEYGSL-----------------------------------------

#HMM          isnFysknvEQAPLEGGQLESPHRDAEAEVmARDGACTTKLHRVEPRLEERVTARRENHVGNVVGR
#MATCH          s F s n
#SEQ          --SCFNSINA--------------------------------------------------------

#HMM          LFRDQSEPsiSltdYlsRiqLLiVSVGLavkycptspsvylvavfLlaliYldRleknYhsrgarc
#MATCH                P+iS  dYl Ri+                +y  +s+          +f+lal+Y+dR+ k
#SEQ          --------PPIS--DYLVRIA----------RYVNCSN-----ECFVLALVYIDRIMK--------

#HMM          lklaitvtslnvHRLliaalrvAtKfLeDfsysWNsyfaKVgGISlrELNkLEidFLflvDFdL<-*
#MATCH           +++v   ln+HRLli+ +++A+Kf +D+ ys Ns +a VgGI ++E+N+LE++FL l+ + L
#SEQ          --MHKFSVSVLNIHRLLITSVMLAAKFSDDVYYS-NSFYAQVGGIKVAEMNILEAQFLMLIKYQL 137
```

Fig. 5 Pfam analysis of predicted *Theileria parva* cyclins. The figure highlights the conserved residues within cyclin-like domains detected within *T. parva* ORFs indicated by their identification numbers following annotation (Gardner et al. 2005) were searched against the Pfam database. Only "Trusted matches" with *E* values above the cut-off were regarded as significant

6.10 Application of Genomics to the Development and Deployment of Vaccines for the Control of *Theileria*

6.10.1 Molecular Tools to Support Live Vaccination Against East Coast Fever

Currently, management of *Theileria* and other tick-borne diseases is primarily effected through control of the tick vector using acaricides, although this is unsustainable in the medium to long term (Tatchell 1987; Willadsen 2004). An "infection and treatment" (ITM) live vaccine based on the injection of a potentially lethal dose of sporozoites together with a long acting dose of tetracycline (reviewed by Radley 1981) has been developed for control of East Coast fever caused by *T. parva*. However, the live vaccine has not been deployed widely throughout the entire affected region due to problems related to use of potentially lethal live parasites and most importantly the requirement for a cold chain and well-trained veterinarians to ensure effective delivery. Application of VNTR and additional molecular markers based on the sequence of the gene encoding the polymorphic immunodominant molecule, acronym PIM (Toye et al. 1996), has proved useful for quality control of *T. parva* sporozoite stabilates and monitoring deployment of infection and treatment live vaccination in the field. The mini- and microsatellites reveal that the Muguga cocktail, the most widely used incarnation of infection and treatment comprising three parasite stocks, contains at least six genetically distinct components (Oura et al. 2007). It is also apparent that two of the three stocks within the cocktail are genetically very similar to one another (Bishop et al. 2001; Oura et al. 2007). There is currently evidence from the work at a vaccination site in Uganda that the Kiambu stock of the parasite induces a long-term carrier state in vaccinated cattle that can last for up to 4 years. This raises the question as to whether all components of the vaccine cocktail are required for long-term immunity. The Ugandan study has also demonstrated that vaccinated animals that remain PCR positive for the Kiambu stock can be super-infected by several distinct local parasite genotypes but remain immune (Oura et al. 2007). The vaccine, therefore, probably protects against disease, rather than infection. A further discovery is that local ticks and cattle can acquire the vaccine strain from immunized animals (Oura et al. 2007). The latter finding was also demonstrated for a different version of the infection and treatment vaccine based on the Marikebuni stabilate at a vaccination trial site in Western Kenya (Bishop et al. unpublished observations). There is strong evidence that the protection induced by infection and treatment is based on class I restricted $CD8^+$ T cells that are specific for the schizont-infected lymphocyte (Goddeeris and Morrison 1987; McKeever and Morrison 1994) and this has recently been exploited in a rationale approach for the development of a subunit vaccine as described later.

6.10.2 Live Attenuated Cell Culture Vaccines for the Control of *T. annulata*

In the case of *T. annulata*, a live vaccine based on inoculation of schizont-infected leukocytes attenuated through passage in vitro culture has been successfully utilized for control and has been widely deployed in North Africa, the Middle East, and South Asia (reviewed by Boulter and Hall 1999). There is evidence that the mechanism of protective immunity induced by the attenuated cell culture vaccine for control of *T. annulata* also involves cytotoxic T cell responses against the infected monocytes (Innes et al. 1989). Animals recovering from *T. lestoquardi* infections are immune to super-infection. Little is known about the role of cell-mediated immunity. Since the clinical course of the malignant theileriosis is comparable to tropical theileriosis and the target host cells are very similar in both diseases, it is believed that analogous mechanisms are operating. Thus, both CD4+ and CD8+ T-cells might be involved. In contrast, nothing is known about the cells involved in immune response to *T.* species (China). Regarding the humoral immune response, specific antibodies against both *T. lestoquardi* and *T.* species (China) have been demonstrated (Leemans et al. 1997; Bakheit et al. 2006; Miranda et al. 2006). Immunization of sheep with attenuated *T. lestoquardi* schizont-infected ovine cells has been carried out successfully in Iraq and Iran (Hawa et al. 1981; Hooshmand-Rad

1985; Hashemi-Fesharki 1997). Although the tissue culture vaccine might be the best method of control of malignant theileriosis of sheep and goats, it remains to be shown whether the immunity induced will be strain-specific (Fiedhoff 1997). Attempts are ongoing to investigate the molecular basis of attenuation of the cell culture vaccine for *T. annulata* and *T. lestoquardi*.

6.10.3 Application of Genomics to Development of Subunit Vaccines for Control of ECF

Because of the difficulties associated with effective delivery of the ITM vaccine, together with the high cost of vaccination (US$7–20 per animal), there has been a long-standing effort to develop a subunit vaccine for control of *T. parva* that would circumvent the drawbacks of ITM. One approach has been to evaluate a recombinant version of the *T. parva* major surface sporozoite protein p67, based on the rationale that antibodies against p67 can "neutralize" sporozoite entry into lymphocytes. A variety of recombinant forms of p67 consistently induce 70–80% protection against an experimental sporozoite stabilate challenge in the laboratory (Musoke et al. 1992; reviewed by Bishop et al. 2004). A degree of efficacy has also been demonstrated by recombinant p67 against field challenge by *T. parva* infected ticks (Musoke et al. 2005). A homologue of p67 from *T. annulata* designated SPAG1 shares neutralizing determinants with p67, and the two recombinant proteins are mutually cross-protective on heterologous challenge (reviewed by Boulter and Hall 1999). A homologous sporozoite surface antigen has also been isolated from *T. lestoquardi* and expressed, but has yet to be tested as a vaccine (Skilton et al. 2000; Bishop et al. 2004). Despite the efficacy of p67 as an experimental vaccine, it is believed by many scientists involved in vaccine development that schizont antigens will also be required to create an effective recombinant vaccine against ECF. The identification of candidate schizont antigens that would form the basis of such a vaccine has been significantly enhanced by the recent application of genomics. Research conducted over a period of 20 years has made the development of an improved subunit vaccine against ECF a reasonable probability. The demonstration that immunization of cattle by ITM engendered solid immunity against *T. parva* provoked a series of studies to elucidate the protective immune mechanisms induced in cattle. This work led to the discovery that MHC class I restricted $CD8^+$ CTLs were responsible for protecting cattle against a lethal challenge with *T. parva* sporozoites (McKeever and Morrison 1994; Taracha et al. 1995). It was also demonstrated that these CTLs were directed at cells infected with the schizont stage of the parasite, which they recognized and lyzed (Morrison et al. 1987; McKeever et al. 1994; Taracha et al. 1995). It was therefore logical to hypothesize that schizont antigens that are recognized by these $CD8^+$ CTLs would be prime candidates for inclusion in a subunit vaccine for ECF. But the major challenge faced was to dissect which of the thousands of proteins expressed by *T. parva* were targeted by $CD8^+$ CTL responses.

The sequencing of the *T. parva* genome (Gardner et al. 2005) provided an opportunity to use bioinformatics to identify candidate molecules that could be assessed for recognition by CTL. It was known that the *T. parva* schizont lies free in the cytosol of infected lymphocytes (Shaw 2003) and it was, therefore, hypothesized that parasite proteins that are either secreted or membrane bound would access the cytosol and be available for processing and presentation by MHC class I molecules to CTLs. The preliminary contigs derived from the *T. parva* genome sequence data were loaded into an annotation database and subjected to an automated process that searched the *T. parva* sequences against a nonredundant database of proteins extracted from GenBank. The search results were used to produce a set of *T. parva* gene models that encoded proteins similar to those in other organisms. This set of gene models were used to train the gene finding programs GlimmerM (Salzberg et al. 1999) and phat (Cawley et al. 2001). Gene models for the entire preliminary genome sequence were subsequently developed from an algorithm based on a combination of these two gene prediction programs. The combiner algorithm was used to analyze all the preliminary contigs. The proteins encoded by the predicted *T. parva* genes were subjected to a variety of analyses, including database searches (blastp), predictions of signal peptides and signal anchors (SignalP 2.0; http://www.cbs.dtu.dk/services/SignalP-2.0/), and transmembrane domains (TMHMM; http://www.cbs.dtu.dk/services/TMHMM). The results from chromosome 1 were reviewed and a set of 55 genes encoding candidate antigens were selected, 36 were successfully cloned into an eukaryotic expression vector and analyzed for their recognition by CTL isolated from cattle experimentally infected with sporozoite stabilates (Graham et al. 2006). Polyclonal parasite-specific $CD8^+$ CTL lines were generated from peripheral blood mononuclear cells from 13 ITM immunized cattle by sequential restimulation with irradiated autologous *T. parva* infected cells. CTL recognition of autologous immortalized skin fibroblasts transiently transfected with the candidate genes was then assessed using an IFN-γ ELISpot assay. CTL from four animals all showed significant IFN-γ ELISpot responses to cells transfected with a single candidate gene that was assigned the name *Tp2* (GenBank accession number XM_760490) (Graham et al. 2006). The *Tp2* gene encoded a 174 amino acid long protein with an N-terminal signal peptide with a predicted cleavage site between residues 23 and 24. Searches of protein and DNA databases using BLAST identified a homolog of *Tp2* in *T. annulata*, a molecule described as a membrane-associated antigen, TaD (Schneider et al. 2004). *Tp2* transcription was demonstrated in the sporozoite, schizont, and piroplasm life-cycle stages (Graham et al. 2007). ^{51}Chromium release assays confirmed that *Tp2* transfected target cells were lysed by CTL in an MHC class I dependent manner and four antigenic regions were mapped using a library of overlapping synthetic peptides covering the full-length of the *Tp2* protein. Significantly, *Tp2* specific $CD8^+$ T cell responses were demonstrated in ITM immunized cattle following challenge with a lethal dose of sporozoites. The kinetics and magnitude of the responses correlated well with those previously described for protective schizont-specific CTL (Morrison et al. 1987; McKeever et al. 1994) and indicated that *Tp2* is recognized by $CD8^+$ T cells during the resolution of the infection.

As a complementary approach to genome-based predictions of candidate antigens, random screening of cDNA libraries has also been undertaken to successfully identify additional CTL target antigens from *T. parva* (Graham et al. 2006). CTL screening of pools of random schizont cDNA clones led to the identification of an additional five antigens, designated Tp1, Tp4, Tp5, Tp7, and Tp8. Interestingly, three of these antigens did not have a predicted secretion signal or transmembrane domain and therefore would not have appeared on the selected gene list for immuno-screening. There are examples of exported proteins that do not contain predicted secretion signals, especially in *P. falciparum* (Nacer et al. 2001), and therefore it is possible that these proteins are externalized via a nonclassical route. The vaccine potential of these proteins has been evaluated in cattle using heterologous prime-boost regimes using plasmid DNA and attenuated poxvirus vectors. In a trial evaluating a cocktail of five antigens, 79% of the vaccinated cattle exhibited antigen-specific $CD8^+$ T cell responses as measured by a direct IFN-γ ELISpot assay, but only 29% of the vaccinated animals demonstrated CTL activity following in vitro restimulation of PBMC with parasite infected cells (Graham et al. 2006). Interestingly, the presence of a lytic CTL response correlated with protection following a lethal challenge infection.

Whether these data reflect responses of T cells with a differing status in terms of activation or memory or alternatively that there may be distinct roles for IFN-γ secreting and lytic $CD8^+$ T cells remains to be determined. Nevertheless, the data confirmed the potential antigens for an in-depth study on improving vaccination regimes to induce protective CTL responses. The extension of this approach to the complete genome of *T. parva* using CTL restricted by additional bovine haplotypes is likely to result in identification of valuable additional vaccine candidates.

Given the number of antigens already identified and the bias in the response imposed by the bovine MHC, it seems likely that additional antigens and epitopes will almost certainly be required to confer protection in the out-bred cattle population exposed to challenge by antigenically diverse parasite populations in the field. It also seems probable that any schizont antigen based vaccine that is ultimately developed will need to be multivalent (i.e., it will need to include epitopes from several different antigens).

6.11 Initial Investigation of *Theileria* Interaction with Ixodid Tick Vectors Using Expressed Sequence Tags Derived from *R. appendiculatus* and *A. variegatum* Tick Salivary Gland Transcripts

In an initial attempt to analyze genes encoding potentially secreted proteins of ixodid tick species, 2,109 nonredundant ESTs from a total dataset of 3,992 sequences were determined from cDNA derived from *A. variegatum* dissected salivary glands (Nene et al. 2004). In addition, 9,844 ESTs from *T. parva*-infected *R. appendiculatus* salivary glands, dissected 4 days after induction of sporogony, and 9,162 ESTs from uninfected salivary glands from ticks fed for the same length of time were determined (Nene et al. 2004). These were clustered and analyzed using the TIGR automated annotation pipeline and used to build a Gene Index (accessible at http://compbio.dfci.harvard.edu/tgi/). The index contains links to annotation data from other organisms and potential roles in metabolic pathways. Excluding sporozoite sequences identified by BLAST searching against the *T. parva* genome, the *R. appendiculatus* dataset contained a total of 18,422 ESTs and equated to 4,971 clusters and 7,359 singletons (Nene et al. 2004). Relative to the *A. variegatum* gene index, this indicates diminishing returns in terms of information and suggests that even for a single tick tissue such as the salivary gland, 20,000 ESTs may not represent adequate coverage to gain a comprehensive snapshot of gene expression. This problem could be solved by construction of normalized libraries. A notable feature of the data was that six out of the eight most highly transcribed *A. variegatum* salivary gland genes potentially encoded glycine/proline rich proteins that may be components of the tick cement cone. Distantly related homologs of a T cell immunosuppressant protein originally described in *Dermacentor* (Bergman et al. 2000) were also identified, suggesting that certain proteins may be conserved between the sialomes of different tick species.

One aim of the *R. appendiculatus* study was to identify transcripts whose expression is modulated by infection with parasite. Few highly expressed transcripts fell into this category, although cDNAs encoding three putative glycine-rich transcripts were significantly up-regulated in the infected salivary glands according to λ^2 analysis ($p < 0.05$). In addition, 653 and 558 tentative consensus sequences, comprising 1–7 clones, were present only in the infected or uninfected salivary glands, respectively. This may indicate parasite modulation of expression of secreted proteins, but a higher level of transcriptome coverage will be required to confirm this.

Six hundred and twenty ESTs were identified as being derived from the sporozoite, although only approximately 200 were absent from a schizont transcriptome data developed using MPSS (Bishop et al. 2005). Automated annotation indicated that the majority represented hypothetical or conserved hypothetical proteins. An interesting finding was the identification of homologs of two major *B. microplus* heme-binding proteins. These exhibited domains were also present in vitellogenins, which are major components of yolk. The apovitellin B sequence homolog in *R. appendiculatus* contained a predicted signal peptide, consistent with secretion. "Signal trap" functional selection was used to isolate additional *R. appendiculatus* salivary gland proteins (Lambson et al. 2005). This indicates that the classical secretory pathway operates in ticks as well as other arthropods, with predicted signal peptides ranging from 15 to 47 amino acids in length. Ten of the functional signal peptides could not be predicted using Hidden Markov or Neural Network algorithms, indicating a requirement for improved algorithms. Both the signal peptide and gene index approaches resulted in identification of homologs (50–60% identity) to the *R. appendiculatus* 64 trp cement antigen that is protective against *R. appendiculatus* tick infestation and also blocks pathogen transmission of *Ixodes*-transmitted encephalitis virus in rodent models (Trimnell et al. 2002; Labuda et al. 2006).

6.12 Opportunities for Future Research

The availability of two *Theileria* genome sequences, combined with high-resolution analysis of parasite transcriptomes will enable new approaches to understand the biology of *Theileria*. Identification of *T. parva* genes that are specifically expressed in sporozoites and piroplasms, combined with mammalian host and vector transcriptome, should result in novel insights into pathogen/vector/host interaction. More specifically, creation of sophisticated genomic resources including oligonucleotide arrays for *Theileria* species and their hosts, combined with arrays derived from ixodid tick EST sequences, have the potential to reveal hitherto unsuspected molecular interactions between host, parasite proteins, and vector proteins.

There are many future possibilities for the application of proteomics to *Theileria* research. For example, a proteome comparison of antigenically different *T. parva* clones would be of great value in identifying shared antigens with vaccine potential. A comparison of the proteins expressed in uninfected and infected bovine T cells might help to reveal the mechanisms of *Theileria* replication and parasite-induced transformation. Similarly, a comparison of the protein expression profiles of *T. parva* and *T. annulata* would potentially provide insight into the different mechanisms of pathogenesis of these closely related parasites.

The comparative analysis of the genome sequences of *T. parva*, *T. annulata*, and *P. falciparum* has provided a highly informative dataset with which to generate hypotheses for further testing in the context of host cell transformation and cell-cycle coordination. The value of these analyses will be further enhanced through the availability of the *T. orientalis* and *Babesia bovis* genomes in the near future. Analysis of these genomes for putative regulators of parasite/host signaling will enable rapid screening of candidate proteins for specific properties. It must be stressed, however, that such predictive tests remain biased by current knowledge and so there is still a strong case for screening all secreted and surface expressed proteins in an unbiased manner. Such a screen is likely to uncover novel protein/protein interactions and improved definition of host cell activation pathways and the as yet unknown parasite proteins that are directly responsible for activating these. The availability of this data will also provide targets for modification by transgenic approaches, if the technical hurdles in development of effective protocols for stable parasite transfection can be overcome.

On a more applied note, new avenues for improving disease control can be explored. The identification of *T. parva* antigens that are the targets of MHC class I restricted cytotoxic T cells, will provide useful input data for the emerging discipline of immunoinformatics. This should enable the development of new algorithms for T cell epitope prediction that can be validated through application to genome sequence data from *Theileria* and other protozoan genera. Additionally, in the short-term homologues of the *T. parva*, antigens can be directly isolated from related protozoa, including *T. annulata*, and evaluated as candidate vaccines.

References

Abrahamsen MS, Templeton TJ, Enomoto S, Abrahante JE, Zhu G, Lancto CA, Deng M, Liu C, Widmer G, Tzipori S, Buck GA, Xu P, Bankier AT, Dear PH, Konfortov BA, Spriggs HF, Iyer L, Anantharaman V, Aravind L, Kapur V (2004) Complete genome sequence of the apicomplexan, *Cryptosporidium parvum*. Science 304:441–445

Aktas M, Altay K, Dumanli N (2006) PCR-based detection of *Theileria ovis* in *Rhipicephalus bursa* adult ticks. Vet Parasitol 140:259–263

Allsopp B, Carrington M, Baylis H, Sohal S, Dolan T, Iams K (1989) Improved characterization of *Theileria parva* isolates using the polymerase chain reaction and oligonucleotide probes. Mol Biochem Parasitol 35:137–147

Allsopp BA, Baylis HA, Allsopp MT, Cavalier-Smith T, Bishop RP, Carrington DM, Sohanpal B, Spooner P (1993) Discrimination between six species of *Theileria* using oligonucleotide probes which detect small subunit ribosomal RNA sequences. Parasitology 107:157–165

Allsopp MT, Cavalier-Smith T, De Waal DT, Allsopp BA (1994) Phylogeny and evolution of the piroplasms. Parasitology 108:147–152

Altay K, Dumanli N, Aktas M (2007) Molecular identification, genetic diversity and distribution of *Theileria* and *Babe-sia* species infecting small ruminants. Vet Parasitol 147:161–165

Anderson NG, Anderson NL (1978) Analytical techniques for cell fractions. XXI. Two-dimensional analysis of serum and tissue proteins: multiple isoelectric focusing. Ann Biochem 85:331–340

Andrews NW, Webster P (1991) Phagolysosomal escape by intracellular pathogens. Parasitol Today 7:335–340

Awadalla P, Walliker D, Babiker H, Mackinnon M (2001) The question of *Plasmodium falciparum* population structure. Trends Parasitol 17:351–353

Babiker HA, Ranford-Cartwright LC, Currie D, Charlwood JD, Billingsley P, Teuscher T, Walliker D (1994) Random mating in a natural population of the malaria parasite Plasmodium falciparum. Parasitology 109:413–421

Bagby S, Kim S, Maldonado E, Tong KI, Reinberg D, Ikura M (1995) Solution structure of the C-terminal core domain of human TFIIB: similarity to cyclin A and interaction with TATA-binding protein. Cell 82:857–867

Bakheit MA, Scholzen T, Ahmed JS, Seitzer U (2006) Molecular characterization of a *Theileria* lestoquardi gene encoding for immunogenic protein splice variants. Parasitol Res 100:161–170

Baldwin CL, Malu MN, Grootenhuis JG (1988) Evaluation of cytotoxic lymphocytes and their parasite strain specificity from African buffalo infected with *Theileria parva*. Parasite Immunol 10:393–403

Barry JD, Ginger ML, Burton P, McCulloch R (2003) Why are parasite contingency genes often associated with telomeres? Int J Parasitol 33:29–45

Bateman A, Coin L, Durbin R, Finn RD, Hollich V, Griffiths-Jones S, Khanna A, Marshall M, Moxon S, Sonnhammer EL, Studholme DJ, Yeats C, Eddy SR (2004) The Pfam protein families database. Nucl Acids Res 32:D138–D141

Baumgartner M, Tardieux I, Ohayon H, Gounon P, Langsley G (1999) The use of nocodazole in cell cycle analysis and parasite purification from *Theileria parva*-infected B cells. Microb Infect 1:1181–1188

Baylis HA, Sohal SK, Carrington M, Bishop RP, Allsopp BA (1991) An unusual repetitive gene family in *Theileria parva* which is stage-specifically transcribed. Mol Biochem Parasitol 49:133–142

Bergman DK, Palmer MJ, Caimano MJ, Radolf JD, Wikel SK (2000) Isolation and molecular cloning of a secreted immunosuppressant protein from *Dermacentor andersoni* salivary gland. J Parasitol 86:516–525

Bishop R, Musoke A, Morzaria S, Sohanpal B, Gobright E (1997) Concerted evolution at a multicopy locus in the protozoan parasite *Theileria parva*: extreme divergence of potential protein-coding sequences. Mol Cell Biol 17:1666–1673

Bishop R, Morzaria S, Gobright E (1998) Linkage of two distinct AT-rich minisatellites at multiple loci in the genome of *Theileria parva*. Gene 216:245–254

Bishop R, Gobright E, Spooner P, Allsopp B, Sohanpal B, Collins N (2000a) Microsequence heterogeneity and expression of the LSU rRNA genes within the two single copy ribosomal transcription units of *Theileria parva*. Gene 257:299–305

Bishop R, Gobright E, Nene V, Morzaria S, Musoke A, Sohanpal B (2000b) Polymorphic open reading frames encoding secretory proteins are located less than 3 kilobases from *Theileria parva* telomeres. Mol Biochem Parasitol 110:359–371

Bishop R, Geysen D, Spooner P, Skilton R, Nene V, Dolan T, Morzaria S (2001) Molecular and immunological characterisation of *Theileria parva* stocks which are components of the 'Muguga cocktail' used for vaccination against East Coast fever in cattle. Vet Parasitol 94:227–237

Bishop R, Geysen D, Skilton R, Odongo D, Nene V, Allsopp BA, Mbogo S, Spooner P, Morzaria S (2002) Genomic polymorphism, sexual recombination and molecular epidemiology of *Theileria parva*. In: Dobbelaere DAE, Mckeever DJ (eds) *Theileria*. Kluwer, Boston, pp 23–40

Bishop R, Musoke A, Morzaria S, Gardner M, Nene V (2004) Theileria: intracellular protozoan parasites of wild and domestic ruminants transmitted by ixodid ticks. Parasitology 129:S271–S283

Bishop R, Shah T, Pelle R, Hoyle D, Pearson T, Haines L, Brass A, Hulme H, Graham SP, Taracha EL, Kanga S, Lu C, Hass B, Wortman J, White O, Gardner MJ, Nene V, de Villiers EP (2005) Analysis of the transcriptome of the protozoan *Theileria parva* using MPSS reveals that the majority of genes are transcriptionally active in the schizont stage. Nucl Acids Res 33:5503–5511

Bishop R, Musoke A, Skilton R, Morzaria S, Gardner M, Nene V *Theileria*: life cycle stages associated with the ixodid tick vector. In: Nuttall (ed.) Ticks: biology, disease and control. Cambridge Univ Press, New York, in press

Bishop RP, Sohanpal BK, Allsopp BA, Spooner PR, Dolan TT, Morzaria SP (1993) Detection of polymorphisms among *Theileria parva* stocks using repetitive, telomeric and ribosomal DNA probes and anti-schizont monoclonal antibodies. Parasitology 107:19–31

Bishop RP, Sohanpal BK, Morzaria SP, Dolan TT, Mwakima FN, Young AS (1994) Discrimination between *Theileria parva* and *T. taurotragi* in the salivary glands of *Rhipicephalus appendiculatus* ticks using oligonucleotides homologous to ribosomal RNA sequences. Parasitol Res 80:259–261

Borst P, Bitter W, McCulloch R, Van Leeuwen F, Rudenko G (1995) Antigenic variation in malaria. Cell 82:1–4

Boulter N, Hall R (1999) Immunity and vaccine development in the bovine theilerioses. Adv Parasitol 44:41–97

Brenner S, Johnson M, Bridgham J, Golda G, Lloyd DH, Johnson D, Luo S, McCurdy S, Foy M, Ewan M, Roth R, George D, Eletr S, Albrecht G, Vermaas E, Williams SR, Moon K, Burcham T, Pallas M, DuBridge RB, Kirchner J, Fearon K, Mao J, Corcoran K (2000) Gene expression analysis by massively parallel signature sequencing (MPSS) on microbead arrays. Nat Biotechnol 18:630–634

Brown CGD (1997) Dynamics and impact of tick-borne diseases of cattle. Trop Anim Health Prod 29:S1–S3

Burton JL, Rosa GJM (2006) Physiological genomics special issue on animal functional genomics. Physiol Genom 28:1–4

Campbell JDM, Spooner RL (1999) Macrophages behaving badly – infected cells and subversion of immune responses to Theileria annulata. Parasitol Today 15:10–16

Carlton JM, Angiuoli SV, Suh BB, Kooij TW, Pertea M et al. (2002) Genome sequence and comparative analysis of the model rodent malaria parasite Plasmodium yoelii yoelii. Nature 419:512–519

Carrington M, Allsopp B, Baylis H, Malu NM, Shochat Y, Sohal S (1995) Lymphoproliferation caused by Theileria parva and Theileria annulata. In: Boothroyd K (ed) Molecular Approaches to Parasitology. Wiley, New York, pp 43–56

Cawley SE, Wirth AI, Speed TP (2001) Phat–a gene finding program for Plasmodium falciparum. Mol Biochem Parasitol 118:167–174

Chae J, Levy M, Hunt J Jr, Schlater J, Snider G, Waghela SD, Holman PJ, Wagner GG (1999) Theileria sp. Infections associated with bovine fatalities in the United States confirmed by small-subunit rRNA gene analyses of blood and tick samples. J Clin Microbiol 37:3037–3040

Chansiri K, Kawazub S, Kamiob T, Teradab Y, Fujisakic K, Philipped H, Sarataphane N (1999) Molecular phylogenetic studies on Theileria parasites based on small subunit ribosomal RNA gene sequences. Vet Parasitol 83:99–105

Chin L, De Pinho RL (2000) Flipping the oncogene switch; illumination of tumor maintenance and regression. Trends Genet 16:147–150

Conrad PA, Stagg DA, Grootenhuis JG, Irvin AD, Newson J, Njamunggeh RE, Rossiter PB, Young AS (1987) Isolation of *Theileria parasites* from African buffalo (*Syncerus caffer*) and characterization with anti-schizont monoclonal antibodies. Parasitology 94:413–423

Coulson RM, Hall N, Ouzounis CA (2004) Comparative genomics of transcriptional control in the human malaria parasite Plasmodium falciparum. Genome Res 14:1548–1554

De Castro JJ, James AD, Minjauw B, Di Giulio GU, Permin A, Pegram RG, Chizyuka GB, Sinyangwe P (1997) Long-term studies on the economic impact of ticks on Sanga cattle in Zambia. Exp Appl Acarol 21:3–19

Dobbelaere D, Heussler V (1999) Transformation of leukocytes by *Theileria parva* and *T. annulata*. Annu Rev Microbiol 53:1–42

Dobbelaere DA, Küenzi P (2004) The strategies of the *Theileria* parasite: a new twist in host-pathogen interactions. Curr Opin Immunol 16:524–530

Dobbelaere DA, Spooner PR, Barry WC, Irvin AD (1984) Monoclonal antibody neutralizes the sporozoite stage of different *Theileria parva* stocks. Parasite Immunol 6:361–370

Fadiel A, Anidi I, Eichenbaum KD (2005) Farm animal genomics and informatics: an update. Nucleic Acids Res 33:6308–6318

Fedele M, Battista S, Manfioletti G, Croce CM, Giancotti V, Fusco A (2001) Role of the high mobility group A proteins in human lipomas. Carcinogenesis 22:1583–1591

Galley Y, Hagens G, Glaser I, Davis W, Eichhorn M, Dobbelaere D (1997) Jun NH_2 -terminal kinase is constitutively activated in T cells transfromed by the intracellular parasite *Theileria parva*. Proc Nat Acad Sci USA 94:5119–5124

Ganesan K, Jiang L, Rathod PK (2002) Stochastic versus stable transcriptional differences on *Plasmodium falciparum* DNA microarrays. Int J Parasitol 32:1543–1550

Gardner MJ, Bishop R, Shah T, de Villiers EP, Carlton JM et al (2005) Genome sequence of *Theileria parva*, a bovine pathogen that transforms lymphocytes. Science 309: 134–137

Gauer M, Mackenstedt U, Mehlhorn H, Schein E, Zapf F, Njenga E, Young A, Morzaria S (1995) DNA measurements and ploidy determination of developmental stages in the life cycles of *Theileria annulata* and *T. parva*. Parasitol Res 81:565–574

Glass EJ, Coussens PM (2005) Functional genomics of host – pathogen interactions in species of veterinary importance. Vet Immunol Immunopathol 105:173–174

Glass EJ, Jensen K (2007) Resistance and susceptibility to a protozoan parasite of cattle – gene expression differences in macrophages from different breeds of cattle. Vet Immunol Immunopathol 120:20–30

Glass EJ, Spooner RL (1990) Parasite-accessory cell interactions in theileriosis. Antigen presentation by *Theileria annulata*-infected macrophages and production of continuously growing antigen-presenting cell lines. Eur J Immunol 20:2491–2497

Glass EJ, Innes EA, Spooner RL, Brown CG (1989) Infection of bovine monocyte/macrophage populations with *Theileria annulata* and *Theileria parva*. Vet Immunol Immunopathol 22:355–368

Glass EJ, Preston PM, Springbett A, Craigmile S, Kirvar E, Wilkie G, Brown CGD (2005) *Bos taurus* and *Bos indicus* (Sahiwal) calves respond differently to infection with *Theileria annulata* and produce markedly different levels of acute phase proteins. Int J Parasit 35:337–347

Goddeeris BM, Morrison WI (1987) The bovine autologous *Theileria* mixed leucocyte reaction: influence of monocytes and phenotype of the parasitized stimulator cell on proliferation and parasite specificity. Immunology 60:63–69

Goddeeris BM, Morrison WI, Toye PG, Bishop R (1990) Strain specificity of bovine *Theileria parva*-specific cytotoxic T cells is determined by the phenotype of the restricting class I MHC. Immunology 69:38–44

Graham SP, Pelle R, Honda Y, Mwangi DM, Tonukari NJ et al. (2006) *Theileria parva* candidate vaccine antigens recognized by immune bovine cytotoxic T lymphocytes. Proc Natl Acad Sci USA 103:3286–3291

Graham SP, Honda Y, Pellé R, Mwangi DM, Glew EJ, de Villiers EP, Shah T, Bishop R, van der Bruggen P, Nene V, Taracha EL (2007) A novel strategy for the identification of antigens that are recognised by bovine MHC class I restricted cytotoxic T cells in a protozoan infection using reverse vaccinology. Immun Res 3:2

Green RD, Qureshi MA, Long JA, Burfening PJ, Hamernik DL (2007) dentifying the future needs for long-term USDA efforts in agricultural animal genomics. Int J Biol Sci 3:185–191

Gubbels MJ, Yin H, Bai Q, Liu G, Nijman IJ, Jongejan F (2002) The phylogenetic position of the *Theileria buffeli* group in relation to other *Theileria* species. Parasitol Res 88: S28–S32

Gunasekera AM, Patankar S, Schug J, Eisen G, Kissinger J, Roos D, Wirth DF (2004) Widespread distribution of antisense transcripts in the *Plasmodium falciparum* genome. Mol Biochem Parasitol 136:35–42

Gunderson JH, Sogin ML, Wollett G, Hollingdale M, de la Cruz VF, Waters AP, McCutchan TF (1987) Structurally distinct, stage-specific ribosomes occur in Plasmodium. Science 238:933–937

Hanks SK, Hunter T (1995) Protein kinases 6. The eukaryotic protein kinase superfamily: kinase (catalytic) domain structure and classification. FASEB J 9:576–596

Hashemi-Fesharki R (1997) Tick-borne diseases of sheep and goats and their related vectors in Iran. Parassitologia 39:115–117

Hawa NJ, Latif BM, Ali SR (1981) Immunization of sheep against *Theileria hirci* infection with schizonts propagated in tissue culture. Vet Parasitol 9:91–97

Heussler V (2002) *Theileria* survival strategies and host cell transformation. In: Dobbelaere DAE, McKeever DJ (eds) World Class Parasites. Kluwer, Dorderecht, The Netherlands, pp 69–84

Heussler VT, Rottenberg S, Schwab R, Küenzi P, Fernandez PC, McKellar S, Shiels B, Chen ZJ, Orth K, Wallach D, Dobbelaere DA (2002) Hijacking of host cell IKK signalosomes by the transforming parasite *Theileria*. Science 298:1033–1036

Hinterberg K, Mattei D, Wellems TE, Scherf A (1994) Interchromosomal exchange of a large subtelomeric segment in a *Plasmodium falciparum* cross. EMBO J 13:4174–4180

Honda Y, Matsubara Y, Morzaria S, Mckeever D (1998) Immunohistochemical detection of parasite antigens in *Theileria parva*-infected bovine lymphocytes. Vet Parasitol 80:137–147

Hooshmand-Rad P (1985) The use of tissue culture attenuated live vaccine for *Theileria hirci*. Dev Biol Stand 62:119–127

Hooshmand-Rad P, Hawa NJ (1973a) Transmission of *Theileria hirci* in sheep with Hyalomma anatolicum. Trop Anim Health Prod 5:103–109

Hooshmand-Rad P, Hawa NY (1973b) Malignant theileriosis of sheep and goats. Trop Anim Health Prod 5:97–102

Hooshmand-Rad PH, Hawa NJ (1975) Cultivation of *Theileria hirci* in sheep lymphoid cells. Trop Anim Health Prod 7:121–122

Huang J, Mullapudi N, Sicheritz-Ponten T, Kissinger JC (2004) A first glimpse into the pattern and scale of gene transfer in Apicomplexa. Int J Parasitol 34:265–274

Hunt T (1991) Cell biology. Destruction's our delight. Nature 349:100–101

Innes EA, Millar P, Brown CG, Spooner RL (1989) The development and specificity of cytotoxic cells in cattle immunized with autologous or allogeneic *Theileria annulata*-infected lymphoblastoid cell lines. Parasite Immunol 11:57–68

Irvin AD, Morrison WI (1987) Immunopathology, immunology and immunoprophylaxis of Theileria infections. In: Soulsby EJL (ed.) Immune responses in parasitic infections: immunology, immunopathology and immunoprophylaxis. CRC, Boca Ratan, Florida, pp 223–274

Irvin AD, Ocama JG, Spooner PR (1982) Cycle of bovine lymphoblastoid cells parasitised by Theileria parva. Res Vet Sci 33:298–304

Jensen K, Speed D, Paxton E, Williams JL, Glass EJ (2006a) Construction of a normalized Bos taurus and Bos indicus macrophage-specific cDNA library. Anim Genet 37:75–77

Jensen K, Talbot R, Paxton E, Waddington D, Glass EJ (2006b) Development and validation of a bovine macrophage specific cDNA microarray. BMC Genom 7:224

Jensen K, de Miranda Santos IK, Glass EJ (2007) Using genomic approaches to unravel livestock (host)-tick-pathogen interactions. Trends Parasitol 23:439–444

Jensen K, Paxton E, Waddington D, Talbot R, Darghouth MA, Glass EJ (2008) Differences in the transcriptional responses induced by *Theileria annulata* infection in bovine monocytes derived from resistant and susceptible cattle breeds. Int J Parasitol 38:313–325

Jura WG, Brown CG, Perry M (1985) Comparative autoradiographic study of parasite-host-cell cyclical relationship in lymphoblastoid cell lines infected with *Theileria annulata* and *Theileria parva* in vitro. Vet Parasitol 18:339–348

Katzer F, Ngugi D, Oura C, Bishop RP, Taracha EL, Walker AR, McKeever DJ (2006) Extensive genotypic diversity in a recombining population of the apicomplexan parasite *Theileria parva*. Infect Immun 74:5456–5464

Labuda M, Trimnell AR, Licková M, Kazimírová M, Davies GM, Lissina O, Hails RS, Nuttall PA (2006) An antivector vaccine protects against a lethal vector-borne pathogen. PLoS Pathog 2:e27

Lambson B, Nene V, Obura M, Shah T, Pandit P, Ole-Moiyoi O, Delroux K, Welburn S, Skilton R, de Villiers E, Bishop R (2005) Identification of candidate sialome components expressed in ixodid tick salivary glands using secretion signal complementation in mammalian cells. Insect Mol Biol 14:403–414

Le Roch KG, Zhou Y, Blair PL, Grainger M, Moch JK, Haynes JD, De La Vega P, Holder AA, Batalov S, Carucci DJ, Winzeler EA (2003) Discovery of gene function by expression profiling of the malaria parasite life cycle. Science 301:1503–1508

Leemans I, Hooshmand-Rad P, Uggla A (1997) The indirect fluorescent antibody test based on schizont antigen for study of the sheep parasite *Theileria lestoquardi*. Vet Parasitol 69:9–18

Li Y, Luo J, Liu Z, Guan G, Gao J, Ma M, Dang Z, Liu A, Ren Q, Lu B, Liu J, Zhao H, Li J, Liu G, Bai Q, Yin H (2007) Experimental transmission of *Theileria* sp. (China 1) infective for small ruminants by *Haemaphysalis longicornis* and *Haemaphysalis qinghaiensis*. Parasitol Res 101:533–538

McCutchan TF, de la Cruz VF, Lal AA, Gunderson JH, Elwood HJ, Sogin ML (1988) Primary sequences of two small subunit ribosomal RNA genes from *Plasmodium falciparum*. Mol Biochem Parasitol 28:63–68

McGuire K, Glass EJ (2005) The expanding role of microarrays in the investigation of macrophage responses to pathogens. Vet Immunol Immunopathol 105:259–275

McGuire K, Manuja A, Russell GC, Springbett A, Craigmile SC, Nichani AK, Malhotra DV, Glass EJ (2004) Quantitative analysis of pro-inflammatory cytokine mRNA expression in *Theileria annulata*-infected cell lines derived from resistant and susceptible cattle. Vet Immunol Immunopathol 99:87–98

McKeever DJ, Morrison WI (1994) Immunity to a parasite that transforms T lymphocytes. Curr Opin Immunol 6:564–567

McKeever DJ, Taracha EL, Innes EL, MacHugh ND, Awino E, Goddeeris BM, Morrison WI (1994) Adoptive transfer of immunity to *Theileria parva* in the CD8+ fraction of responding efferent lymph. Proc Natl Acad Sci USA 91:1959–1963

Mehlhorn H, Schein E (1998) Redescription of Babesia equi Laveran, 1901 as *Theileria* equi. Parasitol Res 84:467–475

Miranda J, Bakheit MA, Liu Z, Yin H, Mu Y, Guo S, Beyer D, Oliva A, Ahmed JS, Seitzer U (2006) Development of a recombinant indirect ELISA for the diagnosis of *Theileria* sp. (China) infection in small ruminants. Parasitol Res 98:561–567

Mooi WJ, Peeper DS (2006) Oncogene-induced cell senescence-halting on the road to cancer. N Eng J Med 355:1037–1046

Morrison WI, Goddeeris BM, Teale AJ, Groocock CM, Kemp SJ, Stagg DA (1987) Cytotoxic T-cells elicited in cattle challenged with *Theileria parva* (Muguga): evidence for restriction by class I MHC determinants and parasite strain specificity. Parasite Immunol 9:563–578

Morrissette NS, Sibley LD (2002) Cytoskeleton of apicomplexan parasites. Microbiol Mol Biol Rev 66:21–38

Morzaria SP, Young JR (1992) Restriction mapping of the genome of the protozoan parasite *Theileria parva*. Proc Natl Acad Sci USA 89:5241–5245

Morzaria SP, Young JR, Spooner PR, Dolan TT, Bishop RP (1993) *Theileria parva*: a restriction map and genetic recombination. In: Mozaria SP (ed) Genome Analysis of Protozoan Parasites. ILRAD, Nairobi, Kenya, pp 67–73

Mukhebi AW, Perry BD (1992) Economic implications of the control of East Coast fever in eastern, central and southern Africa. 20–23 July 1992, Kadona, Zimbabwe. ILCA (International Livestock Centre for Africa), Addis Ababa, Ethiopia, pp 107–112

Musoke A, Morzaria S, Nkonge C, Jones E, Nene V (1992) A recombinant sporozoite surface antigen of *Theileria parva* induces protection in cattle. Proc Natl Acad Sci USA 89:514–518

Musoke A, Rowlands J, Nene V, Nyanjui J, Katende J, Spooner P, Mwaura S, Odongo D, Nkonge C, Mbogo S, Bishop R, Morzaria S (2005) Subunit vaccines based on the p67 major surface protein of *Theileria parva* sporozoites reduce severity of infection derived from field tick challenge. Vaccine 23:3084–3095

Musoke AJ, Nantulya VM, Rurangirwa FR, Buscher G (1984) Evidence for a common protective antigenic determinant on sporozoites of several *Theileria parva* strains. Immunology 52:231–238

Nacer A, Berry L, Slomianny C, Mattei D (2001) Plasmodium falciparum signal sequences: simply sequences or special signals? Int J Parasitol 31:1371–1379

Nagore D, Garcia-Sanmartin J, Garcia-Perez AL, Juste RA, Hurtado A (2004) Identification, genetic diversity and prevalence of *Theileria* and *Babesia* species in a sheep population from Northern Spain. Int J Parasitol 34:1059–1067

Neitz WO (1953) Aureomycin in *Theileria parva* infection. Nature 171:34–35

Neitz WO (1972) The experimental transmission of *Theileria ovis* by *Rhipicephalus evertsi* mimeticus and *R. Bursa*. Onderstepoort J Vet Res 39:83–86

Nene V, Bishop R, Morzaria S, Gardner MJ, Sugimoto C, ole-MoiYoi OK, Fraser CM, Irvin A (2000) *Theileria parva* genomics reveals an atypical apicomplexan genome. Int J Parasitol 30:465–474

Nene V, Lee D, Kang'a S, Skilton R, Shah T, de Villiers E, Mwaura S, Taylor D, Quackenbush J, Bishop R (2004) Genes transcribed in the salivary glands of female Rhipicephalus appendiculatus ticks infected with *Theileria parva*. Insect Biochem Mol Biol 34:1117–1128

Nesvizhskii AI, Aebersold R (2005) Interpretation of shotgun proteomic data: the protein inference problem. Mol Cell Proteomics 4:1419–1440

Ngumi PN, Lesan AC, Williamson SM, Awich JR, Morzaria SP, Dolan TT, Shaw MK, Young AS (1994) Isolation and preliminary characterisation of a previously unidentified Theileria parasite of cattle in Kenya. Res Vet Sci 57:1–9

Norval RA, Lawrence JA, Young AS, Perry BD, Dolan TT, Scott J (1991) Theileria parva: influence of vector, parasite and host relationships on the epidemiology of theileriosis in southern Africa. Parasitology 102:347–356

Norval RAI, Perry BD, Young AS (1992) The Epidemiology of Theileriosis in Africa. Academic Press, London, UK, p 481

Odongo DO, Oura CA, Spooner PR, Kiara H, Mburu D, Hanotte OH, Bishop RP (2006) Linkage disequilibrium between alleles at highly polymorphic mini- and micro-satellite loci of *Theileria parva* isolated from cattle in three regions of Kenya. Int J Parasitol 36:937–946

Osman SA, Al Gaabary MH (2007) Clinical, haematological and therapeutic studies on tropical theileriosis in water buffaloes (*Bubalus bubalis*) in Egypt. Vet Parasitol 146:337–340

Oura CA, Odongo DO, Lubega GW, Spooner PR, Tait A, Bishop RP (2003) A panel of microsatellite and minisatellite markers for the characterisation of field isolates of *Theileria parva*. Int J Parasitol 33:1641–1653

Oura CA, Bishop RP, Wampande EM, Lubega GW, Tait A (2004a) Application of a reverse line blot assay to the study of haemoparasites in cattle in Uganda. Int J Parasitol 34:603–613

Oura CA, Bishop R, Wampande EM, Lubega GW, Tait A (2004b) The persistence of component *Theileria parva* stocks in cattle immunized with the 'Muguga cocktail' live vaccine against East Coast fever in Uganda. Parasitology 129:27–42

Oura CA, Asiimwe BB, Weir W, Lubega GW, Tait A (2005) Population genetic analysis and sub-structuring of *Theileria parva* in Uganda. Mol Biochem Parasitol 140:229–239

Oura CA, Bishop R, Asiimwe BB, Spooner P, Lubega GW, Tait A (2007) *Theileria parva* live vaccination: parasite transmission, persistence and heterologous challenge in the field. Parasitology 134:1205–1213

Pain A, Renauld H, Berriman M, Murphy L, Yeats CA et al (2005) Genome of the host-cell transforming parasite *Theileria annulata* compared with *T. parva*. Science 309:131–133

Paul RE, Day KP (1998) Mating patterns of *Plasmodium falciparum*. Parasitol Today 14:197–202

Paul RE, Packer MJ, Walmsley M, Lagog M, Ranford-Cartwright LC, Paru R, Day KP (1995) Mating patterns in malaria parasite populations of Papua New Guinea. Science 269:1709–1711

Preston PM, Hall FR, Glass EJ, Campbell JD, Darghouth MA, Ahmed JS, Shiels BR, Spooner RL, Jongejan F, Brown CG (1999) Innate and adaptive immune responses co-operate to protect cattle against *Theileria annulata*. Parasitol Today 15:268–274

Radley DE (1981) Infection and treatment method of immunization. In: Irvin AD, Cunningham, MP and Young, AS (eds) Advances in the Control of Theileriosis. Martinus Nijhoff, Hague, The Netherlands, pp 227–237

Rechsteiner M, Rogers SW (1996) PEST sequences and regulation by proteolysis. Trends Biochem Sci 21:267–271

Reeves R (2001) Molecular biology of HMGA proteins: hubs of nuclear function. Gene 277:63–81

Reinartz J, Bruyns E, Lin JZ, Burcham T, Brenner S, Bowen B, Kramer M, Woychik R (2002) Massively parallel signature sequencing (MPSS) as a tool for in-depth quantitative gene expression profiling in all organisms. Brief Funct Genomic Proteomic 1:95–104

Sager HDW, Dobbelaere DA, Jungi TW (1997) Macrophage-parasite relationship in theileriosis. Reversible phenotypic and functional dedifferentiation of macrophages infected with *Theileria annulata*. J Leukoc Biol 61:459–468

Salih DA, ElHussein AM, Hayat M, Taha KM (2003) Survey of *Theileria lestoquardi* antibodies among Sudanese sheep. Vet Parasitol 111:361–367

Salih DA, El Hussein AM, Seitzer U, Ahmed JS (2007) Epidemiological studies on tick-borne diseases of cattle in Central Equatoria State, Southern Sudan. Parasitol Res 101:1035–1044

Salzberg SL, Pertea M, Delcher AL, Gardner MJ, Tettelin H (1999) Interpolated Markov models for eukaryotic gene finding. Genomics 59:24–31

Schneider I, Haller D, Seitzer U, Beyer D, Ahmed JS (2004) Molecular genetic characterization and subcellular localization of a putative *Theileria annulata* membrane protein. Parasitol Res 94:405–415

Schnittger L, Hong Y, Jianxun L, Ludwig W, Shayan P, Rahbari S, Voss-Holtmann A, Ahmed JS (2000) Phylogenetic analysis by rRNA comparison of the highly pathogenic sheep-infecting parasites *Theileria lestoquardi* and a *Theileria* species identified in China. Ann NY Acad Sci 916:271–275

Schnittger L, Yin H, Gubbels MJ, Beyer D, Niemann S, Jongejan F, Ahmed JS (2003) Phylogeny of sheep and goat *Theileria* and *Babesia* parasites. Parasitol Res 91:398–406

Schulman BA, Lindstrom DL, Harlow E (1998) Substrate recruitment to cyclin-dependent kinase 2 by a multipurpose docking site on cyclin A. Proc Natl Acad Sci USA 95:10453–10458

Shaw MK (1997) The same but different: the biology of *Theileria* sporozoite entry into bovine cells. Int J Parasitol 27:457–474

Shaw MK (2002) Theileria development and host cell invasion. In: Dobbelaere DAE, McKeever DJ (eds) World Class Parasites, Kluwer, Boston, London, UK, pp 1–22

Shaw MK (2003) Cell invasion by *Theileria* sporozoites. Trends Parasitol 19:2–6

Shaw MK, Tilney LG (1992) How individual cells develop from a syncytium: merogony in *Theileria parva* (Apicomplexa). J Cell Sci 101:109–123

Shaw MK, Tilney LG (1995) The entry of *Theileria parva* merozoites into bovine erythrocytes occurs by a process similar to sporozoite invasion of lymphocytes. Parasitology 111:455–461

Shaw MK, Tilney LG, Musoke AJ (1991) The entry of *Theileria parva* sporozoites into bovine lymphocytes: evidence for MHC class I involvement. J Cell Biol 113:87–101

Shaw MK, Tilney LG, McKeever DJ (1993) Tick salivary gland extract and interleukin-2 stimulation enhance susceptibility of lymphocytes to infection by *Theileria parva* sporozoites. Infect Immun 61:1486–1495

Shiels BR, McKellar S, Katzer F, Lyons K, Kinnaird J, Ward C, Wastling JM, Swan D (2004) A *Theileria annulata* DNA binding protein localized to the host cell nucleus alters the phenotype of a bovine macrophage cell line. Eukaryot Cell 3:495–505

Shiels B, Langsley G, Weir W, Pain A, McKellar S, Dobbelaere D (2006) Alteration of host cell phenotype by *Theileria annulata* and *Theileria parva*: mining for manipulators in the parasite genomes. Int J Parasitol 36:9–21

Skilton RA, Musoke AJ, Nene V, Wasawo D, Wells CW, Spooner PR, Bishop RP, Osaso J, Nkonge C, Latif A, Morzaria SP (2000) Molecular characterisation of a *Theileria lestoquardi* gene encoding a candidate sporozoite vaccine antigen. Mol Biochem Parasitol 107:309–314

Sohanpal B, Wasawo D, Bishop R (2000) Cloning of telomere-associated DNA using single-specific-primer polymerase chain reaction provides evidence for a conserved sequence directly adjacent to *Theileria parva* telomeric repeats. Gene 255:401–409

Sparagano OA, Spitalska E, Namavari M, Torina A, Cannella V, Caracappa S (2006) Phylogenetics of *Theileria* species in small ruminants. Ann NY Acad Sci USA 1081:505

Stagg DA, Young AS, Leitch BL, Grootenhuis JG, Dolan TT (1983) Infection of mammalian cells with *Theileria* species. Parasitology 86:243–254

Steinfeld H, Gerber P, Wassenaar T, Castel V, Rosales M, de Haan C (2006) Livestock's Long Shadow - Environmental Issues and Options. FAO report

Sugimoto C (1997) Economic importance of Theileriosis in Japan. Trop Anim Health Prod 29:49S

Swan DG, Phillips K, Tait A, Shiels BR (1999) Evidence for localisation of a *Theileria* parasite AT hook DNA-binding protein to the nucleus of immortalised bovine host cells. Mol Biochem Parasitol 101:117–129

Swan DG, Stern R, McKellar S, Phillips K, Oura CA, Karagenc TI, Stadler L, Shiels BR (2001) Characterisation of a cluster of genes encoding *Theileria annulata* AT hook DNA-binding proteins and evidence for localisation to the host cell nucleus. J Cell Sci 114:2747–2754

Tageldin MH, Zakia AM, Nagwa ZG, el Sawi SA (1992) An outbreak of theileriosis in sheep in Sudan. Trop Anim Health Prod 24:15–16

Taracha EL, Goddeeris BM, Morzaria SP, Morrison WI (1995) Parasite strain specificity of precursor cytotoxic T cells in individual animals correlates with cross-protection in cattle challenged with *Theileria parva*. Infect Immun 63:1258–1262

Tatchell RJ (1987) Tick control in the context of ECF immunization. Parasitol Today 3:7–10

Taylor LH, Katzer F, Shiels BR, Welburn SC (2003) Genetic and phenotypic analysis of Tunisian *Theileria annulata* clones. Parasitology 126:241–252

Toye P, Nyanjui J, Goddeeris B, Musoke AJ (1996) Identification of neutralization and diagnostic epitopes on PIM, the polymorphic immunodominant molecule of *Theileria parva*. Infect Immun 64:1832–1838

Toye PG, Goddeeris BM, Iams K, Musoke AJ, Morrison WI (1991) Characterization of a polymorphic immunodominant molecule in sporozoites and schizonts of *Theileria parva*. Parasite Immunol 13:49–62

Trimnell AR, Hails RS, Nuttall PA (2002) Dual action ectoparasite vaccine targeting 'exposed' and 'concealed' antigens. Vaccine 20:3560–3568

Uilenberg G (1981) Theilerial species of domestic livestock. In: Irvin AD, Cunningham MP, Young AS (eds) Advances in the Control of Theileriosis, Martinus Nijhoff, The Hague, The Netherlands, pp 4–37

Vanhee-Brossollet C, Vaquero C (1998) Do natural antisense transcripts make sense in eukaryotes? Gene 211:1–9

Vega-Palas MA, Martin-Figueroa E, Florencio FJ (2000) Telomeric silencing of a natural subtelomeric gene. Mol Gen Genet 263:287–291

Watanabe M, Kikawada T, Minagawa N, Yukuhiro F, Okuda T (2002) Mechanism allowing an insect to survive complete dehydration and extreme temperatures. J Exp Biol 205:2799–2802

Weir W, Ben-Miled L, Karagenç T, Katzer F, Darghouth M, Shiels B, Tait A (2007) Genetic exchange and sub-structuring in *Theileria* annulata populations. Mol Biochem Parasitol 154:170–180

Willadsen P (2004) Anti-tick vaccines. Parasitology 129: S367–S387

Wirth DF (2002) Biological revelations. Nature 419:495–496

Womack JE (2005) Advances in livestock genomics: opening the barn door. Genom Res 15:1699–1705

Wood LJ, Maher JF, Bunton TE, Resar LMS (2000) The oncogenic properties of the HMG-I gene family. Cancer Res 60:4256–4261

Wu CL, Kirley SD, Xiao H, Chuang Y, Chung DC, Zukerberg LR (2001) Cables enhances cdk2 tyrosine 15 phosphorylation by Weel, inhibits cell growth, and is lost in many human colon and squamous cancers. Cancer Res 61:7325–7332

Yin H, Liu G, Luo J, Guan G, Ma M, Ahmed J, Bai Q (2003) Observation on the schizont stage of an unidentified *Theileria sp.* in experimentally infected sheep. Parasitol Res 91:34–39

Yin H, Luo J, Guan G, Gao Y, Lu B, Zhang Q, Ma M, Lu W, Lu C, Yuan Z, Guo S, Wang B, Du H, Schnittger L, Ahmed J, Jongejan F (2002) Transmission of an unidentified *Theileria* species to small ruminants by *Haemaphysalis qinghaiensis* ticks collected in the field. Parasitol Res 88:S25–S27

Yin H, Schnittger L, Luo J, Seitzer U, Ahmed JS (2007) Ovine theileriosis in China: a new look at an old story. Parasitol Res 101(S2):S191–195

Zuckerberg LR, Patrick GN, Nikolic M, Humbert S, Wu CL, Lanier LM, Gertler FB, Vidal M, van Etten RA, Tsai LH (2000) Cables links Cdk5 and c-Abl and facilitates Cdk5 tyrosine phosphorylation, kinase upregulation, and neurite out growth. Neuron 26:633–646

Index

MIX
Papier aus verantwortungsvollen Quellen
Paper from responsible sources
FSC® C105338

If you have any concerns about our products,
you can contact us on
ProductSafety@springernature.com

In case Publisher is established outside the EU,
the EU authorized representative is:
Springer Nature Customer Service Center GmbH
Europaplatz 3, 69115 Heidelberg, Germany

Printed by Libri Plureos GmbH
in Hamburg, Germany